Hormones, Growth Factors, and Oncogenes

Author

Enrique Pimentel, M.D.
Professor and Director
Centro Nacional de Genética
Instituto de Medicina Experimental
Universidad Central de Venezuela
Caracas, Venezuela

CRC Press, Inc.
Boca Raton, Florida

Library of Congress Cataloging-in-Publication Data

Pimentel, Enrique.
    Hormones, growth factors, and oncogenes.

    Bibliography: p.
    Includes index.
    1. Hormones. 2. Growth promoting substances.
3. Oncogenes. 4. Cells—Growth—Regulation.
I. Title. [DNLM: 1. Cell Transformation, Neoplastic.
2. Growth Substances—physiology. 3. Hormones—
physiology. 4. Oncogenes. 5. Peptides—physiology.
6. Proto-Oncogenes. QZ 202 P644h]
QP571.P55   l1987   616.99'4071   86-23220
ISBN 0-8493-5346-7

This book represents information obtained from authentic and highly regarded sources. Reprinted material is quoted with permission, and sources are indicated. A wide variety of references are listed. Every reasonable effort has been made to give reliable data and information, but the author and the publisher cannot assume responsibility for the validity of all materials or for the consequences of their use.

All rights reserved. This book, or any parts thereof, may not be reproduced in any form without written consent from the publisher.

Direct all inquiries to CRC Press, Inc., 2000 Corporate Blvd., N.W., Boca Raton, Florida, 33431.

© 1987 by CRC Press, Inc.

International Standard Book Number 0-8493-5346-7

Library of Congress Card Number 86-23220
Printed in the United States

THE AUTHOR

Enrique Pimentel, M.D., is Professor of General Pathology and Pathophysiology at the School of Medicine. Central University of Venezuela. He was formerly Director of the Institute of Experimental Medicine at the same university and is now Director of the National Center of Genetics in Venezuela.

Born April 7, 1928 in Caracas, Venezuela, he obtained an M.D. degree from the Universities of Madrid, Spain, and Caracas, Venezuela. He is a member of the National Academy of Medicine in Venezuela and a honorary, correspondent or active member of 30 national and international scientific societies. He is Vice President of the International Academy of COGENES (1986). He is the Editor of a recently created journal *CRC Critical Reviews in Oncogenesis*. In many occasions, Dr. Pimentel has been invited to give lectures and seminars at universities and other scientific institutions from North America and Europe. He received several decorations in his country and the Grosse Verdienstkreuz of the Federal Republic of Germany. In 1982, he received in Venezuela the National Award of Science.

ACKNOWLEDGMENTS

I am deeply indebted to the Deutscher Akademischer Austauschdienst (Academic Exchange Service of the Federal Republic of Germany) for supporting my stay at the Deutsches Krebsforschungszentrum (German Cancer Research Center), Heidelberg, in order to complete the manuscript of this book. I also wish to express my gratitude to a number of dear colleagues who have encouraged and assisted me in the course of this endeavor, particularly Drs. Harald zur Hausen, Dietrich Schmähl, Fritz Anders, Janis V. Klavins, Robert C. Gallo, Henry T. Lynch, Peter N. Magee, Avery A. Sandberg, Arthur D. Riggs, and Morley D. Hollenberg.

LIST OF ABBREVIATIONS

AAF	2-Acetylamino-fluorene
ACTH	Adrenocorticotropic hormone
ACV	Avian carcinoma virus
ADP	Adenosine diphosphate
AEV	Avian erythroblastosis virus
AFP	Alpha-fetoprotein
AIDS	Acquired immune deficiency syndrome
ALV	Avian leukemia virus
AMCV	Avian myelocytomatosis virus
AMP	Adenosine 3':5'-monophosphate
A-MuLV	Abelson murine leukemia virus
AMV	Avian myeloblastosis virus
ANF	Atrial natriuretic factor
ARV	Avian reticuloendotheliosis virus
ASMC	Aortic smooth muscle cells
ASV	Avian sarcoma virus
ATP	Adenosine triphosphate
AV	Avian virus
AWTA	Aniridia-Wilms' tumor association
BCDF	B-cell differentiation factor
BCG	Bacillus Calmette-Guérin
BCGF	B-cell growth factor
BDGF	Bone-derived growth factor
BFU-E	Burst-forming units-erythroid
bp	Base pairs
BSF-1	B-cell stimulating factor 1
BSF-2	B-cell stimulating factor 2
cAMP	Cyclic adenosine 3':5'-monophosphate
CAP	Catabolyte gene activator protein
CBGF	Colostrum basic growth factor
CDGF	Cartilage-derived growth factor
cDNA	Complementary deoxyribonucleic acid copy
CEA	Carcinoembryonic antigen
CGF	Chondrocyte growth factor
cGMP	Cyclic guanosine 3':5'-monophosphate
CHE	Chinese hamster embryo
CHO	Chinese hamster ovary
CIL	Colony-inhibiting lymphokine
CLL	Chronic lymphocytic leukemia
CML	Chronic myelogenous leukemia
CNBr-EGF	Cyanide bromide-cleaved epidermal growth factor
CSF	Colony-stimulating factor
CSF-1	Colony-stimulating factor 1
CSF-2	Colony-stimulating factor 2
DENA	Diethylnitrosamine
DM	Double minute chromosome
DMBA	7,12-dimethylbenz(a)anthracene
DMSO	Dimethyl sulfoxide
DNA	Deoxyribonucleic acid

EBPA	Erythroid burst-promoting activity
EBV	Epstein-Barr virus
ECGF	Endothelial cell growth factor
EDF	Eosinophil differentiation factor
EDGF	Eye-derived growth factor
EF	Elongation factor
EGF	Epidermal growth factor
EPA	Erythroid-potentiating activity
ERRP	Epidermal growth factor receptor-related polypeptide
FCS	Fetal calf serum
FDGF	Fibroblast-derived growth factor
FeSV	Feline sarcoma virus
FGF	Fibroblast growth factor
FLV	Feline leukemia virus
F-MuLV	Friend murine leukemia virus
GABA	Gamma-aminobutyric acid
GAD	Glutamic acid decarboxylase
G-CSF	Granulocyte colony-stimulating factor
GDP	Guanosine diphosphate
GGPF	Glial growth-promoting factor
GH	Growth hormone
GM-CSF	Granulocyte-macrophage colony-stimulating factor
GMP	Guanosine monophosphate
GTP	Guanosine triphosphate
HBG	Human beta globin
HBV	Hepatitis B virus
HGF	Hematopoietic growth factor
hGH	Human growth hormone
HMBA	Hexamethylene bisacetamide
HSF	Hepatocyte-stimulating factor
hsp	Heat shock protein
HSR	Homogeneously staining region
HTLV	Human T-cell leukemia virus
HTLV-I	Human T-cell leukemia virus type I
HTLV-II	Human T-cell leukemia virus type II
HTLV-III	Human T-cell leukemia virus type III
IFN	Interferon
IFN-α	Interferon alpha
IFN-β	Interferon beta
IFN-γ	Interferon gamma
Ig	Immunoglobulin
IGF-I	Insulin-like growth factor I
IGF-II	Insulin-like growth factor II
IL	Interleukin
IL-1	Interleukin-1
IL-2	Interleukin-2
IL-3	Interleukin-3
IL-4	Interleukin-4
kbp	Kilobase pairs
kdalton	Kilodalton
K-MuSV	Kirsten murine sarcoma virus

LAV	Lymphadenopathy-associated virus
LDGF	Leukemia-derived growth factor
LDL	Low-density lipoprotein
LTR	Long terminal repeat
MAF	Macrophage-activating factor
MCGF	Mast cell growth factor
MCSF	Macrophage colony-stimulating factor
MEL	Mouse erythroleukemia
MGF	Melanocyte growth factor
MMGF	Myelomonocytic growth factor
MMTV	Mouse mammary tumor virus
M-MuSV	Moloney murine sarcoma virus
MNU	Methylnitrosourea
MOV	Murine osteosarcoma virus
mRNA	Messenger ribonucleic acid
MSA	Multiplication-stimulating activity
MSH	Melanocyte-stimulating hormone
MuLV	Murine leukemia virus
MuSV	Murine sarcoma virus
mol wt	Molecular weight
ND-TGF	Neuronal-derived transforming growth factor
NGF	Nerve growth factor
NHL	Non-Hodgkin lymphoma
NMU	Nitrosomethylurea
NNNG	N-methyl-N'-nitro-N-nitrosoguanidine
NPF	Neurite promoting factor
NRK	Normal rat kidney
ODC	Ornithine decarboxylase
ODGF	Osteosarcoma-derived growth factor
PCNA	Proliferating cell nuclear antigen
PDGF	Platelet-derived growth factor
PDGF-1	Platelet-derived growth factor 1 (or A) chain
PDGF-2	Platelet-derived growth factor 2 (or B) chain
PEM	Peritoneal exudate macrophages
PGE	Prostaglandin E
PHA	Phytohemagglutinin
PL	Placental lactogen
PMA	Phorbol 12-myristate 13-acetate
PRP	Proliferin-related peptide
PTH	Parathyroid hormone
RNA	Ribonucleic acid
RSV	Rous sarcoma virus
SCLC	Small cell lung cancer
SDS	Sodium dodecyl sulfate
SIRS	Soluble immune response suppressor
SLE	Systemic lupus erythematosus
SSV	Simian sarcoma virus
SV40	Simian virus 40
T3	3,3′,5-triiodo-L-thyronine
T4	L-thyroxine

TAF	Tumor angiogenesis factor
TAT	Tyrosine aminotransferase
TCGF	T-cell growth factor
TFP	Trifluoperazine
TGF	Transforming growth factor
TGF-α	Transforming growth factor alpha
TGF-β	Transforming growth factor beta
TIF	Tumor cell growth inhibiting factor
TIF-1	Tumor cell growth inhibiting factor 1
TIF-2	Tumor cell growth inhibiting factor 2
TNF	Tumor necrosis factor
TNF-α	Tumor necrosis factor alpha
TNF-β	Tumor necrosis factor beta
TPA	12-*O*-tetradecanoyl-phorbol 13-acetate
ts	Temperature-sensitive
TSH	Thyroid-stimulating hormone
URO	Urogastrone
VVGF	Vaccinia virus growth factor

TABLE OF CONTENTS

Chapter 1
Introduction
I. Overview 1
II. The Cell Cycle 2
III. Role of Hormones and Growth Factors in the Regulation of DNA Synthesis and Cellular Proliferation 5
    A. Mechanisms of Regulation of DNA Synthesis and Cell Proliferation by Hormones and Growth Factors 5
        1. Cyclin 6
    B. Growth-Inhibiting Factors 6
IV. Oncogenes and Proto-Oncogenes 8
    A. Functions of the Proto-Oncogene Protein Products 9
    B. Proto-Oncogenes and Cancer 9
V. Hormones and Growth Factors in Relation to Oncogenic Processes 11
    A. Animal Tumors 12
        1. The Autocrine Hypothesis of Oncogenesis 12
        2. Cellular Responses to v-*onc* Products 13
        3. Growth Factor Independence and Proto-Oncogene Expression 14
    B. Plant Tumors 14
        1. Hormones and Oncogene Products in Plant Tumors 14
VI. Influence of Hormones and Growth Factors on Proto-Oncogene Expression 15
    A. Mechanisms of Proto-Oncogene Regulation by Hormones and Growth Factors 16
    B. Kinetics of Growth Factor-Regulation Proto-Oncogene Expression 16
    C. Mechanisms of Growth Factor-Regulated Proto-Oncogene Expression 16
    D. Physiological Significance of Growth Factor-Regulated Proto-Oncogene Expression 17
    E. Action of Growth Factors in v-*onc*-Transformed Cells 18
    F. Summary 18
VII. Common Functions of Hormones, Growth Factors, and Oncogene Protein Products 18
References 19

Chapter 2
Cellular Mechanisms of Action of Hormones and Growth Factors and Their Relation to Oncogene Function
I. Introduction 27
II. Receptors 27
    A. Mobility and Internalization of Surface Receptors 28
    B. Presence of Hormone Receptors and Production of Hormones in Microorganisms 28
    C. Relationship between Hormones, Growth Factors, and Their Cellular Receptors 28
    D. Regulation of Receptor Expression 29
III. Postreceptor Transductional Mechanisms 29

		A.	Mediators and Modulators..29
		B.	Ion and pH Changes..30
		C.	Protein Phosphorylation 30
		D.	Regulation of Nuclear Functions 30
IV.	cAMP and the Adenylate Cyclase System 31
		A.	The Adenylate Cyclase System...31
			1. Regulation of the Adenylate Cyclase System 32
			2. cAMP in Relation to Cell Proliferation or Transformation 32
		B.	cAMP-Dependent Protein Kinases..33
			1. Regulatory Actions of cAMP-Dependent Protein Kinases.........34
			2. Structural Homology between the Catalytic Subunit of cAMP-Dependent Protein Kinase and the Protein Products of the *src* Oncogene Family 34
			3. cAMP and the Control of Proto-Oncogene Expression............35
V.	cGMP and the Guanylate System 35
		A.	cGMP and Cell Proliferation 35
		B.	cGMP-Dependent Protein Kinase..35
VI.	GTP-Binding Proteins..35
		A.	G Proteins..36
		B.	Transducin..36
		C.	The *ras* Oncogene Proteins and the Adenylate Cyclase System 37
			1. Structural and Functional Homologies of the *ras* Proteins 37
			2. Evolutionary Aspects of the *ras* Proteins..........................38
			3. Functional Aspects of the *ras* Proteins............................39
			4. Mutant c-*ras* Proteins..40
			5. Influence of *ras* Proteins of Cell Proliferation and Differentiation 41
			6. Other Functions of *ras* Proteins..42
			7. Conclusions..42
VII.	The Calcium-Calmodulin System 43
		A.	Calcium Ions 43
		B.	Calmodulin...44
		C.	Ca^{2+}/Calmodulin-Dependent Protein Kinase 45
		D.	Oncomodulin..45
		E.	Caldesmon 45
VIII.	Phosphoinositides and Protein Kinase C 46
		A.	Ca^{2+}/Phospholipid-Dependent Protein Kinase (Protein Kinase C) 46
		B.	Ca^{2+} Mobilization and Protein Kinase C Activation in Relation to DNA Synthesis and Cell Proliferation..47
		C.	Hormones and Growth Factors in Relation to Phosphoinositide Metabolism and Protein Kinase C..48
		D.	Tumor Promoters, Differentiation-Inducers, Phosphoinositide Metabolism and Protein Kinase C..49
			1. Mechanisms of Action of Phorbol Esters 50
			2. Phorbol Ester-Induced Cell Differentation 50
			3. Mechanisms of Action of Phorbol Esters on Cell Differentiation 51
		E.	Oncogene Protein Products, the Calcium/Calmodulin System, Phosphoinositide Metabolism and Protein Kinase C....................52
			1. The *src* and *ros* Oncogenes 52
			2. The *myc* Oncogene 53

		3.	The *abl* Oncogene 55
		4.	Summary 55
IX.	Protein Phosphorylation 55
	A.	Tyrosine-Specific Protein Kinase Activity 56
		1.	Substrates of Tyrosine-Specific Protein Kinases 57
		2.	Tyrosine-Specific Protein Kinase Activity in Normal Cells 57
		3.	Tyrosine-Specific Protein Kinase Activity in Malignant Cells 58
		4.	Origin and Regulation of Tyrosine-Specific Protein Kinase Activity Present in Normal and Malignant Cells 59
		5.	Phosphotyrosine-Specific Protein Phosphatases 60
		6.	Summary 60
	B.	Nontyrosine-Specific Kinases 60
		1.	Nontyrosine-Specific Protein Kinase Activities in Oncogene Protein Products 61
		2.	Cellular Substrates for Nontyrosine-Specific Protein Kinase Activity 61
	C.	Protein Phosphorylation in Relation to Cellular Proliferation and Cellular Differentiation 62
		1.	DNA Topoisomerases 62
		2.	Relationship between Protein Phosphorylation and Mitogenesis 63
	D.	Protein Phosphorylation and Neoplastic Transformation 63
	E.	Conclusion 64
References 64

Chapter 3
Insulin

I.	Introduction 85
II.	Insulin Structure and the Insulin Gene 85
	A. The Human Insulin Gene and Insulin Biosynthesis 85
	1. Structure of the Insulin Gene 85
	2. Insulin Biosynthesis in Metazoans and Microbes 86
	3. Pancreatic Beta-Cell Tumors in Transgenic Mice Expressing Recombinant Insulin/SV40 Genes 86
III.	Insulin Receptors 87
	A. Structure of Insulin Receptors 87
	B. Insulin Receptor Synthesis 87
	C. Formation and Phosphorylation of the Insulin Receptor Complex 87
	1. Insulin Receptor Phosphorylation 88
	2. Defective Insulin Receptors 88
	D. Relationship between the Insulin Receptor and Oncogene Protein Products 89
	E. Regulation of Insulin Receptor Expression 89
	F. Insulin Receptor Abnormalities in Neoplastic Cells 89
	G. Functional Role of the Insulin Receptor 90
	H. Internalization of the Insulin-Insulin Receptor Complex 90
IV.	Transductional Mechanisms of Insulin Action 90
	A. Insulin Mediators 91
	B. Phosphoinositides as Transducers of Insulin Action 91
V.	Post-Transductional Mechanism of Insulin Action 91

	A.	Insulin Action on Protein Phosphorylation	92
		1. Insulin and Phosphorylation of the v-*ras* Protein	93
	B.	Postreceptor Defects of Insulin Action in Neoplastic Cells	93
	C.	Effects of Insulin of DNA Synthesis, Cell Proliferation, and Cell Differentiation	93
VI.	Insulin Requirement by Neoplastic Cells		94
	A.	Insulin Action and the Induction of Differentiation in Neoplastic Cells	95
VII.	Summary		95
References			95

Chapter 4
Insulin-Like Growth Factors

I.	Introduction		103
II.	The IGFs Genes		103
III.	Structure and Synthesis of IGFs		104
IV.	Functions of IGFs		104
V.	Cellular Receptors for IGFs		105
	A.	IGF-I Receptor	105
	B.	IGF-II Receptor	105
VI.	Insulin and IGFs Receptors in Tranformed Cells		105
	A.	Insulin Receptors in Primary Tumors	105
	B.	Insulin and IGFs Receptors in Neoplastic Cell Lines	106
	C.	Insulin and IGFs Receptors in Induction of Differentiation of Neoplastic Cells	107
	D.	Conclusion	107
VII.	Summary		107
References			108

Chapter 5
Epidermal Growth Factor

I.	Introduction		111
II.	Structure of EGF		111
	A.	EGF and the Vaccinia Virus Growth Factor	111
III.	Biosynthesis of EGF		111
	A.	The EGF Precursor	112
		1. Structural Homologies of the EGF Precursor	112
		2. Homology between the EGF Precursor and the c-*mos* Protein	113
IV.	Production and Biological Actions of EGF and EGF-Related Growth Factors		113
V.	The EGF Receptor		114
	A.	The EGF Receptor Gene	114
	B.	Structure of the EGF Receptor	115
	C.	Biosynthesis of the EGF Receptor	115
	D.	Expression of EGF Receptors in Different Types of Cells	116
		1. Functional Heterogeneity of EGF Receptors in Different Types of Cells	116
	E.	Phosphorylation and Processing of the EGF Receptor	117
	F.	Internalization and Degradation of the EGF Receptor Complex	118
	G.	The *erb*-B Oncogene Protein Product and the EGF Receptor	119

	H.	The c-*neu* Oncogene Protein Product and the EGF Receptor............120
	I.	The v-*src* Oncogene Protein Product and the EGF Receptor............120
	J.	EGF Receptors in Human Tumors...121
	K.	Factors Involved in Changes of EGF Receptors..........................122
	L.	Expression of EGF Receptors in Malignant Cells........................122
		1. Biological Significance of EGF Receptor Alterations in Malignant Cells...123
	M.	Tumor Promoters and EGF Receptors...................................123
V.	Postreceptor Mechanisms of Action of EGF...................................124
	A.	Redistribution of Cell Membrane Components..........................124
	B.	EGF-Stimulated Phosphorylation of Cellular Proteins..................124
	C.	EGF and Phosphorylation of the v-*ras* Protein Product................125
	D.	EGF Action and Phosphoinositide Metabolism.........................126
	E.	EGF Actions at the Nuclear Level.......................................126
	F.	EGF-Mediated Induction of Proto-Oncogene Expression...............127
	G.	EGF Action on Cell Proliferation.......................................128
	H.	EGF Influence on Carcinogenic Processes..............................128
VII.	Summary..129
References...129

Chapter 6
Fibroblast, Melanocyte, and Nerve Growth Factors

I.	Introduction..141
II.	Fibroblast Growth Factor..141
	A. Types of FGFs..141
	B. Functions of FGF...141
	C. Mechanisms of Action of FGF..142
III.	Melanocyte Growth Factor..142
IV.	Nerve Growth Factor..142
	A. The NGF Gene and the Structure of NGF............................142
	B. Mechanisms of Action of NGF.......................................142
	1. NGF-Induced Protein Phosphorylation..........................143
	2. NGF and c-*fos* Proto-Oncogene Expression....................143
	3. Abrogation of NGF Requirement by Oncogene Protein Products...143
References...144

Chapter 7
Transforming Growth Factors

I.	Introduction..147
	A. Cellular Sources of TGFs..147
	B. TGFs Types and TGFs Receptors....................................147
II.	Transforming Growth Factor α..148
	A. The TGF-α Gene...148
	B. Structure of TGF-α...148
	C. Cellular Mechanism of Action of TGF-α............................149
III.	Transforming Growth Factor β..149
	A. Structure of TGF-β...149
	B. Cellular Mechanisms of Action of TGF-β...........................149
	C. Bifunctional Action of TGF-β on Cell Proliferation and/or Transformation...150

IV. Other Transforming Growth Factors 151
V. Summary 151
References.. 151

Chapter 8
Hematopoietic Growth Factors
I. Introduction... 157
    A. Colony-Stimulating Factors... 158
II. Hemopoietins 158
    A. Hemopoietin-1... 159
    B. Hemopoietin-2 (IL-3).. 159
        1. IL-3 Structure and the IL-3 Gene 159
        2. IL-3-Like Proteins 159
        3. IL-3 and Oncogene Expression 160
III. Lymphokines 160
    A. IL-1 161
        1. IL-1 Structure and the IL-1 Gene 161
    B. IL-2 162
        1. The Human Antigen-Specific T-Cell Receptor 162
        2. Structure and Synthesis of IL-2................................. 163
        3. The IL-2 Gene 163
        4. The Cellular IL-2 Receptor 164
        5. IL-2 Receptor in Leukemic Cells 165
        6. Functions of IL-2 in Normal and Neoplastic Cells 165
        7. IL-2 Mediated Changes in Phosphoinositide Metabolism
            and Protein Kinase C Activity 166
        8. IL-2 Action on Monovalent Ion Transport 167
        9. IL-2 Mediated Induction of Proto-Oncogene Expression 167
        10. IL-2 and the Acquired Immune Deficiency Syndrome 167
    C. Glial Growth Promoting Factor....................................... 168
    D. B-Cell Growth Factors 168
IV. Granulocyte-Macrophage Colony Stimulating Factors........................ 169
    A. Colony Stimulating Factor 1... 169
        1. The CSF-1 Receptor and the c-*fms* Proto-Oncogene
            Protein 170
        2. Tumor Promoters and c-*fms*/CSF-1 Receptor Expression......... 170
        3. CSF-1 and Proto-Oncogene Expression 171
    B. Colony-Stimulating Factor 2... 171
        1. CSF-2 and Proto-Oncogene Expression 172
    C. Granulocyte Colony-Stimulating Factor 173
    D. Leukemia-Derived Growth Factor 173
    E. Myelomonocytic Growth Factor 173
    F. Macrophage-Derived Growth Factors................................ 173
V. Interferons 174
    A. IFN Action on Cell Proliferation and Differentiation 174
    B. IFN Action on Proto-Oncogene Expression......................... 175
VI. Cytotoxins 175
    A. Lymphotoxin (TNF-β) 176
    B. Tumor Necrosis Factor α (TNF-α) 176
VII. Erythropoiesis-Stimulating Factors 177
    A. Erythroid-Potentiating Activity 177

		B.	Erythropoietin 177
			1. Structure and Function of Erythropoietin........................ 177
			2. Erythropoietin in Neoplastic Diseases 178
VIII.	Relationships between Hematopoietic Cell Proliferation and Differentiation, HGFs, and Proto-Oncogene Expression 178
	A.	Activation of c-*myc* Expression 178
	B.	Activation of c-*fos* Expression... 179
	C.	Proto-Oncogene Expression and Induction of Differentiation in Hematic Neoplastic Cells 179
IX.	Summary 181
References... 181

Chapter 9
Platelet-Derived Growth Factor

I.	Introduction 195
II.	Structure of PDGF 195
III.	Functions of PDGF 196
IV.	The PDGF Receptor and the Transduction of PDGF Signal 196
	A. Structure of the PDGF Receptor....................................... 196
	B. Phosphorylation of the PDGF Receptor 196
	C. Internalization and Processing of the PDGF Receptor 197
	D. Transductional Mechanisms of Action of PDGF............. 197
	E. Effects of PDGF on EGF Action 197
V.	Post-Transductional Mechanisms of Action of PDGF 198
	A. Extranuclear Effects of PDGF... 198
	B. Nuclear Effects of PDGF... 199
	1. Effects of PDGF on c-*fos* Expression 199
	2. Effects of PDGF on c-*myc* Expression 199
	3. Effects of PDGF on c-*src* Expression 200
	4. Effects of PDGF on the Expression of Non–Proto-Oncogene Genes 200
	5. PDGF as a Competence Factor 200
VI.	PDGF and the *sis* Oncogene Protein Product 201
	A. Comparative Structures of PDGF and $p28^{v\text{-}sis}$............... 201
	B. The c-*sis* Proto-Oncogene 202
	C. v-*sis*- and c-*sis*-Induced Transformation 203
VII.	PDGF and PDGF-Like Factors in Relation to Neoplasia........... 203
	A. Production of PDGF-Like Proteins by Normal Cells 203
	B. Production of PDGF or PDGF-like Proteins by Transformed Cells... 203
VIII.	Non-PDGF-Like Factors Produced by Platelets 205
IX.	Summary 205
References... 205

Chapter 10
Transferrins

I.	Introduction 211
II.	Synthesis and Structure of Transfereins 211
	A. Human Transferrin 211
	B. Rat Transferrin 212
	C. Transferrins and the B-*lym*-1 Proto-Oncogene 212

III. The Transferrin Receptor ... 212
   A. Transferrin Receptor Gene ... 212
   B. Structure of the Transferrin Receptor ... 213
   C. Expression of the Tranferrin Receptor on the Cell Surface ... 213
   D. Transferrin Receptors in Normal and Neoplastic Cells ... 214
IV. Summary ... 215
References ... 215

Chapter 11
Thyroid Hormones

I. Introduction ... 219
II. Thyroid Hormone Receptors ... 219
   A. Regulation of EGF and Insulin Receptors by Thyroid Hormone ... 219
III. Regulation of Gene Expression by Thyroid Hormone ... 220
IV. Thyroid Hormone and Human Cancer ... 220
   A. Thyroid Hormone in Experimental Tumors ... 220
      1. Effects of Thyroid Function on the Growth of Experimental Tumors ... 221
   B. Thyroid Hormone and In Vitro Neoplastic Transformation ... 221
   C. Mechanisms Involved in the Modulation of Neoplastic Transformation by Thyroid Hormone ... 222
V. Summary ... 222
References ... 222

Chapter 12
Steroid Hormones

I. Introduction ... 227
II. Steroid Hormone Receptors ... 227
   A. Phosphorylation of Steriod Hormone Receptors ... 228
   B. Steroid Hormone Receptors in Cancer Cells ... 228
III. Cellular Mechanisms of Action of Steroid Hormones and Neoplasia ... 228
   A. Androgens ... 229
   B. Estrogens ... 229
      1. Estrogens and Proto-Oncogene Expression ... 230
   C. Progesterone ... 230
   D. Glucocorticoids ... 230
      1. Glucocorticoid Receptors in Neoplastic Cells ... 231
      2. Glucocorticoids and Regulation of Chronic Transforming Retrovirus Expression ... 232
      3. Glucocorticoids and Oncogene Expression ... 233
   E. Calcitriol ... 233
      1. Induction of Cell Differentiation by Calcitriol ... 234
      2. Calcitriol and Proto-Oncogene Expression ... 234
      3. Summary ... 234
References ... 234

Index ... 239

Chapter 1

INTRODUCTION

I. OVERVIEW

Hormones and growth factors are important regulatory substances present in metazoan organisms. These substances represent biological signals involved in the regulation of cell growth and differentiation during both pre- and postnatal life. Some hormones and growth factors act in a restricted manner on specific types of responsive cells, whereas others may have a broad, perhaps universal, spectrum of activity for different types of cells and tissues.

Hormones are defined as chemical messengers synthesized in the endocrine glands of multicellular organisms and secreted into the extracellular body fluids. Upon their arrival to the respective target organs, hormones are recognized by and binded to specific cellular receptors. The formation of a hormone-receptor complex determines the cellular responses that are specific for both the hormone and the target cell.[1,2]

Peptide growth factors, commonly called in an abbreviate manner "growth factors", are hormone-related substances with an important role in the control mechanisms of growth and development in a diversity of organs and tissues as well as in cultured cells.[3-10] In contrast to the classical hormones, growth factors are usually not synthesized in specialized endocrine organs but are produced and secreted by cells from different tissues in a steady flow to diffuse to responsive cells, which are frequently located not far from the site of release. Different types of growth factors are produced by a diversity of normal cells as well as by neoplastically transformed cells. Isolation and characterization of growth factors and other tumor-secreted products may be facilitated by culturing cells in appropriate conditions, for example, in a defined protein-free medium.[11]

The biological properties of hormones and growth factors can be studied either in vitro or in vivo. Growth factors are usually defined by their ability to induce stimulation of target cell multiplication and their activity is measured by assays where either the increase of cell populations or the incorporation of labeled thymidine into DNA is determined. The latter can be estimated by autoradiography and counting of labeled cells or by determination of radioactivity in liquid scintillation vials. The most suitable routine test for detection and determination of growth factor activity lies on measuring the incorporation of tritiated thymidine into DNA of target cells with a scintillation counting.[12] A list of the most important hormones, growth factors, and other exogenous agents of cellular origin involved in growth-regulatory phenomena appears in Table 1. Available data corresponding to the chromosomal assignment and location of the genes coding for some of the hormones and growth factors discussed in this volume, as well as those corresponding to the genes coding for their cellular receptors, are indicated in Table 2.

II. THE CELL CYCLE

Serum contains many different hormones and growth factors capable of stimulating the growth of animal cells in vivo and in vitro. Cultured mouse fibroblasts, such as BALB/c 3T3 cells, have been conveniently used as an assay system for evaluating the activity of hormones and growth factors in the stimulation, or inhibition, of cell proliferation. Fibroblasts whose growth has been arrested by serum starvation undergo a synchronous progression through the cell cycle by the addition of serum. A schematic representation of the cell cycle appears in Figure 1.

The cell cycle can be divided into four major phases: G_1, S (DNA synthesis period), G_2,

Table 1
GROWTH FACTORS, HORMONES, AND OTHER GROWTH-REGULATING AGENTS OF EXOGENOUS CELLULAR ORIGIN

Well characterized growth factors
  Insulin-like growth factor I (IGF-I) or somatomedin C
  Insulin-like growth factor II (IGF-II), somatomedin A, or multiplication-stimulating activity (MSA)
  Epidermal growth factor (EGF) or urogastrone (URO)
  Fibroblast growth factor (FGF) or fibroblast-derived growth factor (FDGF)
  Nerve growth factor (NGF)
  Transforming growth factor α (TGF-α)
  Transforming growth factor β (TGF-β)
  Hemopoietin-2, interleukin-3 (IL-3), mast-cell growth factor (MCGF), or erythroid burst-promoting activity (EBPA)
  Interleukin-1 (IL-1)
  Interleukin-2 (IL-2) or T-cell growth factor (TCGF)
  Colony-stimulating factor 1 (CSF-1) or macrophage colony-stimulating factor (MCSF)
  Colony-stimulating factor 2 (CSF-2) or granulocyte-macrophage colony-stimulating factor (GM-CSF)
  Granulocyte colony-stimulating factor (G-CSF)
  Erythropoietin
  Platelet-derived growth factor (PDGF)
Partially characterized growth factors
  EGF-like mitogens
  TGF-like growth factors
  PDGF-like growth factors
  Melanocyte growth factor (MGF)
  Hepatopoietin
  Hepatocyte-stimulating factor (HSF)
  Prostate growth factor
  Cartilage-derived growth factor (CDGF)
  Chondrocyte growth factor (CGF)
  Bone-derived growth factor (BDGF)
  Osteosarcoma-derived growth factor (ODGF)
  Neuronal-derived transforming growth factor (ND-TGF)
  Glial growth-promoting factor (GGPF)
  Colostrum basic growth factor (CBGF)
  Endothelial cell growth factor (ECGF)
  Tumor angiogenesis factor (TAF)
  Hemopoietin-1
  IL-3-like proteins
  Eosinophil differentiation factor (EDF) or interleukin-4 (IL-4)
  B-cell growth factor 1 (BCGF) or B-cell stimulating factor 1 (BSF-1)
  B-cell stimulating factor 2 (BSF-2)
  B-cell differentiation factor (BCDF)
  Leukemia-derived growth factor (LDGF)
  Myelomonocytic growth factor (MMGF)
  Macrophage-derived growth factor (MDGF)
  Macrophage-activating factor (MAF)
  Erythroid-potentiating activity (EPA)
Hormones and other growth-regulating agents
  Insulin
  Growth hormone (GH) or somatotropin
  Thyroid-stimulating hormone (TSH) or thyrotropin
  Adrenocorticotropic hormone (ACTH)
  Gonadotropins
  Prolactin
  Melanocyte-stimulating hormone (MSH)
  Placental lactogen (PL)
  Thyroid hormone

Table 1 (continued)
GROWTH FACTORS, HORMONES, AND OTHER GROWTH-REGULATING AGENTS OF EXOGENOUS CELLULAR ORIGIN

Parathyroid hormone (PTH)
Calcitonin
Steroid hormones
Transferrin
Interferon (IFN)
Tumor necrosis factor α (TNF-α)
Tumor necrosis factor β (TNF-β) or lymphotoxin

Table 2
CHROMOSOMAL ASSIGNMENT AND LOCATION OF THE HUMAN GENES CODING FOR SOME HORMONES AND GROWTH FACTORS AND FOR SOME CELLULAR RECEPTORS

Chromosome	Location	Agent	Ref.
1	1p22—p22.1	NGF	13
2	2p11—p13	TGF-α	14
2	2q13—q21	IL-1	200
3	3q15—q25	Transferrin	15
3	3q26.2	Transferrin receptor	16
4	4q25—q27	EGF	17
4	4q26—q28	IL-2	18
5	5q21—32	CSF-2	19
5		Glucocorticoid receptor	20
6		IFn-α receptor	201
6		TNFs α and β	21
7	7pter—q22	PDGF-A	202
9		IFNs α and β	22
10	10p14—p15	IL-2 receptor	23
11	11p11.21	PTH	24
11	11p15	IGF-II	17
11	11p14.1	Insulin	24
12	12q22—q24.1	IGF-I	17
12		IFN-γ	25
17		GH	26
19	19p13.2—p13.3	Insulin receptor	27
19	19q13.1—q13.3	TGF-β	203
22	22q13.1	PDGF-β	94

and M (mitosis).[28] G_1 is the gap period between mitosis and the initiation of DNA synthesis, and G_2 is the period between S and M. Cells in G_2 contain double amount of DNA than cells in G_1. For most cells growing exponentially in culture, the interval between cell divisions is 10 to 30 hr. Variation in cell cycle between different cell types, or different environmental conditions, is mainly due to variation in the length of G_1, with the duration of S (6 to 8 hr) + G_2 (2 to 6 hr) + M (1 hr) being relatively constant. In addition, there is much variability in the length of G_1 among individual cells in a single population.[28]

Animal cells, both in vivo and in vitro, can also exist in a nongrowing, quiescent state during which they do not divide for long periods. Most frequently, normal cells that have

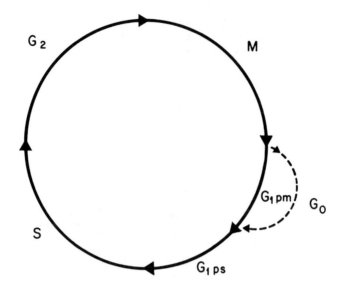

FIGURE 1. Schematic representation of the cell cycle.

ceased to grow have the G_1 content of DNA. Most or all of these quiescent cells are metabolically different from cycling G_1 cells and are considered to be in a qualitatively distinct state, termed G_0.

The crucial control events for the regulation of cell proliferation seem to reside in the G_1 phase of the cycle. Evidence has accumulated for the existence of a restriction or commitment point in mid- to late G_1, at which time a cell "decides" whether to initiate DNA synthesis and undergo division or to cease proliferation.[28] Environmental conditions, including the presence of certain hormones and growth factors, may influence this decision. Suboptimal conditions, like the absence of serum, usually shift normal cells into quiescence. In contrast, neoplastically transformed cells can lose this restriction-point control in whole or in part and may grow even in the absence of serum (or in the absence of specific hormones and growth factors contained in serum).

On the basis of the cell cycle arrest rapidly induced by brief serum starvation or transient inhibition of protein synthesis, the G_1 period of exponentially proliferating 3T3 cells can be subdivided into two physiologically discrete phases: one postmitotic (G_1pm) and one pre-DNA-synthetic (G_1ps).[29] During G_1pm the cells complete the growth factor-dependent processes leading to commitment for proliferation. Thereafter they enter the growth-factor-independent pre-DNA-synthetic part of G_1. The commitment process of G_1pm may be successfully completed in the presence of PDGF as the only supplied growth factor. EGF and insulin are insufficient for the completion of the commitment processes in G_1pm. Under conditions optimal for proliferation, the cells complete the commitment processes of G_1pm within a remarkably constant time period (3 to 4 hr after mitosis). The duration of G_1ps, on the other hand, shows a large intercellular variability, consistent with a transition probability event. In fact, G_1ps accounts for most of the variability in G_1 and cell cycle time.[29]

III. ROLE OF HORMONES AND GROWTH FACTORS IN THE REGULATION OF DNA SYNTHESIS AND CELLULAR PROLIFERATION

Hormones and growth factors are critically involved in the regulation of DNA synthesis and cellular proliferation in the respective target tissues. Hormones and growth factors contained in serum are required for the growth of cells cultured in vitro.[30] Although each of these factors may have a differential effect on DNA, RNA, protein synthesis, and cell division of particular cell types, the combined action of them is required for the optimal growth of cells.[31-34] Some growth factors, called competence factors, act early in the G_1 phase of the cell cycle, rendering cells competent to DNA replication whereas other growth factors, called progression factors, would allow the progression of cells through the prereplicative phase of the cycle.[35] Competence and progression factors would therefore act sequentially and synergistically in the cell cycle. The similar action of some growth factors in the prereplicative phase of cells suggests the operation of certain common intracellular pathways.[36]

The responsiveness of different types of cells to the mitogenic action of hormones and growth factors may show great variation according to the state of development and differentiation. In general, cells from newborn animals have greater mitogenic responsiveness to growth factors than adult cells. The responsiveness of cultured human fibroblasts to hormones and growth factors contained in fetal calves serum declines with age.[37,38] As a general rule, neoplastic cells are less dependent on the exogenous supply of hormones and growth factors than are normal cells.[39] However, the growth of neoplastic cells is not always characterized by an autonomy from the exogenous supply of hormones and growth factors.[40,41] Moreover, the proliferative response of neoplastic cells to growth factors may show great variation according to factors such as the microenvironment and the presence of chromosome abnormalities.[42] The progression, or regression, of hormone-dependent tumors may depend on nonspecific changes in gene expression.[43]

In addition to their characteristic effects on cell proliferation, growth factors may be involved in the control of physiological processes not associated with cell proliferation. For example, PDGF, EGF, and FGF regulate the production of IFN-γ by lymphocytes through a mechanism unrelated to their known proliferative effects on cells.[44] Other secretory processes may also be regulated by hormones and growth factors by mechanisms independent from DNA synthesis and cell proliferation. In general it can be said that hormones and growth factors are crucially involved in controlling the expression of differentiated functions in many types of cells and tissues.

Cultured growing cells, as well as certain normal and tumor cells growing in vivo, may produce hormone- and growth factor-like substances which are apparently useful and/or required for their proliferation. Murine BALB/c 3T3 cells in culture express a mRNA coding for proliferin, which is a 24,000 dalton protein with clear homology to preprolactin.[45] In the mouse, proliferin is a glycoprotein apparently produced only in the placenta, being especially abundant between days 8 and 10 of pregnancy.[46] A proliferin-related protein (PRP) has also been detected in the mouse placenta but, in contrast to proliferin, PRP is not present in proliferating BALB/c 3T3 cells in culture.[47] Structurally, PRP is closely related to both its peak level at the start of DNA synthesis. The high level of expression of the proliferin gene in certain growing cells in culture suggests that it may also serve as an autocrine growth factor in nonplacental cells in vitro, perhaps related to the immortality of these cell lines.

A. Mechanisms of Regulation of DNA Synthesis and Cell Proliferation by Hormones and Growth Factors

The mechanisms by which hormones and growth factors may influence DNA synthesis and cell proliferation are little understood. The mitogenic action of growth factors and the antiproliferative effects of interferon may be independent cell cycle events, as demonstrated in studies with vascular smooth muscle cells and endothelial cells.[48] Changes in cyclic AMP (cAMP) levels and polyamine synthesis may be involved in the regulation of the complex metabolic pathways leading to DNA synthesis and cell division. An increase in intracellular cAMP and polyamine concentration would depend on the activity of adenylate cyclase and ornithine decarboxylase, respectively. The latter enzyme is rate-limiting in the polyamine biosynthesis pathway. Insulin and several growth factors with mitogenic properties (EGF, PDGF, NGF) induce the expression of ornithine decarboxylase activity in the presence of amino acids like asparagine, and this activity is closely related to the synthesis of polyamines and the initiation of DNA synthesis.[49] Inhibition of ornithine decarboxylase without the concurrent administration of polyamine leads to cessation of DNA synthesis and cell proliferation. However, an increase in ornithine decarboxylase activity is necessary but is not sufficient for cell division to occur, and an increase in cAMP concentration may be neither necessary nor sufficient for the induction of cell division in particular types of cells.[50] It seems thus clear that polyamines and cAMP must act synergistically with other agents in the regulation of DNA synthesis and cellular proliferation stimulated by hormones and growth factors.

The level of expression of many different genes may show variation in relation to the different phases of the cell cycle in both normal and transformed proliferating cells.[51] Obviously, the fact that certain genes are expressed in a cell cycle-dependent manner does not mean that they are involved in cell cycle regulatory phenomena. It is thus difficult to recognize which of these genes are directly responsive, for example, for the induction of DNA synthesis. Over 3300 proteins can be distinguished in Swiss 3T3 cells by two dimensional electrophoresis on giant gels; 34 of these proteins are consistently induced more than threefold by growth factors (PDGF, CDGF) or fetal bovine serum after stimulation for 3 hr.[52] An additional 30 inductions are variable present and some of these inductions are inhibited by dexamethasone. However, only some of these proteins show a constant linear relationship to DNA synthesis, suggesting that they play an important role in early control of the cell cycle. Nuclear proteins that are differentially expressed in the cell cycle may be good candidates for playing a role in cell cycle control. A human histone H4 gene exhibits cell cycle-dependent changes in chromatin structure that correlate with its expression.[53] Identification of sets of genes that are specifically expressed in transition stages of the cell cycle, for example, in G_0/G_1 transition,[54] may be important for understanding the genetic mechanisms involved in the molecular controls of the cell cycle.

1. Cyclin

Cyclin is an acidic nuclear protein of 36,000 mol wt which may be involved in the control of DNA synthesis and cell proliferation. Cyclin has been identified in various types of normal or transformed cells from different animal species.[55,56] Cyclin is identical to a previously described protein called proliferating cell nuclear antigen (PCNA).[57-59] Isolated human epidermal basal cells synthesize very little cyclin, whereas simian virus 40 (SV40)-transformed keratinocytes synthesize this protein constitutively.[60] Induction of cyclin in serum-stimulated cultured cells is independent of DNA synthesis and cell transformation but it precedes DNA synthesis and correlates with cell proliferation.[61,62] Moreover, the synthesis of cyclin is not triggered by DNA replication, but changes in the nuclear distribution of cyclin are observed in relation to the S phase of the cell cycle.[63] Immunofluorescence analysis of synchronously growing transformed human amnion cells using autoantibodies specific for cyclin reveals remarkable changes in the nuclear distribution of cyclin during the S phase of the cell cycle.[64] Moreover, individual nuclei in polykaryons produced by polyethylene glycol-induced fusion of populations of transformed human amnion cells can control cyclin distribution and DNA synthesis in spite of the fact that they share a common cytoplasm.[60] These results suggest that cyclin may be an important component of the events leading to DNA replication and cell division.

B. Growth-Inhibiting Factors

Most frequently, hormones and growth factors have stimulatory actions on the respective target cells but some of these agents may have inhibitory effects when acting on particular types of cells. The growth inhibitory properties of glucocorticoids, for example, are profited for the treatment of many different types of inflammatory or neoplastic diseases.

Particular types of growth-inhibiting factors may be produced by either normal or neoplastic cells. An antimitogenic peptide factor with capability for inhibiting growth of cells in several normal and neoplastic cell lines has been detected in acetic acid extracts of the porcine hypothalamus.[66] A growth inhibitory factor is apparently produced by human monocytic leukemia cells and can be detected in the serum-free culture medium.[67] Another inhibitory factor, termed tumor-inhibitory factor (TIF), has been identified in rat ascites fluid and a human colon carcinoma cell line.[68,69] TIF acts in a partially reversible, nontoxic manner and coexists in the same malignant cells with TGF activity, which suggests that control of cell growth, either normal or neoplastic, may involve a balance between positive and negative factors.

Two types of tumor cell growth-inhibiting factors, termed TIF-1 and TIF-2, have been isolated and partially purified from the A673 human rhabdomyosarcoma cell line.[70,71] TIF-1 is a low-molecular-weight polypeptide (10,000 to 16,000 mol wt) that is inactivated by trypsin and dithiothreitol. TIF-1 inhibits the growth of a wide range of human tumor cells and virally transformed human and rat cells in soft agar and monolayer culture but stimulates the growth of normal human fibroblasts and a normal epithelial cell line.[70] TIF-1 blocks the stimulatory effect of both TGF-α and EGF in a soft agar assay with human tumor cells. TIF-2, purified from the same human rhabdomyosarcoma cell line, shares many properties in common with TIF-1 but can be distinguished by its molecular weight, chromatographic properties, and its effects on certain target cells.[71] TIF-1 and TIF-2 have no antiviral activity and their growth-inhibiting effects are reversible when the affected cells are no longer exposed to the factor. TIFs may be a family of related polypeptides which selectively inhibit the growth of tumor cells.

The physiological role of inhibitory growth factors in the regulation of cell proliferation is unknown. Regulation of many physiological phenomena and biological activities is probably governed by processes that activate or inhibit specific pathways. In the immune system, for example, the existence of antigen-specific and antigen-nonspecific regulatory proteins that either promote or inhibit immune functions has been well documented. Soluble immune response suppressor (SIRS) is a protein of 14,000 mol wt produced by mitogen- or interferon-activated suppressor T-cells that inhibits antibody secretion by B-lymphocytes and cell division by neoplastic cell lines.[72] IL-1, IL-2, and EGF each inhibit SIRS-mediated suppression of antibody secretion by cultured mouse spleen cells.[73] These results indicate that growth factors may act in some biological systems by interfering the action of growth-inhibiting factors.

Factors classically considered as stimulators of cell proliferation and/or transformation may act in certain types of cells, or even in the same type of cell under different conditions, as inhibitors of these phenomena. EGF, for example, stimulates growth in a wide variety of cells in vivo and in vitro but inhibits the growth of human A431 cells despite the fact that these cells express a very high number of EGF receptors.[74] TGF-β is capable of inducing anchorage-independent growth in mouse embryo AKR-2B cells as well as in normal rat kidney (NRK) cells, although in the latter case either EGF of TGF-α is required in association with TGF-β. However, TGF-β is a bifunctional factor because it is capable of inhibiting the anchorage-independent growth of many human tumor cell lines at concentrations in the same range as those than enhance the anchorage-independent growth of NRK or AKR-2B cells.[75] Moreover, in Fischer rat 3T3 fibroblasts transfected with a c-*myc* proto-oncogene, TGF-β can function as either an inhibitor or an enhancer of anchorage-independent growth, depending on the particular set of growth factors operant in the cell together with TGF-β.[75] These results suggest that transformation may result not only from the action of some positive, stimulating factors in susceptible cells, but also from the failure of some cells to respond

Table 3
ACUTE TRANSFORMING RETROVIRUSES AND THEIR RESPECTIVE ONCOGENES

Oncogene	Isolate origin	Prototype virus strain
abl	Rodent	Abelson murine leukemia virus (A-MuLV)
erb-A	Avian	Avian erythroblastosis virus (AEV)
erb-B	Avian	Avian erythroblastosis virus (AEV)
ets	Avian	Avian leukemia virus E26 (E26-ALV)
fes	Feline	Snyder-Theilen feline sarcoma virus (ST-FeSV)
fgr	Feline	Gardner-Rasheed feline sarcoma virus (GR-FeSV)
fms	Feline	McDonough feline sarcoma virus (MD-FeSV)
fos	Rodent	FBJ murine osteosarcoma virus (FBJ-MOV)
mht	Avian	Mill Hill 2 avian carcinoma virus (MH2-ACV)
mos	Rodent	Moloney murine sarcoma virus (M-MuSV)
myb	Avian	Avian myeloblastosis virus (AMV)
myc	Avian	MC29 avian myelocytomatosis virus (MC29-AMCV)
raf	Rodent	3611 murine sarcoma virus (3611-MuSV)
H-ras	Rodent	Harvey murine sarcoma virus (H-MuSV)
K-ras	Rodent	Kirsten murine sarcoma virus (K-MuSV)
rel	Avian	Avian reticuloendotheliosis virus (ARV)
ros	Avian	Rochester URII avian sarcoma virus (URII-ASV)
sis	Primate	Simian sarcoma virus (SSV)
ski	Avian	SKV 770 avian virus (SKV 770 AV)
src	Avian	Rous sarcoma virus (RSV)
yes	Avian	Yamaguchi avian sarcoma virus (Y-ASV)

to specific factors with inhibitory properties. Moreover, the general or local environmental conditions may critically modify the type of response of particular cells to different hormones or growth factors.

IV. ONCOGENES AND PROTO-ONCOGENES

Recent evidence strongly suggests the existence of important relationships between different types of growth-regulating agents, including several hormones and growth factors, proto-oncogene protein products, and the appearance and/or development of neoplastic diseases.[76,77] Oncogenes have been defined as genes with potential properties for the induction of neoplastic transformation in either natural or experimental conditions.[78] Most oncogenes have been isolated from acute transforming retroviruses, which act as oncogene transducers (Table 3). Acute transforming retroviruses are defective, nonreplicating viruses. These viruses are not infectious agents under natural conditions but have been isolated from different types of animal tumors and are originated from recombination events that may occur between cellular sequences and sequences derived from infectious chronic transforming retroviruses (Figure 2). No transforming sequences (oncogenes) are contained in chronic transforming retroviruses. The viral oncogenes (v-*onc* genes) contained in acute transforming retroviruses are responsible for the oncogenic potential exhibited by these viruses under experimental conditions and are derived from sequences of cellular origin, which are termed proto-oncogenes or c-*onc* genes. Some proto-oncogenes are apparently not present in the genomes of acute transforming retroviruses but have been isolated directly from the cellular genome. Proto-oncogenes have been detected in all vertebrate species studied so far and are also present in the genome of invertebrates.

Apparently, proto-oncogenes are normal constituents of the genome of all multicellular organisms, and at least some types of proto-oncogenes are present in unicellular eukaryotes.

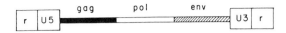
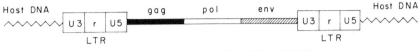

FIGURE 2. Schematic representation of the general structure of chronic and acute transforming retroviruses.

DNA sequences related to several oncogenes (v-*ras*, v-*mos*, and v-*abl*) are present in a variety of fungi and some of them are linked to actin-related sequences.[79] Moreover, DNA sequences related to the oncogene v-*myb* have been detected in primitive living organisms like archaebacteria.[80]

According to the criteria currently used for their definition, the total number of proto-oncogenes is probably limited to about 2 dozens. Most of the known proto-oncogenes have already been assigned to particular chromosomes, and particular regions within chromosomes, of the human karyotype and research work is under way in several laboratories for the chromosomal assignment and chromosomal sublocalization of proto-oncogenes in several nonhuman animal species. A list of the human proto-oncogenes with their respective chromosome assignment and chromosome location appears in Table 4 and is also indicated in the human idiogram represented in Figure 3.

A. Functions of the Proto-Oncogene Protein Products

The normal functions of proto-oncogene protein products are only partially understood but there is evidence that many of these products have important roles in the control mechanisms of metabolic processes as well as in the processes of cell differentiation and cell proliferation, which suggests *a priori* the existence of important interactions between proto-oncogene protein products and the cellular mechanisms of action of hormones and growth factors. The expression of many proto-oncogenes is subjected to developmental and tissue-specific regulation, and different types of hormones and growth factors may participate in these complex regulatory phenomena. Moreover, some proto-oncogene products are structurally and functionally related to either particular types of hormones and growth factors or to their cellular receptors.

B. Proto-Oncogenes and Cancer

Although viral oncogenes are genes with clear oncogenic potential in susceptible hosts, the possible role of proto-oncogenes in neoplastic diseases occurring in humans and other animals is currently a subject of high controversy.[104] Both quantitative and qualitative changes

Table 4
CHROMOSOMAL LOCATION OF HUMAN PROTO-ONCOGENES

Chromosome	Location	Proto-oncogene	Ref.
1	1p11—p13	c-N-*ras*	81, 82
1	1p32	c-B-*lym*-1	83
1	1p34—p36	c-*src*-2	84
1	1p36.1—p36.2	c-*fgr*	85
2	2p23—p24	c-N-*myc*	86
3	3p25	c-*raf*	87
5	5q34	c-*fms*	88
6	6p11—p12	c-K-*ras*-1	82
6	6q22—q24	c-*myb*	89
7	7pter—q22	c-*erb*-B	90
7	7q21—q31	c-*met*	91
8	8q11	c-*mos*	92
8	8q24	c-*myc*	93
9	9q34.1	c-*abl*	94
11	11q23—q24	c-*ets*-1	95
11	11p14.1	c-H-*ras*-1	96
12	12p11.1—p12.1	c-K-*ras*-2	82
12	12pter—q14	c-*int*-1	97
14	14q21—q31	c-*fos*	98
15	15q26.1	c-*fes*	94
17	17q21	c-*neu*	99
17	17q21—q22	c-*erb*-A	100
18	18q21.3	c-*yes*	101
20	20q12—q13	c-*src*-1	84
21	21q22.1—22.3	c-*ets*-2	102
22	22q13.1	c-*sis*	94
X		c-H-*ras*-2	103

of proto-oncogene expression could, in principle, be involved in oncogenic phenomena under either natural or experimental conditions. Four basic types of mechanisms could be responsible for a role of proto-oncogene alterations in carcinogenic processes:

1. Increased transcriptional activity of particular proto-oncogenes at unscheduled sites or times
2. Amplification and overexpression of proto-oncogenes
3. Translocation and/or rearrangement of proto-oncogenes
4. Mutation of proto-oncogenes

From these possible mechanisms, only translocation of particular types of proto-oncogenes is regularly detected in specific hematologic malignant diseases, but such translocations and rearrangements are heterogeneous at the molecular level. Thus, there is no strong evidence in favor of the hypothesis that proto-oncogene alterations constitute a kind of universal, indispensable and critical pathway leading to the origin of cancer, especially to the common forms of human solid tumors.[104] However, proto-oncogenes may be involved, in conjunction with other genes, in the multistage phenomena contributing to the origin and/or development of tumors. A detailed discussion on oncogenes and proto-oncogenes is contained in *Oncogenes*.[78]

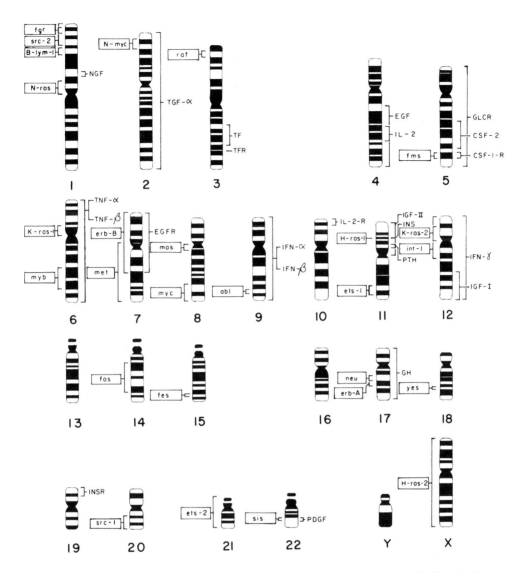

FIGURE 3. Human idiogram with indication of the location of proto-oncogenes (on the left side of each chromosome) and the location of the genes coding for some hormones and growth factors or their cellular receptors (on the right side of each chromosome). TF = transferrin, TFR = transferrin receptor, GLCR = glucocorticoid receptor, EGFR = EGF receptor, INSR = insulin receptor. Other symbols are indicated in the general list of abbreviations.

V. HORMONES AND GROWTH FACTORS IN RELATION TO ONCOGENIC PROCESSES

Hormones and growth factors are the most important environmental factors regulating cellular proliferation and differentiation, two processes that are critically altered in malignant neoplastic diseases. Thus, it is evident by itself that the hormonal environment of the whole organism, as well as the local action of hormones and growth factors, are of great importance for the origin and/or development of neoplastic diseases, not only under experimental conditions,[105-109] but also in human cancer.[110,111] Moreover, hormonal changes occur in specific types of tumors, which may allow the use of hormones, or their respective precursors and/or fragments, as potential tumor markers for the diagnosis and following-up of patients with these tumors.[112,113]

A. Animal Tumors

Animal tumors are constituted by a relatively autonomous population of transformed cells and are characterized, as a general rule, by a reduced dependence on the exogenous supply of hormones and growth factors in comparison with the respective normal tissue. Mouse cells transformed by chemical carcinogens may be less dependent on the exogenous supply of growth factors like EGF.[114] The requirements of mouse 3T3 cells for exogenous insulin and PDGF are diminished or abolished after SV40-induced transformation.[115] Cultured cells infected by acute transforming retroviruses may become independent of the presence of serum. Rat adrenal cells fully transformed by K-MuSV, which transduces the v-K-*ras* oncogene, loss the differential response to growth factors contained in serum.[116] Viral oncogene protein products may act as potent mitogens, inducing a relative autonomy of the transformed cells, which is reflected in their capacity to grow in the absence of exogenous growth factors. For example, the protein product of the v-*src* oncogene of RSV is a potent and complete mitogen which is able to stimulate host cell proliferation without the help of exogenous growth factors.[117,118]

The sensitivity of different tumors to particular types of hormones and growth factors may show great variation according to the type of tumor. Even a given type of tumor may respond in different ways to these agents according to the histological subtype and biological behavior. Human nonHodgkin lymphomas (NHLs), for example, may derive from both T- and B-cell precursors. The neoplastic lymphoid cells from these tumors respond normally (increased thymidine incorporation) to the respective specific interleukin growth factor, TCGF, or BCGF, when the tumors are histologically classified as small T-cell and small B-cell NHLs, respectively. Large-cell NHLs, on the other hand, are quite different tumors both clinically and biologically. Both the T- and B-cell types of large-cell NHLs tend to be rather aggressive tumors and the cells from these tumors do not show a normal response (i.e., do not show increased thymidine incorporation) when they are exposed to the specific interleukin growth factor.[119]

The cellular sensitivity of cells to different types of growth factors (TGF, EGF, PDGF) may show variation even among different subclones of the same cultured cells.[120] Clones from the rat fibroblastoid cell line F2408 that respond most efficiently to one growth factor are also the most sensitive to the other growth factor, and resistant clones are resistant to all of the growth factors examined. Moreover, clones responsive to growth factors are the more susceptible to transformation induced by two acute transforming retroviruses, K-MuSV and the A-MuLV, which transduce the oncogenes v-K-*ras* and v-*abl*, respectively. Complex interactions between different hormones and growth factors may be important for the cloning of human tumor cells.[121]

1. The Autocrine Hypothesis of Oncogenesis

Cells transformed by acute retroviruses transducing different types of oncogenes may become independent from the exogenous supply of growth factors and may produce different types of these factors, especially TGFs.[122-130] Similar factors may be present in human tumor cells.[131,132] The possibility that such factors are involved in the expression of a transformed phenotype is reinforced by the observation that the expression of a growth factor cDNA in a factor-dependent cell line may result in autonomous growth and tumorigenicity.[133] According to the autocrine hypothesis of oncogenesis, transformed cells maintained in culture would produce factors that, upon interaction with the specific receptors on the cell surface, would stimulate their own growth in monolayer or in suspension culture.[134] A similar phenomenon could be responsible for the autonomous growth of malignant tumor cell populations in vivo.

Evidence in favor of the autocrine hypothesis of neoplasia has been obtained, for example, with the study of in vitro growth of human small cell lung cancer (SCLC) cell lines, which produce and secrete bombesin-like peptides and express high-affinity receptors for these peptides.[135-137] A monoclonal antibody to bombesin binds to the carboxyterminal region of bombesin-like peptides, blocks the binding of the hormone to cellular receptors, and inhibits the clonal growth of SCLC cells in vitro and the growth of SCLC xenografts in vivo.[138] Avian retroviruses containing oncogenes of the v-*src* family render v-*myb*-transformed myeloblasts and, in most cases, v-*myc*-transformed macrophages, independent of exogenous growth factors.[139] In v-*myb*-transformed myeloid cells, this independence is achieved via the induced secretion of a specific growth factor activity.

However, the general validity of the autocrine hypothesis of neoplasia is questionable. There is evidence that transformation induced by viral oncogenes is not necessarily dependent on the endogenous production of growth factors according to an autocrine type of mechanism. For example, A-MuLV is capable of transforming growth factor-dependent, nontumorigenic myeloid cell lines into growth factor-independent, tumorigenic cell lines but the v-*abl*-induced transformation is not associated with an endogenous synthesis of growth factors capable of determining autocrine stimulation.[140-142] A-MuLV-derived transformants of mast cells do not express or secrete detectable levels of the specific growth factor nor is their growth inhibited by an antiserum against the growth factor.[142] Such results, as well as the results obtained in other systems,[143] argue strongly against the general validity of the autocrine hypothesis of neoplastic transformation.

2. Cellular Responses to v-onc Products

Acute transforming retroviruses carrying v-*onc* genes are capable of inducing malignant transformation in a diversity of cells both in vivo and in vitro. However, the production of both a viral oncogene protein and a transforming growth factor may not be sufficient for the maintenance of the transformed phenotype. Flat revertants of K-MuSV-transformed cells, although expressing elevated levels of the v-K-*ras* p21 protein product and producing transforming growth factors, nevertheless express a majority of the properties associated with the nontransformed state.[144] In murine myeloid leukemic myeloblastic cells maintained in suspension cultures, no relationship could be demonstrated between the capacity of a cell line to grow autonomously and its capacity to release autostimulating growth factors.[145] It may be concluded that the expression of a viral oncogene protein product is not necessarily associated with the expression of a transformed phenotype and that transformation induced by viral oncogenes may be independent from the endogenous production of growth factors by an autocrine type of growth stimulation.

The avian acute leukemia retrovirus E26 contains two oncogenes, v-*myb* and v-*ets*, and is capable of inducing malignant transformation of various cell types in vitro (myeloblasts, erythroblasts, fibroblasts) and causes a mixed erythroid/myeloid leukemia in chicks.[146] E26 virus mutants that are temperature-sensitive (ts) for myeloblast transformation have been isolated.[147] At the permissive temperature, ts E26-transformed myeloid cells resemble macrophage precursors and proliferate rapidly, provided the medium contains a specific growth factor (chicken myelomonocytic growth factor). When shifted to the nonpermissive temperature, the cells stop growing, differentiate into macrophage-like cells, and lose the responsiveness to the particular growth factor but secrete the factor or a substance similar to the factor.[147]

Chick heart mesenchymal cells transformed with a c-*myc* proto-oncogene and treated with heparin become proliferative quiescent and assume a normal morphology but are hypersensitive to stimulation with EGF, FGF, and PDGF.[148] These cells, however, do not respond to insulin. On the other hand, the same chicken cells transformed with a v-*ras* oncogene and treated with heparin are refractory to mitogenic stimulation with EGF or FGF but are hypersensitive to insulin or IGFs. It is thus apparent that transformation by different viral oncogenes may result in specific altered responses of the cells to specific types of hormones and growth factors and that in some cases the v-*onc*-transformed cells become hypersensitive to the stimulation by particular types of hormones and growth factors.

3. Growth Factor Independence and Proto-Oncogene Expression

Progression of tumor cells through stages of increased tumorigenic properties is frequently accompanied by a decreased ability of the cells to differentiate as well as by a change from growth factor dependence to growth factor independence. It may be argued that the growth factor-independent proliferation of malignant cells may be due to the differential expression of proto-oncogenes.[149] In order to test this hypothesis, the differential expression of a series of 18 different proto-oncogenes was examined in a group of cultured murine T-cell lymphomas that were induced following virus inoculation into, or X-irradiation of, C57BL/6 mice.[149] Two classes of T-lymphoma cell lines were studied: growth factor-dependent cells and growth factor-independent cells. Of the 18 proto-oncogenes examined, 5 (c-*myc*, c-*myb*, c-*abl*, c-H-*ras*, and c-K-*ras*) were consistently expressed in all cell lines tested. The results indicated no significant differences in proto-oncogene expression between growth factor-dependent and growth factor-independent cells, which demonstrates that proto-oncogene expression is in itself not sufficient to induce progression of T-lymphoma cells to growth factor independence.[149] The results further suggest that expression of other cellular genes, which may not be proto-oncogenes, is required for the progression of tumors from growth factor dependence to growth factor independence.

B. Plant Tumors

A phenomenon that is something similar to the autonomous, growth-independent proliferation of animal tumor cells is observed in tumors occurring in plants. Dicotyledonous plants may be affected by a neoplastic disease, crown gall, which is induced by the bacterium *Agrobacterium tumefaciens*.[150] The bacteria transfer a specific portion of a large plasmid, Ti, to the plant cell and the transferred DNA may be stably integrated into the plant nuclear DNA, which results in autonomous growth in the absence of the phytohormones, auxin, and cytokinin.[151,152] The tumor morphology locus, *tms*, of the Ti plasmid is transcribed into two mRNA species, and one of these species code for a protein which has significant homology to the adenine binding region of an enzyme (*p*-hydroxybenzoate hydroxylase from *Pseudomonas fluorescens*), suggesting that the protein binds adenine either as substrate or cofactor.[153]

1. Hormones and Oncogene Products in Plant Tumors

The origin of the sequences at the *tms* locus has not been determined but the possibility of an eukaryotic origin exists, and if this origin is confirmed such sequences contributing to the origin of neoplastic diseases in plants and conferring hormone- and growth-factor-independent growth to plant tissues, would be analogous to the oncogenes described in animal cells. At least three genes, or oncogenes, of the Ti plasmid are responsible for the transformed plant cell phenotype: oncogenes 1 and 2, which have roles in auxin independence, and gene 4, which is involved in cytokinin independence. Gene 4 from the T-region of Ti plasmids is responsive for cytokinin effects in crown gall cells and it has been demonstrated that this gene codes for an enzyme of cytokinin biosynthesis.[154] However, there are genetic differences between Ti plasmids from different *A. tumefaciens* strains, and differences in the oncogene complements of the Ti plasmids as well as other Ti oncogenes and undefined host factures may influence their host range characteristics and tumorigenic effects.[155,156] In particular, it is not known if specific plant hormones are involved in the regulation of plant tumor growth and if plant tumors are characterized by hormone independency. Certain plant hormones, like auxin, act very rapidly at the nuclear level in the regulation of transcriptional functions in plant tissues.[157] Phosphotyrosine is more abundant in normal plant tissues than in normal animal tissues but, in contrast to animal cells where transformation may result in a severalfold increase in phosphotyrosine content, no increase of the modified amino acid is detected in Ti plasmid-transformed crown gall tissue.[158] Further studies are required for a better characterization of the relationships between hormones, growth factors, and putative oncogene products in the tumorigenic processes occurring in plants.

VI. INFLUENCE OF HORMONES AND GROWTH FACTORS ON PROTO-ONCOGENE EXPRESSION

The levels of expression of some proto-oncogenes, proto-oncogene-like sequences, or other cellular genes may show important fluctuation in association with the different stages of the cell cycle and the proliferative stimuli acting on the susceptible cells. In general, proto-oncogenes should be considered, like other cellular genes, as cell cycle-dependent genes.[46,159,160] However, there are exceptions to this general rule and the levels of expression of certain proto-oncogenes, like c-*myc*, are constant throughout the cell cycle.[161,162] However, the level of c-*myc* proteins may show transient variation when resting cells are stimulated by serum.[163] In any case, it is clear that the level of expression of some proto-oncogenes will show variation according to the action of agents that, like hormones and growth factors, are involved in the control of cell proliferation. Moreover, the results of important research work performed during the last few years in several laboratories has demonstrated the existence of close relationships between hormones, growth factors, the cellular receptors for hormones and growth factors, and the protein products of proto-oncogenes. However, it seems obvious that not all the genes (including proto-oncogenes) regulated by hormones and growth factors in a cell cycle-dependent manner are necessarily involved in the control mechanisms of the cell cycle. The biochemical phenomena related to the expression of cell cycle-dependent genes are also not clear. These genes may be fully expressed, or may be even overexpressed, in the presence of cycloheximide,[164] which indicates that protein synthesis is not required for the induction of at least some genes (including proto-oncogenes) regulated during the cell cycle by hormones and growth factors. The molecular mechanisms involved in this induction remain unknown.

The influence of hormones and growth factors on proto-oncogene expression may be conveniently ascertained by measuring the mRNA levels for different types of proto-oncogene protein products after stimulation of quiescent cultured cells with serum (which contains a complex mixture of hormones and growth factors) or known amounts of each one of the purified hormones and growth factors. In general, the agents capable of initiating the first phase of a proliferative response, termed competence, can enhance the expression of specific proto-oncogenes. For example, a strong induction of c-*myc* expression occurs very soon following the activation of T-lymphocytes with concanavalin A, B-lymphocytes with lipopolysaccharide (LPS), and 3T3 fibroblasts in PDGF.[165] A large increase in c-*myc* expression is also induced in human T-lymphocytes stimulated with phytohemagglutinin (PHA).[166] Early increase of c-*myc*-encoded proteins is observed in synchronized mouse 3T3 cells upon stimulation by growth factors contained in serum.[167] Thyrotropic hormone (TSH) and cAMP enhance the expression of c-*myc* and c-*fos* in cultured thyroid cells.[168]

The c-*myc* protein may act as an intracellular competence factor, rendering cells competent to progress into S phase of the cycle. Upon microinjection into nuclei of quiescent mouse fibroblasts (Swiss 3T3 cells), the c-*myc* protein cooperates with platelet-poor plasma in the stimulation of cellular DNA synthesis.[169] Extracellular competence factors include growth factors like PDGF and FGF.

A. Mechanisms of Proto-Oncogene Regulation by Hormones and Growth Factors

The regulation of proto-oncogene expression by hormones or growth factors may involve activation of phosphoinositide turnover and Ca^{2+} mobilization. In Swiss 3T3 cells, PDGF induces Ca^{2+} mobilization and c-*myc* expression, and a similar expression is stimulated by the calcium ionophores, A23187 and ionomycin.[170] These results suggest that Ca^{2+} may serve as a messenger for expression of the c-*myc* proto-oncogene induced by PDGF and other growth factors, which probably involves stimulation of phosphoinositide turnover to activate protein kinase C and to generate inositol trisphosphate, the latter being responsible for Ca^{2+} mobilization from intracellular stores to the cytoplasm.[171]

B. Kinetics of Growth Factor-Regulated Proto-Oncogene Expression

An analysis has been performed with quiescent mouse fibroblasts (NIH/3T3 cells) stimulated after 3 days of serum deprivation with 20% fetal calf serum (FCS), which leads to synchronous cell cycle initiation.[172] The results clearly demonstrated that:

1. The expression of the proto-oncogene c-*fos* increases as early as 5 min after stimulation and reaches at 1 hr levels that are two orders of magnitude higher than those existing before serum stimulation, decreasing to basal levels 1 hr later.
2. Transcripts of the c-*myc* gene are significantly elevated after 30 min of stimulation and are induced more than 20-fold 1 hr after the addition of serum, followed by a slow decrease until reaching the basal level of quiescent cells after about 18 hr.
3. The levels of c-H-*ras* and c-K-*ras* mRNA increase no more than threefold in serum-stimulated cells at any of the time points analyzed.
4. Transcripts of c-*abl* and c-*raf* do not show significant changes in the level of expression.
5. Transcripts of c-*sis* are not detectable in this cell system.

When the action of purified growth factors was studied in the same cellular system the following findings were obtained:[172]

1. Expression of c-*fos* increases markedly 1 hr after stimulation with FGF or PDGF and is then abolished rapidly, whereas in cells treated with EGF the induction of c-*fos* is much less pronounced.
2. The strong initial induction of c-*myc* expression by FCS is reproduced using either FGF, PDGF, or EGF alone but it is much more stable in FCS- or FGF-treated cells than in PDGF- or EGF-stimulated cells, where the concentration of c-*myc* decreases to relatively low levels 2 hr after stimulation.
3. No changes in c-H-*ras* expression are observed after stimulation with FGF, PDGF, or EGF.

C. Mechanisms of Growth Factor-Regulated Proto-Oncogene Expression

The action of the protein synthesis inhibitor cycloheximide in serum-stimulated quiescent NIH/3T3 cells yields interesting results:[172]

1. Cultivation of the quiescent cells in 0.05% FCS for 4 hr in the presence of cycloheximide results in a slightly increased expression of both c-*myc* and c-*fos* mRNA.
2. Cultivation of the quiescent cells for 4 hr with 20% FCS and cycloheximide leads to the accumulation of c-*fos* and c-*myc* mRNA concentrations severalfold higher than the maximum observed in the absence of cycloheximide.

These results suggest that c-*fos* and c-*myc* genes may be modulated through repression by an unstable protein or by autoregulation. The mRNAs of these proto-oncogenes could become more stable when a protein normally causing their rapid degradation is absent as a consequence of cycloheximide treatment. Results obtained in other studies suggest that c-*myc* expression is regulated mainly at the post-transcriptional level.[173,174] The proto-oncogene is transcribed at a high rate in G_0-arrested hamster lung fibroblasts, although the level of mature c-*myc* mRNA is barely detectable. Stimulation by growth factors is not accompanied by any appreciable change in the transcription rate of c-*myc*, which supports a model of regulation of c-*myc* expression at the level of mRNA degradation.[174]

D. Physiological Significance of Growth Factor-Regulated Proto-Oncogene Expression

The rapid induction of c-*fos* and c-*myc* expression after stimulation by either serum or purified growth factors, in contrast to the slight response or lack of response of other proto-oncogenes, has been detected in an independent study.[175] A transcriptional enhancer element is contained upstream from the c-*fos* proto-oncogene.[176] The results suggest that c-*fos* and c-*myc* may have a primary role in the control of proliferation in at least certain types of cells. In general, expression of c-*fos* and c-*myc* is higher in rapidly proliferating than in quiescent cells.[172] However, persistence of the competent state in cells like mice fibroblasts may be independent from c-*fos* and c-*myc* expression. Cells can remain competent in spite of expressing c-*fos* and c-*myc* genes at very low levels (i.e., at levels similar to those of quiescent cells), which suggests that these proto-oncogene products are not directly involved in maintenance of competence.[177] Apparently, cells rendered competent by growth factors like PDGF do not require high levels of c-*fos* and c-*myc* gene expression to progress through G_1. However, the transient induction of c-*fos* and c-*myc* in quiescent stimulated cells could play an important role in making the cells competent.[177]

The c-*fos* gene would participate, in association with other cellular genes, in the regulatory phenomena related to expression of cellular differentiation.[140] The highest levels of c-*fos* and c-*myc* expression are detected in NIH/3T3 cells transformed by either a long terminal repeat (LTR)-activated c-*raf* proto-oncogene or the McDonough strain of FeSV, which carries the v-*fms* oncogene. However, the levels of both c-*myc* and c-*fos* may be as high or even higher at certain times in serum-stimulated normal cells than in NIH/3T3 transformed cells. Transient accumulation of c-*fos* mRNA following serum stimulation requires a conserved 5' element and c-*fos* 3' sequences.[179]

As a general rule, in the NIH/3T3 system there is a correlation between the expression of c-*fos* and c-*myc* proto-oncogenes and cellular proliferative activity.[172] Studies performed in other types of cells or tissues (fetal membranes and placenta, bone marrow, early fetal liver, and differentiated macrophages) suggest a correlation between c-*fos* expression and cell differentiation.[180-182] Rapid increase in c-*fos* expression is observed when PC12 rat pheochromocytoma cells are induced to differentiate by treatment with NGF.[183] This induction is enhanced more than 100-fold in the presence of peripherally active benzodiazepines. Expression of the c-*fos* proto-oncogene and a c-*fos*-related gene is increased in quiescent BALB/c 3T3 mouse fibroblasts upon treatment with purified PDGF.[184] Rapid and early expression of c-*fos* proto-oncogene also occurs when the promyelocytic cell line HL-60 or the monocytic cell line U-937 are induced to differentiation with phorbol ester.[185] Expression of c-*fos* is inducible when quiescent terminally differentiated macrophages are treated with the macrophage-specific growth factor CSF-1, although the kinetics of c-*fos* induction in this system is entirely different from that in growth factor-stimulated fibroblasts, which supports the view that the c-*fos* gene protein product may serve different functions in different cell types.[186] Moreover, rapid induction of c-*fos* is not observed when the HL-60 cell line is induced to differentiation by either calcitriol or retinoic acid,[186] which indicates that c-*fos* induction is not universally related to cell differentiation phenomena.

Growth control may be associated with hormone- and growth factor-dependent cell-cycle-timed proto-oncogene expression. Hypothetically, a relaxed regulation of proto-oncogene expression may contribute to loss of growth control in neoplastic transformation but such alteration should have some kind of specificity for both the cell type and the proto-oncogene. Although expression of both c-*myc* and c-K-*ras* is dependent upon the cellular growth state, c-*myc*, but not c-K-*ras*, is expressed constitutively in chemically transformed mouse fibroblasts.[187] Expression of c-*myc* is, however, independent of the different phases of the cell cycle in both normal and transformed cells.[161-163]

E. Action of Growth Factors in v-*onc*-Transformed Cells

The growth of oncogene-transformed cells may not be completely autonomous but may depend on the presence of particular growth factors. Fischer rat 3T3 fibroblasts transfected with a c-*myc* gene linked to an SV40 promoter can be induced to grow and form colonies in soft agar by treatment either with EGF alone or with a combination of PDGF and TGF-β.[75] The mechanism involved in this induction, however, may be different, as ascertained by the action of a differentiation-inducing agent like retinoic acid. Colony formation induced by the combined action of PDGF and TGF-β in c-*myc* transformed cells is 100-fold more sensitive to inhibition by retinoic acid than is colony formation induced by treatment of the same cells with EGF.[188]

F. Summary

Expression of certain proto-oncogenes, especially c-*fos* and c-*myc* is closely related to cell proliferation and/or differentiation induced by particular types of growth factors in certain types of cells responsive to these factors. The same or other proto-oncogenes may have a similar role in the responses elicited by different hormones and growth factors in different types of cells and tissues. The possible role of hormones, growth factors and proto-oncogenes in the origin and/or development of malignant diseases is not immediately apparent on the basis of such results. However, it seems likely that hormones, growth factors, and proto-oncogene protein products participate through complex interactions in the multi-stage processes leading to tumorigenesis.

VII. COMMON FUNCTIONS OF HORMONES, GROWTH FACTORS, AND ONCOGENE PROTEIN PRODUCTS

Hormones, growth factors, and oncogene protein products may share some common cellular mechanisms of action, exerting similar or identical functional changes at the cellular level. A particular mRNA species, termed pTR1 RNA, is present in significantly high levels in rat fibroblasts transformed by polyoma virus, RSV, or the EJ mutant human c-H-*ras* oncogene, and the same RNA is rapidly induced by the addition of EGF to the culture medium of rat fibroblasts.[189] Both the growth factor and the oncogene proteins control expression of the corresponding gene at the transcriptional level. pTR1 RNA is not present in serum-stimulated cells and is apparently not a participant of the "normal" or general cellular growth response. The function of the pTR1-derived protein has not been established but the results indicate that both growth factors and oncogene proteins may act by controlling the transcriptional activity of similar or identical genes.

Another protein that may be induced by both growth factor(s) and an oncogene protein product is the mammalian heat shock protein 70 (hsp 70). Expression of the hsp 70 gene may be induced by both serum stimulation and the c-*myc* protein.[190,191] Protein hsp 70 could be involved in the control of cell growth, acting during the S phase of the cell cycle. Heat shock proteins usually have a nuclear location, suggesting a role in the regulation of genomic functions. In parasitic protozoa like *Trypanosoma* and *Leishmania*, expression of heat shock genes may be responsible for differentiation processes involving extensive morphological and physiological changes upon transmission of the parasite from the insect vector to the mammalian host, which implicates a marked difference in temperature (from 25°C in the vector to 37°C in the mammalian host).[193]

It is possible that some protein products of proto-oncogenes are involved in the regulation of differentiated functions of the nervous system, including the generation of neurotransmitters. This possibility is suggested by the observation that in neuronal cells (quail embryo neuroretina cultures) the protein product of the v-*src* oncogene, pp60 $^{v\text{-}src}$, can selectively stimulate the activity of the enzyme glutamic acid decarboxylase (GAD), which is responsible for the synthesis of the neurotransmitter gamma-aminobutyric acid (GABA).[193] GAD activity is also stimulated in the same cells by retroviruses carrying the v-*mil* oncogene. The possible physiological significance of these results is suggested by the fact that the c-*src* proto-oncogene is expressed at high levels specifically during organogenesis of neural tissues in humans and other vertebrates.[178,194-197] In both chick neural retina and cerebellum, expression of the c-*src* proto-oncogene occurs in developing neurons at the onset of differentiation, at a stage when cell proliferation ceases.[198,199] In general, expression of the c-*src* gene is associated more with cell differentiation than with cell proliferation.[197] Further studies are required for a better characterization of the respective roles of hormones, growth factors, and proto-oncogene protein products in the regulation of the functional properties of different organs and tissues at different stages of development.

REFERENCES

1. **Pimentel, E.**, Cellular mechanisms of hormone action. I. Transductional events, *Acta Cient. Venez.*, 29, 73, 1978.
2. **Pimentel, E.**, Cellular mechanisms of hormone action. II. Posttransductional events, *Acta Cient. Venez.*, 29, 147, 1978.
3. **Gospodarowicz, R. A. and Moran, J. S.**, Growth factors in mammalian cell culture, *Ann. Rev. Biochem.*, 45, 531, 1976.
4. **Golde, D. W., Herschman, H. R., Lusis, A. J., and Groopman, J. E.**, Growth factors, *Ann. Intern. Med.*, 92, 650, 1980.
5. **Gospodarowicz, D.**, Growth factors and their action in vivo and in vitro, *J. Pathol.*, 141, 201, 1983.
6. **Carpenter, G. and Cohen, S.**, Peptide growth factors, *Trends Biochem. Sci.*, 9, 169, 1984.
7. **James, R. and Bradshaw, R. A.**, Polypeptide growth factors, *Ann. Rev. Biochem.*, 53, 259, 1984.
8. **Courty, J., Courtois, Y., and Barritault, D.**, Propriétés in vitro de quelques facteurs de croissance et effets in vivo, *Biochimie*, 66, 419, 1984.
9. **Heinemann, V. and Jehn, U.**, Wachstumfaktoren: Eine neue Dimension im Verständnis der Onkogenese, *Klin. Wochenschr.*, 63, 740, 1985.
10. **Goustin, A. S., Leof, E. B., Shipley, G. D., and Moses, H. L.**, Growth factors and cancer, *Cancer Res.*, 46, 1015, 1986.
11. **Alderman, E. M., Lobb, R. R., and Fett, J. W.**, Isolation of tumor-secreted products from human carcinoma cells maintained in a defined protein-free medium, *Proc. Natl. Acad. Sci. U.S.A.*, 82, 5771, 1985.
12. **Plouët, J., Courty, J., Olivié, M., Courtois, Y., and Barritault, D.**, A highly reliable and sensitive assay for the purification of cellular growth factors, *Cell. Mol. Biol.*, 30, 105, 1984.
13. **Zabel, B. U., Eddy, R. L., Lalley, P. A., Scott, J., Bell, G. I., and Shows, T. B.**, Chromosomal locations of the human and mouse genes for precursors of epidermal growth factor and the beta subunit of nerve growth factor, *Proc. Natl. Acad. Sci. U.S.A.*, 82, 469, 1985.
14. **Brissenden, J. E., Derynck, R., and Francke, U.**, Mapping of transforming growth factor alpha gene on human chromosome 2 close to the breakpoint of the Burkitt's lymphoma 7(2;8) variant translocation, *Cancer Res.*, 45, 5593, 1985.
15. **Huerre, C., Uzan, G., Grzeschik, K. H., Weil, D., Levin, M., Hors-Cayla, M.-C., Boué, J., Kahn, A., and Junien, C.**, The structural gene for transferrin (TF) maps to 3q21 → 3qter, *Ann. Genet. (Paris)*, 27, 5, 1984.
16. **Rabin, M., McClelland, A., Kühn, L., and Ruddle, F. H.**, Regional localization of the human transferrin receptor gene to 3q26.2 → ter, *Am. J. Human Genet.*, 37, 1112, 1985.
17. **Morton, C. C., Byers, M. G., Naka, H., Bell, G. I., and Shows, T. B.**, Human genes for insulin-like growth factors I and II and epidermal growth factors are located on 12q 22 → q 24.1, 11p 15, and 4q 25 → q 27, respectively, *Cytogenet. Cell Genet.*, 41, 245, 1986.
18. **Seigel, L. J., Harper, M. E., Wong-Staal, F., Gallo, R. C., Nash, W. G., and O'Brien, S. J.**, Gene for T-cell growth factor: location on human chromosome 4q and feline chromosome B1, *Science*, 223, 175, 1984.

19. **Huebner, K., Isobe, M., Croce, C. M., Golde, D. W., Kaufman, S. E., and Gasson, J. C.,** The human gene encoding GM-CSF is at 5q21-q32, the chromosome region deleted in the 5q- anomaly, *Science*, 230, 1282, 1985.
20. **Gehring, U., Segnitz, B., Foellmer, B., and Francke, U.,** Assignment of the human gene for the glucocorticoid receptor to chromosome 5, *Proc. Natl. Acad. Sci. U.S.A.*, 82, 3751, 1985.
21. **Nedwin, G. E., Jarrett-Nedwin, J., Smith, D. H., Naylor, S. L., Sakaguchi, A. Y., Goeddel, D. V., and Gray, P. W.,** Structure and chromosomal localization of the human lymphotoxin gene, *J. Cell. Biochem.*, 29, 171, 1985.
22. **Owerbach, D., Rutter, W. J., Shows, T. B., Gray, P. W., Goeddel, D. V., and Lawn, R. M.,** Leukocyte and fibroblast interferon genes are located on human chromosome 9, *Proc. Natl. Acad. Sci. U.S.A.*, 78, 3123, 1981.
23. **Leonard, W. J., Donlon, T. A., Lebo, R. V., and Greene, W. C.,** Localization of the gene encoding the human interleukin-2 receptor on chromosome 10, *Science*, 228, 1547, 1985.
24. **Chaganti, R. S. K.,** Germ-line chromosomal localization of genes in chromosome 11p linkage: parathyroid hormone, beta-globin, c-Ha-ras-1, and insulin, *Somat. Cell Mol. Genet.*, 11, 197, 1985.
25. **Sagar, A. D., Sehgal, P. B., May, L. T., Inouye, M., Slate, D. L., Shulman, L., and Ruddle, F. H.,** Interferon-beta-related DNA is dispersed in the human genome, *Science*, 223, 1312, 1984.
26. **Owerbach, D., Rutter, W. J., Martial, J. A., Baxter, J. D., and Shows, T. B.,** Genes for growth hormone, chorionic somatomammotropin, and growth hormone-like gene on chromosome 17 in humans, *Science*, 209, 289, 1980.
27. **Yang-Feng, T. L., Francke, U., and Ullrich, A.,** Gene for human insulin receptor: localization to site on chromosome 19 involved in pre-B-cell leukemia, *Science*, 228, 728, 1985.
28. **Pardee, A. B., Dubrow, R., Hamlin, J. L., and Kletzien, R. F.,** Animal cell cycle, *Ann. Rev. Biochem.*, 47, 715, 1978.
29. **Zetterberg, A. and Larsson, O.,** Kinetic analysis of regulatory events in G_1 leading to proliferation or quiescence of Swiss 3T3 cells, *Proc. Natl. Acad. Sci. U.S.A.*, 82, 5365, 1985.
30. **Holley, R. W. and Kiernan, J. A.,** Control of the initiation of DNA synthesis in 3T3 cells: serum factors, *Proc. Natl. Acad. Sci. U.S.A.*, 71, 2908, 1974.
31. **Olashaw, N. E., Harrington, M., and Pledger, W. J.,** The regulation of the cell cycle by multiple growth factors, *Cell Biol. Int. Rep.*, 7, 489, 1983.
32. **Shipley, G. D., Childs, C. B., Volkenant, M. E., and Moses, H. L.,** Differential effects of epidermal growth factor, transforming growth factor, and insulin on DNA and protein synthesis and morphology in serum-free cultures of AKR-2B cells, *Cancer Res.*, 44, 710, 1984.
33. **McKeehan, W. L., Adams, P. S., and Rosser, M. P.,** Direct mitogenic effects of insulin, epidermal growth factor, glucocorticoid, cholera toxin, unknown pituitary factors and possibly prolactin, but not androgen, on normal rat prostate epithelial cells in serum-free, primary cell culture, *Cancer Res.*, 44, 1998, 1984.
34. **Balk, S. D., Morisi, A., Gunther, H. S., Svoboda, M. F., Van Wyk, J. J., Nissley, S. P., and Scanes, C. G.,** Somatomedins (insulin-like growth factors), but not growth hormone, are mitogenic for chicken heart mesenchymal cells and act synergistically with epidermal growth factor and brain fibroblast growth factor, *Life Sci.*, 35, 335, 1984.
35. **O'Keefe, E. J. and Pledger, W. J.,** A model of cell cycle control: sequential events regulated by growth factors, *Mol. Cell. Endocrinol.*, 31, 167, 1983.
36. **Westermark, B. and Heldin, C.-H.,** Similar action of platelet-derived growth factor and epidermal growth factor in the prereplicative phase of human fibroblasts suggests a common intracellular pathway, *J. Cell. Physiol.*, 124, 43, 1985.
37. **Plisko, A. and Gilchrest, B. A.,** Growth factor responsiveness of cultured human fibroblasts declines with age, *J. Gerontol.*, 38, 513, 1983.
38. **Phillips, P. D., Kaji, K., and Cristofalo, V. J.,** Progressive loss of the proliferative response of senescing WI-38 cells to platelet-derived growth factor, epidermal growth factor, insulin, transferrin, and dexamethasone, *J. Gerontol.*, 39, 11, 1984.
39. **Scher, C. D. and Todaro, G. J.,** Selective growth of human neoplastic cells in medium lacking serum growth factor, *Exp. Cell Res.*, 68, 479, 1972.
40. **Metcalf, D., Moore, M. A. S., Sheridan, J. W., and Spitzer, G.,** Responsiveness of human granulocytic leukemia cell to colony-stimulating factor, *Blood*, 43, 847, 1974.
41. **Greenberg, B. R., Hirasuna, J. D., and Woo, L.,** In vitro response to erythropoietin in erythroblastic transformation of chronic myelogenous leukemia, *Exp. Hematol.*, 8, 52, 1980.
42. **Haas, M., Altman, A., Rothenberg, E., Bogart, M. H., and Jones, O. W.,** Mechanism of T-cell lymphomagenesis: transformation of growth-factor-dependent T-lymphoblastoma cells to growth-factor-independent T-lymphoma cells, *Cancer Res.*, 81, 1742, 1984.
43. **Huang, F. L. and Cho-Chung, Y. S.,** Alteration in gene expression at the onset of hormone-dependent mammary tumor regression, *Cancer Res.*, 43, 2138, 1983.

44. Johnson, H. M. and Torres, B. A., Peptide growth factors PDGF, EGF, and FGF regulate interferon-gamma production, *J. Immunol.*, 134, 2824, 1985.
45. Linzer, D. I. H. and Nathans, D., Growth-related changes in specific mRNAs of cultured mouse cells, *Proc. Natl. Acad. Sci. U.S.A.*, 80, 4271, 1983.
46. Linzer, D. I. H., Lee, S.-J., Ogren, L., Talamantes, F., and Nathans, D., Identification of proliferin mRNA and protein in mouse placenta, *Proc. Natl. Acad. Sci. U.S.A.*, 82, 4356, 1985.
47. Linzer, D. I. H. and Nathans, D., A new member of the prolactin-growth hormone gene family expressed in mouse placenta, *EMBO J.*, 4, 1419, 1985.
48. Heyns, A. du P., Eldor, A., Vlodavsky, I., Kaiser, N., Fridman, R., and Panet, A., The antiproliferative effect of interferon and the mitogenic action of growth factors are independent cell cycle events: studies with vascular smooth muscle cells and endothelial cells, *Exp. Cell Res.*, 161, 297, 1985.
49. Rinehart, C. A., Jr. and Canellakis, E. S., Induction of ornithine decarboxylase activity by insulin and growth factors is mediated by amino acids, *Proc. Natl. Acad. Sci. U.S.A.*, 82, 4365, 1985.
50. Willey, J. C., Laveck, M. A., McClendon, I. A., and Lechner, J. F., Relationship of ornithine decarboxylase activity and cAMP metabolism to proliferation of normal human bronchial epithelial cells, *J. Cell. Physiol.*, 124, 1985.
51. Bravo, R. and Celis, J. E., Gene expression in normal and virally transformed mouse 3T3B and hamster BHK21 cells, *Exp. Cell Res.*, 127, 249, 1980.
52. Levenson, R., Iwata, K., Klagsbrun, M., and Young, D. A., Growth factor- and dexamethasone-induced proteins in Swiss 3T3 cells: relationship to DNA synthesis, *J. Biol. Chem.*, 260, 8056, 1985.
53. Chrysogelos, S., Riley, D. E., Stein, G., and Stein, J., A human histone H4 gene exhibits cell cycle-dependent changes in chromatin structure that correlate with its expression, *Proc. Natl. Acad. Sci. U.S.A.*, 82, 7535, 1985.
54. Lau, L. F. and Nathans, D., Identification of a set of genes expressed during the G_0/G_1 transition of cultured mouse cells, *EMBO J.*, 4, 3145, 1985.
55. Bravo, R., Fey, S. J., Bellatin, J., Larsen, P. M., Arevalo, J., and Celis, J. E., Identification of a nuclear and of a cytoplasmic polypeptide whose relative proportions are sensitive to changes in the rate of cell proliferation, *Exp. Cell Res.*, 136, 311, 1981.
56. Bravo, R. and Celis, J. E., Updated catalogue of HeLa cell proteins: percentages and characteristics of the major cell polypeptides labeled with a mixture of 16 (14C)-labeled amino acids, *Clin. Chem.*, 28, 766, 1982.
57. Miyachi, K., Fritzler, M. J., and Tan, E. M., Autoantibody to a nuclear antigen in proliferating cells, *J. Immunol.*, 121, 2228, 1978.
58. Tan, E. M., Autoantibodies to nuclear antigens: their immunobiology and medicine, *Adv. Immunol.*, 33, 167, 1982.
59. Mathews, M. B., Bernstein, R. M., Franza, B. R., Jr., and Garrels, J. I., Identity of the proliferating cell nuclear antigen and cyclin, *Nature (London)*, 309, 374, 1984.
60. Celis, J. E., Fey, S. J., Larsen, P. M., and Celis, A., Expression of the transformation-sensitive protein "cyclin" in normal human epidermal basal cells and simian virus 40-transformed keratinocytes, *Proc. Natl. Acad. Sci. U.S.A.*, 81, 3128, 1984.
61. Bravo, R. and Graf, T., Synthesis of the nuclear protein cyclin does not correlate directly with transformation in quail embryo fibroblasts, *Exp. Cell Res.*, 156, 450, 1985.
62. Macdonald-Bravo, H. and Bravo, R., Induction of the nuclear protein cyclin in serum-stimulated quiescent 3T3 cells is independent of DNA synthesis, *Exp. Cell Res.*, 156, 455, 1985.
63. Bravo, R. and Macdonald-Bravo, H., Changes in the nuclear distribution of cyclin (PCNA) but not its synthesis depend on DNA replication, *EMBO J.*, 4, 655, 1985.
64. Celis, J. E. and Celis, A., Cell cycle-dependent variations in the distribution of the nuclear protein cyclin (proliferating cell nuclear antigen) in cultured cells: subdivision of S phase, *Proc. Natl. Acad. Sci. U.S.A.*, 82, 3262, 1985.
65. Celis, J. E. and Celis, A., Individual nuclei in polykaryons can control cyclin distribution and DNA synthesis, *EMBO J.*, 4, 1187, 1985.
66. Redding, T. W. and Schally, A. V., Inhibition of cell growth by a hypothalamic peptide, *Proc. Natl. Acad. Sci. U.S.A.*, 79, 7014, 1982.
67. Gaffney, E. V., Tsai, S.-C., Dell'Aquila, M. L., and Lingenfelter, S. E., Production of growth-inhibitory activity in serum-free medium by human monocytic leukemia cells, *Cancer Res.*, 43, 3668, 1983.
68. Levine, A. E., Hamilton, D. A., Yeoman, L. C., Busch, H., and Brattain, M. G., Identification of a tumor inhibitory factor in rat ascites fluid, *Biochem. Biophys. Res. Comm.*, 119, 76, 1984.
69. Levine, A. E., McRae, L. J., Hamilton, D. A., Brattain, D. E., Yeoman, L. C., and Brattain, M. G., Identification of endogenous inhibitory growth factors from a human colon carcinoma cell line, *Cancer Res.*, 45, 2248, 1985.
70. Iwata, K. K., Fryling, C. M., Knott, W. B., and Todaro, G. J., Isolation of tumor cell growth-inhibiting factors from a human rhabdomyosarcoma cell line, *Cancer Res.*, 45, 2689, 1985.

71. **Fryling, C. M., Iwata, K. K., Johnson, P. A., Knott, W. B., and Todaro, G. J.**, Two distinct tumor cell growth-inhibiting factors from a human rhabdomyosarcoma cell line, *Cancer Res.*, 45, 2695, 1985.
72. **Aune, T. M., Webb, D. R., and Pierce, C. W.**, Purification and initial characterization of the lymphokine soluble immune response suppressor, *J. Immunol.*, 131, 2848, 1983.
73. **Aune, T. M.**, Inhibition of soluble immune response suppressor activity by growth factors, *Proc. Natl. Acad. Sci. U.S.A.*, 82, 6240, 1985.
74. **Gill, G. N. and Lazar, C. S.**, Increased phosphotyrosine content and inhibition of proliferation in epidermal growth factor-treated A431 cells, *Nature (London)*, 293, 305, 1981.
75. **Roberts, A. B., Anzano, M. A., Wakefield, L. M., Roche, N. S., Stern, D. F., and Sporn, M. B.**, Type beta transforming growth factor: a bifunctional regulator of cellular growth, *Proc. Natl. Acad. Sci. U.S.A.*, 82, 119, 1985.
76. **Heldin, C.-H. and Westermark, B.**, Growth factors: mechanism of action and relation to oncogenes, *Cell*, 37, 9, 1984.
77. **Deuel, T. F. and Huang, J. S.**, Roles of growth factor activities in oncogenesis, *Blood*, 64, 951, 1984.
78. **Pimentel, E.**, *Oncogenes*, CRC Press, Boca Raton, Fla., 1986.
79. **Prakash, K. and Seligy, V. L.**, Oncogene related sequences in fungi: linkage of some to actin, *Biochem. Biophys. Res. Commun.*, 133, 293, 1985.
80. **Perbal, B. and Kohiyama, M.**, Existence de séquences homologues de l'oncogene V-MYB dans le génome des archaebactéries, *C. R. Acad. Sci. (Paris)*, 300, 177, 1985.
81. **Rabin, M., Watson, M., Barker, P. E., Ryan, J., Breg, W. R., and Ruddle, F. H.**, N-*ras* transforming gene maps to region p11 → p13 on chromosome 1 by in situ hybridization, *Cytogenet. Cell Genet.*, 38, 70, 1984.
82. **Popescu, N. C., Amsbaugh, S. C., DiPaolo, J. A., Tronick, S. R., Aaronson, S. A., and Swan, D. C.**, Chromosomal localization of three human *ras* genes by in situ molecular hybridization, *Somat. Cell Mol. Genet.*, 11, 149, 1985.
83. **Morton, C. C., Taub, R., Diamond, A., Lane, M. A., Cooper, G. M., and Leder, P.**, Mapping of the human *Blym*-1 transforming gene activated in Burkitt lymphomas to chromosome 1, *Science*, 223, 173, 1984.
84. **Le Beau, M. M., Westbrook, C. A., Diaz, M. O., and Rowley, J. D.**, Evidence for two distinct c-*src* loci on human chromosomes 1 and 20, *Nature (London)*, 312, 70, 1984.
85. **Tronick, S. R., Popescu, N. C., Cheah, M. S. C., Swan, D. C., Amsbaugh, S. C., Lengel, C. R., DiPaolo, J. A., and Robbins, K. C.**, Isolation and chromosomal localization of the human *fgr* porotooncogene, a distinct member of the tyrosine kinase gene family, *Proc. Natl. Acad. Sci. U.S.A.*, 82, 6595, 1985.
86. **Schwab, M., Varmus, H. E., Bishop, J. M., Grzeschik, K. H., Naylor, S. L., Sakaguchi, A. Y., Brodeur, G., and Trent, J.**, Chromosome localization in normal cells and neuroblastomas of a gene related to c-*myc*, *Nature (London)*, 308, 288, 1984.
87. **Bonner, T., O'Brien, S. J., Nash, W. G., Rapp, U. R., Morton, C. C., and Leder, P.**, The human homologous of *raf (mil)* oncogene are located on human chromosomes 3 and 4, *Science*, 223, 71, 1984.
88. **Groffen, J., Heisterkamp, N., Spurr, N., Dana, S., Wasmuth, J. J., and Stephenson, J. R.**, Chromosomal localization of the human c-*fms* oncogene, *Nucleic Acids Res.*, 11, 6331, 1983.
89. **Harper, M. E., Franchini, G., Love, J., Simon, M. I., Gallo, R. C., and Wong-Staal, F.**, Chromosomal sublocalization of human c-*myb* and c-*fes* cellular *onc* genes, *Nature (London)*, 304, 169, 1983.
90. **Spurr, N. K., Solomon, E., Jansson, M., Sheer, D., Goodfellow, P. N., Bodmer, W. F., and Vennström, B.**, Chromosomal localisation of the human homologues to the oncogenes *erb* A and B, *EMBO J.*, 3, 159, 1984.
91. **Dean, M., Park, M., LeBeau, M. M., Robins, T. S., Diaz, M. O., Rowley, J. D., Blair, D. G., and Vande Woude, G. F.**, The human *met* oncogene is related to the tyrosine kinase oncogene, *Nature (London)*, 318, 385, 1985.
92. **Caubet, J.-F., Mathieu-Mahul, D., Bernheim, A., Larsen, C.-J., and Berger, R.**, Human protooncogene c-*mos* maps to 8q11, *EMBO J.*, 4, 2245, 1985.
93. **Neel, B. G., Jhanwar, S. C., Chaganti, R. S. K., and Hayward, W. S.**, Two human c-*onc* genes are located on the long arm of chromosome 8, *Proc. Natl. Acad. Sci. U.S.A.*, 79, 7842, 1982.
94. **Jhanwar, S. C., Neel, B. G., Hayward, W. S., and Chaganti, R. S. K.**, Localization of the cellular oncogenes ABL, SIS, and FES on human germ line chromosomes, *Cytogenet. Cell Genet.*, 38, 73, 1984.
95. **de Taisne, C., Gegonne, A., Stehelin, D., Bernheim, A., and Berger, R.**, Chromosomal localization of the human protooncogene c-*ets*, *Nature (London)*, 310, 581, 1984.
96. **de Martinville, B., Giacalone, J., Shih, C., Weinberg, R. A., and Francke, U.**, Oncogene from human EJ bladder carcinoma is located on the short arm of chromosome 11, *Science*, 219, 498, 1983.
97. **van't Veer, L. J., van Kessel, A. G., van Heerikhuizen, H., van Ooyen, A., and Nusse, R.**, Molecular cloning and chromosomal assignment of the human homolog of *int*-1, a mouse gene implicated in mammary tumorigenesis, *Mol. Cell. Biol.*, 4, 2532, 1984.

98. **Barker, P. E., Rabin, M., Watson, M., Breg, W. R., Ruddle, F. H., and Verma, I. M.**, Human c-*fos* oncogene mapped within chromosomal region 14q21-q31, *Proc. Natl. Acad. Sci. U.S.A.*, 81, 5826, 1984.
99. **Schechter, A. L., Hung, M-C., Vaidyanathan, L., Yang-Feng, T. L., Francke, U., Ullrich, A., and Coussens, L.**, The *neu* gene: an *erb*B-homologous gene distinct from and unlinked to the gene encoding the EGF receptor, *Science*, 229, 976, 1985.
100. **Jhanwar, S. C., Chaganti, R. S. K., and Croce, C. M.**, Germ-line chromosomal localization of human c-*erb*-A oncogene, *Somat. Cell Mol. Genet.*, 11, 99, 1985.
101. **Yoshida, M. C., Sasaki, M., Mise, K., Semba, K., Nishizawa, M., Yamamoto, T., and Toyoshima, K.**, Regional mapping of the human proto-oncogene c-*yes*-1 to chromosome 18 at band q21.3, *Jpn. J. Cancer Res.*, 76, 559, 1985.
102. **Watson, D. K., McWilliams-Smith, M. J., Nunn, M. F., Duesberg, P. H., O'Brien, S. J., and Papas, T. S.**, The *ets* sequence from the transforming gene of avian erythroblastosis virus, E26, has unique domains on human chromosomes 11 and 21: both loci are transcriptionally active, *Proc. Natl. Acad. Sci. U.S.A.*, 82, 7294, 1985.
103. **O'Brien, S. J., Nash, W. G., Goodwin, J. L., Lowy, D. R., and Chang, E. H.**, Dispersion of the *ras* family of transforming genes to four different chromosomes in man, *Nature (London)*, 302, 839, 1983.
104. **Pimentel, E.**, Oncogenes and human cancer, *Cancer Genet. Cytogenet.*, 14, 347, 1985.
105. **Pierpaoli, W., Haran-Ghera, N., Bianchi, E., Mueller, J., Meshorer, A., and Bree, M.**, Endocrine disorders as contributory factor to neoplasia in SJL/J mice, *J. Natl. Cancer Inst.*, 53, 731, 1974.
106. **Pierpaoli, W., Haran-Ghera, N., and Kopp, H. G.**, Role of host endocrine status in murine leukaemogenesis, *Br. J. Cancer*, 35, 621, 1977.
107. **Pierpaoli, W. and Meshorer, A.**, Host endocrine status mediates onocogenesis: leukemia virus-induced carcinomas and reticulum cell sarcomas in acyclic or normal mice, *Eur. J. Cancer*, 18, 1181, 1982.
108. **Mishkin, S. Y., Farber, E., Ho, R. K., Mulay, S., and Mishkin, S.**, Evidence for the hormone dependency of hepatic hyperplastic nodules: inhibition of malignant transformation after exogenous 17 beta-estradiol and tamoxifen, *Hepatology*, 3, 308, 1983.
109. **Lupulescu, A.**, Glucagon control of carcinogenesis, *Endocrinology*, 113, 527, 1983.
110. **Henderson, B. E., Ross, R. K., Pike, M. C., and Casagrande, J. T.**, Endogenous hormones as a major factor in human cancer, *Cancer Res.*, 42, 3232, 1982.
111. **de Waard, F.**, Hormonal factors in human carcinogenesis, *J. Cancer Res. Clin. Oncol.*, 108, 177, 1984.
112. **Pimentel, E.**, Hormones as tumor markers, *Cancer Detect. Prevent.*, 6, 87, 1983.
113. **Pimentel, E.**, Peptide hormone precursors, subunits and fragments as human tumor markers, *Ann. Clin. Lab. Sci.*, 15, 381, 1985.
114. **Moses, H. L. and Robinson, R. A.**, Growth factors, growth factor receptors, and cell cycle control mechanisms in chemically transformed cells, *Fed. Proc.*, 41, 3008, 1982.
115. **Powers, S., Fisher, P. B., and Pollack, R.**, Analysis of the reduced growth factor dependency of simian virus 40-transformed 3T3 cells, *Mol. Cell. Biol.*, 4, 1572, 1984.
116. **Auersperg, N. and Calderwood, G. A.**, Development of serum independence in Kirsten murine sarcoma virus-infected rat adrenal cells, *Carcinogenesis*, 5, 175, 1984.
117. **Durkin, J. P. and Whitfield, J. F.**, Partial characterization of the mitogenic action of $pp60^{v-src}$, the oncogenic protein product of the *src* gene of avian sarcoma virus, *J. Cell. Physiol.*, 120, 135, 1984.
118. **Durkin, J. P. and Whitfield, J. F.**, The mitogenic activity of $pp60^{v-src}$, the oncogenic protein product of the *src* gene of avian sarcoma virus, is independent of external serum growth factors, *Biochem. Biophys. Res. Comm.*, 123, 411, 1984.
119. **Ford, R. J., Davis, F., and Ramirez, I.**, Growth factors for human lymphoid neoplasms, in *Mediators in Cell Growth and Differentiation*, Ford, R. J. and Maizel, A. L., Eds., Raven Press, New York, 1985, 233.
120. **Kaplan, P. L. and Ozanne, B.**, Cellular responsiveness to growth factors correlates with a cell's ability to express the transformed phenotype, *Cell*, 33, 931, 1983.
121. **Singletary, S. E., Tomasovic, B., Spitzer, G., Tucker, S. L., Hug, V., and Drewinko, B.**, Effects and interactions of epidermal growth factor, insulin, hydrocortisone, and estradiol on the cloning of human tumor cells, *Int. J. Cell Cloning*, 3, 407, 1985.
122. **De Larco, J. E. and Todaro, G. J.**, Growth factors from murine sarcoma virus-transformed cells, *Proc. Natl. Acad. Sci. U.S.A.*, 75, 4001, 1978.
123. **Kryceve-Martinerie, C., Lawrence, D. A., Crochet, J., Julien, P., and Vigier, P.**, Cells transformed by Rous sarcoma virus release transforming growth factors, *J. Cell Physiol.*, 113, 365, 1982.
124. **Roberts, A. B., Anzano, M. A., Lamb, L. C., Smith, J. M., Frolik, C. A., Marquardt, M., Todaro, G. J., and Sporn, M. B.**, Isolation from the murine sarcoma cells of novel transforming growth factors potentiated by EGF, *Nature (London)*, 295, 417, 1982.
125. **Marquardt, H., Hunkapiller, M. W., Hood, L. E., Twardzik, D. R., De Larco, J. E., Stephenson, J. R., and Todaro, G. J.**, Transforming growth factors produced by retrovirus-transformed rodent fibroblasts and human melanoma cells: amino acid sequence homology with epidermal growth factor, *Proc. Natl. Acad. Sci. U.S.A.*, 80, 4684, 1983.

126. **Anzano, M. A., Roberts, A. B., Smith, J. M., Sporn, M. B., and De Larco, J. E.,** Sarcoma growth factor from conditioned medium of virally transformed cells is composed of both type alpha and type beta transforming growth factors, *Proc. Natl. Acad. Sci. U.S.A.,* 80, 6264, 1983.
127. **Hirai, R., Yamaoka, K., and Mitsui, H.,** Isolation and purification of a new class of transforming growth factors from an avian sarcoma virus-transformed rat cell line, *Cancer Res.,* 43, 5742, 1983.
128. **Yamaoka, K., Hirai, R., Tsugita, A., and Mitsui, H.,** The purification of an acid- and heat-labile transforming growth factor from an avian sarcoma virus-transformed rat cell line, *J. Cell. Physiol.,* 119, 307, 1984.
129. **Massagué, J.,** Type beta transforming growth factor from feline sarcoma virus-transformed rat cells: isolation and biological properties, *J. Biol. Chem.,* 259, 9756, 1984.
130. **Anzano, M. A., Roberts, A. B., De Larco, J. E., Wakefield, L. M., Assoian, R. K., Roche, N. S., Smith, J. M., Lazarus, J. E., and Sporn, M. B.,** Increased secretion of type beta transforming growth factor accompanies viral transformation of cells, *Mol. Cell. Biol.,* 5, 242, 1985.
131. **Hamburger, A. W. and White, C. P.,** Autocrine growth factors for human tumor clonogenic cells, *Int. J. Cell Cloning,* 3, 399, 1985.
132. **Richmond, A., Lawson, D. H., Nixon, D. W., and Chawla, R. K.,** Characterization of autostimulatory and transforming growth factors from human melanoma cells, *Cancer Res.,* 45, 6390, 1985.
133. **Lang, R. A., Metcalf, D., Gough, N. M., Dunn, A. R., and Gonda, T. J.,** Expression of a hemopoietic growth factor cDNA in a factor-dependent cell line results in autonomous growth and tumorigenicity, *Cell,* 43, 531, 1985.
134. **Sporn, M. B. and Todaro, G. J.,** Autocrine secretion and malignant transformation, *N. Engl. J. Med.,* 303, 878, 1980.
135. **Moody, T. W., Russell, E. K., O'Donohue, T. L., Linden, C. D., and Gazdar, A. F.,** Bombesin-like peptides in small cell lung cancer: biochemical characterization and secretion from a cell line, *Life Sci.,* 32, 487, 1983.
136. **Moody, T. W., Carney, D. N., Cuttitta, F., Quattrocchi, K., and Minna, J. D.,** High affinity receptors for bombesin/GRP-like peptides on human small cell lung cancer, *Life Sci.,* 37, 105, 1985.
137. **Miller, Y. E.,** Growth factors, oncogenes, and lung cancer, *Am. Rev. Respir. Dis.,* 132, 178, 1985.
138. **Cuttitta, F., Carney, D. N., Mulshine, J., Moody, T. W., Fedorko, J., Fischler, A., and Minna, J. D.,** Bombesin-like peptides can function as autocrine growth factors in human small-cell lung cancer, *Nature (London),* 316, 823, 1985.
139. **Adkins, B., Leutz, A., and Graf, T.,** Autocrine growth induced by *src*-related oncogenes in transformed chicken myeloid cells, *Cell,* 39, 439, 1984.
140. **Oliff, A., Agranovsky, O., McKinney, M. D., Murty, V. V. V. S., and Bauchwitz, R.,** Friend murine leukemia virus-immortalized myeloid cells are converted into tumorigenic cell lines by Abelson leukemia virus, *Proc. Natl. Acad. Sci. U.S.A.,* 82, 3306, 1985.
141. **Cook, W. D., Metcalf, D., Nicola, N. A., Burgess, A. W., and Walker, F.,** Malignant transformation of a growth factor-dependent myeloid cell line by Abelson virus without evidence of an autocrine mechanism, *Cell,* 41, 677, 1985.
142. **Pierce, J. H., Di Fiore, P. P., Aaronson, S. A., Potter, M., Pumphrey, J., Scott, A., and Ihle, J. N.,** Neoplastic transformation of mast cells by Abelson-MuLV: abrogation of IL-3 dependence by a non-autocrine mechanism, *Cell,* 41, 685, 1985.
143. **Rein, A., Keller, J., Schultz, A. M., Holmes, K. L., Medicus, R., and Ihle, J. N.,** Infection of immune mast cells by Harvey sarcoma virus: immortalization without loss of requirement for interleukin-3, *Mol. Cell. Biol.,* 5, 2257, 1985.
144. **Salomon, D. S., Zwiebel, J. A., Noda, M., and Bassin, R. H.,** Flat revertants derived from Kirsten murine sarcoma virus-transformed cells produce transforming growth factors, *J. Cell. Physiol.,* 121, 22, 1984.
145. **Fichelson, S., Heard, J.-M., and Levy, J.-P.,** The in vitro autocrine secretion of CSFs alone does not account for the longterm growth of murine myeloid leukemic cells in suspension cultures, *J. Cell. Physiol.,* 124, 487, 1985.
146. **Moscovici, M. G., Jurdic, P., Samarut, J., Gazzolo, L., Mura, C. V., and Moscovici, C.,** Characterization of the hemopoietic target cells for the avian leukemia virus E26, *Virology,* 129, 65, 1983.
147. **Beug, H., Leutz, A., Kahn, P., and Graf, T.,** *Ts* mutants of E26 leukemia virus allow transformed myeloblasts, but not erythroblasts or fibroblasts, to differentiate at the nonpermissive temperature, *Cell,* 39, 579, 1984.
148. **Balk, S. D., Riley, T. M., Gunther, H. S., and Morisi, A.,** Heparin-treated, v-*myc*-transformed chicken heart mesenchymal cells assume a normal morphology but are hypersensitive to epidermal growth factor (EGF) and brain fibroblast growth factor (bFGF); cells transformed by the v-Ha-*ras* oncogene are refractory to EGF and bFGF but are hypersensitive to insulin-like growth factors, *Proc. Natl. Acad. Sci. U.S.A.,* 82, 5781, 1985.
149. **Mally, M. I., Vogt, M., Swift, S. E., and Haas, M.,** Oncogene expression in murine splenic T cells and murine T-cell neoplasms, *Virology,* 144, 115, 1985.

150. **Smith, E. F. and Townsend, C. O.**, A plant tumor of bacterial origin, *Science*, 25, 671, 1907.
151. **Nester, E. W. and Kosuge, T.**, Plasmids specifying plant hyperplasias, *Ann. Rev. Microbiol.*, 35: 531, 1981.
152. **Bevan, M. W. and Chilton, M.-D.**, T-DNA of *Agrobacterium* Ti and Ri plasmids, *Ann. Rev. Genet.*, 16, 357, 1982.
153. **Klee, H., Montoya, A., Horodyski, F., Lichtenstein, C., Garfinkel, D., Fuller, S., Flores, C., Peschon, J., Nester, E., and Gordon, M.**, Nucleotide sequence of the *tms* genes of the pTiA6NC octopine Ti plasmid: two gene products involved in plant tumorigenesis, *Proc. Natl. Acad. Sci. U.S.A.*, 81, 1728, 1984.
154. **Buchmann, I., Marner, F.-J., Schröder, G., Waffenschmidt, S., and Schröder, J.**, Tumour genes in plants: T-DNA encoded cytokinin biosynthesis, *EMBO J.*, 4, 853, 1985.
155. **Buchholz, W. C. and Thomashow, M. F.**, Comparison of T-DNA oncogene complements of *Agrobacterium tumefaciens* tumor-inducing plasmids with limited and wide host ranges, *J. Bacteriol.*, 160, 319, 1984.
156. **Buchholz, W. C. and Thomashow, M. F.**, Host range encoded by the *Agrobacterium tumefaciens* tumor-inducing plasmid ptiAg63 can be expanded by modification of its T-DNA oncogene complement, *J. Bacteriol.*, 160, 327, 1984.
157. **Hagen, G. and Guilfoyle, T. J.**, Rapid induction of selective transcription by auxins, *Mol. Cell Biol.*, 5, 1197, 1985.
158. **Elliott, D. C. and Geytenbeck, M.**, Identification of products of protein phosphorylation in T37-transformed cells and comparison with normal cells, *Biochim. Biophys. Acta*, 845, 317, 1985.
159. **Calabretta, B., Kaczmarek, L., Mars, W., Ochoa, D., Gibson, C. W., Hirschhorn, R. R., and Baserga, R.**, Cell-cycle-specific genes differentially expressed in human leukemias, *Proc. Natl. Acad. Sci. U.S.A.*, 82, 4463, 1985.
160. **Kaczmarek, L., Calabretta, B., and Baserga, R.**, Expression of cell-cycle-dependent genes in phytohemagglutinin-stimulated human lymphocytes, *Proc. Natl. Acad. Sci. U.S.A.*, 82, 5375, 1985.
161. **Thompson, C. B., Challoner, P. B., Neiman, P. E., and Groudine, M.**, Levels of c-*myc* oncogene mRNA are invariant throughout the cell cycle, *Nature (London)*, 314, 363, 1985.
162. **Hann, S. R., Thompson, C. B., and Eisenman, R. N.**, c-*myc* oncogene protein synthesis is independent of the cell cycle in human and avian cells, *Nature (London)*, 314, 366, 1985.
163. **Eisenman, R. N. and Hann, S. R.**, Proteins expressed by the c-*myc* oncogene in lymphomas of human and avian origin, *Proc. R. Soc. London Ser. B*, 226, 73, 1985.
164. **Rittling, S. R., Gibson, C. W., Ferrari, S., and Baserga, R.**, The effect of cycloheximide on the expression of cell cycle dependent genes, *Biochem. Biophys. Res. Comm.*, 132, 327, 1985.
165. **Kelly, K., Cochran, B. H., Stiles, C. D., and Leder, P.**, Cell-specific regulation of the c-*myc* gene by lymphocyte mitogens and platelet-derived growth factor, *Cell*, 35, 603, 1983.
166. **Persson, H., Hennighausen, L., Taub, R., De Grado, W., and Leder, P.**, Antibodies to human c-*myc* oncogene product: evidence of an evolutionarily conserved protein induced during cell proliferation, *Science*, 225, 687, 1984.
167. **Persson, H., Gray, H. E., and Godeau, F.**, Growth-dependent synthesis of c-*myc*-encoded proteins: early stimulation by serum factors in synchronized mouse 3T3 cells, *Mol. Cell. Biol.*, 5, 2903, 1985.
168. **Tramontano, D., Chin, W. W., Moses, A. C., and Ingbar, S. H.**, Thyrotropi and dibutyryl cyclic AMP increase levels of c-myc and c-fos MRNAs in cultured rat thyroid cells, *J. Biol. Chem.*, 261, 3919, 1986.
169. **Kaczmarek, L., Hyland, J., Watt, R., Rosenberg, M., and Baserga, R.**, Microinjected c-*myc* as a competence factor, *Science*, 228, 1313, 1985.
170. **Tsuda, T., Kaibuchi, K., West, B., and Takai, Y.**, Involvement of Ca^{2+} in platelet-derived growth factor-induced expression of c-*myc* oncogene in Swiss 3T3 fibroblasts, *FEBS Lett.*, 187, 43, 1985.
171. **Kaibuchi, K., Tsuda, T., Kikuchi, A., Tanimoto, T., Yamashita, T., and Takai, Y.**, Possible involvement of protein kinase C and calcium ion in growth factor-induced expression of c-*myc* oncogene in Swiss 3T3 fibroblasts, *J. Biol. Chem.*, 261, 1187, 1986.
172. **Müller, R., Bravo, R., Burckhardt, J., and Curran, T.**, Induction of c-*fos* gene and protein by growth factors precedes activation of c-myc, *Nature (London)*, 312, 716, 1984.
173. **Dani, C., Blanchard, J. M., Piechaczyk, M., El Sabouty, S., Marty, L., and Jeanteur, P.**, Extreme instability of *myc* mRNA in normal and transformed human cells, *Proc. Natl. Acad. Sci. U.S.A.*, 81, 7046, 1984.
174. **Blanchard, J.-M., Piechaczyk, M., Dani, C., Chambard, J.-C., Franchi, A., Pouyssegur, J., and Jeanteur, P.**, c-*myc* gene is transcribed at high rate in G_0-arrested fibroblasts and is post-transcriptionally regulated in response to growth factors, *Nature (London)*, 317, 443, 1985.
175. **Greenberg, M. E. and Ziff, E. B.**, Stimulation of 3T3 cells induces transcription of the c-*fos* proto-oncogene, *Nature (London)*, 311, 433, 1984.
176. **Deschamps, J., Meijlink, F., and Verma, I. M.**, Identification of a transcriptional enhancer element upstream from protooncogene *fos*, *Science*, 230, 1174, 1985.

177. **Bravo, R., Burckhardt, J., and Müller, R.**, Persistence of the competent state in mouse fibroblasts is independent of c-*fos* and c-*myc* expression, *Exp. Cell Res.*, 160, 540, 1985.
178. **Rüther, U., Wagner, E., and Müller, R.**, Analysis of the differentiation-promoting potential of inducible c-*fos* genes introduced into embryonal carcinoma cells, *EMBO J.*, 4, 1775, 1985.
179. **Treisman, R.**, Transient accumulation of c-*fos* RNA following serum stimulation requires a conserved 5' element and c-*fos* 3' sequences, *Cell*, 42, 889, 1985.
180. **Müller, R., Tremblay, J. M., Adamson, E. D., and Verma, I. M.**, Tissue and cell type-specific expression of two human c-*onc* genes, *Nature (London)*, 304, 454, 1983.
181. **Müller, R., Müller, D., and Guilbert, L.**, Differential expression of c-*fos* in hematopoietic cells: correlation with differentiation of monomyelocytic cells, *in vitro*, *EMBO J.*, 3, 1887, 1984.
182. **Gonda, T. J. and Metcalf, D.**, Expression of *myb*, *myc* and *fos* proto-oncogenes during the differentiation of a murine myeloid leukaemia, *Nature (London)*, 310, 249, 1984.
183. **Curran, T. and Morgan, J. I.**, Superinduction of c-*fos* by nerve growth factor in the presence of peripherally active benzodiazepines, *Science*, 229, 1265, 1985.
184. **Cochran, B. H., Zullo, J., Verma, I. M., and Stiles, C. D.**, Expression of the c-*fos* gene and of an *fos*-related gene is stimulated by platelet-derived growth factor, *Science*, 226, 1080, 1984.
185. **Mitchell, R. L., Zokas, L., Schreiber, R. D., and Verma, I. M.**, Rapid induction of proto-oncogene *fos* during human monocytic differentiation, *Cell*, 40, 209, 1985.
186. **Müller, R., Curran, T., Müller, D., and Guilbert, L.**, Induction of c-*fos* during myelomonocytic differentiation and macrophage proliferation, *Nature (London)*, 314, 546, 1985.
187. **Todaro, G. J., De Larco, J. E., and Cohen, S.**, Transformation by murine and feline sarcoma viruses specifically blocks binding of epidermal growth factor to cells, *Nature (London)* 264, 26, 1976.
188. **Usui, T., Moriyama, N., Ishibe, T., and Nakatsu, H.**, Loss of epidermal growth factor receptor on the renal neoplasm induced in vivo with xenotropic pseudotype Kirsten muring sarcoma virsus, *Biochem. Biophys. Res. Comm.*, 120, 879, 1984.
189. **Robinson, R. A., Branum, E. L., Volkenant, M. E., and Moses, H. L.**, Cell cycle variation in ^{125}I-labeled epidermal growth factor binding in chemically transformed cells, *Cancer Res.*, 42, 2633, 1982.
190. **Wakshull, E., Kraemer, P. M., and Wharton, W.**, Multistep change in epidermal growth factor receptors during spontaneous neoplastic progression in Chinese hamster embryo fibroblasts, *Cancer Res.*, 45, 2070, 1985.
191. **Chua, C. C., Geiman, D. E., Schreiber, A. B., and Ladda, R. L.**, Nonfunctional epidermal growth factor receptor in cells transformed by Kirsten sarcoma virus, *Biochem. Biophys. Res. Comm.*, 118, 538, 1984.
192. **LaX, I., Kris, R., Sasson, I. Ullrich, A., Hayman, M. J., Beug, H., and Schlessinger, J.**, Activation of c-*erb*-B in avian leukosis virus-induced erythroblastosis leads to the expression of a truncated EGF receptor kinase, *EMBO J.*, 4, 3179, 1985.
193. **Bodine, P. V. and Tupper, J. T.**, Calmodulin antagonists decrease binding of epidermal growth factor to transformed, but not to normal human fibroblasts, *Biochem. J.*, 218, 629, 1984.
194. **Cotton, P. C. and Brugge, J. S.**, Neural tissues express high levels of the cellular *src* gene product pp60^{c-src}, *Mol. Cell. Biol.*, 3, 1157, 1983.
195. **Jacobs, C. and Rübsamen, H.**, Expression of pp60^{c-src} protein kinase in adult and fetal human tissue: high activities in some sarcomas and mammary carcinomas, *Cancer Res.*, 43, 1696, 1983.
196. **Levy, B. T., Sorge, L. K., Meymandi, A., and Maness, P. F.**, pp60^{c-src} kinase is in chick and human embryonic tissues, *Dev. Biol.*, 104, 9, 1984.
197. **Schartl, M. and Barnekow, A.**, Differential expression of the cellular *src* gene during vertebrate development, *Dev. Biol.*, 105, 415, 1984.
198. **Sorge, L. K., Levy, B. T., and Maness, P. F.**, pp60 $^{c-src}$ is developmentally regulated in the neural retina, *Cell*, 36, 249, 1984.
199. **Fults, D. W., Towle, A. C., Lauder, J. M., and Maness, P. F.**, pp60^{c-src} in the developing cerebellum, *Mol. Cell. Biol.*, 5, 27, 1985.
200. **Webb, A. C., Collins, K. L., Auron, P. E., Eddy, R. L., Nakai, H., Byers, M. G., Haley, L. L., Henry, W. M., and Shows, T. B.**, Interleukin-1 gene (IL1) assigned to long arm of human chromosome 2, *Lymphokine Res.*, 5, 77, 1986.
201. **Rashidbaigi, A., Langer, J. A., Jung, V., Jones, C., Morse, H. G., Tischfield, J. A., Trill, J. J., Kung, H., and Pestka, S.**, The gene for the human immune interferon receptor is located on chromosome 6, *Proc. Natl. Acad. Sci. U.S.A.*, 83, 384, 1986.
202. **Betsholtz, C., Johnsson, A., Heldin, C. H., Westermark, B., Lind, P., Urdea, M. S., Eddy, R., Shows, T. B., Philpott, K., Mellor, A. L., Knott, T. J., and Scott, J.**, cDNA sequence and chromosomal localization of human platelet-derived growth factor A-chain and its expression in tumour cell lines, *Nature (London)*, 320, 695, 1986.
203. **Fujii, D., Brissenden, J. E., Derynck, R., and Francke, U.**, Transforming growth factor beta gene maps to human chromosome 19 long arm and to mouse chromosome 7, *Somat. Cell Mol. Genet.*, 12, 281, 1986.

Chapter 2

CELLULAR MECHANISMS OF ACTION OF HORMONES AND GROWTH FACTORS AND THEIR RELATION TO ONCOGENE FUNCTIONS

I. INTRODUCTION

The cellular mechanisms of action of hormones and peptide growth factors are exceedingly complex.[1-4] Hormones and growth factors may be considered as essential signal molecules and their action mechanisms are intimately and tightly related to the most delicate events occurring in the molecular machinery of cells and particularly to the regulatory phenomena at this level. The biochemical events occurring in the target cell under the stimulus of such signal molecules may be classified in two sequential steps: transductional and post-transductional. In turn, the transductional events may be constituted by two sequential steps: binding of the hormone or growth factor to a specific cellular receptor and production and/or translocation of mediators (second messengers). In the post-transductional stages, specific types of regulatory changes can occur at either the genome level or at extragenomic sites like the cell membrane, the cytoplasm, and different organelles.

The protein products of cellular oncogenes may share some common biochemical pathways with the transductional and post-transductional mechanisms of hormone and growth factor action. Moreover, hormones and growth factors are signal molecules that may be involved in the regulation of proto-oncogene expression at both the transcriptional and post-transcriptional levels.

Cell differentiation and cell proliferation are subjected to complex, multifactorial regulatory mechanisms. Each phase of the cell cycle (G_1, S, G_2, and M), as well as the transition from one phase to another, is controlled by a combination of endogenous (intracellular) and exogenous (extracellular) factors.[5] Complex interactions between the action of hormones, growth factors, and proto-oncogene protein products are of critical importance for the proper development of such intricate regulatory phenomena. In the adult multicellular organism a delicate balance between the processes responsible for cell differentiation and cell proliferation is continuously operating a loss or damage in some of these control mechanisms may result in neoplastic transformation and tumorigenesis.

II. RECEPTORS

Hormones and growth factors interact with specific cellular proteins, termed receptors, which discriminate between the various types of signal molecules. These molecules can be considered as receptor ligands, capable of establishing a highly specific and reversible interaction, which occurs with high affinity for the particular receptor.[6-10] After formation of the ligand-receptor complex, the receptor would undergo a specific conformational change, which in some manner induces the generation of intracellular transducing signals leading to a cellular response.

According to their respective functions, receptors are located at different sites in the cell. The receptors for peptide hormones and growth factors are located in the plasma membrane. Thyroid hormone receptors have been identified in several cellular components, including the plasma membrane, the cytosol, the mitochondria, and the nucleus. Receptors for steroid hormones would be located primarily in the nucleus, although translocation of "activated" cytoplasmic steroid hormone receptors into the nucleus has been described in many reports.

A. Mobility and Internalization of Surface Receptors

Lateral mobility and aggregation of hormone- and growth factor-receptor complexes at the cell surface may be important for the physiological activities of these agents. Plant lectins such as concanavalin A or wheat germ agglutinin do not significantly affect DNA synthesis in cultures human skin fibroblasts, but these lectins decrease EGF- and insulin-stimulated DNA synthesis.[11] This inhibition cannot be overcome by increasing the concentration of peptide added. Since these plant lectins are multivalent molecules which bind to specific carbohydrates on the cell surface and can cross-link a variety of cell surface receptors, the noncompetitive inhibition of the cellular response to hormones and growth factors induced by lectins may be attributed to effects at a postreceptor level.

Desensitization of cellular responses to hormonal, growth factor, or drug ligands is correlated with the loss of available receptor sites on the cell surface. The receptors for peptide hormones and growth factors are concentrated on coated pits at the level of the cell membrane and are internalized by endocytosis, which involves invagination of the plasma membrane, fusion of the neck of the invagination, and detachment of the fused vesicle.[7,12] Internalization of ligands bound to the surface receptors might proceed by one of two mechanisms: in an "escalator model", receptors are continuously being taken into the cell via coated pits and the ligand is a passive passenger, and in an "elevator model", receptors are internalized only after ligand has given a "signal" for uptake.[13]

Internalized receptors may become associated with different forms of lysosomes as well as with the nuclear membrane, the endoplasmic reticulum and the Golgi apparatus. Membrane receptors may be degraded by lysosomal proteases or may be recycled back to the cell surface.[8] There is evidence indicating that peptide hormones may have intracellular receptors and that interaction of the hormone with these receptors may be followed by possible physiological actions.[14] However, the definite physiological meaning of the possible direct intracellular actions of peptide hormones remains uncertain.

B. Presence of Hormone Receptors and Production of Hormones in Microorganisms

Proteins with affinity for hormones and growth factors have been detected in microorganisms, including protozoa, bacteria, and algae.[15] Although the biological significance of the presence of such "receptors", or receptor-like molecules, in microorganisms not usually exposed to hormones or growth factors is not understood, the findings indicate that these signal receivers are phylogenetically very ancient. Moreover, some unicellular organisms may be able to produce substances which are similar or identical to the hormones and growth factors of higher animal species. For example, the yeast *Saccharomyces cerevisiae* possesses a high-affinity estrogen-binding protein and is capable of synthesizing 17 beta-estradiol.[16] Insulin or a molecule closely related to insulin, is produced by microbes such as *Escherichia coli* and *Tetrahymena*.[17] The possible role of hormones or hormone-like materials produced by microorganisms in unknown.

C. Relationship between Hormones, Growth Factors, and Their Cellular Receptors

The molecular relationship between hormones and growth factors and their respective cellular receptors can be evaluated at both the nucleic acid level and the peptide level. An interesting pattern in the genetic code has been observed recently: codons for hydrophilic and hydrophobic amino acids on one strand of nucleic acid (DNA or RNA) are complemented by codons for hydrophobic and hydrophilic amino acids on the other strand, respectively.[18] The average tendency of codons for "uncharged" (slightly hydrophylic) amino acids is complemented by codons for "uncharged" amino acids. Furthermore, regions of complementarity exist between the mRNAs for growth factors, including EGF, transferrin and IL-2, and the mRNAs for the respective cellular receptors.[19] Naturally occurring peptides such as ACTH and γ-endorphin bind specifically derived counterparts that are specified by RNA sequences complementary to the mRNA for ACTH and γ-endorphin, respectively.[20] This binding might result from one peptide being an "internal image" of the other, as demonstrated by the fact that antibody to the peptide that was encoded by the complementary RNA for ACTH recognizes the adrenal cell ACTH receptor.[20] These findings are important from the evolutionary point of view.

D. Regulation of Receptor Expression

In multicellular organisms receptors are subjected to quantitative, and perhaps also qualitative, ontogenic regulatory changes. Many physiologic factors acting at different cellular levels can be involved in modifying the cellular concentration and/or affinity of receptors, especially in the intact animal.[21-23] The cooperation of different types of cells may be important in the regulation of receptor expression. Receptor regulation can be either homospecific (or self-regulatory), whereby a ligand can regulate the function and/or number of its own receptors, or heterospecific, wherein the activation of one receptor system leads to changes in a second distinct receptor system. Heterospecific receptor regulation may occur via mechanisms that are localized either at the level of the plasma membrane or at levels comparatively remote from the initial ligand-receptor interaction.

III. POSTRECEPTOR TRANSDUCTIONAL MECHANISMS

The cellular reactions occurring after stimulation by signal molecules can be classified in three groups, which differ in the principal mechanisms and the time involved in the development of cellular responses:[24]

1. Highly lipophilic hormones like steroids easily permeate the cell membrane and interact with cytoplasmic and/or nuclear receptors. The activated cytoplasmic hormone-receptor complexes would be translocated into the nucleus, whereas particular nuclear receptors would be directly activated upon binding of other types of ligands, for example, thyroid hormone. These phenomena may result in an increased transcription of specific DNA sequences and a subsequent increased formation of proteins that lead to the cellular response, usually after a time delay of one or more hours.
2. Some signal molecules, especially neurotransmitters, can interact with membrane-bound receptors that are closely connected with the ion channels of the plasma membrane. The interaction may determine ion exchanges that occur within a few milliseconds, leading to an almost immediate cellular response.
3. Binding of a peptide hormone or growth factor to its membrane-bound receptor triggers the formation of intracellular signals, consisting in the production and/or translocation of particular mediators. The best characterized mediators are cyclic nucleotides (cAMP and cGMP), calcium ions, and phosphoinositide metabolites. The mediators interact with specific intracellular proteins, causing modification and subsequent activation of particular enzymes, including protein kinases, which may result in functional changes, including specific phosphorylation/dephosphorylation processes. These effects require a few seconds to develop and reach a maximum within one or a few minutes after exposure to the signal molecule.

A. Mediators and Modulators

Mediators are more clearly associated with the action mechanisms of peptide hormones and growth factors than with the actions of thyroid hormone and steroid hormones. The latter two types of hormones are apparently able to produce in a direct manner regulatory changes in genomic functions upon interaction of the activated hormone-receptor complex with the DNA or chromatin proteins. In addition to mediators or second messengers, substances like prostaglandins and polyamines may act as modulators or regulators of hormone and growth factor action at different biochemical levels. The polyamines (putrescine, spermidine, and spermine) are ubiquitous in vertebrate cells and have been implicated in the regulation of cellular proliferation and differentiation.[25,26]

B. Ion and pH Changes

The regulation of cell growth and differentiation by peptide hormones and growth factors is associated with changes in the flux rates and intracellular levels of several monovalent and divalent cations, including H^+, Na^+, K^+, and Ca^{2+}.[27] Mobilization of Ca^{2+} from intracellular stores and activation of the Na^+/H^+ antiport with rise in cytoplasmic pH and alkalinization of the cytoplasm may result in profound physiological changes in cells responding to hormones or growth factors.[28] Growth factors involved in lymphocyte functions, like IL-2, may induce a rapid increase in intracellular pH through activation of a Na^+/H^+ antiport, but cytoplasmic alkalinization is not required for lymphocyte proliferation.[29] However, regulatory changes induced by growth factors on membrane transport may be associated with stimulation of DNA synthesis in particular types of cells under specific physiological conditions.[30] A growth factor-induced increase in cytoplasmic pH can stimulate the entry of mitogen-stimulated quiescent fibroblasts into the S phase of the cycle.[31] Prevention of mitogen-stimulated Na^+ influx and cytoplasmic pH rise in a fibroblast mutant lacking Na^+/H^+ antiport activity suppresses growth factor-induced DNA synthesis at neutral and acidic pH.[31]

C. Protein Phosphorylation

Reactions involving protein phosphorylation and dephosphorylation, controlled by different types of protein kinases, participate in crucial ways in the cellular mechanisms of action of hormones and growth factors.[32-45] Protein kinases involved in such reactions may be either cAMP-dependent or cAMP-independent, the latter including a diversity of enzymes, for example, cGMP-dependent, Ca^{2+}-calmodulin-dependent, Ca^{2+}-phospholipid-dependent, and polyamine-dependent protein kinases. Substrates for phosphorylation/dephosphorylation phenomena may be located on different cellular compartments, including the plasma membrane, the cytoplasm, and the nucleus. Different types of cellular proteins are important substrates for these processes and some of these proteins are histones or other types of nuclear proteins.[46] It is widely accepted that these changes may have important consequences in the regulation of genomic functions but, in general, the biochemical consequences of hormone-induced changes in the phosphorylation of cellular proteins remain a subject for speculation. There is little solid evidence in support of a role for any specific protein kinase in mediating hormone- or growth factor-induced modification in the phosphorylation of specific, identified nuclear proteins.[46]

D. Regulation of Nuclear Functions

DNA binding proteins with the ability to discriminate between distinctive DNA sequences found in the promoter 5' regions of different genes are involved in the regulation of gene expression mediated in part by promoter-specific transcription factors.[47] In turn, the activity of many promoters is modulated by an enhancer, which is a regulatory element contained in separate and more distant DNA sequences. Both hormones and growth factors may influence specific transcriptional processes through mechanisms involving the activation or inactivation of promoter and/or enhancer elements. Eventually, such regulatory changes may be reflected in stimulation of DNA synthesis and mitosis.

IV. cAMP AND THE ADENYLATE CYCLASE SYSTEM

cAMP (adenosine 3':5'-monophosphate) has been recognized as an important mediator in the mechanisms of action of many hormones and growth factors.[48-50] Adenylate cyclase, a plasma membrane-bound enzyme, catalyzes the synthesis of cAMP from ATP, whereas a phosphodiesterase inactivates cAMP to the noncyclic compound adenosine 5'-monophosphate (5'-AMP). The equilibrium between the activities of the enzymes adenylate cyclase and phosphodiesterase determines the intracellular concentration of cAMP under specific physiological conditions, including hormone and growth factor action. These signal molecules may change the equilibrium mainly through modification of adenylate cyclase activity, although influences on phosphodiesterase have been described.

The specificity of the adenylate cyclase system is determined not by the relatively unspecific cAMP molecule but by the particular type of cell surface receptor involved in the activation of the system and the particular endogenous substrate for cAMP-dependent phosphorylation. In turn, both the receptors expressed by a given cell and the substrates available for phosphorylation depend on the specific genetic program of the particular cell.

cAMP is a universal second messenger which influences a great diversity of biochemical and physiological phenomena. The cyclic nucleotide has been detected in both animals and plants and is an important regulator of transcriptional processes occurring in prokaryotes, where it acts upon binding to a specific receptor protein, the catabolite gene activator protein (CAP).[51] In bacteria the cAMP-CAP complex binds to specific DNA sequences and enhances the rate of transcription of several operons. In eukaryotes cAMP may act as either an activator or an inhibitor of gene transcription.[52] Hypothetically, the expression of cAMP-responsive genes in higher eukaryotes may be controlled, as in *E. coli*, by proteins that would form complexes with cAMP and then would show sequence-specific DNA-binding properties.[53] There is evidence that cAMP can stimulate transcription of discrete subsets of eukaryotic genes by at least two independent molecular mechanisms.[54] cAMP would stimulate growth hormone (GH) gene transcription and phosphorylation of a 19,000-dalton nuclear protein by the direct activation of a cAMP-dependent protein kinase. In contrast, the stimulation of prolactin gene transcription by cAMP would reflect the activation of a discrete calcium-dependent event. In any case, it is clear that cAMP and cAMP-dependent biochemical processes are intimately involved in the control of cell proliferation and differentiation.

A. The Adenylate Cyclase System

Adenylate cyclase is a multicomponent system consisting of at least three physically distinct units.

1. The hormone receptor (R component), located in the outer plasma membrane and containing a specific site for a given hormone.
2. The catalytic moiety (C component) of adenylate cyclase, located on the inner face of the plasma membrane and bearing the site responsive for catalysis of the cyclizing reaction.
3. The guanine nucleotide-binding regulatory subunit (G component), which binds guanine nucleotide and is mainly located within the plasma membrane.[55]

The catalytic subunit (C component) of adenylate cyclase has been purified from rabbit myocardial membranes and is approximately 150,000 mol wt.[56] The same subunit purified from brain tissue has been identified as a calmodulin-binding protein of 135,000 kdaltons.[57] Receptors are molecularly separated from the rest of the adenyl cyclase system and "float" freely in the phospholipid bilayer of the plasma membrane, independently of the adenylate cyclase system.

Guanosine triphosphate (GTP) exerts three actions in the adenylate cyclase system.

1. GTP stimulates basal activity in the absence of the hormone.
2. GTP accelerates the rate of hormone binding to and release from its receptor.
3. GTP promotes coupling of receptors to adenylate cyclase.[55]

A Mg^{2+}-dependent GTPase activity is associated with the adenylate cyclase system. GTPase is stimulated by catecholamines and inhibited by cholera toxin. Stimulation of adenylate cyclase activity by cholera toxin is due to accumulation of enzyme-GTP complex because of a decreased rate of hydrolysis of bound GTP. Thus, adenylate cyclase is regulated by GTPase activity and hormone- and peptide growth factor-induced stimulation is associated with an exchange process between GTP and guanosine diphosphate (GDP), which promotes dissociation of inhibitory effects of GDP and allows association of stimulatory GTP.[55] Mg^{2+}-dependent GTPase-like activity is apparently responsive for the turn off of adenylate cyclase activity and the formation of an inactive enzyme-GDP complex.

1. Regulation of the Adenylate Cyclase System

Guanyl nucleotides are importantly involved in the regulation of the adenylate cyclase system because they affect not only the basic activity of the system in the absence of added hormone or growth factor, but also hormone- and growth factor-receptor interaction. Interaction of the signal molecules with their specific cellular receptors lead to a decrease in the affinity of the receptors for guanyl nucleotide. In general, the role of guanine nucleotide regulation of hormone and growth factor action on adenylate cyclase constitutes a GTP-mediated amplification and GTP-mediated dampening system whose degree of amplification and dampening varies with the receptor type as well as from tissue to tissue.[58]

The primary site of action of receptors is at the guanine nucleotide-binding (G) regulatory subunit of the adenylate cyclase system.[59] Hormone- and growth factor-induced activation of the G component would determine a reduced requirement for Mg^{2+}, resulting in subunit dissociation of the complex. Subunit dissociation constitutes a rate-limiting step in the functional activity of the adenylate cyclase system. Changes in the guanyl nucleotide regulatory protein could be responsible for alterations in the hormonal response of the adenylate cyclase system observed in tumor tissues when hormone receptors are found to be normal, for example, in certain neoplastic human thyroid tissues.[60] Uncoupling between the effector binding and catalytic stimulation of adenylate and guanylate cyclase may have occurred in a spontaneous murine thyroid tumor which did not respond to TSH by increasing cAMP and cGMP levels in spite of the presence of apparently intact TSH receptors and adenylate and guanylate cyclase activity.[61]

In addition to the regulatory action of guanyl nucleotides, the adenylate cyclase system is also regulated by other influences, in particular by the calcium-calmodulin transductional system.[62] The Ca^{2+}-calmodulin complex could activate either adenylate cyclase or phosphodiesterase, thus regulating the intracellular concentration of cAMP. Many actions of the Ca^{2+}-calmodulin system are exerted through activation of phosphorylation/dephosphorylation processes regulated by particular types of protein kinases and there is evidence that phosphorylation and dephosphorylation phenomena are involved in regulating the activity of the catalytic (C) component of the adenylate cyclase system.[63] Activation of a calmodulin-sensitive phosphodiesterase by a rise of intracellular free calcium may contribute to regulate the cAMP level.[64]

2. cAMP in Relation to Cell Proliferation or Transformation

In multicellular animals there is frequently an inverse relationship between intracellular cAMP levels and the rate of cell proliferation. However, numerous studies concerning variations of cAMP levels in normal or tumor cells have yielded conflicting results with either an increase, a decrease, or no changes in the intracellular levels of cAMP.[65-70] The action of cAMP on proliferative growth appears variable in different cell types and in different experimental protocols. cAMP maybe involved in contact inhibition of cell proliferation,[71,72] and there is evidence that treatment of either cultured neoplastic cells (Reuber H35 hepatoma cells) or tumors growing in vivo (rat mammary carcinomas) with active cAMP derivatives may result in inhibition of the growth of neoplastic cells and regression of tumors.[73-75]

Changes in cAMP may not be related directly to the origin or maintenance of the transformed phenotype. However, decreased intracellular levels of cAMP (usually as a consequence of decreased adenylate cyclase activity rather than increased phosphodiesterase activity) are observed in different types of transformed cells, including epithelial cells or fibroblasts transformed by RNA viruses (K-MuSV and RSV) or DNA viruses (SV40 and polyoma virus).[76] On the other hand, increased cAMP content directly correlates with morphological transformation of certain types of cells, like normal rat kidney (NRK) cells transformed with M-MuSV, which contains the v-mos oncogene.[77] In these cells, agents capable of elevating endogenous levels of cAMP, like prostaglandin E_1 (PGE_1) and cholera toxin, stimulate a dose- and time-dependent accumulation of endogenous cAMP that is paralleled by morphological signs of transformation.

Transformed cells may exhibit altered response to cAMP-mediated modulation of protein phosphorylation.[78] cAMP metabolism may also be involved in the complex processes of tumor cell spreading in vivo. Experimental metastasis from B16 murine melanoma cell clones correlates with higher levels of cAMP accumulation induced by melanocyte-stimulating hormone (MSH) or the diterpene, forskolin.[79] Although the physiological role of cAMP in the regulation of cellular growth remains unresolved,[80] it might be safe to conclude that cAMP can potentiate the effects of polypeptide mitogenic factors.

Complex inter-relationships exist between the intracellular levels of cAMP and the expression of viral and cellular oncogenes. While nontransformed NRK cells show increased cAMP levels and decreased growth rate at confluency, the transformation of chicken embryo fibroblasts by acute transforming retroviruses (RSV and K-MuSV) is accompanied by a failure of adenylate cyclase to rise in the growing cells.[81] The adenylate cyclase system may have a role in transformation induced by either v-ras proteins or mutant c-ras proteins. Adenylate cyclase activity is depressed in C127 mouse fibroblasts infected with wild type K-MuSV but the activity increases when cells infected with a temperature-sensitive mutant of K-MuSV are shifted from the permissive temperature to the nonpermissive temperature.[82] Treatment of H-MuSV-transformed NIH/3T3 cells with cAMP analogs which are selective for two different cAMP-binding sites of type II protein kinase results in inhibition of both $p21^{v-ras}$ protein synthesis and phenotypic transformation.[83] Transformation of BALB/3T3 cells with EJ/T24 mutant c-H-ras proto-oncogene inhibits adenylate cyclase response to beta-adrenergic agonist while increasing muscarine receptor-dependent hydrolysis of inositol lipids.[84] However, the p21 product of the v-K-ras oncogene can induce changes in growth, morphology, and the relative rate of collagen production independent of cAMP.[85]

B. cAMP-Dependent Protein Kinases

The effect of cAMP in eukaryotic cells are mediated exclusively through the activation of a cAMP-dependent protein kinase, also termed A-kinase.[43] The enzyme exists as an inactive complex of two regulatory (R) and two catalytic (C) subunits. When cAMP is generated under the action of hormones, growth factors or other stimuli, the complex dissociates to liberate active catalytic subunits, which are apparently identical. The complete amino acid sequence of the catalytic (C) subunit of bovine cardiac muscle cAMP-dependent protein kinase has been determined.[86]

Two types of cAMP-dependent protein kinases (type-I and type-II) are distinguished according to differences in the regulatory components, functional properties, and tissue-specific functional activities.[87,88] The two isoenzymes are identical in the structure and function of the catalytic (C) subunit. Two forms of regulatory (R) subunits of cAMP-dependent protein kinases, termed RI and RII, have been identified in erythroleukemic cells transformed by the Friend murine leukemia virus (F-MuLV).[89] Differentiation and treatment with a cAMP analog (8-bromo-cAMP) elicit a large and selective increase in the rate of biosynthesis of only one type of these R subunits. A cAMP-resistant Chinese hamster ovary (CHO) cell line containing both wild type and mutant species of the RI subunit of cAMP-dependent protein kinase has been characterized.[90] The results suggested the possibility that the two forms of RI detected in these cells occurred by mutation in one of two separate alleles of a single gene coding for RI.

1. Regulatory Actions of cAMP-Dependent Protein Kinases

The activation of cAMP-dependent protein kinases may have important effects at various intracellular levels, including the cell membrane, the cytoplasm, the mitochondria, the lysosomes, the microtubules, and the nucleus. Unfortunately, the substrates responsible for these pleiotropic effects of cAMP-dependent protein kinase remain poorly characterized. At the cell membrane level, some components of the membrane skeleton, acting as accessory proteins of the spectrin-actin complex, maybe phosphorylated by cAMP-dependent protein kinases.[91] At the nuclear level, phosphorylation and dephosphorylation of histone and non-histone proteins may contribute, in principle, to the regulation of gene expression.[46] However, the precise mechanisms of cAMP-dependent regulatory phenomena at the nuclear level remain largely unknown. Although there is much indirect evidence indicating that the adenylate cyclase system is involved in the regulation of eukaryotic genome expression by hormones and growth factors, in no instance this phenomenon has been characterized at the molecular level. A possible mechanism is modification of the degree of DNA supercoiling or relaxation, processes which are controlled by DNA topoisomerases.[92] There is evidence that the phospho form of the type II regulatory subunit of cAMP-dependent protein kinase (phospho-RII-cAMP) from rat liver possesses intrinsic topoisomerase activity towards several DNA substrates.[93] However, the precise functional and biological significance of the topoisomerase activity exhibited by phospho-RII-cAMP is unknown. In general, the physiologic role of cAMP in growth regulation remains unresolved.[80]

2. Structural Homology between the Catalytic Subunit of cAMP-Dependent Protein Kinase and the Protein Products of the src Oncogene Family

Structural homology has been detected between the catalytic (C) subunit of mammalian cAMP-dependent protein kinase and the protein products of some members of the *src* oncogene family, including the v-*src* oncogene. Considerable homology (28 of 96 amino acid residues) exists between the human c-*abl* protein and the C subunit of bovine cAMP-dependent protein kinase, although the tyrosine acceptor residue is replaced by a tryptophan residue in the bovine kinase.[94] Interestingly, this substitution is directly adjacent to a threonine residue known to represent one of two phosphorylation sites previously identified in the C subunit of the kinase. Other members of the *src* oncogene family (the oncogenes v-*fes*, v-*fps*, and v-*yes*) are more distantly related to the same mammalian kinase.[94]

A somewhat distant degree of structural homology has been detected between the C subunit of mammalian cAMP-dependent protein kinase and the protein product of the v-*mos* oncogene.[95] The v-*mos* oncogene is also a member of the *src* family but has not acquired in evolution, as most of the other members, tyrosine-specific protein kinase activity. Both the C subunit of mammalian cAMP-dependent protein kinase and the protein product of the v-*mos* oncogene are structurally related to the protein product of the CDC28 gene of the yeast *Saccharomyces cerevisiae*, the degrees of identity being 25 and 22%, respectively.[95] The protein product of the CDC28 gene is involved in the control of yeast cell division.

3. cAMP and the Control of Proto-Oncogene Expression

Intracellular concentration of cAMP may have influence, either directly or indirectly, on the transcriptional activity of particular proto-oncogenes. Treatment of PY815 mouse mastocytoma cells maintained in culture with a dibutyryl cAMP analog (which increases the intracellular concentration of cAMP) induces a temporary increase in the level of c-*myc* transcripts.[96] The levels of c-*myc* mRNA are increased by 1.8-fold 3 hr after the treatment, but the increase is not sustained despite the continued presence of the dibutyryl cAMP analog, and between 6 and 15 hr c-*myc* expression is reduced to half that in untreated cells. c-*myc* expression again increases briefly to 1.5-fold that in untreated cells before decreasing again to low levels 3 hr after removing the analog. The peaks of c-*myc* mRNA that occurs during dibutyryl cAMP treatment, or when the cAMP analog is removed, correspond to times when the majority of cells are in late G_1 or early S phase of the cycle.[96]

V. cGMP AND THE GUANYLATE CYCLASE SYSTEM

cGMP (guanosine 3':5'-monophosphate) is generated from guanosine triphosphate (GTP) by the activity of the enzyme guanylate cyclase.[97,98] The intracellular concentration of cGMP depends on the equilibrium existing between guanylate cyclase and cGMP phosphodiesterase. The activity of guanylate cyclase is, at least partially, regulated by hormones and growth factors via cAMP.[99] The activity of cGMP phosphodiesterase is under separate genetic control from cAMP phosphodiesterase.[100,101]

A. cGMP and Cell Proliferation

The effects of cGMP on the control of the proliferation of normal or transformed cells are, as those of cAMP, difficult to evaluate.[102,103] High concentrations of cGMP have been found in the blood and urine of guinea pigs during the development of a transplantable leukemia.[104,105] In some cellular systems an increase in the intracellular levels of cAMP is associated with inhibition of DNA synthesis and mitosis whereas cGMP may have opposite effects, acting in a positive manner in the mediation of cellular responses to mitogenic agents.[106-109] However, in other cellular systems the situation is much less clear and a general conclusion on this subject cannot be reached on the basis of the available data.

B. cGMP-Dependent Protein Kinase

The physiological actions of cGMP are mediated by a cGMP-dependent protein kinase which is present in a diversity of mammalian tissues.[36,110] cAMP- and cGMP-dependent kinases are homologous proteins, having many characteristics in common, but there are some differences between them, especially in the mechanisms of activation of the enzymes by cyclic nucleotides and the effects of different stimulatory and inhibitory modulators on their activity. Unfortunately, the substrates of cGMP-dependent protein kinases at different cellular sites are, as those of the other protein kinases, little known. cGMP may stimulate the incorporation of phosphate into specific nuclear proteins of peripheral blood lymphocytes subjected to proliferative stimuli whereas cAMP would exert an opposing effect.[111] There is evidence that cGMP is associated with the regulation of particular transcriptional processes, for example, those occurring in the polytene chromosomes of *Drosophila melanogaster* after heat-shock treatment.[112] It has also been suggested that a protein kinase modulator may participate in cGMP stimulation of histone phosphorylation processes occurring in the liver,[113] but the hypothetical modulator has not been characterized.

VI. GTP-BINDING PROTEINS

An important system in the mediation of transduction processes across the cell membrane involves an ubiquitous family of guanosine triphosphate (GTP)-binding proteins.[114-116] These proteins transduce signals generated by ligand interactions with specific cell surface receptors into changes in intracellular levels of cyclic nucleotides.

The molecular mechanisms involved in the transductional processes of extracellular signals into intracellular messages involve stimulation and inhibition of the adenylate cyclase system. These mechanisms are constituted by three interacting components: a receptor, a GTP-binding protein, and an effector. In hormonally regulated adenylate cyclase systems the signal is the hormone, which is recognized by the specific hormone receptor, and the GTP-binding proteins are called G proteins. In the visual system, the signals are photons which are captured by the specific receptor, rhodopsin, and the GTP-binding protein is called transducin. In both the hormonally regulated system and the visual system, the effector is a cGMP phosphodiesterase.

A. G Proteins

In a diversity of cell types the cell surface receptors for hormones and peptide growth factors communicate with a pair of homologous GTP-binding proteins, called G proteins. One of these proteins, termed G_s, mediates the stimulation of adenylate cyclase activity, and the other, termed G_i, is responsible for its inhibition. The G proteins are associated with the catalytic unit of adenylate cyclase and may represent evolutionary branch points for transduction of information across the plasma membrane. G proteins are distinct components of the adenylate cyclase system on the cell membrane and both G_s and G_i are oriented toward the inside of the cell.

Both the stimulatory (G_s) and the inhibitory (G_i) coupling G proteins of the hormone-sensitive adenylate cyclase system consist of three distinct subunits termed alpha, beta, and gamma. Whereas the beta subunits of G_s and G_i are identical in their molecular weights (35,000 daltons) and amino acid compositions and proteolytic peptide maps, the alpha subunits are disimilar in molecular weight (45,000 daltons for G_s and 41,000 daltons for G_i) and other characteristics. The gamma subunits (approximate 10,000 mol wt) are less well characterized but are probably identical. Activated cell surface receptors stimulate the alpha subunit to bind GTP, and the alpha subunit-GTP complex interacts with the effector of the transductional system, i.e., with the hormone-sensitive adenylate cyclase. The interaction is terminated when the bound GTP is hydrolyzed to guanosine diphosphate (GDP) by the alpha subunit, which is thereafter recoupled with the beta-gamma complex. In addition to the guanine nucleotide binding sites, the alpha subunits of G proteins contain an ADP-ribosylation site. The alpha subunit of G_s is ADP-ribosylated by pertussis toxin. A cDNA clone encoding the alpha subunit of G_s has been isolated and identified.[117]

Cellular receptors on the cell surface are functionally coupled with G proteins.[118] The beta-adrenergic receptor is stoichiometrically phosphorylated by the cAMP-dependent protein kinase predominantly in two serine residues and this phosphorylation induces a direct inhibition of the coupling of the receptor to the G_s guanine nucleotide regulatory protein.[119] Thus, there exists a direct correlation between beta-adrenergic receptor phosphorylation and decreased receptor-G_s coupling, which contributes to regulation of the receptor functional properties.

B. Transducin

GTP-binding proteins are involved in the regulation of proteins other than the catalytic unit of adenylate cyclase.[115] In the visual system the rod outer segment in the retina converts light stimulus into a change in membrane polarization. The concentration of cGMP in rod cells is regulated by light and is importantly involved in visual excitatory phenomena. Rhodopsin, a receptor for light, contains a phosphodiesterase that hydrolyzes intracellular cGMP. A G protein analog, termed transducin, is found in the disk membranes of the outer segment of the retinal rod, where it regulates GMP concentrations by activating a cGMP phosphodiesterase in response to photoexcitation of rhodopsin. When photon capture causes rhodopsin to isomerize, each rhodopsin molecule interacts with hundreds of transducing molecules in a way that permits transducin to exchange bound GDP with GTP. Upon GTP binding transducin dissociates from rhodopsin and reacts with phosphodiesterase, which in turn hydrolyzes cGMP. The cycle ends when GTP-bound transducin is hydrolyzed to GDP, phosphodiesterase activation ceases, and transducin reassociates with rhodopsin.[116]

Transducin exhibits striking structural and functional homologies to the G proteins and is a heterodimer composed, as G_s and G_i, of alpha, beta, and gamma subunits. The beta subunit of transducin has a molecular weight of about 35,000 and is indistinguishable from that of G_s or G_i. The alpha subunit of transducin, as deduced from the bovine cDNA sequence, has a molecular weight of 39,971 and is composed of 350 amino acids.[120-122] It can be ADP-ribosylated at distinct sites by pertussis toxin in the dark and by cholera toxin in the light. The gamma subunit of transducin has a molecular weight of about 8,000 and its structure is different from that of G_s and G_i gamma subunits. The sequence of the 74 amino acids in the gamma subunit of transducin from bovine rod outer segments has been deduced from the cDNA sequence.[123] Binding of guanine nucleotide causes dissociation of the alpha subunit from the other components of the complex (beta and gamma subunits), which results in activation of phosphodiesterase, causing hydrolysis of cGMP to 5'-guanosine monophosphate. After hydrolysis of GTP bound to the alpha subunit, the beta-gamma complex recouples the alpha subunit to the signal detector. The beta-gamma complexes derived from either transducin or G protein dissociation inhibit membrane-bound adenylate cyclase.[124]

C. The *ras* Oncogene Proteins and the Adenylate Cyclase System

As stated above, G proteins (G_s and G_i) are GTP-binding proteins which are associated with the catalytic subunit of adenylate cyclase and are involved in the regulation of the generation of cAMP. A striking degree of structural and functional homology has been detected between the G proteins and the protein products of the *ras* family of oncogenes. The protein products of the c-*ras* proto-oncogenes (c-H-*ras*, c-K-*ras*, and c-N-*ras*) are, like the G proteins, membrane-bound proteins that bind guanine nucleotide, hydrolyze GTP, and reside at the inner surface of the cell membrane.[125-128] Homology exists between the amino acid sequences of the alpha subunit of G proteins and the middle and amino-terminal portions of the protein products of the c-*ras* oncogene family, $p21^{c-ras}$, which suggests that both protein families may be derived from a common ancestor molecule.[129-131] The products of both gene families, the *ras* proteins and the G proteins, are apparently involved in the transduction of regulatory signals across the plasma membrane.

1. Structural and Functional Homologies of the ras Proteins

Comparison of amino acid sequences has revealed the presence of multiple homologous regions common to all members of the human c-*ras* family and the bacterial translation elongation factors Tu (EF-Tu) and G (EF-G), which also contain in their amino-terminal region a binding site for GTP.[132] During protein biosynthesis in bacteria *(Escherichia coli)*, the elongation factor EF-Tu recognizes, transports, and positions the codon-specified aminoacyl-tRNA onto the A site of the ribosome. In this role, EF-Tu interacts with GDP and GTP, which act as allosteric effectors to control the protein conformation required during the elongation cycle.[133] High-resolution X-ray diffraction analysis of EF-Tu reveals that four regions of the amino acid sequence that are homologous to $p21^{ras}$ are located in the vicinity of the GDP-binding site, and most of the invariant amino acids shared by the proteins interact directly with the GDP ligand.[133] A GTP-binding protein from *E. coli*, called LepA, is possibly involved in bacterial protein secretion and shows sequence homology to initiation factor 2 (IF-2) as well as to EF-Tu and EF-G.[134]

The $p21^{ras}$ product is structurally homologous to both the bovine brain G protein and the alpha subunit of transducin.[129-131] The amino acid sequence of the amino-terminal part of transducin alpha subunit is 59% homologous to p21^{ras}. The amino acid sequence of bovine transducin alpha subunit as deduced from the cDNA sequence, includes four regions, ranging from 11 to 19 residues in length, that exhibit significant homology to sequences of GTP-binding proteins, including G proteins, the c-*ras* proteins of man and yeast, and the elongation factors of ribosomal protein synthesis in bacteria, EF-G and EF-Tu.[120,121] All four regions probably interact with GTP. Homology also exists between transducin alpha subunit c-*ras* proteins, and the bacterial translation initiation factor, IF2.[121]

In the retina, transducin functions to couple photolysis of rhodopsin to changes in intracellular levels of GMP. Light stimulation of photoreceptors determines an exchange of GDP bound to the alpha subunit of transducin for GTP. The complex formed by this reaction activates a cGMP phosphodiesterase and the activation is terminated when the intrinsic GTPase activity of transducin alpha subunit hydrolyzes the bound GTP, restoring the transducin-GDP complex.

The structural similarities existing between mammalian and yeast *ras* proto-oncogene products, bacterial translation elongation and initiation factors, G proteins, and transducin alpha subunit reflect their involvement in guanine nucleotide binding and hydrolysis activities and the requirements for an alternation between GDP- and GTP-bound conformations. The portions of these proteins that are not directly involved in nucleotide binding or hydrolysis may govern the specificity for interaction with different subcellular components.[130] The carboxy-terminal region of G proteins alpha subunits and c-*ras* proteins may represent a functional domain involved in receptor-coupling and thus conferring relative functional specificity to the respective proteins.[135] A model for the tertiary structure of p21^{ras} has been constructed.[136] This model predicts p21^{ras} sequences responsible for guanine nucleotide specificity and for binding and hydrolysis of GTP, and allows to propose mechanisms by which specific mutations may lead to oncogenic activation of the protein.

In spite of the structural similarities existing between the p21 products of *ras* oncogenes and the G proteins, no direct evidence has been found that p21^{ras} regulates adenylate cyclase activity in a manner similar to G proteins.[137] Thus, the decreased adenylate cyclase activity frequently observed in virally transformed cells would not depend on a direct action of p21^{v-ras} on the activity of the enzyme. However, there is evidence that a mutant p21 K-*ras* protein stimulates adenylate cyclase activity early in G_1 phase of the cycle of NRK cells.[138] There is also evidence that cAMP and its receptor protein, type II protein kinase, may be involved in a quantitative regulation of *ras* oncogene expression. Treatment of H-MuSV-infected NIH/3T3 cells with two classes of cAMP analogs, which are selective for two different binding sites of type II protein kinase, in combination, synergistically inhibits both p21^{ras} synthesis and phenotype transformation. Thus, at least some of the effects of cAMP on cell proliferation could be mediated by regulation of p21^{c-ras} synthesis and expression.

2. Evolutionary Aspects of the ras Proteins

The general biological importance of the c-*ras* protein products is indicated by the fact that genes related to this family are present in all of the vertebrate species studied so far,[139-141] as well as in insects like *Drosophila melanogaster*.[142-145] Three c-*ras* functional genes are contained in different locations on chromosome 3 of this insect and their nucleotide sequences show a high degree of homology to the vertebrate c-*ras* counterparts.[144] All three *Drosophila* c-*ras* genes are expressed in all developmental stages during the life cycle of the insect.[146] However, the expression of c-*ras* genes of *Drosophila* is subjected to developmental regulation.[146,147] Each *Drosophila* c-*ras* gene codes for one large transcript and two smaller transcripts. The larger transcript of each gene is present at a constant abundance in all stages of development, whereas the shorter transcripts are found mainly in embryos. These results suggest the possible participation of c-*ras* genes in processes related to cell differentiation and/or proliferation.

DNA sequences related to c-*ras* are also present in the yeast *Saccharomyces cerevisiae*.[148-151] The protein products of the yeast c-*ras*-related genes are functionally homologous to mammalian c-*ras* genes, being involved in the control of adenylate cyclase and exhibiting guanine nucleotide binding and GTPase activities.[152-155] Yeast cells that lack functional *ras*-related genes (RAS-1 and RAS-2 genes) are ordinarily nonviable,[156,157] but may remain viable if they carry an intact c-*ras* gene of mammalian origin.[158] RAS-1 and RAS-2 encode proteins of 309 and 322 amino acids, respectively. RAS-2 gene of *S. cerevisiae* is required for gluconeogenic growth and proper response to nutrient limitation.[159] The reserve carbohydrates in yeast are glycogen and tetralose and, as with glycogen in higher cells, their degradative enzymes are activated by cAMP-dependent phosphorylation processes. cAMP also affects other functions in yeast and RAS-2 mutants might be affected in reserve carbohydrates and growth when placed on media with pyruvate or other noncarbohydrate carbon sources.[160] Variant yeast RAS-1 protein, which are structurally analogous to oncogenic human c-*ras* proteins, are characterized by reduced GTPase activity.[154] These results suggest that yeast and mammalian c-*ras* proteins have similar biochemical and physiological properties.

A *ras*-related gene exists also in an eukaryotic microbe, the slime mold *Dictyostelium discoideum*, where it is developmentally regulated.[161-163] The c-*ras*-related product of *D. discoideum* is synthesized as a 23,000 dalton protein (p23) which is specifically immunoprecipitated by an anti-p21^{ras} monoclonal antibody and is closely related to mammalian p21^{c-ras} in structure as defined by tryptic peptide analysis. There is a clear down-regulation of p23 protein expression and level during *D. discoideum* development, suggesting that the synthesis of the c-*ras* gene product is closely correlated with the extent of cellular proliferation. A novel *ras*-related gene family has been described recently in the marine mollusc, *Aplysia*.[164]

3. Functional Aspects of the ras Proteins

The viral v-*ras* p21 protein products are susceptible to autophosphorylation by transferring the gamma-phosphate of GTP (but not ATP) to threonine in position 59.[126,165] p21^{v-ras} products expressed in *E. coli* after cloning of v-H-*ras* and v-K-*ras* in molecular vectors may retain both guanine nucleotide binding and autophosphorylation activities.[166-173] The purified p21^{v-ras} product obtained in *E. coli* transfected with a molecular vector contains stoichiometric amounts of noncovalently associated GDP, which acts as a competitive inhibitor of the interaction of added guanine nucleotides with p21^{v-ras}.[169,170] EGF stimulates guanine nucleotide binding activity and phosphorylation of *ras* oncogene products.[174] In contrast to p21^{v-ras}, normal p21^{c-ras} proteins are not phosphorylated due to substitution for the 59-threonine with alanine. The three activities displayed by p21^{v-ras} proteins (GTP/GDP binding, autophosphorylation, and GTPase) are related and may depend on a single active center within the p21 molecule since they are specifically affected by a monoclonal antibody.[172] The v-H-*ras* p21 product is tightly bound to cellular lipid on the inner surface of the plasma membrane,[175,176] and the study of carboxy-terminal point mutations of the protein indicates that a simple cysteine residue at position 186 is involved in both the lipid binding and the membrane localization of p21^{v-ras}.[177,178] A synthetic tetrapeptide of the p21 carboxy terminus has been used to demonstrate that the acylation site of p21 is at cysteine-186 and that the lipid moiety of the hydrophobic peptide is palmitic acid.[179] Since cysteine occupies the same location in all known *ras* oncogene proteins of either viral or cellular origin, it may serve a similar function in the other *ras* proteins as well. This function is probably associated with both the membrane localization and the oncogenic potential of the *ras* oncogene proteins.

There is evidence that the v-*ras* protein consists of at least two functional domains, one which specifies the p21 guanine nucleotide binding activity and the other which is involved in the p21 membrane association in transformed cells.[180] The functional domains of p21^{ras} proteins can be characterized by using chimeric genes or molecular vectors containing deletion mutants of *ras* oncogene sequences.[171,181] The amino acid 12 of *ras* proteins is involved in GTP binding. An antibody that specifically recognize an epitope of p21^{ras} including amino acid 12 blocks binding of GTP to p21 and, conversely, guanine nucleotides prevent interaction of the antibody with p21^{ras}.[182] Microinjection of antibodies specific for amino acid 12 of the v-K-*ras* protein into cells transformed by this viral oncogene causes a transient reversion of the cells to a normal phenotype.[183] The fact that this antibody inhibits binding of GTP to the v-K-*ras* p21 product supports the notion that GTP binding is essential to the transforming function of this oncogene product.

Expression vectors have been constructed which encode the authentic wild-type human c-H-*ras* protein and its T24 oncogenic variant containing a single point mutation.[184] Both the normal (glycine-12) and the oncogenic (valine-12) versions of the p21 product expressed from the cloned human c-H-*ras* gene were able to bind approximately the same quantities of guanine nucleotides but a marked impairment was observed in the GTPase activity of the oncogenic p21 product as compared with the normal product.

Expression vectors have also been constructed that encode two truncated forms of the T24 variant protein.[171] One of the truncated c-H-*ras* proteins was deleted between residues 1 and 23 at the amino terminus, whereas the other truncated protein was deleted for 23 residues at the carboxy terminus. The full length c-H-*ras* wild-type and T24 variant proteins produced by the respective vectors in *E. coli* retained guanine nucleotide binding properties in essential equimolar amounts. The carboxy-terminal truncated *ras* protein also retained the ability to bind GTP and, although the extent of binding was somewhat reduced, it was apparent that the carboxy-terminal 23 amino acid residues are not a crucial part of the guanine nucleotide binding site. The carboxy-terminal truncated molecule, however, had lost its transforming ability, probably as a result of the inability to localize properly to the inner face of the plasma membrane. In contrast, the amino terminal truncated *ras* protein was completely deficient in guanine nucleotide binding, which suggests that amino acids of this region are part of the nucleotide binding site.[171] This result is in accordance with the observation that antibodies specific for amino acid 12 of p21^{ras} peptides inhibit binding of GTP to the p21 molecule.[182,183] The GTP binding site is also required for the transforming capacity of p21^{v-ras} products because such antibodies produce a transient reversion to the normal phenotype in the treated cells. The carboxy-terminal region of *ras* gene products, however, may be involved in some way with guanine nucleotide-binding function as demonstrated by using antibodies of predetermined specificity to this region.[185] In conclusion, both the carboxy-terminal part of p21^{ras} molecules, involved in the location of the protein on the inner aspect of the cell surface, and the amino-terminal part of the same protein, involved in guanine nucleotide binding, would be necessary for the expression of transforming capacity.

The cellular targets for p21^{ras} have not been identified. Aside from the intramolecular autophosphorylation in threonine-59 of p21^{v-ras}, no protein kinase activity has been detected in the protein. GDP bound to intracellular p21^{ras} would have to be released and replaced by GTP to activate the p21 protein.[170] Glycerol and Mn^{2+} strongly stimulate the GTPase activity of normal c-*ras* proteins.[186]

4. Mutant c-ras Proteins

Mutant c-*ras* proteins containing amino acid substitutions at positions either 12 or 61 of the polypeptide chain have enhanced oncogenic potential in the NIH/3T3 assay system.[187,188] An example of such mutant proteins in the EJ/T24 c-H-*ras* gene product identified in several types of human tumors and tumor cell lines. The structural mutational changes which activate the transforming capacity of c-*ras* proteins do not alter the protein's known biochemical parameters and, in particular, do not affect the localization of the protein at the cell membrane or the intrinsic ability of the protein to bind guanine nucleotides.[189] Rather, mutational activation of p21$^{c\text{-}ras}$ may be a consequence of alterations in the interaction of p21 with other cellular proteins. Augmented expression of certain proto-oncogenes, especially c-*myc*, is important for the cotransformation of susceptible cells (rat embryo fibroblasts) with a mutant *ras* gene.[190]

The level of intrinsic GTPase activity distinguishes normal and mutant forms of p21$^{c\text{-}ras}$ molecules.[191] A severalfold reduction of GTPase activity is present in the activated, mutant versions of p21$^{c\text{-}ras}$ products, which may determine a persistent activation of the GTP-associated transductional system, possibly leading to uncontrolled cell proliferation and neoplastic transformation of susceptible cells.[191-193] Recombinant plasmids carrying the human c-H-*ras* gene with two point mutations in codons 12 and 61 have the same transforming activity in the NIH/3T3 assay as the genes with single mutations.[194] Thus, a single point mutation of c-*ras* at either codon 12 or 61 could be enough to reduce the GTPase activity of p21$^{c\text{-}ras}$ and activate the oncogenic potential of the gene. However, the available evidence indicates that the solely presence of an activated *ras* oncogene is not sufficient to induce the complete range of phenomena associated with the carcinogenic process, which requires the operation of other genetic changes such as aneuploidy.[195] Moreover, reversions to a nontransformed phenotype can occur in cells transformed by H-MuSV when they are submitted to particular types of environmental changes.[196] There is evidence for cell-mediated resistance in the molecular mechanisms responsible for such reversions.[197]

5. Influence of ras Proteins on Cell Proliferation and Differentiation

The structural homologies existing between the G proteins located on the cell membrane and c-*ras* oncogene proteins suggest that they may share some similar functional properties, especially in functions related to cell proliferation or differentiation. The protein products of c-*ras* proto-oncogenes are apparently involved in the control mechanisms of cell proliferation in certain tissues, like hemopoietic cells.[198] Retroviruses containing the v-*ras* oncogene can alter the growth of cultured cells from the erythroid lineage, inducing colony formation and erythropoietin-dependent proliferation but the cells are not irreversibly blocked in their differentiation and they can eventually show extensive hemoglobinization and can reach a terminal stage of erythroid differentiation.[199] A similar phenomenon is observed in cells from the myeloid lineage, which after infection with BALB and Harvey murine sarcoma viruses from diffuse colonies but the colonies are made up of relatively mature macrophages which, although exhibit increased self-renewal capacity, can eventually undergo terminal differentiation in culture.[200]

Microinjection of the mutated (T24) oncogenic form of the c-*ras* protein, but not of the normal c-*ras* protein, into some PC12 rat pheochromocytoma cells results in the induction of morphological differentiation.[201] This differentiation occurs in the absence of the specific growth factor (NGF) which is usually required for the differentiation of PC12 cells. These results suggest that the mutated c-*ras* protein is able to eliminate the need for an exogenous signal represented by NGF. Sarcoma viruses carrying *ras* oncogenes also induce differentiation-associated properties in a neuronal cell line.[202] In contrast, an irreversible block of cell differentiation is induced by K-MuSV in cultured rat thyroid cells, with persistent suppression of thyroglobulin synthesis and iodide uptake in spite of reversion of the transformed phenotype as assessed by using temperature-sensitive (ts) mutants of the virus.[203] Whereas the malignant properties of rat thyroid cells are suppressed at the nonpermissive temperature, the blockage of differentiated thyroid functions persist at this temperature. The latter phenomenon due to an alteration at the level of transcription but the mechanism of the dissociation between expression of transformation and expression of differentiated functions is not understood.

The oncogene protein products of murine sarcoma viruses carrying different types of v-*ras* can produce marked enhancement in the proliferation of susceptible cells but the mechanisms involved in such effects are not yet clear. The mitogenic activity of p21^{ras} could be mediated, at least partially, through changes in adenylate cyclase.[138] Different types of transforming growth factors produced by murine sarcoma virus-infected cells may contribute to the stimulation of DNA synthesis and cell proliferation.[204,205] The p21$^{v\text{-}ras}$ products of murine sarcoma viruses can block calcium-induced terminal differentiation of BALB/MK epidermal keratinocytes and can abrogate completely their requirement for EGF.[206] Thus, v-*ras* proteins confer to epidermal cells the rapid acquisition of EGF-independent growth. The purified p21$^{v\text{-}ras}$ protein made in *E. coli* from the oncogene of BALB murine sarcoma virus is sufficient to induce a morphologically transformed phenotype and to stimulate cell proliferation when it is administered by microinjection into quiescent mouse NIH/3T3 fibroblasts.[207] The normal, nonmutated human c-*ras* p21 protein produced by a similar procedure can also induce the same actions but only at higher concentrations and the effects are not as pronounced as with the v-H-*ras* protein. Interaction of tumor promoters with *ras* oncogene-transformed cells may be important for the determination of specific types of growth response.[208] The immediate-early enhancer element of herpes simplex virus type 1 (HSV-1) can replace a regulatory region of the c-H-*ras*-1 proto-oncogene required for transformation.[209]

Microinjection of cloned oncogenes of the *ras* family stimulates DNA synthesis in quiescent mammalian cells.[210,211] Microinjection of an anti-*ras* antibody into NIH/3T3 cells induced to divide by adding serum to the culture medium results in inability of the induced cells to enter the S phase of the cell cycle.[212] These results suggest that p21$^{c\text{-}ras}$ is required for the action of a variety of growth factors as no growth factor present in serum can promote efficiently cell division in cells containing anti-*ras* antibody. The c-*ras* protein may, therefore, represent a common element in the molecular sequence of cellular events initiated by numerous growth factors.[212]

Transfection of a normal human bronchial epithelial (NHBE) cell line with a plasmid vector carrying a cloned v-H-*ras* oncogene changes the growth requirements, terminal differentiation, and tumorigenic properties of the recipient cells.[213] The transfectants acquire anchorage independence, indefinite lifespan (immortality), and tumorigenicity upon implantation into susceptible hosts (athymic nude mice). Myeloid cell transformation is induced by v-*ras*-containing murine sarcoma viruses.[200] The v-K-*ras* oncogene is also able to transform differentiated rat thyroid epithelial cell lines.[214] Transformation by K-MuSV blocks the typical differentiated functions of the thyroid gland, since neither thyroglobulin nor thyroglobulin-specific mRNA are synthesized, nor is iodide concentrated in the thyroid cells after transformation. However, p21$^{v\text{-}ras}$ expression does not seem to be continuously required in order to cause this block in the differentiated properties of the transformed cell lines since the block is not reversed when cells infected with a ts mutant of K-MuSV are shifted to the temperature nonpermissive for transformation.[203]

G. Other Functions of ras Proteins

The possible involvement of c-*ras* proteins in functions other than the control of cell proliferation is suggested by their presence in the rat brain.[215] The portions of G proteins or c-*ras* proteins that are not directly involved in nucleotide binding or hydrolysis may determine the specificity for interaction with different cellular macromolecules and the respective specific functional properties of these proteins in each tissue.

7. Conclusion

The *ras* p21 products are proteins that specifically bind GTP and GDP, resembling G regulatory proteins. These proteins are involved in the modulation of other cellular proteins by a cycle of GTP binding and GTP hydrolysis to GDP. Acting in close association with the plasma membrane, p21^{ras} proteins may be part of a signal-transducing complex similar to the membrane-bound G components of the adenylate cyclase and transducin systems. However, the biochemical function of p21^{ras} has not been definitely identified. In particular, a rigorous test is needed of the hypothesis that p21^{ras} functions like a G protein.[128] As yet, there is not direct evidence that p21 functions by a cycle of GTP binding and hydrolysis, and the relative biological effects of p21^{ras}-GTP vs. p21^{ras}-GDP are ignored. The possible biological consequences of the decreased GTP hydrolytic activity caused by c-*ras* gene mutations are also a subject of speculation. The biological significance of p21^{v-ras} autophosphorylation at a threonine-59 acceptor site is unknown and the attempts to demonstrate protein kinase activity with exogenous substrates for p21^{v-ras} have been unsuccessful.[128] The molecular mechanisms involved in the enhanced oncogenic potential of mutant p21^{c-ras} proteins have not been elucidated. Further studies are required to make clearer the function of normal and mutant p21^{ras} proteins.

VII. THE CALCIUM-CALMODULIN SYSTEM

The role of calcium ions in regulatory phenomena occurring in living organisms, including cell proliferation and cell differentiation, was established many years ago, although the transducing mechanisms responsible for the generation of a Ca^{2+}-mediated signal have been characterized only in the last few years.[216-222] In particular, it has been recognized that calcium ions act usually, if not always, through their binding to specific cellular proteins.[223] The best characterized and the most widely distributed of these proteins is calmodulin.[62,224-233] The biological actions of calcium are frequently mediated by activation of calcium-dependent protein kinases. Two major classes of these enzymes have been identified: Ca^{2+}/calmodulin-dependent protein kinase and Ca^{2+}/phospholipid-dependent protein kinase (protein kinase C).[234]

A. Calcium Ions

Calcium ions play an important role in the regulation of cellular activity and metabolism, and calcium has been considered, in conjunction with cyclic nucleotides, as an important second messenger to external stimuli, including hormone and growth factor action.[235] Treatment of Ehrlich ascites tumor cells with micromolar concentrations of extracellular ATP induces a rapid and transient increase in cytosolic Ca^{2+} by simultaneously mobilizing a nonmitochondrial pool of intracellular Ca^{2+} and selectively increasing the Ca^{2+} permeability of the plasma membrane.[236] Other nucleotide triphosphates, including GTP, induce Ca^{2+} transient increase which are identical to those produced by ATP. Addition of serum to serum-deprived, quiescent mouse fibroblasts results in an immediate and transient elevation of the intracellular free Ca^{2+} concentration, which may also be observed by addition of some purified growth factors, like PDGF and FGF, but not EFG.[237] In another study, it was demonstrated that the addition of PDGF, EGF, or fetal calf serum to quiescent cultured human fibroblasts causes an immediate, up to threefold, rise in cytoplasmic free Ca^{2+} concentration.[238] Apparently, growth factors may act by triggering the release of Ca^{2+} from intracellular stores (e.g., mitochondria, endoplasmatic reticulum, and plasma membrane) rather than by stimulating influx of extracellular Ca^{2+}. The Ca^{2+} growth requirement of normal prostatic epithelial and foreskin fibroblastic cells are markedly affected by EGF. A 50- to 120-fold reduction in the half-maximal requirement for Ca^{2+} is noted in epithelial cells stimulated by EGF.[239,240] These results suggest that Ca^{2+} may be a primary messenger in the action of growth factors.

Extracellular calcium ions have an important role in the control of proliferation of normal and neoplastic cells.[221,241-247] The calcium ionophore A23187, which selectively transfers calcium, magnesium, and other divalent cations across biologic membranes, is highly effective in inducing blast transformation, DNA synthesis and mitosis in cultured human lymphocytes.[248] Nontransformed cells require rather large amounts of Ca^{2+} in the media, and Ca^{2+} seems to be necessary for progression of cell cycle from G_1 into S phase as well as for re-entry of plateau (G_0) cells into the cell cycle and for mitosis. In contrast, transformed cells (in particular, virally transformed cells) require much less Ca^{2+} for growing.[230,249,250] However, the precise role of calcium ions in mitogenic processes occurring under either natural or experimental conditions remains to be elucidated.

B. Calmodulin

The cellular effects of Ca^{2+} are mediated by a specific calcium binding protein, calmodulin. This protein is widely distributed in nature and has been found in all eukaryotic cells so far examined, including plant cell,[251] but it has not been detected in prokaryotes. Calmodulin is a highly conserved protein in evolution and calmodulins isolated from diverse organisms are remarkably similar in biological, chemical, and physical properties.

The primary structure of calmodulin is constituted by a single polypeptide chain of 148 amino acids containing four Ca^{2+}-binding sites. The three-dimensional structure of calmodulin has been determined by crystallography at 3.0 Å resolution.[252] The molecule consists of two globular lobes connected by a long exposed alpha-helix. Each lobe binds two calcium ions through helix-loop-helix domains. The long helix between the lobes may be involved in interactions of calmodulin with drugs and various proteins. The molecule is stabilized by multiple interactions between the helices. It is possible that the regulation of different enzymes by calmodulin is partially a function of the number of Ca^{2+} bound, but controversy exists regarding the order in which the four sites are filled and the discrete structural changes associated with Ca^{2+} binding. Calmodulin exhibits a high degree of homology with other calcium-binding proteins like parvalbumin and troponin-C. Variation of intracellular calmodulin levels are regulated mainly by transcriptional mechanisms, i.e., by changes in the synthesis of calmodulin mRNA.

Calcium is an important regulator of cell division and there is much evidence supporting the hypothesis that the Ca^{2+}-calmodulin system is critically involved in the regulation of cell proliferation and cell differentiation. Ca^{2+}-calmodulin complexes may be active in the calcium-dependent processes leading to the initiation of DNA synthesis. Calcium stimulates DNA synthesis in calcium-deprived cultured rat liver cells through the formation of Ca^{2+}-calmodulin complexes which may undergo a brief intracellular redistribution in the later prereplicative phase of the cell cycle.[253] Calmodulin levels are elevated twofold at late G_1 and/or early S phase during the cell cycle in actively growing cells.[254] Changes in intracellular calmodulin levels are observed in the re-entry of growth-arrested cells into the cell cycle, when the cells have been rendered competent by the action of mitogenic agents, hormones, and growth factors.[255] Calmodulin may have a role in at least two distinct phases of the cell cycle. First, release from plateau is associated with a decrease in calmodulin content within the first hour, and second, a rapid increase in calmodulin levels is associated with the entry into S phase 5 to 8 hr after release.[255] Treatment of cells with anticalmodulin drugs may produce important effects on the growth of normal or malignant cells, eliciting specific and reversible progression delays into and through S phase of the cycle.[254,256]

In general, calmodulin levels are higher in transformed cells and rapidly dividing cells than in their quiescent normal counterparts, which suggests that calmodulin levels are involved in the regulation of cell division.[221,257] Calmodulin itself may be an exogenous substrate for phosphorylation on tyrosine by the purified protein kinase product of Rous sarcoma virus, pp60^{v-src}.[258] The possible biological significance of this phenomenon is not understood because different types of exogenous polypeptides, including angiotensin, can also be phosphorylated by the pp60^{v-src}-associated kinase activity.[259] Transformed cells do not obligatorily contain elevated levels of calmodulin and many factors, intrinsic or extrinsic to these cells, may contribute to produce wide variation in the calmodulin levels present in either normal or transformed cells.

Initiation of carcinogenesis in the mouse skin may be associated with the evolution of cells which resist Ca^{2+}-induced terminal differentiation.[260] It has been postulated that differentiation induced by substances like dimethyl sulfoxide (DMSO) in certain types of transformed cells, for example, in Friend murine erythroleukemic cells, is critically associated with an early increase in Ca^{2+} influx.[261] However, the measurement of cytosolic Ca^{2+} using a fluorescent indicator shows that, contrary to expectation, a small decrease occurs upon treatment of Friend erythroleukemia cells with DMSO.[262]

C. Ca^{2+}/Calmodulin-Dependent Protein Kinase

The calcium-calmodulin complex can act as an integral subunit of an enzyme or through direct interaction with specific enzymes. A frequent mechanism of action of this system is through activation of Ca^{2+}-calmodulin-dependent protein kinases. There is definite evidence that calmodulin regulates the activity of the adenylate cyclase system in a wide range of structurally different tissues and species.[263] Unfortunately, the molecular mechanisms by which calmodulin activates adenylate cyclase are not yet understood. Other questions to be answered are whether both calmodulin-dependent and calmodulin-independent forms of adenylate cyclase exist within any one tissue and whether some tissues lack a calmodulin-dependent enzyme entirely.[263] In any case, the available evidence indicates that the calcium-calmodulin system modulates, either directly or indirectly, most if not all of the protein phosphorylation occurring in eukaryotic cells. Ca^{2+}-calmodulin-dependent protein kinase activities contribute to the regulation of both cytosolic and nuclear protein phosporylation.[264-267] The multifunctional Ca^{2+}/calmodulin-dependent protein kinase purified from rat brain cytosol is composed of 51,000- and 60,000-dalton subunits, both of which are autophosphorylated by an intramolecular reaction.[268] Such autophosphorylation is probably important for modulating the enzyme activity.

D. Oncomodulin

Oncomodulin is a calcium-binding protein which has been detected in many rodent tumors induced by various chemical agents as well as in the blood of tumor-bearing animals.[269] Oncomodulin is not present in normal fetal or adult tissues. Oncomodulin, however, is not a tumor-specific protein but has been found in human and rodent extraembrionic tissues descended from both lineages of the blastocyst, including placenta, parietal yolk sac, trophectoderm, parietal endoderm, and amnion.[270,271] Oncomodulin is different from other oncodevelopmental proteins, like alpha-fetoprotein (AFP) and carcinoembryonic antigen (CEA), in several ways, the most important one being that oncomodulin expression is not linked to cellular proliferation but may be linked to the invasive properties of either extraembryonic tissues or tumor tissues. Certain proto-oncogenes, in particular c-*fos*, follow a similar pattern of expression during gestation.[272-274]

E. Caldesmon

Calcium may bind to complexes of calmodulin and several classes of calmodulin-binding proteins.[275] Caldesmon is a calmodulin-binding and F-actin-binding protein that is considered as a key protein through which the Ca^{2+}-dependent regulatory action of calmodulin is transmitted to F-actin filaments.[276,277] Caldesmon has been detected in bovine aorta and uterus and in human platelets as well as in cultured fibroblast cells where it is present as a 77,000 mol wt protein and is localized in the cellular stress fibers and leading edges in close association with actin filaments. In nonmuscle cells, the calmodulin-caldesmon system may play a regulatory role in cellular functions associated with actin-containing microfilaments, such as maintenance of cell morphology, cell locomotion, membrane ruffling, and cell adhesion to a substratum. In transformed cells, caldesmon is decreased to about one third of that in the untransformed cells and its intracellular distribution is changed to a diffuse and blurred appearance.[277] These changes may be important in relation to the morphological and functional expression of a transformed phenotype. Caldesmon may be phosphorylated by protein kinase C.[278]

Calcineurin has been recognized as a major calmodulin-binding protein in the bovine brain and human placental membranes.[279] Calcineurin can catalyze the complete dephosphorylation of phosphotyrosine and phosphoserine residues in the human placental EGF receptor.

VIII. PHOSPHOINOSITIDES AND PROTEIN KINASE C

Phospholipids are essential components of the cell membranes but, until recently, they were usually considered as relatively inert substances. In the last years, however, it has been recognized that inositol phospholipids (phosphoinositides) are important participants in transductional mechanisms related to the action of hormones and other chemical messengers.[280-288] The general biological importance of these compounds is suggested by the fact that polyphosphoinositides are present not only in animal cells but have also been found in plant cells.[289] Phospholipase C, which is involved in the hydrolysis of phosphatidylinositol, is regulated partially by Ca^{2+}. Two forms of soluble phospholipase C with different requirements for calcium have been isolated from rat brain and liver.[290] One of the hydrolysis products of phosphatidylinositol is the polyphosphoinositol inositol 1,4,5-trisphosphate, which may function to mobilize calcium from intracellular reservoirs and to regulate enzymic reactions or the permeability of the plasma membrane to monovalent cations. Inositol 1,4,5-trisphosphate is stepwise dephosphorylated by the action of two or more enzymes in the liver and other organs and tissues.[291] Another of the primary products of phosphoinositides breakdown is 1,2-diacylglycerol (Figure 1).

Both inositol 1,4,5-trisphosphate and 1,2-diacylglycerol are generated by activation of the enzyme polyphosphoinositide (PPI) phosphodiesterase at the level of the plasma membrane. There is evidence that the coupling factor linking receptors at the cell surface and PPI phosphodiesterase is a guanine nucleotide-binding protein analogous to the G_s and G_i proteins involved in the activation and inhibition of adenylate cyclase.[292] Inositol 1,4,5-trisphosphate and 1,2-diacylglycerol may function as second messengers to activate signal pathways that may also be responsible for releasing arachidonic acid and for activating guanylate cyclase. Degradation of 1,2-diacylglycerol by the enzyme diacylglycerol lipase results in production of monoacylglycerol and arachidonic acid. The later substance is a polyunsaturated fatty acid that is oxygenated to produce a family of compounds, called icosanoids, which includes prostaglandins, thromboxane, and leukotrienes.

Cyclic phosphate esters may be formed, in addition to noncyclic products, as a result of phosphoinositide degradation by phospholipase C.[293] The physiological role of these cyclic phosphate intermediates is unknown but the cyclic phosphate esters contain a reactive bond that may play a role in phosphoinositide-derived signal transduction.

A. Ca^{2+}/Phospholipid-Dependent Protein Kinase (Protein Kinase C)

1,2-diacylglycerol is an important messenger that functions within the plane of the plasma membrane to induce a marked increase in the affinity for Ca^{2+} and an activation of protein kinase C.[234,294] This enzyme is a multifunctional Ca^{2+}/calmodulin-dependent protein kinase and its activity is associated with transmembrane signaling and autophosphorylation processes.[268,295,296] Protein kinase C has been partially purified and characterized from several animal sources, including the chick oviduct.[297] Protein kinase C may be present even in plants.[298] The enzyme is a membrane-associated protein involved in the phosphorylation of serine and threonine residues in specific cellular proteins, including hormone and growth factor receptors as well as the glucose transporter.[299,300] Interaction of protein kinase C with

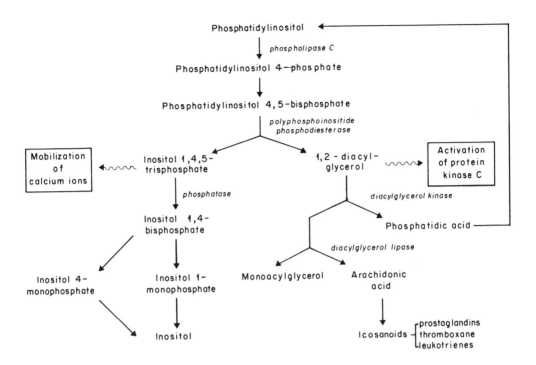

FIGURE 1. Phosphoinositide metabolic pathways.

membranes is regulated by Ca^{2+}, ATP, and phorbol esters.[301] In particular, certain Ca^{2+}-binding proteins may be involved in regulating the activity of protein kinase C. A novel 21,000-dalton Ca^{2+}-binding protein from bovine brain is the most potent inhibitor of protein kinase C among this group of proteins.[302]

In addition to the receptor-mediated phospholipase C and 1,2-diacylglycerol formation sequence, there may be other mechanisms of protein kinase C activation. Protein kinase C can be activated independently of phospholipid and Ca^{2+} by *cis* unsaturated fatty acids (oleate and arachidonate).[303] Although the physiological role of this type of activation is unknown, unsaturated fatty acids are normally present at very small concentrations and are liberated from membrane phospholipids by phospholipase A_2, which may serve as a signal for protein kinase C activation.

Protein kinase C and its fragments have been quantitated by immunological methods, which allowed an estimation of tissue levels, subcellular distribution, and ontogenic changes of the enzyme in brain and heart.[304]

B. Ca^{2+} Mobilization and Protein Kinase C Activation in Relation to DNA Synthesis and Cell Proliferation

Generation of 1,2-diacylglycerol and activation of protein kinase C is important in the control of many physiological phenomena, including DNA synthesis, cell proliferation, wound healing, and tissue and organ regeneration.[305] Calcium ions may influence cell proliferation, at least partially, through changes in the activity of protein kinase C. Extracellular calcium deprivation inhibits the proliferation of BALB/c 3T3 cells and this inhibition correlates with a loss of protein kinase C activity from the particulate fraction.[306] Addition of calcium to the calcium-deprived cells induces DNA synthesis and cell proliferation with a parallel increase in protein kinase C activity. Extracellular Ca^{2+} may not be required for the action of hormones like angiotensin II, which induces an elevation in cytosolic Ca^{2+} concentration and enhances phosphoinositol turnover, a result that suggests that accumulation of cytosolic Ca^{2+} can result solely from mobilization of intracellular stores of Ca^{2+}.[307]

However, this independence from intracellular Ca^{2+} is observed only with high doses of angiotensin II and responses to low doses of the hormone are totally dependent on the supply of external Ca^{2+}.

Elevated intracellular Ca^{2+} levels may have a synergistic effect on protein kinase C activity by acting on the enzyme itself.[308,309] Moreover, activation of protein kinase C may be involved in the cell cycle G_0 to S and G_1 to S transitions.[310] These molecular mechanisms may lead to induction of DNA synthesis and cell proliferation. Sea urchin eggs, for example, can be activated by microinjection of inositol 1,4,5-trisphosphate, which produces Ca^{2+} mobilization.[311] Activators of protein kinase C stimulate meiotic maturation of rat oocytes.[312] In contrast to the uniformity in the tissue concentrations of cyclic AMP-dependent protein kinase and calmodulin, the levels of protein kinase C activity among adult rat organs range over orders of magnitude.[313]

Ion fluxes and redistribution are thought to play a role in mediating the action of mitogenic agents and many mitogens stimulate rapid increase in the rate of H^+, Na^+, and K^+ fluxes across the plasma membrane and induce alterations in the intracellular levels of H^+, Na^+, K^+, and Ca^{2+}.[314-316] Activation of phosphatidylinositol turnover and protein kinase C by agents such as LPS and phorbol esters may lead either directly or indirectly to enhanced rates of monovalent cation fluxes in quiescent cells.[316,317] Protein kinase C increases the activity of the Na^+/H^+ antiport system, which in turn promotes Na^+ influx, enhances intracellular pH, and stimulates Na^+/K^+ pump activity. In turn, protein kinase C may inhibit the activity of phospholipase C, thus reducing the rate of phosphoinositide metabolism turnover. However, activation of protein kinase C may not be a sufficient signal to activate the Na^+/H^+ exchange in some types of cells, which would require some cooperative effects of Ca^{2+} elevation.[318] The precise association between such cellular changes and the stimulus to enter DNA synthesis and cell proliferation is not understood in molecular terms.

An important substrate of protein kinase C activity is the ribosomal protein S6,[319] whose function may be related in some manner to the control of DNA synthesis. Another substrate of protein kinase C is vinculin,[320] a cytoskeletal protein which is also phosphorylated in response to Ca^{2+} and phorbol esters in intact cells.[321] Nonhistone nuclear proteins may also constitute substrates for protein kinase C in organs such as the testis.[322] This phosphorylation is inhibited by adriamycin, an inhibitor of spermatogenesis.

The precise role of Ca^{2+} in the processes leading to cell proliferation through activation of phosphoinositide breakdown, however, is not totally clear. Thrombin produced at the site of a wound may be an important regulator in wound healing and tissue repair upon its interaction with high-affinity cell surface thrombin receptors, and it has been shown that addition of highly purified human thrombin to hamster fibroblasts in serum-free medium stimulates phosphoinositide turnover.[323] The stimulation results in phosphorylation of both phosphatidylinositol 4-phosphate and phosphatidylinositol 4,5-bisphosphate, as well as in release of inositol 1,4,5-trisphosphate and Ca^{2+} mobilization. However, inositol 1,4,5-trisphosphate-induced Ca^{2+} mobilization may not be required for thrombin-induced mitogenesis since neomycin can block inositol 1,4,5-trisphosphate release without inhibiting initiation of DNA synthesis.[323] In platelets, addition of thrombin or ADP may lead to increased cytosolic-free calcium independently of phosphatidylinositol 4,5-bisphosphate degradation.[324] Breakdown of phosphatidylinositol 4,5-bisphosphate in a T-cell leukemia line stimulated by mitogen (PHA) is apparently not dependent on Ca^{2+} mobilization.[325]

C. Hormones and Growth Factors in Relation to Phosphoinositide Metabolism and Protein Kinase C

Many hormones and growth factors have been shown to elicit striking effects on phosphoinositide metabolism, stimulating phosphatidylinositol breakdown with activation of phospholipase C, generation of 1,2-diacylglycerol and activation of protein kinase C. In turn, the receptors for several peptide hormones and growth factors may constitute substrates for protein kinase C activity, including the receptors for insulin, IGF-I, EGF, PDGF, and transferrin. It is thus likely that changes in phosphoinositide turnover play an important role in regulating the function of receptors through alteration of protein kinase C activity, which would result in phosphorylation/dephosphorylation of the receptors.[326] Such changes may precede, follow, or amplify changes in the adenylate cyclase and calcium-calmodulin systems. Early changes in phosphatidylinositol and arachidonic acid metabolism are observed, for example, in quiescent Swiss 3T3 cells stimulated to divide by platelet-derived growth factor.[327] Enhancement of calcium uptake and phosphatidylinositol turnover is induced by EGF in A431 human carcinoma cells.[328]

In rat pinealocytes, activation of protein kinase C is involved in the alpha$_1$-adrenergic amplification of beta-adrenergic stimulation of cAMP accumulation.[329] Thus, there is a synergistic interaction between protein kinase C and neurotransmitter-dependent stimulation of cAMP.

In the adipose tissue, insulin increases the concentration of phosphatidic acid, phosphatidylinositol, and phosphoinositides, as well as the incorporation of ^{32}P-labeled orthophosphate into phosphatidic acid, phosphatidylinositol and 1,2-diacylglycerol.[330,331] However, it seems unlikely that activation of phosphoinositide metabolism constitutes a universal pathway for the cellular mechanisms of action of hormones and growth factors. There is a lack of association of EGF-, insulin-, and serum-induced mitogenesis with stimulation of phosphoinositide degradation in BALB/c 3T3 fibroblasts.[332] Neither insulin nor EGF stimulate the synthesis of phosphatidylinositol, phosphatidylinositol-4-phosphate, or phosphatidylinositol-4,5-bisphosphate in rat liver plasma membranes or intact hepatocytes, which suggests that the insulin- and EGF-stimulated receptor tyrosine kinases do not act on phosphoinositides in the liver.[333] The tyrosine-specific protein kinase and phosphatidylinositol kinase activities which copurify with the EGF receptor reside on different molecules and the EGF receptor is devoided of the latter type of activity.[334]

Phosphatidylcholine biosynthesis from phosphatidic acid in cultured Swiss mouse 3T3 fibroblasts is increased in varying degrees by several types of mitogenic factors, including bovine serum, TPA, EGF, FGF, IGF-I, IGF-II, and vasopressin.[335] There is a clear synergy between growth factors in the stimulation of phosphatidylcholine biosynthesis and the stimulation is associated with a two- to threefold increase in intracellular choline kinase activity.[336]

In general, the molecular mechanisms responsible for phosphoinositide metabolism regulation by hormones are not understood. The role of phosphoinositide metabolism in carcinogenic processes is also unknown. Enhanced phosphatidylinositol kinase activity is associated with early stages of hepatocarcinogenesis and hepatocellular carcinoma.[337] Further studies are required for a better characterization of the complex inter-relationships between hormones, growth factors, phosphoinositide metabolism, and the control of proliferation in normal or transformed cells.

D. Tumor Promoters, Differentiation-Inducers, Phosphoinositide Metabolism, and Protein Kinase C

Tumor promoters are potent mitogenic substances that can be involved in the formation of tumors through stimulation of the proliferation of transformed cells. Tumor promoters may display mitogenic properties both in vivo and in vitro. In quiescent mouse 3T3 fibroblasts, tumor promoters such as the phorbol ester 12-O-tetradecanoyl-phorbol 13-acetate (TPA) induce a pleiotropic mitogenic response which include changes that, although reversible upon their removal from the medium, are reminiscent of the transformed phenotype. In vitro, phorbol 12-myristate 13-acetate (PMA) triggers phenomena mimicking angiogenic processes that occur in vivo in association with the invasion of the perivascular extracellular matrix by sprouting endothelial cells.[338] Angiogenesis is necessary for the continuous growth of solid tumors.[339]

1. Mechanisms of Action of Phorbol Esters

Phorbol esters like TPA acts through activation of protein kinase C.[234,341,342] Moreover, protein kinase C purified to homogeneity has been identified with the phorbol ester receptor in bovine brain.[294,343] Phorbol diesters are substrates for diacylglycerol lipase,[344] and it has been suggested that phosphatidylinositol phosphorylation might be a key event in the mechanism of action of phorbol esters.[345,346]

The molecular mechanisms of phorbol ester-induced activation of protein kinase C are not understood. Incubation of platelets or peripheral blood lymphocytes from patients with chronic lymphocytic leukemia with TPA causes a rapid decrease in soluble protein kinase C and an increase in protein kinase C associated with the membrane-associated particulate cell fraction.[347,348] The results suggest that TPA induces an irreversible activation of protein kinase C by proteolytic processing to a protein of molecular weight about 50,000 which is active in the absence of Ca^{2+} and phospholipid.

PMA causes a complete block of hormone-stimulated production of inositol 1,4,5-trisphosphate and a concomitant loss of Ca^{2+} mobilization in cultured human 1321N1 astrocytoma cells.[349] As phorbol esters interact with protein kinase C, protein phosphorylation may be the mechanism by which phorbol esters modulate phosphoinositide metabolism. Addition of phorbol ester or a calcium ionophore to quiescent rat embryo fibroblasts results, in part, in the phosphorylation of two 28,000-dalton proteins which lack methionine and are members of the mammalian stress protein family.[350] Proteins of this family show increased rates of synthesis during different conditions of physiological stress and could represent substrates for protein kinase C-induced phosphorylation. Stress genes of this family, termed heat-shock genes, are highly conserved in evolution and are expressed in insects following heat treatment.[351,352] In *Drosophila melanogaster* the transcription of heat-shock genes is regulated, at least partially, by the protein product of the c-*myc* proto-oncogene.[353] Moreover, a heat-shock protein produced by *Drosophila* is bound to the protein product of the c-*src* proto-oncogene in form of a relatively stable complex.[354]

Tumor promoters not structurally related to TPA, such as teleocidin and debromoaplysiatoxin, also activate protein kinase C.[355] It has been suggested that for each molecule of tumor promoters intercalated into the membrane to modify the phospholipid microenvironment, one molecule of protein kinase C moves to it and then produces a quaternary complex of phospholipid, Ca^{2+}, tumor promoter and protein kinase C, which is fully active for protein phosphorylation.[356] Variation in the levels of protein kinase C is observed in mouse strains showing different susceptibility to tumor promoting agents.[357]

In addition to the stimulation of protein kinase C, phorbol esters like TPA stimulate the adenylate cyclase system causing increased phosphorylation of endogenous substrates by cAMP-dependent protein kinase.[358] The TPA-activated cAMP-dependent protein kinase can inhibit phosphorylation, or can enhance dephosphorylation, of a specific subset of TPA-dependent phosphoproteins. Activation of Na^+/H^+ exchange is also observed in cultured fibroblasts exposed to phorbol esters, calcium ionophores, and growth factors.[318] It may be concluded that the cellular mechanisms of action of phorbol esters are exceedingly complex and not understood at present.

2. Phorbol Ester-Induced Cell Differentiation

Paradoxically, tumor promoters can induce in vitro terminal differentiation of several types of cells, including some leukemic cell lines, with disappearance of the transformed phenotype.[341] For example, the human promyelocytic leukemia cell line HL60 is induced to differentiate into macrophage-like cells by low TPA concentrations.[359] The molecular mechanisms involved in this phenomenon are not understood but different molecular forms of diacylglycerol (*sn*-1,2-diacylglycerols with fatty acids 4 to 10 carbons long) may have effects on the differentiation of HL-60 cells similar to those of tumor cell promoters like TPA.[360]

sn-1,2-diacylglycerols are potent activators of protein kinase C and are competitive inhibitors of phorbol dibutyrate binding to the soluble phorbol diester receptor. Moreover, treatment of HL-60 cells with either PMA or *sn*-1,2-diacylglycerol produces identical phosphoprotein changes.[360] Substances with capability for inducing cell differentiation in vitro, like DMSO and retinoic acid may also act through mechanisms involving protein kinase C.[361,362] Incubation of HL-60 human leukemia cells with PMA for only 30 min results in an increased phosphorylation of at least ten different cellular proteins, which is probably due to activation of protein kinase C.[363] Enhanced phosphorylation of the ribosomal protein S6 is also observed in response to TPA treatment.[364]

The possible role of protein kinase C in phorbol-ester-induced differentiation of HL-60 cells, however, is not clear because treatment of these cells with TPA results in the rapid disappearance of protein kinase C activity.[348,365] Moreover, the most remarkable and rapid change induced by TPA in HL60 cells is a decrease in phosphorylation of a 75,000-dalton (75 k) protein,[366] which suggests that mechanisms other than activation of protein kinase C may be involved in such differentiation phenomena. There is evidence that protein kinase C activation may be dissociated from phorbol ester-induced maturation of HL-60 leukemic cells.[367]

Phorbol esters may produce important effects in the expression and/or maintenance of differentiated functions in a diversity of cell types. Addition of TPA to neonatal rat pancreatic islets in long-term culture results in the long-term maintenance of beta-cell differentiated function, as characterized by increases in both the release and tissue content of insulin.[368] Another phorbol ester, PMA, causes dose-dependent increments of testosterone production by purified rat Leydig cells, probably through activation of protein kinase C.[369]

3. Mechanisms of Action of Phorbol Esters on Cell Differentiation

The effects of phorbol esters on various cell functions, including the experimentally induced differentiation of neoplastic cells in vitro, are mediated by a cell membrane alteration. However, the events linking the alteration of cell membrane to changes in the expression of genetic program of the cell, including the possibility of changes in the expression of proto-oncogenes, are not understood.

The biological actions of phorbol esters may be associated with important effects on the receptors for growth factors, including their phosphorylation or transient internalization, without degradation of unoccupied receptors.[370,372] Some actions of tumor promoters are probably mediated by changes in the expression of hormone and growth factor receptors at the cell surface, as well as by changes in postreceptor mechanisms of action of hormones and growth factors. A correlation has been demonstrated between the cytosolic phorbol ester receptor and hormone dependency in several mammary carcinoma cell lines.[373] Differential effects of phorbol ester have been observed on the beta-adrenergic response of normal and *ras*-transformed NIH/3T3 cells.[374] However, at least in part the mitogenic effect of phorbol esters may be independent of modification of hormone/growth factor mechanisms of action at either receptor or postreceptor levels. A variant of Swiss 3T3 cells, the 3T3-TNR9 cell line, does not respond mitogenically to tumor promoters like TPA, but does respond mitogenically to EGF, FGF, and serum. Despite the lack of mitogenic response to TPA, 3T3-TNR9 cells display levels of TPA binding and protein kinase C activity which are at least equal to those of the parental 3T3 cells.[375]

Transcriptional phenomena maybe importantly involved in the mechanisms of action of phorbol esters. Phosphorylation of the histones H2B and H4 on serine and threonine residues is specifically and rapidly stimulated in primary mouse lymphocyte cultures by phorbol ester.[376] The biological role of H2B and H4 phosphorylation is unclear, but may be involved in the processes by which phorbol esters affect gene activity. A countertranscript has been detected in HL-60 cells during early stages of differentiation induced by TPA.[377] The presence of countertranscripts may reflect the operation of key phenomena in the molecular processes leading to cell differentiation.

Ornithine decarboxylase may be induced by phorbol esters through activation of protein kinase C.[378] Systemic administration of TPA rapidly increases the activities of ornithine decarboxylase and plasminogen activator in rat liver and this increase is attenuated by pretreatment with either cycloheximide or actinomycin D.[379] Ornithine decarboxylase is present in all cells studied to date and the polyamines, the products of its activity, are essential metabolites.[25,26] Inhibition of ornithine decarboxylase without the concurrent administration of polyamines leads to cessation of DNA synthesis and cell proliferation. Several growth factors (EGF, PDGF, and NGF) are capable of inducing ornithine decarboxylase activity, acting synergistically with some amino acids like asparagine.[380] Probably, these growth factors induce ornithine decarboxylase activity and stimulate polyamine synthesis through activation of protein kinase C. These phenomena may be involved in the mechanisms responsible for the mitogenic action of growth factors.

In conclusion, the mechanisms involved in the mitogenic actions of phorbol esters are unknown. Phorbol esters like TPA are theirselves not mitogenic and it may be assumed that some particular signals must be generated within cells responding to the mitogenic action of phorbol esters. A putative signal would consist in Ca^{2+} mobilization induced by activation of phosphoinositide turnover rate but it has been shown that the effect of TPA on lymphocyte proliferation may occur in the absence of extracellular Ca^{2+} or detectable changes in Ca^{2+} concentrations within the cells when mitogens such as PHA are present in the incubation medium.[381] Thus, calcium-independent signal(s) may be generated by the mitogen in the presence of phorbol ester. The nature of this signal is unknown but another interesting possibility may be constituted by Na^+/K^+ exchange and elevation of cytoplasmic pH, which are observed in cells exposed to phorbol ester or synthetic diacylglycerol.[382] Further studies are required to make more clear this important address.

E. Oncogene Protein Products, the Calcium/Calmodulin System, Phosphoinositide Metabolism and Protein Kinase C

The protein products of several oncogenes (v-*src*, v-*abl*, and v-H-*ras*) contain tightly bound lipid.[175] The possibility that phosphatidylinositol metabolism can be involved in the mechanisms of action of oncogene protein products arose from the observation that the turnover of phosphoinositides is stimulated in RSV-transformed cells and that the protein product of the RSV oncogene can phosphorylate glycerol.[383-385]

1. The src and ros Oncogenes

Evidence was obtained that the purified RSV oncogene protein product, $pp60^{v-src}$, stimulates phosphatidylinositol phosphorylation to form both mono- and diphosphate derivatives and that 1,2-diacylglycerol generated by phosphatidylinositol 4,5-bisphosphate breakdown was phosphorylated to form phosphatidic acid.[386] The results suggested that $pp60^{v-src}$ may be involved in both the generation and removal of 1,2-diacylglycerol, thus contributing to the regulation of protein kinase C activity. Similar results were obtained with the polyoma virus middle-T antigen-$pp60^{c-src}$ complexes as well as with the v-*ros* oncogene protein product from avian sarcoma virus UR2.[387,388] These results, however, are not compatible with the clear demonstration that phosphorylations of phosphatidylinositol, phosphatidylinositol 4-phosphate, and 1,2-diacylglycerol are catalyzed by kinases that are distinct from the tyrosine-specific protein kinases existing in RSV-infected and RSV-noninfected cells.[389] Thus, any changes in phosphoinositide turnover or in the phosphorylation of phospholipid metabolites occurring in v-*src*-transformed cells should be considered as an indirect consequence of the primary action of the $pp60^{v-src}$ tyrosine-specific protein kinase. Moreover, phosphatidylkinase activity in chicken embryo fibroblasts is not immunoprecipitated by an antibody that recognizes pp60$^{c\text{-}src}$, and overproduction of pp60$^{c\text{-}src}$ do not increase the phosphatidylinositol kinase level in the cells.[390] These data indicate that the phosphatidylinositol kinase activity is not encoded by the v-*src* or c-*src* genes. Other oncogene product with tyrosine-specific protein kinase activity, p140$^{gag\text{-}fps}$, which is the transforming protein of Fujinami sarcoma virus, is also devoided of phosphatidylinositol kinase activity.[390] It is unlikely that this type of activity is encoded by other oncogenes whose products possess tyrosine-specific protein kinase activity, like v-*ros*, v-*mil*, and v-*raf*. Phosphoinositides are not phosphorylated by the very active tyrosine-specific protein kinase from the LSTRA murine lymphosarcoma cell line.[391]

In conclusion, it seems clear that the elevated phosphatidylinositol turnover detected in some cells transformed by viral oncogenes cannot be attributed to a direct effect of oncogene protein products possessing tyrosine-specific kinase activity but depend on specific cellular enzymes.

Treatment of RSV-transformed vole cells with TPA results in increased phosphorylation of pp60$^{v\text{-}src}$ on serine residues located in the amino terminus of the molecule.[392] Such phosphorylation may depend on protein kinase C activity, and it has been demonstrated that protein kinase C phosphorylates pp60^{src} at serine-12.[393]

As a general rule, neoplastic cells of all kinds replicate their DNA and proliferate despite an extracellular Ca^{2+} deficiency that would stop proliferation of the corresponding normal cells. However, the reasons for this reduced extracellular Ca^{2+} requirement by transformed cells are not clear.[245,249] The transforming action of viral oncogenes is usually associated with reduced requirement for the exogenous supply of calcium ions. Normal rat kidney (NRK) cells transformed by avian sarcoma virus (ASV) containing the v-*src* oncogene have a significantly reduced requirement for extracellular Ca^{2+}. ASV-transformed cells need, however, a Ca^{2+}/calmodulin effector mechanism to trigger the G1/S transition phase of the cell cycle.[249] NRK cells infected with a transformation-defective, temperature-sensitive (ts) mutant of ASV cannot proliferate in a Ca^{2+}-deficient medium at the nonpermissive (40°C) that inactivates pp60$^{v\text{-}src}$ and renders the cells phenotypically untransformed. The arrested cells can be stimulated to initiate DNA replication by adding Ca^{2+} or calmodulin at the nonpermissive temperature or by reducing the temperature to the permissive level (36°C), which restores the transformed phenotype by rapidly reactivating pp60$^{v\text{-}src}$. The G1/S transition triggered by restoring the transformed phenotype is suppressed by three different anticalmodulin drugs and this suppression can be overcome by adding calmodulin. These results indicate that v-*src*-induced transformation sharply reduces the initiation of DNA replication without bypassing a Ca^{2+}/calmodulin-dependent mechanism required for the G1/S phase transition.[249] To do this, pp60$^{v\text{-}src}$ might raise the cytosol Ca^{2+} concentration sufficiently to produce the Ca^{2+}/calmodulin complexes needed for the G1/S transition by restraining internal Ca^{2+}-sequestering mechanisms, mobilizing Ca^{2+} from the endoplasmic reticulum and/or promoting the uptake of residual extracellular Ca^{2+}.

2. The myc Oncogene

The c-*myc* protein is highly conserved in evolution and its expression is closely associated with cell proliferation.[394,395] Quiescent BALB/3T3 cells show a rapid and marked induction of c-*myc* messenger RNA subsequent to treatment with mitogenic agents like PDGF, and a similar induction is produced in the same cells by the phorbol ester TPA.[396] A marked increase in the expression of c-*myc* in the liver is observed in rats 1 to 3 hr after partial hepatectomy, preceding the proliferative process associated with liver regeneration.[397] Regulatory changes in the levels of c-*myc* messenger RNA are induced in normal human lymphocytes by modulation of cell proliferation.[398] Coexpression of the c-*myc* and c-*sis* protooncogenes occurs in human placenta, which suggests an autocrine control of trophoblastic growth.[399]

The v-*myc*- and c-*myc*-encoded proteins are localized in the nucleus and are associated with the nuclear matrix.[400-404] Expression of the c-*myc* gene can be studied by determining the cellular levels of c-*myc* mRNA or protein. The c-*myc* protein can be measured specifically in human cells by using monoclonal antibodies that have been produced against the human protein.[405] The isolation and characterization of the human c-*myc* protein has been reported recently.[406] The major human c-*myc* product is a protein of 64,000 mol wt which differs from the v-*myc* protein by 18 amino acid deletions and 42 additions, being longer by 24 amino acids.

The synthesis, turnover, and modification of c-*myc* proteins is constant throughout the cell cycle, although transient increases in level occur upon serum stimulation of resting cells.[404] Several modulators of lymphocyte proliferation produce changes in c-*myc* expression in human peripheral blood mononuclear cells.[398,407] Stimulation of these cells with the plant lectin PHA, the phorbol ester PMA, the calcium ionophore ionomycin, or the monoclonal antibody OKT3 (anti-antigen-receptor complex) produces marked increases in c-*myc* mRNA levels within 3 hr. Recombinant IL-2 has little effect on c-*myc* expression but, in combination with PHA, it augments the levels of c-*myc* transcripts measured at 24 hr but not at 3 hr. Addition of various immunosuppressors and inhibitors of lymphocyte proliferation to PHA-stimulated cultures reveals that cyclosporin A, dexamethasone, and OKT11A antibody (anti-sheep erythrocyte receptor antibody) diminishes levels of c-*myc* mRNA measured at 3 and 24 hr, whereas anti-Tac (anti-IL-2 receptor antibody) inhibits at 24 hr but not at 2 hr. These data indicate that c-*myc* expression is regulated in normal lymphocytes at several points of the cell cycle.[398] The data further suggest that protein kinase C may represent a critical molecule that regulates levels of c-*myc* mRNA in normal T-cells.[398,407,408]

Expression of c-*myc* is inversely correlated with the state of cell differentiation, remaining elevated in growth-inhibited undifferentiated cells.[409-411] Density-dependent arrest of DNA replication is accompanied by decreased levels of c-*myc* mRNA in myogenic but not in differentiation-defective myoblasts.[412] Human growth hormone (hGH) induces and maintain c-*myc* gene expression in Nb2 lymphoma cells.[413] Phorbol esters, and other agents capable of inducing differentiation of neoplastic cells in vitro can produce striking effects on both the activation of protein kinase C and the expression of c-*myc*. Maturation of human promyelocytic leukemia cells (HL-60 cells) induced by phorbol ester may be mediated by activation of protein kinase C.[414] Induction of differentiation in HL-60 cells is accompanied by decreased c-*myc* transcription.[415] This decrease is not associated with detectable changes in the primary structure or methylation of the c-*myc* gene but following the differentiation one of four S_1 nuclease-sensitive sites contained in c-*myc* is not detectable and the remaining three sites appear to be qualitatively decreased.[416] Such changes are suggestive of alterations in the physical state of the c-*myc* gene, which may be closely related to its transcriptional activation.

Alterations in c-*myc* mRNA levels are also observed in other cellular systems after chemically induced differentiation in vitro. An early decrease in phosphatidylinositol turnover, with diminished levels of inositol-trisphosphate and diacylglycerol, occurs in Friend erythroleukemia cells upon induction of differentiation by DMSO.[417] This decrease precedes decreased expression of the c-*myc* proto-oncogene and its protein product. Phorbol esters decrease c-*myc* messenger RNA levels by about 80% and enhance the expression of differentiated cell functions like calcitonin gene transcription in human medullary thyroid carcinoma cells.[418] Treatment of human neuroblastoma cells with retinoic acid induces cellular differentiation, which is associated with decreased expression of the proto-oncogene N-*myc*.[419,420] Treatment of certain human cell lines (Daudi cells) with interferon results in a marked and selective reduction of c-*myc* mRNA levels but other cell lines (HL-60 and U937 cells) do not show this reduction in spite that they display normal responses to other interferon-regulated activities and show a decline in c-*myc* gene expression when they become arrested in the G_0/G_1 phase of the cycle as part of their terminal differentiation.[421]

Calcium may be involved in the expression of proto-oncogenes induced by hormones and growth factors. In Swiss 3T3 cells, PDGF induces both Ca^{2+} mobilization and c-*myc* expression, and this expression is stimulated by the Ca^{2+} ionophores, A23187 and ionomycin.[422] These results suggest that Ca^{2+} can serve as a messenger for PDGF-induced expression of the c-*myc* proto-oncogene. PDGF has been shown to stimulate phosphoinositide turnover to produce inositol trisphosphate, which then mobilizes Ca^{2+} from the intracellular stores to the cytoplasm,[28] and its seems likely that this mechanism is involved in the activation of the c-*myc* proto-oncogene by PDGF. The expression of other proto-oncogenes might also be mediated by a similar mechanism.

3. The abl Oncogene

The protein product of the v-*abl* oncogene is a tyrosine protein kinase,[423] but it does not modulate the activity of protein kinase C by direct phosphorylation of the enzyme on tyrosine residues.[424] The direct phosphorylation of phosphatidylinositol by the v-*abl* oncogene product has no physiological role, but phosphoinositide breakdown appears to be constitutively activated in A-MuLV-transformed cells.[424]

4. Summary

Several lines of evidence suggest that stimulation of phosphatidylinositol breakdown and activation of protein kinase C and Ca^{2+} mobilization by hormones, growth factors, tumor promoters, and oncogene protein products may be involved in processes controlling cell proliferation and cell differentiation. Moreover, protein kinase C activation and/or Ca^{2+} mobilization may be involved in regulating the expression of several proto-oncogenes at either the transcriptional or post-transcriptional levels. The same, or similar, biochemical mechanisms could also be involved in processes leading to malignant transformation. It has been postulated that an uncontrollable production of active protein kinase C may be related to the promotion of oncogenic processes.[356] However, growth factor-stimulated protein phosphorylation is not only protein kinase C-dependent but also protein kinase C-independent.[427] Moreover, no simple correlation is apparent at present between protein kinase C activity and the malignant transformation of cells or the artificially induced differentiation of transformed cells. Further studies are required in order to achieve a better characterization of the relationships between phosphoinositide metabolism, protein kinase C activation, and Ca^{2+} mobilization in relation to the expression of particular oncogenes and the induction of proliferation and/or differentiation in either normal or neoplastic cells.

IX. PROTEIN PHOSPHORYLATION

Reversible phosphorylation of cellular proteins is a general type of regulatory mechanism involved in the control of many metabolic functions. Protein kinases are enzymes involved in the phosphorylation of different types of cellular proteins.[32-45] Most frequently, the phosphorylation occurs at serine or threonine residues but in the last few years it has become increasingly apparent that tyrosine phosphorylation, although occurring less frequently, is very important for the functional modification of different cellular proteins. Phosphotyrosine accounts for only about 0.01 to 0.05% of total acid-stable phosphoamino acids in normal cells.[426]

Changes in the patterns of phosphorylation of cellular proteins may occur in malignant cells, and secreted phosphoproteins may constitute markers for neoplastic transformation of human epithelial and fibroblastic cells.[427-429] However, all the protein kinase modifications observed in human leukemic cells seem to be related to cell proliferation and immaturity rather than to malignancy itself or to the different origin of the malignant cells.[430]

A. Tyrosine-Specific Protein Kinase Activity

In 1978 it was discovered that the protein product of the v-*src* oncogene of RSV is a protein kinase.[431] This product 2 years later identified the protein pp60[v-src] as a tyrosine-specific protein kinase.[426] Subsequently, tyrosine kinase activity has been found to be present in the protein products of other oncogenes, including the products of v-*yes*, v-*fps*, v-*fes*, v-*fgr*, v-*fms*, v-*ros*, v-*erb*-B, v-*abl*.[432-438] These oncogenes show variable degrees of sequence homology with the v-*src* oncogene and are considered as members of the *src* oncogene family.[141] At least some of the classic transformation hallmarks of malignant transformation in RSV-infected cells are absolutely dependent on the tyrosine-specific protein kinase activity of pp60[v-src].[439] Inhibition of the pp60[v-src]-associated protein kinase may result in reversion of RSV-transformed cells to a normal phenotype.[440] The c-*src* proto-oncogene protein product, as well as the protein products of several c-*onc* counterparts of the v-*src* oncogene family, possesses intrinsic tyrosine-specific protein kinase activity. Several proteins are phosphorylated on tyrosine residues by the pp60[v-src] protein kinase in vitro and in vivo. Increased pp60[c-src] tyrosyl kinase activity has been found in human neuroblastoma cells and the alteration is associated with amino-terminal phosphorylation of the c-*src* gene product at tyrosine residues.[441]

Antibodies to phosphotyrosine can be produced by immunization with v-*abl*-encoded protein.[442] The protein product of the c-*abl* proto-oncogene would not possess tyrosine-specific protein kinase activity but can acquire this type of activity upon chromosome translocation. The 9:22 translocation occurring in human chronic myelogenous leukemia (CML), constituting the Philadelphia chromosome (Ph chromosome), may result in the creation of a chimeric gene leading to the production of a hybrid c-*abl*/*bcr* protein possessing tyrosine-specific protein kinase activity.[443-449] The possible relationship between the acquisition of this activity by the c-*abl* protein and the origin of CML is not yet understood.

The human c-*met* proto-oncogene is related to the *src* gene family and its sequenced domains show significant homology to both the v-*abl* oncogene and the human insulin receptor.[450] The protein products of two members of the *src* family, the oncogenes v-*raf* (= v-*mil* or v-*mht*) and v-*mos*, have not acquired in evolution tyrosine kinase activity but may possess kinase activity with specificity for serine and threonine residues.[451-454] Structural and functional homologies strongly suggest that the proteins of the *src* oncogene family, irrespective of their amino acid substrate specificity, comprise a single divergent gene family that also includes the catalytic subunit of cyclic AMP-dependent protein kinase.[455]

The protein products of proto-oncogenes from the *src* family maybe involved in important cellular functions, including developmental processes and expression of differentiated functions in a diversity of tissues. RNA sequences from each of the three c-*src* genes contained in the genome of *Drosophila* are present in preblastoderm embryos, indicating their maternal origin.[456] The *Drosophila* c-*src* genes are differentially expressed in the tissues of the insect,[457] the majority of c-*src* transcripts in the adult insect being contained in the ovaries.[456]

Tyrosine-specific protein kinase activity was found to be associated with the receptor and postreceptor cellular mechanisms of action of several hormones and growth factors, including insulin,[458,459] IGF-I,[460] EGF,[461] TGFs,[462,463] PDGF,[464] and estradiol.[465] Phosphorylation of specific proteins on tyrosine may also be stimulated in cultured human cells (foreskin fibroblasts) by bacterial components like lipoteichoic acid.[466]

Changes in tyrosine-specific protein kinase activity may be associated with both tumorigenic processes and differentiation of transformed cells. Activation of liver tyrosine-specific protein kinase occurs during the early stages of chemically induced hepatocarcinogenesis.[467] Rats receiving a complete carcinogenic regimen (one dose of diethylnitrosamine (DENA) followed by two weeks of dietary 0.02% 2-acetylamino-fluorene (AAF) starting at day 14 after DENA, followed by partial hepatectomy on day 21) showed a 2.6-fold increase in their liver tyrosine kinase activity as compared to sham controls. In contrast, rats that received a partial regimen did not have elevated tyrosine kinase activity nor did they have hyperplastic nodules.[467] It was not determined whether the activation of tyrosine protein kinases is unique to the complete carcinogenic regimen or if it also occurs during noncarcinogenic liver cell regeneration. It was also not determined whether or not the increase in tyrosine kinase activity was specifically associated with persistent liver nodules that undergo progression to malignancy.

Phorbol esters may act as either tumor promoters or inducers of differentiation of transformed cells. Phorbol esters induce rapid phosphorylation on tyrosine residues in specific cellular proteins as well as in tyrosine-containing artificial substrates and exogenous proteins such as casein.[468,469] Tyrosine kinase activity is stimulated in a specific manner by other substances with ability for inducing differentiation of transformed cells, for example, DMSO.[470] Calmodulin may also be involved in stimulating the phosphorylation of proteins on tyrosine residues.[471] The effects on tyrosine phosphorylation suggest further analogies between the action mechanisms of hormones, growth factors, oncogene protein products, tumor promoters, and differentiation-inducing agents.

1. Substrates of Tyrosine-Specific Protein Kinases

Many membrane and cytosolic proteins, including receptors for several peptide hormones and growth factors, may become phosphorylated on tyrosine residues as a result of tyrosine kinase activity elicited by these substances. Substrate specificity may be an important mechanism to preserve the specificity of the hormonal signal at the postreceptor level, after the formation of the activated hormone-receptor complex. Such specificities can be examined by comparison of the abilities of the different kinases to phosphorylate various exogenous substrates.[472] Sharing of common substrates suggest the existence of some common mechanisms by which the hormonal signals are processed. The insulin and EGF receptor stimulated kinases have similar but not identical substrate specificities.[473] Similarly, the substrate specificity of pp60^{v-src}, as determined in rat and mouse cells transformed by RSV, is not shared with mammalian cells infected with retroviruses transducing other oncogenes of the *src* family, like *fps, fes,* and *abl*.[474]

According to the results obtained with immunofluorescence microscopy experiments using a high-affinity polyclonal antibody against phosphotyrosine, permanent lines of fibroblasts, and epithelial cells with nontransformed phenotypes show low, but detectable, levels of phosphotyrosine.[475] In such cells phosphotyrosine-containing proteins are concentrated at sites of microfilament-membrane and membrane-substratum interaction, namely, the focal contacts and intercellular junctions. The presence of phosphotyrosine is particularly evident in A-MuLV-transformed cells.[475] The important identification and purification of protein substrates for tyrosine kinase activity may be facilitated by using monoclonal antibodies to phosphotyrosine.[476] A low-abundance cellular protein maybe a major substrate for growth factor-activated tyrosine-specific protein kinases.[477] Tyrosine-specific protein kinases are not involved in the phosphorylation of phospholipid metabolites such as phosphatidylinositol, phosphatidylinositol 4-phosphate, and 1,2-diacylglycerol, which depends on separate enzymatic activities.[389,390]

2. Tyrosine-Specific Protein Kinase Activity in Normal Cells

Tyrosine-specific kinase activity may not be an exclusive property of oncogene products, hormones, growth factors, or tumor promoter actions. A similar activity is also present in normal adult tissues, such as liver,[478-480] placenta,[481] testis (especially the Leydig cells,[482] spleen,[483] bone marrow,[484] lymphoid cells,[485-487] and nonproliferating, terminally differentiated human blood cells.[488] The normal rat spleen contains tyrosine kinase activity comparable to that of RSV-transformed duck embryo fibroblasts.[483] Tyrosine phosphorylation in the mouse spleen is provided by the distinct molecular weights of the endogenous phosphotyrosine substrates in normal T- and B-lymphocytes.[487] The T-cell substrates are proteins of 58,000 and 64,000 mol wt, termed p58 and p64, respectively. Specific tyrosine phosphorylation is associated with particular stages of B-cell differentiation in human lymphoid leukemias.[489]

Most probably, proto-oncogene protein products possessing tyrosine-specific protein kinase activity participate in tyrosine phosphorylations occurring in normal cells. The highest levels of pp60^{c-src}-associated tyrosine-specific protein kinase activity are present in the central nervous system, where they are subjected to developmental regulation.[490-494] pp60^{c-src} is specifically expressed at high levels in neurones and astrocytes but the structure of the proto-oncogene product in these cells is qualitatively different in two aspects: the pp60^{c-src} protein from neurones exhibits a 6- to 12-times higher tyrosine-specific protein kinase activity than the protein of astrocytes, and the pp60^{c-src} protein from neurones contains a structural change the protein of astrocytes, and the pp60^{c-src} protein from neurones contains a structural change within the amino-terminal domain of the molecule.[495] The functions of pp60^{c-src} in the central postmitotic neurones indicates that activation of this kinase does not correlate with cell proliferation. There is experimental evidence in favor of a role of pp60^{c-src} in the differentiation of nervous tissue, where the proto-oncogene protein may have a function similar to that of NGF.[496]

Phosphotyrosine is a rare modified amino acid in all normal cells, in contrast to phosphoserine and phosphothreonine which may be approximately 3000-fold more abundant in proteins from normal cells.[497] Moreover, the turnover of phosphate in phosphotyrosine is rapid, which is in accordance with the suggestion for a critical role of tyrosine phosphorylation in the regulation of cellular protein functions. Phosphotyrosine has been identified in yeast (*Saccharomyces cerevisiae*) proteins and tyrosine-specific protein kinase activity is associated with the plasma membrane of yeast cells.[498,499] Protein kinases with homology to the products of members of the *src* gene family may be involved in the initiation of the cell cycle in yeast.[398] Tyrosine kinase activity may exist even in bacteria. Phosphotyrosine and tyrosine kinase activity have been detected, for example, in the photosynthetic bacterium *Rhodospirillum rubrum* growing under photo-autotrophic conditions.[500]

3. Tyrosine-Specific Protein Kinase Activity in Malignant Cells

High levels of tyrosine-specific protein kinase activity have been detected in some, but not all, malignant cells.[501,502] The LSTRA cell line, originally derived from a mouse infected with M-MuLV, contains an elevated level of tyrosine kinase activity, which is associated with rearrangement and overexpression of a lymphocyte-specific protein tyrosine kinase gene.[503] Vesicular stomatitis virus produced by infected LSTRA cells may bear tyrosine-specific protein kinase activity.[504] Another murine lymphoma-derived cell line, YAC-1, has a tyrosine-specific protein kinase activity similar to that of normal mouse T-lymphocytes.[505] One substrate for this activity is a cellular protein with 58,000 mol wt termed p58, which is present in both normal and malignant T-lymphocytes. Myristic acid, a rare fatty acid, is covalently bound to tyrosine kinase in LSTRA lymphoma cells.[506] Conjugation of proteins to fatty acids is potentially important for guiding interactions with membranes and other hydrophobic surfaces in cells.

High levels of phosphotyrosine were detected in two proteins (55,000 and 35,000 mol wt) from three different human T-cell lymphoma cell lines, whereas an additional protein (78,000 mol wt) was present only in the human lymphoma cell line Ke 37.[507] Similar size proteins weakly phosphorylated in normal lymphocytes. High levels of the same enzymatic activity were observed in human leukemic cells by using immunofluorescent staining with a monoclonal antibody which identified phosphotyrosine-containing proteins.[508] No immunofluorescence was detected in peripheral blood cells and bone marrow cells from normal persons while 40 to 90% of the malignant cells from six patients with different types of leukemia gave positive results with the same method. It was suggested that the method may be useful for identifying leukemic cells in peripheral blood of patients. Further studies are required for a proper evaluation of these findings. In another study with a different method it was found that higher levels of tyrosine-specific protein kinase activity are present in resting human lymphocytes than in proliferating or leukemic blood cells.[509]

4. Origin and Regulation of Tyrosine-Specific Protein Kinase Activity Present in Normal and Malignant Cells

The origin and regulation of tyrosine-kinase activity present in normal adult tissues is not clear at present. Aside from tyrosine phosphorylation depending on proto-oncogene products and cellular receptors for hormones and growth factors, phosphorylation of cellular proteins on tyrosine may depend on the specific kinase activity of distinct cellular tyrosine-specific protein kinases.[478,479] A tyrosine-specific protein kinase of 56,000 mol wt has been purified from rat spleen.[510] Incubation of this kinase preparation with ATP and Mg^{2+} results in about tenfold increase in the protein kinase activity. These and other results suggest that the activation of the cellular tyrosine protein kinase by ATP is due to phosphorylation of the enzyme.[510] A tyrosine protein kinase with a similar or identical molecular weight, termed p56, was detected at high levels in the LSTRA murine thymoma/lymphoma cell line and at low levels in microsomes from most, but not all, T-lymphoma cell lines as well as in normal thymus tissue.[511] The p56 tyrosine protein kinase is distinct from the protein kinases encoded by the c-*src*, c-*yes*, c-*fgr*, c-*abl*, c-*fes*, and c-*ros* proto-oncogenes and contain covalently bound fatty acid.

Proteins with tyrosine kinase activity antigenically related to pp60^{src} (and more particularly to the carboxy-terminal sequence of pp60^{src} of avian origin) are present in human lymphoblastoid cells of either malignant origin (Raji cell line) or normal origin (Priess cell line).[512] Myristic acid is attached to the oncogene protein product pp60^{src} of either viral or cellular origin,[513] and protein tyrosine kinase activity is present in the same oncogene product.[426] Tyrosine-specific kinase activity of T-cell plasma membrane fractions is inhibited by *N*-alpha-*p*-tosyl-L-lysine chloromethyl ketone (TLCK), an inhibitor of the EGF receptor kinase and pp60^{src}-associated tyrosine-specific kinase activities.[487] In contrast, tyrosine-specific kinase activity of B-cells is unaffected by TLCK, which suggests that it depends on a different enzyme. These results suggest that the protein product of the c-*src* proto-oncogene, and/or the products of other members of the c-*src* proto-oncogene family, may be involved in certain tyrosine phosphorylation processes occurring in normal animal tissues.

Structural differences existing between v-*src* and c-*src* may be responsible for their different biological properties. In general, the c-*src* proto-oncogene, even if it is overexpressed, is unable to induce malignant transformation of cells.[514-516] Phosphorylation of cellular proteins on tyrosine is only slightly increased when the cells are infected with RSV in which the v-*src* oncogene has been replaced by a normal c-*src* proto-oncogene.[517] These results suggest that the low level of tyrosine phosphorylation by overproduced pp60^{c-src} accounts for its inability to transform cells.

The possibility that protein products of the c-*src* proto-oncogene family participate in the biochemical and functional alterations occurring in neoplastic cells is suggested by the fact that tyrosine-specific protein kinase activity of unspecified origin is increased severalfold in various types of human and nonhuman cancer cells.[501] However, high levels of tyrosine-specific protein kinase activity are not specific of malignant transformation because such high levels are also present during embryogenesis,[518] and a similar increase in tyrosine phosphorylation occurs in cells stimulated by different types of mitogenic agents.[435,519] A better characterization of the substrates for phosphorylation in normal and transformed cells is required for understanding the biological significance of tyrosine phosphorylation.

The response of endogenous protein phosphorylation on tyrosine residues to growth factor stimulation may be different in malignant cells when compared with the respective normal cells. A retinoblastoma-derived growth factor stimulates significantly more endogenous protein tyrosine phosphorylation in the human retinoblastoma Y-79 cell line than in normal retina.[520] Amplification and overexpression of the N-*myc* proto-oncogene has been detected in Y-79 cells,[521] but the possible relation, if any, between this alteration and the increased growth factor-dependent phosphorylation is not understood.

5. Phosphotyrosine-Specific Protein Phosphatases

The amount, and possibly also the specificity, of phosphotyrosine present in different cellular proteins would depend not only on the specific protein kinases but also on the activity of phosphotyrosine-specific protein phosphatases.[522-526] These enzymes produce dephosphorylation of tyrosine residues on cellular proteins and at least three different enzymes with this type of activity have been identified in cultured chicken cells.[527] A major phosphotyrosyl phosphatase in the bovine brain and the human placenta is calcineurin, a calmodulin-binding protein.[279]

Orthovanadate selectively inhibits phosphatases acting on phosphotyrosine.[523,524,527] Addition of vanadate to the culture medium of NRK cells results in a maximal 40-fold increase in the level of phosphotyrosine in cell protein,[528] which is greater than the 5- to 10-fold increase observed in cells transformed with viral oncogenes.[497] Vanadate reversibly induces transformation of cultured cells in a dose-dependent manner, so the degree of transformation can be controlled by the concentration of vanadate added to the tissue culture medium.[528] Apparently, vanadate-induced transformation does not depend on changes in phospholipid metabolism, which remains almost unaltered, but on the phosphorylation of one or more protein targets.

6. Summary

In general, the physiological significance of tyrosine phosphorylation and its possible relation to cell transformation are not understood.[529] An intriguing possibility is that protein substrates involved in tyrosine kinase activity related to cell transformation may be different from the substrates for the same activity when it is related to regulatory processes occurring in normal cells. However, the existence of proteins involved specifically in malignant transformation of cells is unlikely. More probably, phosphorylation (or other modifications) of proteins involved in normal processes of cell differentiation may result in a transformed cell phenotype when they occur in unduly high amounts or at inappropriate stages of differentiation.

B. Nontyrosine-Specific Protein Kinases

In addition to the phosphorylation on tyrosine residues, hormones, growth factors, oncogene products, and tumor promoters may stimulate phosphorylation of proteins on serine and threonine residues, and this modification is also important for regulation of the functional activity of the different substrates. Nontyrosine-specific phosphorylation reactions may be mediated by protein kinases dependent on either activation of the adenylate cyclase system (cAMP-protein kinases) or activation of other types of protein kinases (cAMP-independent protein kinases), the latter including Ca^{2+}-phospholipid-dependent protein kinase (protein kinase C). At least two types of cAMP-dependent protein kinases, type-I and type-II, are involved in these reactions. Differences between type-I and type-II isoenzymes reside not in the catalytic subunits, which are identical, but in the regulatory components of the enzymes.[87,88] The type-I and type-II isoenzymes may have different substrates and different functions, and may respond to cAMP concentrations in a hormone- and tissue-specific manner. For example, in a human lung cancer cell line (BEN), calcitonin stimulates selectively type-I isoenzyme whereas in two human breast cancer cell lines (T47D and MCF7), the same hormone activates exclusively type-II isoenzyme.[530] A cAMP-independent, Ca^{2+}-

independent protein kinase, termed protease-activated kinase II(PAK II), may be involved in the phosphorylation of the ribosomal protein S6 by EGF and other mitogenic agents.[531] Substances capable of inducing differentiation of cultured neoplastic cells may act through changes in protein phosphorylations mediated by the activities of cAMP-dependent and cAMP-independent protein kinases.[532]

No differences in total protein kinase activity could be observed between nonneoplastic human gastric mucosa and human gastric carcinomas, but a selective elevation of type-I cAMP-dependent isoenzyme was detected in gastric carcinomas independently of the histological type.[533] In xenotransplantable human gastric carcinomas in nude mice, type-I isoenzyme was significantly elevated. The results suggested that type-I cAMP-dependent protein kinase activity may be a biochemical marker for malignant transformation and transplantability of gastric tumors.[533]

1. Nontyrosine-Specific Protein Kinase Activities in Oncogene Protein Products

Protein kinase activity with specificity not for tyrosine but for serine and threonine residues is present in the purified protein products of two viral oncogenes, v-*mos* and v-*raf*/v-*mil*.[451-454] Although these oncogenes are considered, on the basis of structural homology, as members of the *src* oncogene family, they do not possess, as the other members of the family, an associated tyrosine-specific protein kinase activity.

The product of the Moloney murine sarcoma virus (M-MuSV) oncogene v-*mos* is a soluble cytoplasmic protein, p37 $^{v\text{-}mos}$.[534] p37 $^{v\text{-}mos}$ is structurally related to both pp60 $^{v\text{-}src}$ and the catalytic subunit of mammalian cAMP-dependent protein kinase.[535] A sequence located at amino acids 115 to 128 of the predicted v-*mos* protein sequence shows homology with the ATP-binding site of bovine cAMP-dependent protein kinase as well as with the protein products of the oncogenes v-*abl*, v-*yes*, and v-*fms*, which are members of the *src* oncogene family and possess intrinsic tyrosine-specific protein kinase activity.[455] In all of these oncogene proteins, including p37$^{v\text{-}mos}$, a group of glycines with the sequence Gly-x-Gly-x-x-Gly lies 16 to 28 residues to the amino-terminal side of a lysine residue implicated in binding ATP.[455] Homology has also been detected between p37 $^{v\text{-}mos}$ and the EGF precursor polypeptide.[536]

cAMP-independent protein kinase activity with specificity for serine and threonine residues is associated with the protein product of the v-*mos* oncogene.[453,454] It was initially believed that the c-*mos* proto-oncogene was universally silent in all normal tissues. However, by using a sensitive S_1 nuclease assay to screen RNA preparations from mouse tissues, c-*mos*-related transcripts of different sizes have been detected in normal mouse embryos as well as in mouse testes and ovaries.[537] The c-*mos* transcripts are of different size in the testis (1.7 kb) and the ovary (1.4 kb), and at least two major c-*mos* transcripts (2.3 and 1.3 kb) are present in the mouse embryo. The physiological significance of this variation as well as the function(s) of c-*mos* protein product(s) are unknown.

2. Cellular Substrates for Nontyrosine-Specific Protein Kinase Activity

Phosphorylation of cellular proteins on serine and/or threonine residues maybe a consequence of the effects of hormones, growth factors, or oncogene protein products. The phosphorylated proteins may include receptors for hormones and growth factors, which may result in changes in the functional properties of the receptors. Phosphorylation of the ribosomal protein S6 on serine may play an important role in the regulation of cell proliferation. Phosphorylation of S6 is constitutively stimulated, in the presence or absence of serum, by the v-*abl* oncogene protein product, as demonstrated by microinjection of the purified protein into *Xenopus* oocytes.[538] Since the v-*abl* protein is a tyrosine-specific protein kinase, the phosphorylation of S6 protein on serine maybe an indirect consequence of the primary action of the oncogene protein. S6 phosphorylation is stimulated by the phorbol ester PMA, serum, or pp60 $^{v\text{-}src}$, probably through common pathways.[539,540] The same serine sites of the S6 protein can be nearly maximally phosphorylated under the influence of the different agents. pp60^{src} itself is a substrate for protein kinase C-induced phosphorylation at serine-12.[393]

The precise role of S6 phosphorylation in cell proliferation is unknown. The yeast *Saccharomyces cerevisiae* has a ribosomal protein, called S10, which is equivalent to the S6 protein of animal cells, but the results obtained with oligonucleotide mutagenesis studies indicate that the phosphorylation of S10 is not essential for growth.[541] In vertebrate cells, many other cellular proteins, including membrane and nuclear proteins, may be substrates for the activity of nontyrosine-specific protein kinases present in normal or transformed cells but the exact role of these protein modifications on the respective protein functions remains to be established.

C. Protein Phosphorylation in Relation to Cellular Proliferation and Cellular Differentiation

There is an impressive body of evidence indicating that protein phosphorylation is critically involved in the control of DNA synthesis and cellular proliferation, as well as in the control of cellular differentiation, but the molecular mechanisms responsible for this control are not understood. Hormones and growth factors such as EGF and insulin may act synergistically in resting cultured cells to increase ATP turnover, which may lead to DNA synthesis. However, the presence for only a few minutes of growth factors in the culture medium of quiescent cells is usually not sufficient to induce DNA synthesis, which may explain the existence of a correlation between the total increase in ATP synthesis in the first 5 hr after growth factor addition and the increase in DNA synthesis determined between 6 and 24 hr after the addition.[542] Phosphorylation of particular nuclear proteins, including DNA topoisomerases, maybe is of paramount importance for the control of the molecular processes leading to DNA synthesis and cell division.

1. DNA Topoisomerases

DNA topoisomerases are involved in controlling the topological state of DNA and may play a crucial role in determining the function of DNA in cells, including DNA replication, transcription, and genetic recombination.[92] In eukaryotes, DNA supercoiling is an essential step in forming the DNA-histone complexes of chromatin and this supercoiling is controlled by topoisomerases. Some of these enzymes carry out the relaxation of supercoiled DNA and have been found in all cell types that have been examined. Other group of topoisomerases, the DNA gyrases, carry out the reverse reaction of converting relaxed closed-circular DNA to a superhelical form, a reaction which is coupled to the hydrolysis of ATP. All topological interconversions of DNA require the transient breakage and rejoining of DNA strands.[92]

The activity of DNA topoisomerases is affected by phosphorylation/dephosphorylation processes occurring on tyrosine and/or serine residues of the enzymes.[543,544] EGF, which is a potent mitogen, stimulates a topoisomerase activity in mouse and human cells, and this activity in the nucleus corresponds with DNA synthesis in the cells.[545] It is thus most interesting that proteins with kinase activity like the purified EGF receptor and the oncogene product pp60^{src} are able to interact with and nick supercoiled double-stranded DNA in a ATP-stimulated manner.[546] Incubation of *E. coli* or calf thymus type-I topoisomerases with either pp60$^{v\text{-}src}$ or a tyrosine-specific protein kinase purified from normal rat liver results in a tenfold loss of topoisomerase activity.[547]

These results suggest that retroviral- or cellular-encoded protein kinases could affect genomic functions, including DNA replication and transcription, through direct or indirect effects on phosphorylation/dephosphorylation reactions involving DNA topoisomerases whose respective functions will determine the balance between DNA supercoiling or relaxation.

However, the mechanisms involved in regulation of DNA topoisomerases by tyrosine-specific protein kinases are not yet understood. EGF receptor elicits DNA-nicking activity but this activity is not intrinsic to the EGF receptor kinase.[548] Apparently, the DNA-nicking activity associated with the EGF receptor is due to a distinct molecular species. The phosphoform of the type-II regulatory subunit of cAMP-dependent protein kinase possesses intrinsic topoisomerase activity towards several DNA substrates.[93]

The possible role of DNA topoisomerases alteration in neoplastic transformation or tumor progression is little understood but it has been suggested that DNA topoisomerases may constitute cellular targets for the antitumor properties of particular agents in cells transformed by oncogenic viruses.[549] Increased topoisomerase activity and intracellular distribution has been found in several human leukemic and lymphoblastoid cell lines in comparison to normal human peripheral blood lymphocytes.[550] Moreover, there is some evidence of topoisomerase structural abnormality in these malignant cell lines.

2. Relationship between Protein Phosphorylation and Mitogenesis

It is generally believed that protein phosphorylation is critically involved in the molecular events leading to DNA synthesis and cell division but, at least in certain cases, there seems to exist an apparent dissociation between phosphorylation of cellular proteins and mitogenic events. The 3T3-TNR9 cell line is a variant of Swiss 3T3 mouse fibroblast which does not respond mitogenically to tumor promoters (phorbol esters), but does respond mitogenically to EGF, FGF, and serum. Early phosphorylation reactions in these cells, however, are intact, including the phosphorylation of the EGF receptor and of 80- and 20-kdalton proteins, as well as the phosphorylation of a 22-kdalton protein on tyrosine.[375] These results suggest that, although these phorphorylations may be necessary, they are insufficient to trigger mitogenesis.

D. Protein Phosphorylation and Neoplastic Transformation

Although protein phosphorylation reactions are of great importance for the regulation of many cellular functions, their precise role in the complex phenomena associated with neoplastic transformation is still not understood. Phosphorylations of specific cellular proteins are probably involved in the transforming mechanisms of action of many oncogene protein products but, at present, it is difficult to evaluate their relative importance with respect to other biochemical changes, including Ca^{2+}-calmodulin alterations, monovalent ion exchanges, and phosphoinositide phosphorylation. In an extensive study of the biochemical changes associated with transformation induced by the oncogene protein products of two acute transforming retroviruses (RSV and UR2 sarcoma virus), only small changes in protein phosphorylation were found, and the changes affected only proteins already phosphorylated in nontransformed cells.[551] In contrast, large changes in the synthesis of nonphosphorylated proteins were found in the transformed cells and some of them appear to be specifically related to the maintenance of the transformed state.

It is evident that tyrosine phosphorylation can no longer be considered as a common, universal pathway for the transforming mechanisms of action of oncogene protein products because some of these products possess kinase activity with specificity for other amino acid residues (serine and threonine) or may lack any type of kinase activity. Moreover, both tyrosine-specific and nontyrosine-specific activities are present in normal cells, which would depend not only on the effects of hormones and growth factors but also on the actions of proto-oncogene products normally synthesized in the same cells. However, quantitative differences existing in the levels of protein phosphorylation on tyrosine residues, as well as differences in substrate specificities, could account for the transforming action of at least some viral oncogene proteins possessing tyrosine-specific protein kinase activity.

E. Conclusion

In conclusion, hormones, growth factors, and proto-oncogene protein products are all involved in the control of tyrosine-sepcific and nontyrosine-specific phosphorylations occurring in both normal and transformed cells but there is a lot of ignorance with respect to the specific substrates involved in such phosphorylations as well as to their possible role in the malignant transformation of cells.

REFERENCES

1. **Pimentel, E.**, Transcriptional and translational effects of hormones, *Ann. Endocrinol. (Paris)*, 39, 117, 1978.
2. **Pimentel, E.**, Cellular mechanisms of hormone action. I. Transductional events, *Acta Cient. Venez.*, 29, 73, 1978.
3. **Pimentel, E.**, Cellular mechanisms of hormone action. II. Posttransductional events, *Acta Cient. Venez.*, 29, 147, 1978.
4. **James, R. and Bradshaw, R. A.**, Polypeptide growth factors, *Ann. Rev. Biochem.*, 53, 259, 1984.
5. **LoBue, J. and LoBue, P. A.**, Control of cell growth, *Transplant. Proc.*, 16, 341, 1984.
6. **Levey, G. S. and Robinson, A. G.**, Introduction to the general principles of hormone-receptor interactions, *Metabolism*, 31, 639, 1982.
7. **Gorden, P., Carpentier, J.-L., Fan, J-Y., and Orci, L.**, Receptor mediated endocytosis of polypeptide hormones: mechanism and significance, *Metabolism*, 31, 664, 1982.
8. **Brown, M. S., Anderson, A. G. W., and Goldstein, J. L.**, Recycling receptors: the round-trip itinerary of migrant membrane proteins, *Cell*, 32, 663, 1983.
9. **Lefkowitz, R. J. and Michel, T.**, Plasma membrane receptors, *J. Clin. Invest.*, 72, 1185, 1983.
10. **Valentine, K. A. and Hollenberg, M. D.**, Membrane receptors and hormone action, in *Cell Biology of the Secretory Process*, S. Karger, Basel, 1984, 1.
11. **Kaplowicz, P. B.**, Wheat germ agglutinin and concanavalin A inhibit the response of human fibroblasts to peptide growth factors by a post-receptor mechanism, *J. Cell. Physiol.*, 124, 474, 1985.
12. **Wileman, T., Harding, C., and Stahl, P.**, Receptor-mediated endocytosis, *Biochem. J.*, 232, 1, 1985.
13. **Larrick, J. W., Enns, C., Raubitschek, A., and Weintraub, H.**, Receptor-mediated endocytosis of human transferrin and its cell surface receptor, *J. Cell. Physiol.*, 124, 283, 1985.
14. **Goldfine, I. D.**, Interaction of insulin, polypeptide hormones, and growth factors with intracellular membranes, *Biochim. Biophys. Acta*, 650, 53, 1981.
15. **Csaba, G.**, The present state in the phylogeny and ontogeny of hormone receptors, *Horm. Metab. Res.*, 16, 329, 1984.
16. **Feldman, D., Tökés, L. G., Stathis, P. A., Miller, S. C., Kurz, W., and Harvey, D.**, Identification of 17 beta-estradiol as the estrogenic substance in *Saccharomyces cerevisiae*, *Proc. Natl. Acad. Sci. U.S.A.*, 81, 4722, 1984.
17. **LeRoith, D., Shiloach, J., Heffron, R., Rubinovitz, C., Tanenbaum, R., and Roth, J.**, Insulin-related material in microbes: similarities and differences from mammalian insulins, *Can. J. Biochem. Cell Biol.*, 63, 839, 1985.
18. **Blalock, J. E. and Smith, E. M.**, Hydropathic anticomplementarity of amino acids based on the genetic code, *Biochem. Biophys. Res. Comm.*, 121, 203, 1984.
19. **Bost, K. L., Smith, E. M., and Blalock, J. E.**, Regions of complementarity between the messenger RNAs for epidermal growth factor, transferrin, interleukin-2, and their respective receptors, *Biochem. Biophys. Res. Comm.*, 128, 1373, 1985.
20. **Bost, K. L., Smith, E. M., and Blalock, E. J.**, Similarity between the corticotropin (ACTH) receptor and a peptide encoded by an RNA that is complementary to ACTH mRNA, *Proc. Natl. Acad. Sci. U.S.A.*, 82, 1372, 1985.
21. **Hollenberg, M. D.**, Examples of homospecific and heterospecific receptor regulation, *Trends Pharmacol. Sci.*, 6, 242, 1985.
22. **Hollenberg, M. D.**, Biochemical mechanisms of receptor regulation, *Trends Pharmacol. Sci.*, 6, 299, 1985.
23. **Hollenberg, M. D.**, Pathophysiological and therapeutic implications of receptor regulation, *Trends Pharmacol. Sci.*, 6, 334, 1985.
24. **Schultz, G., Aktories, K., Böme, E., Gerzer, R., and Jakobs, K. H.**, Signal transformation mediated by membrane receptors for hormones and neurotransmitters, *J. Immunol.*, 19, 1207, 1982.

25. **Jänne, J., Pösö, H., and Raina, A.**, Polyamines in rapid growth and cancer, *Biochim. Biophys. Acta*, 473, 241, 1978.
26. **Pegg, A. E.**, Recent advances in the biochemistry of polyamines in eukaryotes, *Biochem. J.*, 234, 249, 1986.
27. **Rozengurt, E.**, Stimulation of Na influx, Na-K pump activity and DNA synthesis in quiescent cultured cells, *Adv. Enzyme Regul.*, 19, 61, 1981.
28. **Berridge, M. J.**, Inositol trisphosphate and diacylglycerol as second messengers, *Biochem. J.*, 220, 345, 1984.
29. **Mills, G. B., Cragoe, E. J., Jr., Gelfand, E. W., and Grinstein, S.**, Interleukin 2 induces a rapid increase in intracellular pH through activation of a Na^+/H^+ antiport: cytoplasmic alkalinization is not required for lymphocyte proliferation, *J. Biol. Chem.*, 260, 12500, 1985.
30. **Tupper, J. T. and Smith, J. W.**, Growth factor regulation of membrane transport in human fibroblasts and its relationship to stimulation of DNA synthesis, *J. Cell. Physiol.*, 125, 443, 1985.
31. **Pouysségur, J., Franchi, A., L'Allemain, G., and Paris, S.**, Cytoplasmic pH, a key determinant of growth factor-induced DNA synthesis in quiescent fibroblasts, *FEBS Lett.*, 190, 115, 1985.
32. **Nimmo, H. G. and Cohen, P.**, Hormonal control of protein phosphorylation, *Adv. Cyclic Nucleotide Res.*, 8, 145, 1977.
33. **Jungmann, R. A. and Kranias, E. G.**, Nuclear phosphoprotein kinases and the regulation of gene transcription, *Int. J. Biochem.*, 819, 1977.
34. **Jungmann, R. A. and Russell, D. H.**, Cyclic AMP, cyclic AMP-dependent protein kinase, and the regulation of gene expression, *Life Sci.*, 20, 1787, 1977.
35. **Greengard, P.**, Phosphorylated proteins as physiological effectors, *Science*, 199, 146, 1978.
36. **Kuo, J. F., Shoji, M., and Kuo, W-N.**, Molecular and physiopathologic aspects of mammalian cyclic GMP-dependent protein kinase, *Ann. Rev. Pharmacol. Toxicol.*, 18, 341, 1978.
37. **Lincoln, T. M. and Corbin, J. D.**, On the role of cAMP and cGMP-dependent protein kinases in cell function, *J. Cyclic Nucleotide Res.*, 4, 3, 1978.
38. **Manning, D. R., DiSalvo, J., and Stull, J. T.**, Protein phosphorylation: quantitative analysis in vivo and in intact cell systems, *Mol. Cell. Endocrinol.*, 19, 1, 1980.
39. **Granot, J., Mildvan, A. S., and Kaiser, E. T.**, Studies on the mechanism of action and regulation of cAMP-dependent protein kinase, *Arch. Biochem. Biophys.*, 205, 1, 1980.
40. **Cohen, P.**, The role of protein phosphorylation in neural and hormonal control of cellular activity, *Nature (London)*, 296, 613, 1982.
41. **Hollenberg, M. D.**, Receptor mediated phosphorylation reactions, *Trends Pharmacol. Sci.*, 3, 271, 1982.
42. **Hunt, T.**, Phosphorylation and the control of protein synthesis, *Philos. Trans. R. S. London, Ser. B*, 302, 127, 1983.
43. **Fischer, E. H.**, Cellular regulation by protein phosphorylation, *Bull. Inst. Pasteur (Paris)*, 81, 7, 1983.
44. **Krebs, E. G.**, The phosphorylation of proteins: a major mechanism for biological regulation, *Biochem. Soc. Trans.*, 13, 813, 1985.
45. **Hunter, T. and Cooper, J. A.**, Protein tyrosine kinases, *Ann. Rev. Biochem.*, 54, 897, 1985.
46. **Cooper, E. and Spaulding, S. W.**, Hormonal control of the phosphorylation of histones, HMG proteins and other nuclear proteins, *Mol. Cell. Endocrinol.*, 39, 1, 1985.
47. **Dynan, W. S. and Tjian, R.**, Control of eukaryotic messenger RNA synthesis by sequence-specific DNA-binding proteins, *Nature (London)*, 316, 774, 1985.
48. **Robison, G. A., Butcher, R. W., and Sutherland, E. W.**, *Cyclic AMP*, Academic Press, New York, 1971.
49. **Earp, H. S. and Steiner, A. L.**, Compartmentalization of cyclic nucleotide-mediated hormone action, *Ann. Rev. Pharmacol. Toxicol.*, 18, 431, 1978.
50. **Gunzburg, J., de, Le** mode d'action de l'AMP cyclique chez les procaryotes et les eucaryotes: CAP et protéine kinases AMPc dépendantes, *Biochimie*, 67, 563, 1985.
51. **Ullmann, A. and Danchin, A.**, Role of cyclic AMP in bacteria, *Adv. Cyclic Nucleotide Res.*, 15, 1, 1983.
52. **Vaulont, S., Munnich, A., Marie, J., Reach, G., Pichard, A-L., Simon, M-P., Besmond, C., Barbry, P., and Kahn, A.**, Cyclic AMP as a transcriptional inhibitor of upper eukaryotic gene transcription, *Biochem. Biophys. Res. Comm.*, 125, 135, 1984.
53. **Nagamine, Y. and Reich, E.**, Gene expression and cAMP, *Proc. Natl. Acad. Sci. U.S.A.*, 82, 4606, 1985.
54. **Waterman, M., Murdoch, G. H., Evans, R. M., and Rosenfeld, M. G.**, Cyclic AMP regulation of eukaryotic gene transcription by two discrete molecular mechanisms, *Science*, 229, 267, 1985.
55. **Abramowitz, J., Iyengar, R., and Birnbaumer, L.**, Guanyl nucleotide regulation of hormonally-responsive adenylyl cyclases, *Mol. Cell. Endocrinol.*, 16, 129, 1979.
56. **Pfeuffer, E., Dreher, R-M., Metzger, H., and Pfeuffer, T.**, Catalytic unit of adenylate cyclase: purification and identification by affinity crosslinking, *Proc. Natl. Acad. Sci. U.S.A.*, 82, 3086, 1985.

57. **Coussen, F., Haiech, J., d'Alayer, J., and Monneron, A.,** Identification of the catalytic subunit of brain adenylate cyclase: a calmodulin binding protein of 135 KDa., *Proc. Natl. Acad. Sci. U.S.A.*, 82, 6736, 1985.
58. **Iyengar, R., Abramowitz, J., Bordelon-Riser, M., Blume, A. J., and Birnbaumer, L.,** Regulation of hormone-receptor coupling to adenylate cyclase: effects of GTP and GDP, *J. Biol. Chem.*, 255, 10312, 1980.
59. **Iyengar, R. and Birnbaumer, L.,** Hormone receptor modulates the regulatory component of adenylyl cyclase by reducing its requirement for Mg^{2+} and enhancing its extent of activation by guanine nucleotides, *Proc. Natl. Acad. Sci. U.S.A.*, 79, 5179, 1982.
60. **Clark, O. H., Gerend, P. L., Goretzki, P., and Nissenson, R. A.,** Characterization of the thyrotropin receptor-adenylate cyclase system in neoplastic human thyroid tissue, *J. Clin. Endocrinol. Metab.*, 57, 140, 1983.
61. **Piot, J-M. and Jacquemin, C.,** Lack of adenylate and guanylate cyclases responsiveness to hormones in a spontaneous murine thyroid tumor, *Biochem. Biophys. Res. Comm.*, 95, 357, 1980.
62. **Stevens, F. C.,** Calmodulin: an introduction, *Can. J. Biochem. Cell Biol.*, 61, 906, 1983.
63. **Akiyama, T., Gotho, E., and Ogawara, H.,** Alteration of adenylate cyclase activity by phosphorylation and dephosphorylation, *Biochem. Biophys. Res. Comm.*, 112, 250, 1983.
64. **Erneux, C., Van Sande, J., Miot, F., Cochaux, P., Decoster, C., and Dumont, J. E.,** A mechanism in the control of intracellular cAMP level: the activation of a calmodulin-sensitive phosphodiesterase by a rise of intracellular free calcium, *Mol. Cell. Endocrinol.*, 43, 123, 1985.
65. **Ryan, W. L. and Heidrick, M. L.,** Role of cyclic nucleotides in cancer, *Adv. Cyclic Nucleotide Res.*, 4, 87, 1974.
66. **MacManus, J. P. and Whitfield, J. F.,** Cyclic AMP, prostaglandins, and the control of cell proliferation, *Prostaglandins*, 6, 475, 1974.
67. **Eker, P.,** Inhibition of growth and DNA synthesis in cell cultures by cyclic AMP, *J. Cell Sci.*, 16, 301, 1974.
68. **Minton, J. P., Matthews, R. H., and Wisenbaugh, T. W.,** Elevated adenosine 3',5'-cyclic monophosphate levels in human and animal tumors in vivo, *J. Natl. Cancer Inst.*, 57, 39, 1976.
69. **Rechler, M. M., Bruni, C. B., Podskalny, J. M., and Carchman, R. A.,** DNA synthesis in cultured human fibroblasts: regulation by 3':5'-cyclic AMP, *J. Supramol. Struct.*, 4, 199, 1976.
70. **Wang, T., Sheppard, J. R., and Foker, J. E.,** Rise and fall of cyclic AMP required for onset of lymphocyte DNA synthesis, *Science*, 201, 155, 1978.
71. **Johnson, G. S., and Pastan, I.,** Role of 3',5'-adenosine monophosphate in regulation of morphology and growth of transformed and normal fibroblasts, *J. Natl. Cancer Inst.*, 48, 1377, 1972.
72. **Froehlich, J. E. and Rachmeler, M.,** Effect of adenosine 3',5'-cyclic monophosphate on cell proliferation, *J. Cell Biol.*, 55, 19, 1972.
73. **van Wijk, R., Wicks, W. D., and Clay, K.,** Effects of derivatives of cyclic 3',5'-adenosine monophosphate on the growth, morphology, and gene expression of hepatoma cells in culture, *Cancer Res.*, 32, 1905, 1972.
74. **Cho-Chung, Y. S. and Redler, B. H.,** Dibutyryl cyclic AMP mimics ovariectomy: nuclear protein phosphorylation in mammary tumor regression, *Science*, 197, 272, 1977.
75. **Huang, F. L. and Cho-Chung, Y. S.,** Dibutyryl cyclic AMP treatment mimics ovariectomy: new genomic regulation in mammary tumor regression, *Biochem. Biophys. Res. Comm.*, 107, 411, 1982.
76. **Beckner, S. K.,** Decreased adenylate cyclase responsiveness of transformed cells correlates with the presence of a viral transforming protein, *FEBS Lett.*, 166, 170, 1984.
77. **Somers, K. D.,** Increased cyclic AMP content directly correlated with morphological transformation of cells infected with a temperature-sensitive mutant of mouse sarcoma virus, *In Vitro*, 161, 851, 1980.
78. **Rieber, M. S. and Rieber, M.,** Transformed cells exhibit altered response to DB cyclic AMP-mediated modulation of protein phosphorylation and different endogenous phosphoprotein acceptors, *Cancer Biochem. Biophys.*, 5, 163, 1981.
79. **Sheppard, J. R., Koestler, T. P., Corwin, S. P., Buscarino, C., Doll, J., Lester, B., Greig, R. G., and Poste, G.,** Experimental metastasis correlates with cyclic AMP accumulation in B16 melanoma clones, *Nature (London)*, 308, 544, 1984.
80. **O'Keefe, E. J. and Pledger, W. J.,** A model of cell cycle control: sequential events regulated by growth factors, *Mol. Cell. Endocrinol.*, 31, 167, 1983.
81. **Anderson, W. B., Russell, T. R., Charchman, R. A., and Pastan, I.,** Interrelationship between adenylate cyclase activity, adenosine 3':5' cyclic monophosphate phosphodiesterase activity, adenosine 3':5' cyclic monophosphate levels, and growth of cells in culture, *Proc. Natl. Acad. Sci. U.S.A.*, 70, 3802, 1973.
82. **Saltarelli, D., Fischer, S., and Gacon, G.,** Modulation of adenylate cyclase by guanine nucleotides and Kirsten sarcoma virus mediated transformation, *Biochem. Biophys. Res. Comm.*, 127, 318, 1985.

83. Tagliaferri, P., Clair, T., DeBortoli, M. E., and Cho-Chung, Y. S., Two classes of cAMP analogs synergistically inhibit p21 *ras* protein synthesis and phenotypic transformation of NIH/3T3 transfected with Ha-MuSV DNA, *Biochem. Biophys. Res. Comm.*, 130, 1193, 1985.
84. Chiarugi, V., Porciatti, F., Pasquali, F., and Bruni, P., Transformation of Balb/3T3 cells with EJ/T24/H-*ras* oncogene inhibits adenylate cyclase response to beta-adrenergic agonist while increasing muscarinic receptor-dependent hydrolysis of inositol lipids, *Biochem. Biophys. Res. Comm.*, 132, 900, 1985.
85. Majmudar, G. and Peterkofsky, B., Cyclic AMP-independent processes mediate Kirsten sarcoma virus-induced changes in collagen production and other properties of cultured cells, *J. Cell. Physiol.*, 122, 113, 1985.
86. Shoji, S., Parmelee, D. C., Wade, R. D., Kumar, S., Ericsson, L. H., Walsh, K. A., Neurath, H., Long, G. L., Demaille, J. G., Fischer, E. H., and Titani, K., Complete amino acid sequence of the catalytic subunit of bovine cardiac muscle cyclic AMP-dependent protein kinase, *Proc. Natl. Acad. Sci. U.S.A.*, 78, 848, 1981.
87. Corbin, J. D., Keely, S. L., and Park, C. R., The distribution and dissociation of cyclic adenosine 3′:5′-monophosphate-dependent protein kinases in adipose, cardiac, and other tissues, *J. Biol. Chem.*, 250, 218, 1975.
88. Hofmann, F., Beavo, J. A., Bechtel, P., and Krebs, E. G., Comparison of adenosine 3′:5′-monophosphate-dependent protein kinases from rabbit skeletal and bovine heart muscle, *J. Biol. Chem.*, 250, 7795, 1975.
89. Schwartz, D. A. and Rubin, C. S., Regulation of cAMP-dependent protein kinase subunit levels in Friend erythroleukemic cells, *J. Biol. Chem.*, 258, 777, 1983.
90. Singh, T. J., Hochman, J., Verna, R., Chapman, M., Abraham, I., Pastan, I. H., and Gottesman, M. M., Characterization of a cyclic AMP-resistant Chinese hamster ovary cell mutant containing both wild-type and mutant species of type I regulatory subunit of cyclic AMP-dependent protein kinase, *J. Biol. Chem.*, 260, 13927, 1985.
91. Horne, W. C. et al., Leto, T. L., and Marchesi, V. T., Differential phosphorylation of multiple sites in protein 4.1 and protein 4.9 by phorbol ester-activated and cyclic AMP-dependent protein kinases, *J. Biol. Chem.*, 260, 9073, 1985.
92. Gellert, M., DNA topoisomerases, *Ann. Rev. Biochem.*, 50, 879, 1981.
93. Constatinou, A. I., Squinto, S. P., and Jungmann, R. A., The phosphoform of the regulatory subunit RII of cyclic AMP-dependent protein kinase possesses intrinsic topoisomerase activity, *Cell*, 42, 429, 1985.
94. Groffen, J., Heisterkamp, N., Reynolds, F. H., Jr., and Stephenson, J. R., Homology between phosphotyrosine acceptor site of human c-*abl* and viral oncogene products, *Nature (London)*, 304, 167, 1983.
95. Lörincz, A. T. and Reed, S. I., Primary structure homology between the product of yeast cell division control gene CDC28 and vertebrate oncogenes, *Nature (London)*, 307, 183, 1984.
96. Le Gros, J., De Feyter, R., and Ralph, R. K., Cyclic AMP and c-myc gene expression in PY815 mouse mastocytoma cells, *FEBS Lett.*, 186, 13, 1985.
97. Hardman, J. G. and Sutherland, E. W., Guanyl cyclase, an enzyme catalyzing the formation of guanosine 3′:5′-monophosphate from guanosine triphosphate, *J. Biol. Chem.*, 249, 6363, 1969.
98. Kimura, H. and Murad, F., Subcellular localization of guanylate cyclase, *Life Sci.*, 17, 837, 1976.
99. Earp, H. S., The role of insulin, glucagon, and cAMP in the regulation of hepatocyte guanylate cyclase activity, *J. Biol. Chem.*, 255, 8079, 1980.
100. Russell, T. R. and Pastan, I. H., Cyclic adenosine 3′:5′-monophosphate and cyclic guanosine 3′:5′-monophosphate phosphodiesterase activities are under separate genetic control, *J. Biol. Chem.*, 249, 7764, 1974.
101. Beavo, J. A., Hansen, R. S., Harrison, S. A., Hurwitz, R. L., Martins, T. J., and Mumby, M. C., Identification and properties of cyclic nucleotide phosphodiesterases, *Mol. Cell. Endocrinol.*, 28, 386, 1982.
102. Goldberg, N. D. and Haddox, M. K., cGMP metabolism and the involvement in biological regulation, *Ann. Rev. Biochem.*, 46, 823, 1977.
103. Coffey, R. G., Hadden, E. M., Lopez, C., and Hadden, J. W., Cyclic GMP and calcium in the initiation of cellular proliferation, *Adv. Cyclic Nucleotide Res.*, 9, 661, 1978.
104. Wood, P. J., Pao, G., and Cooper, A., Changes in guinea pig plasma cyclic nucleotide levels during the development of a transplantable leukemia, *Cancer*, 53, 79, 1984.
105. Williams, A. C. and Light, P. A., Alterations in concentrations of cyclic guanosine 3′,5′-monophosphate in guinea pig urine during the development of a transplantable leukaemia, *Cancer Lett.*, 28, 93, 1985.
106. Hadden, J. W., Hadden, E. M., Haddox, M. K., and Goldberg, N. D., Guanosine 3′:5′-cyclic monophosphate: a possible intracellular mediator of mitogenic influences on lymphocytes, *Proc. Natl. Acad. Sci. U.S.A.*, 69, 3024, 1972.
107. Gillette, R. W., McKenzie, G. O., and Swanson, M. H., Modification of the lymphocyte response to mitogens by cyclic AMP and GMP, *J. Reticuloendothelial Soc.*, 16, 289, 1974.

108. **Diamantstein, T. and Ulmer, A.**, Regulation of DNA synthesis by guanosine-5'-monophosphate, cyclic guanosine-3',5'-monophosphate, and cyclic adenosine-3',5'-monophosphate in mouse lymphoid cells, *Exp. Cell Res.*, 93, 309, 1975.
109. **Miller, Z., Lovelace, E., Gallo, M., and Pastan, I.**, Cyclic guanosine monophosphate and cellular growth, *Science*, 190, 1213, 1975.
110. **Kuo, J. F.**, Guanosine 3':5'-monophosphate-dependent protein kinases in mammalian tissues, *Proc. Natl. Acad. Sci. U.S.A.*, 71, 4037, 1974.
111. **Johnson, E. M. and Hadden, J. W.**, Phosphorylation of lymphocyte nuclear acidic proteins: regulation by cyclic nucleotides, *Science*, 187, 1198, 1975.
112. **Spruill, W. A., Hurwitz, D. R., Lucchesi, J. C., and Steiner, A. L.**, Association of cyclic GMP with gene expression of polytene chromosomes of *Drosophila melanogaster*, *Proc. Natl. Acad. Sci. U.S.A.*, 75, 1480, 1978.
113. **Mackenzie, C. W., III and Donnelly, T. E., Jr.**, Variable dependence on protein kinase stimulatory modulator for cyclic GMP stimulation of histone phosphorylation by rat liver cyclic GMP-dependent protein kinase, *Biochem. Biophys. Res. Comm.*, 88, 462, 1979.
114. **Hughes, S. M.**, Are guanine nucleotide binding proteins a distinct class of regulatory proteins?, *FEBS Lett.*, 164, 1, 1983.
115. **Gilman, A. G.**, G proteins and dual control of adenylate cyclase, *Cell*, 577, 1984.
116. **Spiegel, A. M., Gierschik, P., Levine, M. A., and Downs, R. W., Jr.**, Clinical implications of guanine nucleotide-binding proteins as receptor-effector couplers, *N. Engl. J. Med.*, 312, 26, 1985.
117. **Harris, B. A., Robishaw, J. D., Mumby, S. M., and Gilman, A. G.**, Molecular cloning of complementary DNA for the alpha subunit of the G protein that stimulates adenylate cyclase, *Science*, 229, 1274, 1985.
118. **De Wit, R. J. W. and Snaar-Jagalska, B. E.**, Folate and cAMP modulate GTP binding to isolated membranes of *Dictyostelium discoideum*. Functional coupling between cell surface receptors and G-proteins, *Biochem. Biophys. Res. Comm.*, 129, 11, 1985.
119. **Benovic, J. L., Pike, L. J., Cerione, R. A., Staniszewski, C., Yoshimasa, T., Codina, J., Caron, M. G., and Lefkowitz, R. J.**, Phosphorylation of the mammalian beta-adrenergic receptor by cyclic AMP-dependent protein kinase: regulation of the rate of receptor phosphorylation and dephosphorylation by agonist occupancy and effects on coupling of the receptor to the stimulatory guanine nucleotide regulatory protein, *J. Biol. Chem.*, 260, 7094, 1985.
120. **Medynski, D. C., Sullivan, K., Smith, D., Van Dop, C., Chang, F-H., Fung, B. K-K., Seeburg, P. H., and Bourne, H. R.**, Amino acid sequence of the alpha subunit of transducin deduced from the cDNA sequence, *Proc. Natl. Acad. Sci. U.S.A.*, 82, 4311, 1985.
121. **Yatsunami, K. and Khorana, H. G.**, GTPase of bovine rod outer segments: the amino acid sequence of the alpha subunit as derived from the cDNA sequence, *Proc. Natl. Acad. Sci. U.S.A.*, 82, 4316, 1985.
122. **Sugimoto, K., Nukada, T., Tanabe, T., Takahashi, H., Noda, M., Minamino, N., Kangawa, K., Matsuo, H., Hirose, T., Inayama, S., and Numa, S.**, Primary structure of the beta-subunit of bovine transducint deduced from the cDNA sequence, *FEBS Lett.*, 191, 235, 1985.
123. **Yatsunami, K., Pandya, B. V., Oprian, D. D., and Khorana, H. G.**, cDNA-derived amino acid sequence of the gamma subunit of GTPase from bovine rod outer segments, *Proc. Natl. Acad. Sci. U.S.A.*, 82, 1936, 1985.
124. **Bockaert, J., Deterre, P., Pfister, C., Guillon, G., and Chabre, M.**, Inhibition of hormonally regulated adenylate cyclase by the beta-gamma subunit of transducin, *EMBO J.*, 4, 1413, 1985.
125. **Willingham, M. C., Pastan, I., Shih, T. Y., and Scolnick, E. M.**, Localization of the *src* gene product of the Harvey strain of MSV to plasma membrane of transformed cells by electron microscopic immunocytochemistry, *Cell*, 19, 1005, 1980.
126. **Shih, T. Y., Papageorge, A. G., Stokes, P. E., Weeks, M. O., and Scolnick, E. M.**, Guanine nucleotide-binding and autophosphorylating activities associated with the p21[src] protein of Harvey murine sarcoma virus, *Nature (London)*, 287, 686, 1980.
127. **Papageorge, A., Lowy, D., and Scolnick, E. M.**, Comparative biochemical properties of the p21 *ras* molecules coded for by viral and cellular *ras* genes, *J. Virol.*, 44, 509, 1982.
128. **Gibbs, J. B., Sigal, I. S., and Scolnick, E. M.**, Biochemical properties of normal and oncogenic *ras* p21, *Trends Biochem. Sci.*, 10, 350, 1985.
129. **Hurley, J. B., Simon, M. I., Teplow, D. B., Robishaw, J. D., and Gilman, A. G.**, Homologies between signal transducing G proteins and *ras* gene products, *Science*, 226, 860, 1984.
130. **Lochrie, M. A., Hurley, J. B., and Simon, M. I.**, Sequence of the alpha subunit of photoreceptor G protein: homologies between transducin, *ras*, and elongation factors, *Science*, 228, 96, 1985.
131. **Tanabe, T., Nukuda, T., Nishikawa, Y., Sugimoto, K., Suzuki, H., Takahashi, H., Noda, M., Haga, T., Ichiyama, A., Kangawa, K., Minamino, N., Matsuo, H., and Numa, S.**, Primary structure of the alpha-subunit of transducing and its relationship to *ras* proteins, *Nature (London)*, 315, 242, 1985.
132. **Halliday, K. R.**, Regional homology in GTP-binding protooncogene products and elongation factors, *J. Cyclic Nucleotide Prot. Res.*, 9, 435, 1983.

133. **Jurnak, F.**, Structure of the GDG domain of EF-Tu and location of the amino acids homologous to *ras* oncogene proteins, *Science,* 230, 32, 1985.
134. **March, P. E. and Inouye, M.**, GTP-binding membrane protein of *Escherichia coli* with sequence homology to initiation factor 2 and elongation factors Tu and G, *Proc. Natl. Acad. Sci. U.S.A.,* 82, 7500, 1985.
135. **Pines, M., Gierschik, P., Milligan, G., Klee, W., and Spiegel, A.**, Antibodies against the carboxyl-terminal 5-kDa peptide of the alpha subunit of transducin crossreact with the 40-kDa but not the 39-kDa guanine nucleotide binding protein from brain, *Proc. Natl. Acad. Sci. U.S.A.,* 82, 4095, 1985.
136. **McCormick, F., Clark, B. F. C., La Cour, T. F. M., Kjeldgaard, M., Norskov-Lauritsen, L., and Nyborg, J.**, A model for the tertiary structure of p21, the product of the *ras* oncogene, *Science,* 230, 78, 1985.
137. **Beckner, S. K., Hattori, S., and Shih, T. Y.**, The *ras* oncogene product p21 is not a regulatory component of adenylate cyclase, *Nature (London),* 317, 71, 1985.
138. **Franks, D. J., Whitfield, J. F., and Durkin, J. P.**, Mitogenic/oncogenic p21 Ki-*ras* protein stimulates adenylate cyclase activity early in the G_1 phase of NRK rat kidney cells, *Biochem. Biophys. Res. Comm.,* 132, 780, 1985.
139. **Langbeheim, H., Shih, T. Y., and Scolnick, E. M.**, Identification of a normal vertebrate cell protein related to the p21 *src* of Harvey murine sarcoma virus, *Virology,* 106, 292, 1980.
140. **Ellis, R. W., DeFeo, D., Shih, T. Y., Gonda, M. A., Young, H. A., Tsuchida, N., Lowy, D. R., and Scolnick, E. M.**, The p21 *src* genes of Harvey and Kirsten sarcoma viruses originate from divergent members of a family of normal vertebrate genes, *Nature (London),* 292, 506, 1981.
141. **Shilo, B-Z.**, Evolution of cellular oncogenes, *Adv. Viral Oncol.,* 4, 29, 1984.
142. **Shilo, B. and Weinberg, R. A.**, DNA sequences homologous to vertebrate oncogenes are conserved in *Drosophila melanogaster, Proc. Natl. Acad. Sci. U.S.A.,* 78, 6789, 1981.
143. **Hoffman-Falk, H., Einat, P., Shilo, B-Z., and Hoffmann, F. M.**, Drosophila melanogaster DNA clones homologous to vertebrate oncogenes: evidence for a common ancestor to the *src* and *abl* cellular genes, *Cell,* 32, 589, 1983.
144. **Neuman-Silberberg, F. S., Schejter, E., Hoffmann, F. M., and Shilo, B-Z.**, The Drosophila *ras* oncogenes: structure and nucleotide sequence, *Cell,* 37, 1027, 1984.
145. **Shilo, B-Z. and Hoffmann, F. M.**, *Drosophila melanogaster* cellular oncogenes, *Cancer Surv.,* 3, 299, 1984.
146. **Lev, Z., Kimchie, Z., Hessel, R., and Segev, O.**, Expression of *ras* cellular oncogenes during development of *Drosophila melanogaster, Mol. Cell. Biol.,* 5, 1540, 1985.
147. **Mozer, B., Marlor, R., Parkhurst, S., and Corces, V.**, Characterization and developmental expression of a *Drosophila ras* oncogene, *Mol. Cell. Biol.,* 5, 885, 1985.
148. **DeFeo-Jones, D., Scolnick, E. M., Koller, R., and Dhar, R.**, *ras*-Related gene sequences identified and isolated from *Saccharomyces cerevisiae, Nature (London),* 306, 707, 1983.
149. **Papageorge, A. G., DeFeo-Jones, D., Robinson, P. S., Temeles, G., and Scolnick, E. M.**, *Saccharomyces cerevisiae* synthesizes proteins related to the p21 product of *ras* genes found in mammals, *Mol. Cell. Biol.,* 4, 23, 1984.
150. **Temeles, G. L., DeFeo-Jones, D., Tatchell, K., Ellinger, M. S., and Scolnick, E. M.**, Expression and characterization of *ras* mRNAs from *Saccharomyces cerevisiae, Mol. Cell. Biol.,* 4, 2298, 1984.
151. **Dhar, R., Nieto, A., Koller, R., DeFeo-Jones, D., and Scolnick, E. M.**, Nucleotide sequence of two ras^H-related genes isolated from the yeast *Saccharomyces cerevisiae, Nucleic Acids Res.,* 12, 3611, 1984.
152. **Tamanoi, F., Walsh, M., Kataoka, T., and Wigler, M.**, A product of yeast *RAS2* gene is a guanine nucleotide binding protein, *Proc. Natl. Acad. Sci. U.S.A.,* 81, 6924, 1984.
153. **Kataoka, T., Powers, S., Cameron, S., Fasano, O., Goldfarb, M., Broach, J., and Wigler, M.**, Functional homology of mammalian and yeast *RAS* genes, *Cell,* 40, 19, 1985.
154. **Temeles, G. L., Gibbs, J. B., D'Alonzo, J. S., Sigal, I. S., and Scolnick, E. M.**, Yeast and mammalian *ras* proteins have conserved biochemical properties, *Nature (London),* 313, 700, 1985.
155. **Toda, T., Uno, I., Ishikawa, T., Powers, S., Kataoka, T., Broek, D., Cameron, S., Broach, J., Matsumoto, K., and Wigler, M.**, In yeast, *RAS* proteins are controlling elements of adenylate cyclase, *Cell,* 40, 27, 1985.
156. **Kataoka, T., Powers, S., McGill, C., Fasano, O., Strathern, J., Broach, J., and Wigler, M.**, Genetic analysis of yeast *RAS1* and *RAS2* genes, *Cell,* 37, 437, 1984.
157. **Tatchell, K., Chaleff, D. T., DeFeo-Jones, D., and Scolnick, E. M.**, Requirement of either of a pair of *ras*-related genes of *Saccharomyces cerevisiae* for spore viability, *Nature (London),* 309, 523, 1984.
158. **DeFeo-Jones, D., Tatchell, K., Robinson, L. C., Sigal, I. S., Vass, W. C., Lowy, D. R., and Scolnick, E. M.**, Mammalian and yeast *ras* gene products: biological function in their heterologous systems, *Science,* 228, 179, 1985.
159. **Tatchell, K., Robinson, L. C., and Breitenbach, M.**, *RAS2* of *Saccharomyces cerevisiae* is required for gluconeogenic growth and proper response to nutrient limitation, *Proc. Natl. Acad. Sci. U.S.A.,* 82, 3785, 1985.

160. **Fraenkel, D. G.**, On *ras* gene function in yeast, *Proc. Natl. Acad. Sci. U.S.A.*, 82, 4740, 1985.
161. **Pawson, T. and Weeks, G.**, Expression of *ras*-encoded proteins in relation to cell growth and differentiation, in *Genes and Cancer*, Alan R. Liss, New York, 1984, 461.
162. **Pawson, A. T., Hinze, E., Auersperg, N., Neave, N., Sobolewski, A., and Weeks, G.**, Regulation of a *ras*-related protein during development of *Dictyostelium discoideum*, *Mol. Cell. Biol.*, 5, 33, 1985.
163. **Reymond, C. D., Nellen, W., and Firtel, R. A.**, Regulated expression of *ras* gene constructs in *Dictyostellium* transformants, *Proc. Natl. Acad. Sci. U.S.A.*, 82, 7005, 1985.
164. **Madaule, P. and Axel, R.**, A novel *ras*-related gene family, *Cell*, 41, 31, 1985.
165. **Shih, T. Y., Stokes, P. E., Smythers, G. W., Dhar, R., and Oroszlan, S.**, Characterization of the phosphorylation and surrounding amino acid sequences of the p21 transforming proteins coded for by the Harvey and Kirsten strains of sarcoma viruses, *J. Biol. Chem.*, 257, 11767, 1982.
166. **Lautenberger, J. A., Ulsh, L., Shih, T. Y., and Papas, T. S.**, High-level expression in *Escherichia coli* of enzymatically active Harvey murine sarcoma virus 21^{ras} protein, *Science*, 221, 858, 1983.
167. **Stein, R. B., Robinson, P. S., and Scolnick, E. M.**, Photoaffinity labeling with GTP of viral p21 *ras* protein expressed in *Escherichia coli*, *J. Virol.*, 50, 343, 1984.
168. **Lacal, J. C., Santos, E., Notario, V., Barbacid, M., Yamazaki, S., Kung, H., Seamans, C., McAndrew, S., and Crowl, R.**, Expression of normal and transforming H-*ras* genes in *Escherichia coli* and purification of their encoded proteins, *Proc. Natl. Acad. Sci. U.S.A.*, 81, 5305, 1984.
169. **Manne, V., Yamazaki, S., and Kung, H-F.**, Guanosine nucleotide binding by highly purified Ha-*ras*-encoded p21 protein produced in *Escherichia coli*, *Proc. Natl. Acad. Sci. U.S.A.*, 81, 6953, 1984.
170. **Poe, M., Scolnick, E. M., and Stein, R. B.**, Viral Harvey *ras* p21 expressed in *Escherichia coli* purifies as a binary one-to-one complex with GDP, *J. Biol. Chem.*, 260, 3906, 1985.
171. **Gross, M., Sweet, R. W., Sathe, G., Yokoyama, S., Fasano, O., Goldfarb, M., Wigler, M., and Rosenberg, M.**, Purification and characterization of human H-*ras* proteins expressed in *Escherichia coli*, *Mol. Cell. Biol.*, 5, 1015, 1985.
172. **Hattori, S., Ulsh, L. S., Halliday, K., and Shih, T. Y.**, Biochemical properties of a highly purified v-ras^H p21 protein overproduced in *Escherichia coli* and inhibition of its activities by a monoclonal antibody, *Mol. Cell. Biol.*, 5, 1449, 1985.
173. **Tamaoki, T., Mizukami, T., Perucho, M., and Nakano, H.**, Expression of intact Ki-*ras* p21 protein in *Escherichia coli*, *Biochem. Biophys. Res. Comm.*, 132, 126, 1985.
174. **Kamata, T. and Feramisco, J. R.**, Epidermal growth factor stimulates guanine nucleotide binding activity and phosphorylation of *ras* oncogene proteins, *Nature (London)*, 310, 147, 1984.
175. **Sefton, B. M., Trowbridge, I. S., Cooper, J. A., and Scolnick, E. M.**, The transforming proteins of Rous sarcoma virus, Abelson sarcoma virus, and Harvey sarcoma virus contain tightly-bound lipid, *Cell*, 31, 465, 1982.
176. **Willumsen, B. M., Papageorge, A. G., Hubbert, N., Bekesi, E., Kung, H-F., and Lowy, D. R.**, Transforming $p21^{ras}$ protein: flexibility in the major variable region linking the catalytic and membrane anchoring domains, *EMBO J.*, 4, 2893, 1985.
177. **Willumsen, B. M., Christensen, A., Hubbert, N. L., Papageorge, A. G., and Lowy, D. R.**, The p21 *ras* C-terminus is required for transformation and membrane association, *Nature (London)*, 310, 583, 1984.
178. **Willumsen, B. M., Norris, K., Papageorge, A. G., Hubbert, N. L., and Lowy, D. R.**, Harvey murine sarcoma virus p21 *ras* protein: biological and biochemical significance of the cysteine nearest the carboxy terminus, *EMBO J.*, 3, 2581, 1984.
179. **Chen, Z., Ulsh, L. S., Du Bois, G., and Shih, T. Y.**, Posttranslational processing of p21 *ras* proteins involves palmitylation of the C-terminal tetrapeptide containing cysteine-186, *J. Virol.*, 56, 607, 1985.
180. **Weeks, M. O., Hager, G. L., Lowe, R., and Scolnick, E. M.**, Development and analysis of a transformation-defective mutant of Harvey murine sarcoma *tk* virus and its gene product, *J. Virol.*, 54, 586, 1985.
181. **Schejter, E. D. and Shilo, B.-Z.**, Characterization of functional domains of p21 *ras* by use of chimeric genes, *EMBO J.*, 4, 407, 1985.
182. **Clark, R., Wong, G., Arnheim, N., Nitecki, D., and McCormick, F.**, Antibodies specific for amino acid 12 of the *ras* oncogene product inhibit GTP binding, *Proc. Natl. Acad. Sci. U.S.A.*, 82, 5280, 1985.
183. **Feramisco, J. R., Clark, R., Wong, G., Arnheim, N., Milley, R., and McCormick, F.**, Transient reversion of *ras* oncogene-induced cell transformation by antibodies specific for amino acid 12 of *ras* protein, *Nature (London)*, 314, 639, 1985.
184. **McGrath, J. P., Capon, D. J., Goeddel, D. V., and Levinson, A. D.**, Comparative biochemical properties of normal and activated human *ras* p21 protein, *Nature (London)*, 310, 644, 1984.
185. **Srivastava, S. K., Lacal, J. C., Reynolds, S. H., and Aaronson, S. A.**, Antibody of predetermined specificity to a carboxy-terminal region of H-*ras* gene products inhibits their guanine nucleotide-binding function, *Mol. Cell. Biol.*, 5, 3316, 1985.
186. **Manne, V. and Kung, H.**, Effect of divalent metal ions and glycerol on the GTPase activity of H-*ras* proteins, *Biochem. Biophys. Res. Comm.*, 128, 1440, 1985.

187. **Weinberg, R. A.**, *ras* Oncogenes and the molecular mechanisms of carcinogenesis, *Blood,* 64, 1143, 1984.
188. **Balmain, A.**, Transforming *ras* oncogeneses and multistage carcinogenesis, *Br. J. Cancer,* 51, 1, 1985.
189. **Finkel, T., Der, C. J., and Cooper, G. M.**, Activation of *ras* genes in human tumors does not affect localization, modification, or nucleotide binding properties of p21, *Cell,* 37, 151, 1984.
190. **Lee, W. M. F., Schwab, M., Westaway, D., and Varmus, H. E.**, Augmented expression of normal c-*myc* is sufficient for cotransformation of rat embryo cells with a mutant *ras* gene, *Mol. Cell. Biol.,* 5, 3345, 1985.
191. **Gibbs, J. B., Sigal, I. S., Poe, M., and Scolnick, E. M.**, Intrinsic GTPase activity distinguishes normal and oncogenic *ras* p21 molecules, *Proc. Natl. Acad. Sci. U.S.A.,* 81, 5704, 1984.
192. **Sweet, R. W., Yokoyama, S., Kamata, T., Feramisco, J. R., Rosenberg, M., and Gross, M.**, The product of *ras* is a GTPase and the T24 oncogenic mutant is deficient in this activity, *Nature (London),* 311, 273, 1984.
193. **Manne, V., Bekesi, E., and Kung, H-F.**, Ha-*ras* proteins exhibit GTPase activity: point mutations that activate Ha-*ras* gene products result in decreased GTPase activity, *Proc. Natl. Acad. Sci. U.S.A.,* 82, 376, 1985.
194. **Sekiya, T., Tokunaga, A., and Fushimi, M.**, Essential region for transforming activity of human c-Ha-*ras*-1, *Jpn. J. Cancer Res.,* 76, 787, 1985.
195. **Yuspa, S. H., Kilkenny, A. E., Stanley, J., and Lichti, U.**, Keratinocytes blocked in phorbol ester-responsive early stage of terminal differentiation by sarcoma viruses, *Nature (London),* 314, 459, 1985.
196. **Flatow, U., Willingham, M. C., and Rabson, A. S.**, Butyrate prevents Harvey sarcoma virus focus formation but permits oncogene expression, *Cancer Lett.,* 22, 203, 1984.
197. **Norton, J. D., Cook, F., Roberts, P. C., Clewley, J. P., and Avery, R. J.**, Expression of Kirsten murine sarcoma virus in transformed nonproducer and revertant NIH/3T3 cells: evidence for cell-mediated resistance to a viral oncogene in phenotypic reversion, *J. Virol.,* 50, 439, 1984.
198. **Scolnick, E. M., Weeks, M. O., Shih, T. Y., Ruscetti, S. K., and Dexter, T. M.**, Markedly elevated levels of an endogenous *sarc* protein in a hemopoietic precursor cell line, *Mol. Cell. Biol.,* 1, 66, 1981.
199. **Hankins, W. D. and Scolnick, E. M.**, Harvey and Kirsten sarcoma viruses promote the growth and differentiation of erythroid precursor cells *in vitro, Cell,* 26, 91, 1981.
200. **Pierce, J. H. and Aaronson, S. A.**, Myeloid cell transformation by *ras*-containing murine sarcoma viruses, *Mol. Cell. Biol.,* 5, 667, 1985.
201. **Bar-Sagi, D. and Feramisco, J. R.**, Microinjection of the *ras* oncogene protein into PC12 cells induces morphological differentiation, *Cell,* 42, 841, 1985.
202. **Noda, M., Ko, M., Ogura, A., Liu, D., Amano, T., Takano, T., and Ikawa, Y.**, Sarcoma viruses carrying *ras* oncogenes induce differentiation-associated properties in a neuronal cell line, *Nature (London),* 318, 73, 1985.
203. **Colletta, G., Pinto, A., Di Fiore, P. P., Fusco, A., Ferrentino, M., Avvedimento, V. E., Tsuchida, N., and Vecchio, G.**, Dissociation between transformed and differentiated phenotype in rat thyroid epithelial cells after transformation with a temperature-sensitive mutant of the Kirsten murine sarcoma viruses, *Mol. Cell. Biol.,* 3, 2099, 1983.
204. **De Larco, J. E. and Todaro, G. J.**, Growth factors from murine sarcoma virus-transformed cells, *Proc. Natl. Acad. Sci. U.S.A.,* 75, 4001, 1978.
205. **Chua, C. C., Geiman, D., and Ladda, R. L.**, Transforming growth factors released from Kirsten sarcoma virus transformed cells do not compete for epidermal growth factor membrane receptors, *J. Cell. Physiol.,* 117, 116, 1983.
206. **Weissman, B. E. and Aaronson, S. A.**, BALB and Kirsten murine sarcoma viruses alter growth and differentiation of EGF-dependent BALB/c mouse epidermal keratinocyte lines, *Cell,* 32, 599, 1983.
207. **Stacey, D. W. and Kung, H-F.**, Transformation of NIH 3T3 cells by microinjection of Ha-*ras* p21 protein, *Nature (London),* 310, 508, 1984.
208. **Dotto, G. P., Parada, L. F., and Weinberg, R. A.**, Specific growth response of *ras*-transformed embryo fibroblasts to tumour promoters, *Nature (London),* 318, 472, 1985.
209. **Puga, A., Gomez-Marquez, J., Brayton, P. R., Cantin, E. M., Long, L. K., Barbacid, M., and Notkins, A. L.**, The immediate-early enhancer elements of herpes simplex virus type 1 can replace a regulatory region of the c-Ha-*ras*1 oncogene required for transformation, *J. Virol.,* 54, 879, 1985.
210. **Feramisco, J. R., Gross, M., Kamata, T., Rosenberg, M., and Sweet, R. W.**, Microinjection of the oncogene form of the human H-*ras* (T24) protein results in rapid proliferation of quiescent cells, *Cell,* 38, 109, 1984.
211. **Hyland, J. K., Rogers, C. M., Scolnick, E. M., Stein, R. B., Ellis, R. B., and Baserga, R.**, Microinjected *ras* family oncogenes stimulate DNA synthesis in quiescent mammalian cells, *Virology,* 141, 333, 1985.
212. **Mulcahy, L. S., Smith, M. R., and Stacey, D. W.**, Requirement for *ras* proto-oncogene function during serum-stimulated growth of NIH 3T3 cells, *Nature (London),* 313, 241, 1985.

213. **Yoakum, G. H., Lechner, J. F., Gabrielson, E. W., Korba, B. E., Malan-Shibley, L., Willey, J. C., Valerio, M. G., Shamsuddin, A. M., Trump, B. F., and Harris, C. C.**, Transformation of human bronchial epithelial cells transfected by Harvey *ras* oncogene, *Science*, 227, 1174, 1985.
214. **Ferrentino, M., Di Fiore, P. P., Fusco, A., Colletta, G., Pinto, A., and Vecchio, G.**, Expression of the *onc* gene of the Kirsten murine sarcoma virus in differentiated rat thyroid epithelial cell lines, *J. Gen. Virol.*, 65, 1955, 1984.
215. **Scheinberg, D. A. and Strand, M.**, A brain membrane protein similar to the rat *src* gene product, *Proc. Natl. Acad. Sci. U.S.A.*, 78, 55, 1981.
216. **Whitfield, J. F., Rixon, R. H., MacManus, J. P., and Balk, S. D.**, Calcium, cyclic adenosine 3′,5′-monophosphate, and the control of cell proliferation: a review, *In Vitro*, 8, 257, 1973.
217. **Rasmussen, H. and Goodman, D. B.**, Relationships between calcium and cyclic nucleotides in cell activation, *Physiol. Rev.*, 57, 421, 1977.
218. **Whitfield, J. F., Boynton, A. L., MacManus, J. P., Sikorska, M., and Tsang, B. K.**, The regulation of cell proliferation by calcium and cyclic AMP, *Mol. Cell. Biochem.*, 27, 155, 1979.
219. **Williamson, J. R., Cooper, R. H., and Hoek, J. B.**, Role of calcium in the hormonal regulation of liver metabolism, *Biochim. Biophys. Acta*, 639, 243, 1981.
220. **Whitfield, J. P.**, The role of calcium and magnesium in cell proliferation: an overview, in *Ions, Cell Proliferation, and Cancer*, Boynton, A. L., McKeehan, W. L., and Whitfield, J. F., Eds., Academic Press, New York, 1982, 283.
221. **Veigl, M. L., Vanaman, T. C., and Sedwick, W. D.**, Calcium and calmodulin in cell growth and transformation, *Biochim. Biophys. Acta*, 738, 21, 1984.
222. **Rasmussen, H.**, The calcium messinger system, *N. Engl. J. Med.*, 314, 1094, 1986.
223. **Moore, P. B. and Dedman, J. R.**, Calcium binding proteins and cellular regulation, *Life Sci.*, 31, 2937, 1982.
224. **Cheung, W. Y.**, Calmodulin plays a pivotal role in cellular regulation, *Science*, 207, 19, 1980.
225. **Means, A. R. and Dedman, J. R.**, Calmodulin in endocrine cells and its multiple roles in hormone action, *Mol. Cell. Endocrinol.*, 19, 215, 1980.
226. **Scharff, O.**, Calmodulin and its role in cellular activation, *Cell Calcium*, 2, 1, 1981.
227. **Stoclet, J-C.**, Calmodulin, an ubiquitous protein which regulates calcium-dependent cellular functions and calcium movements, *Biochem. Pharmacol.*, 30, 1723, 1981.
228. **Cheung, W. Y.**, Calmodulin and the adenylate cyclase-phosphodiesterase system, *Cell Calcium*, 2, 263, 1981.
229. **Means, A. R., Tash, J. S., and Chafouleas, J. G.**, Physiological implications of the presence, distribution, and regulation of calmodulin in eukaryotic cells, *Physiol. Rev.*, 62, 1, 1982.
230. **Means, A. R., Lagace, L., Guerriero, V., Jr., and Chafouleas, J. G.**, Calmodulin as a mediator of hormone action and cell regulation, *J. Cell. Biochem.*, 20, 317, 1982.
231. **Oldham, S. B.**, Calmodulin: its role in calcium-mediated cellular regulation, *Miner. Electrolyte Metab.*, 8, 1, 1982.
232. **Cheung, W. Y.**, Calmodulin: an overview, *Fed. Proc.*, 41, 2253, 1982.
233. **Lin, Y. M.**, Calmodulin, *Mol. Cell. Biochem.*, 45, 101, 1982.
234. **Kuo, J. F., Schatzman, R. C., Turner, R. S., and Mazzei, G. J.**, Phospholipid-sensitive Ca^{2+}-dependent protein kinase: a major protein phosphorylation system, *Mol. Cell. Endocrinol.*, 35, 65, 1984.
235. **Rebhum, L. I.**, Cyclic nucleotides, calcium and cell division, *Int. Rev. Cytol.*, 49, 1, 1977.
236. **Dubyak, G. R. and De Young, M. B.**, Intracellular Ca^{2+} mobilization activated by extracellular ATP in Ehrlich as cites tumor cells, *J. Biol. Chem.*, 260, 10653, 1985.
237. **McNeil, P. L., McKenna, M. P., and Taylor, D. L.**, A transient rise in cytosolic calcium follows stimulation of quiescent cells with growth factors and is inhibitable with phorbol myristate acetate, *J. Cell Biol.*, 101, 372, 1985.
238. **Moolenaar, W. H., Tertoolen, L. G. J., and de Laat, S. W.**, Growth factors immediately raise cytoplasmic free Ca^{2+} in human fibroblasts, *J. Biol. Chem.*, 259, 8066, 1984.
239. **McKeehan, W. L. and McKeehan, K. A.**, Epidermal growth factor modulates extracellular Ca^{2+} requirement for multiplication of normal human skin fibroblasts, *Exp. Cell Res.*, 123, 397, 1979.
240. **Lechner, J. F.**, Interdependent regulation of epithelial cell replication by nutrients, hormones, growth factors, and cell density, *Fed. Proc.*, 43, 116, 1984.
241. **Boynton, A. L., Whitfield, J. F., Isaacs, R. J., and Morton, H. J.**, Control of 3T3 cell proliferation by calcium, *In Vitro*, 10, 12, 1974.
242. **Dulbecco, R. and Elkington, J.**, Induction of growth in resting fibroblastic cell cultures by Ca^{++}, *Proc. Natl. Acad. Sci. U.S.A.*, 72, 1584, 1975.
243. **Swierenga, S. H. H., Whitfield, J. F., and Gillan, D. J.**, Alteration by malignant transformation of the calcium requirements for cell proliferation in vitro, *J. Natl. Cancer Inst.*, 57, 125, 1976.
244. **Hennings, H., Michael, D., Cheng, C., Steinert, P., Holbrook, K., and Yuspa, S. H.**, Calcium regulation of growth and differentiation of mouse epidermal cells in culture, *Cell*, 19, 245, 1980.

245. **Boynton, A. L., McKeehan, W. L., and Whitfield, J. F.,** Eds., *Ions, Cell Proliferation and Cancer,* Academic Press, New York, 1982.
246. **Lichtman, A. H., Segel, G. B., and Lichtman, M. A.,** The role of calcium in lymphocyte proliferation (an interpretative review), *Blood,* 61, 413, 1983.
247. **Miyaura, C., Abe, E., and Suda, T.,** Extracellular calcium is involved in the mechanism of differentiation of mouse myeloid leukemia cells (M1) induced by 1 alpha, 25-dihydroxyvitamin D_3, *Endocrinology,* 115, 1891, 1984.
248. **Luckasen, J. R., White, J. G., and Kersey, J. H.,** Mitogenic properties of a calcium ionophore, A23187, *Proc. Natl. Acad. Sci. U.S.A.,* 71, 5088, 1974.
249. **Durkin, J. P. and Whitfield, J. F.,** Transforming NRK cells with avian sarcoma virus reduces the extracellular Ca^{2+} requirement without affecting the calcicalmodulin requirement for the G1/S transition, *Exp. Cell Res.,* 157, 544, 1985.
250. **Parsons, P. G., Moss, D. J., Morris, C., Musk, P., Maynard, K., and Partridge, R.,** Decreased calcium dependence of lymphoblastoid cell lines compared with Burkitt lymphoma cell lines, *Int. J. Cancer,* 35, 743, 1985.
251. **Dieter, P.,** Calmodulin and calmodulin-mediated processes in plants, *Plant Cell Environ.,* 7, 371, 1984.
252. **Babu, Y. S., Sack, J. S., Greenhough, T. J., Bugg, C. E., Means, A. R., and Cook, W. J.,** Three-dimensional structure of calmodulin, *Nature (London),* 315, 37, 1985.
253. **Boynton, A. L., Whitfield, J. F., and MacManus, J. P.,** Calmodulin stimulates DNA synthesis by rat liver cells, *Biochem. Biophys. Res. Comm.,* 95, 745, 1980.
254. **Chafouleas, J. G., Bolton, W. E., Hidaka, Y., Boyd, A. E., III, and Means, A. R.,** Calmodulin and the cell cycle: involvement in regulation of cell-cycle progression, *Cell,* 28, 41, 1982.
255. **Chafouleas, J. G., Lagace, L., Bolton, W. E., Boyd, A. E., III, and Means, A. R.,** Changes in calmodulin and its mRNA accompany reentry of quiescent (GO) cells into the cell cycle, *Cell,* 36, 73, 1984.
256. **Kikuchi, Y., Iwano, I., and Kato, K.,** Effects of calmodulin antagonists on human ovarian cancer cell proliferation in vitro, *Biochem. Biophys. Res. Comm.,* 123, 385, 1984.
257. **MacManus, J. P., Braceland, B. M., Rixon, R. H., Whitfield, J. F., and Morris, H. P.,** An increase in calmodulin during growth of normal and cancerous liver in vivo, *FEBS Lett.,* 133, 99, 1981.
258. **Fukami, Y. and Lipmann, F.,** Purification of the Rous sarcoma virus src kinase by casein-agarose and tyrosine-agarose affinity chromatography, *Proc. Natl. Acad. Sci. U.S.A.,* 82, 321, 1985.
259. **Wong, T. W. and Goldberg, A. R.,** Kinetics and mechanism of angiotensin phosphorylation by the transforming gene product of Rous sarcoma virus, *J. Biol. Chem.,* 259, 3127, 1984.
260. **Yuspa, S. H. and Morgan, D. L.,** Mouse skin cells resistant to terminal differentiation associated with initiation of carcinogenesis, *Nature (London),* 293, 72, 1981.
261. **Levenson, R., Housman, D., and Cantley, L.,** Amiloride inhibits erythroleukemia cell differentiation: evidence for a Ca^{2+} requirement for commitment, *Proc. Natl. Acad. Sci. U.S.A.,* 77, 5948, 1980.
262. **Faletto, D. L. and Macara, I. G.,** The role of Ca^{2+} in dimethyl sulfoxide-induced differentiation of Friend erythroleukemia cells, *J. Biol. Chem.,* 260, 4884, 1985.
263. **Mac Neil, S., Lakey, T., and Tomlinson, S.,** Calmodulins regulation of adenylate cyclase activity, *Cell Calcium,* 6, 213, 1985.
264. **Wolff, D. J., Ross, J. M., Thompson, P. N., Brostrom, M. A., and Brostrom, C. O.,** Interaction of calmodulin with histones: alteration of histone dephosphorylation, *J. Biol. Chem.,* 256, 1846, 1981.
265. **Iwasa, Y., Iwasa, T., Higashi, K., Matsui, K., and Miyamoto, E.,** Modulation by phosphorylation of interaction between calmodulin and histones, *FEBS Lett.,* 133, 95, 1981.
266. **Maizels, E. T. and Jungmann, R. A.,** Ca^{2+}-calmodulin-dependent phosphorylation of soluble and nuclear proteins in the rat ovary, *Endocrinology,* 112, 1895, 1983.
267. **Pearson, R. B., Woodgett, J. R., Cohen, P., and Kemp, B. E.,** Substrate specificity for a multifunctional calmodulin-dependent protein kinase, *J. Biol. Chem.,* 260, 14471, 1985.
268. **Kuret, J. and Schulman, H.,** Mechanism of autophosphorylation of the multifunctional Ca^{2+}/calmodulin-dependent protein kinase, *J. Biol. Chem.,* 260, 6427, 1985.
269. **MacManus, J. P., Whitfield, J. F., Boynton, A. L., Durkin, J. P., and Swierenga, S. H. H.,** Oncomodulin: a widely distributed tumour-specific, calcium-binding protein, *Oncodevel. Biol. Med.,* 3, 79, 1982.
270. **MacManus, J. P., Brewer, L. M., and Whitfield, J. F.,** The widely-distributed tumour protein, oncomodulin, is a normal constituent of human and rodent placentas, *Cancer Lett.,* 27, 145, 1985.
271. **Brewer, L. M. and MacManus, J. P.,** Localization and synthesis of the tumor protein oncomodulin in extraembryonic tissues of the fetal rat, *Develop. Biol.,* 112, 49, 1985.
272. **Müller, R., Slamon, D. J., Tremblay, J. M., Cline, M. J., and Verma, I. M.,** Differential expression of cellular oncogenes during pre- and postnatal development of the mouse, *Nature (London),* 299, 640 1982.

273. **Müller, R., Verma, I. M., and Adamson, E. D.,** Expression of c-*onc* genes: c-*fos* transcripts accumulate to high levels during development of mouse placenta, yolk sac and amnion, *EMBO J.*, 2, 679, 1983.
274. **Mason, I., Murphy, D., and Hogan, B. L. M.,** Expression of c-*fos* in parietal endoderm, amnion and differentiating F9 teratocarcinoma cells, *Differentiation*, 30, 76, 1985.
275. **Olwin, B. B. and Storm, D. R.,** Calcium binding to complexes of calmodulin and calmodulin binding proteins, *Biochemistry*, 24, 8081, 1985.
276. **Kakiuchi, R., Inui, M., Morimoto, K., Kanda, K., Sobue, K., and Kakiuchi, S.,** Caldesmon, a calmodulin-binding F actin-interacting protein, is present in aorta, uterus and platelets, *FEBS Lett.*, 154, 351, 1983.
277. **Owada, M. K., Hakura, A., Iida, K., Yahara, I., Sobue, K., and Kakiuchi, S.,** Occurrence of caldesmon (a calmodulin-binding protein) in cultured cells: comparison of normal and transformed cells, *Proc. Natl. Acad. Sci. U.S.A.*, 81, 3133, 1984.
278. **Umekawa, H. and Hidaka, H.,** Phosphorylation of caldesmon by protein kinase C, *Biochem. Biophys. Res. Comm.*, 132, 56, 1985.
279. **Pallen, C. J., Valentine, K. A., Wang, J. H., and Hollenberg, M. D.,** Calcineurin-mediated dephosphorylation of the human placental membrane receptor for epidermal growth factor urogastrone, *Biochemistry*, 24, 4727, 1985.
280. **Michell, R. H.,** Inositol phospholipids and cell surface receptor function, *Biochim. Biophys. Acta*, 415, 81, 1975.
281. **Farese, R. V.,** Phosphoinositide metabolism and hormone action, *Endocrine Rev.*, 4, 78, 1983.
282. **Farese, R. V.,** The phosphatidate-phosphoinositide cycle: an intracellular messenger system in the action of hormones and neurotransmitters, *Metabolism*, 32, 628, 1983.
283. **Farese, R. V.,** Phospholipids as intermediates in hormone action, *Mol. Cell. Endocrinol.*, 35, 1, 1984.
284. **Berridge, M. J.,** Inositol trisphosphate and diacylglycerol as second messengers, *Biochem. J.*, 220, 345, 1984.
285. **Majerus, P. W., Neufeld, E. J., and Wilson, D. B.,** Production of phosphoinositide-derived messengers, *Cell*, 37, 701, 1984.
286. **Macara, I. G.,** Oncogenes, ions, and phospholipids, *Am. J. Physiol.*, 248, C3, 1985.
287. **Majerus, P. W., Neufeld, E. J., and Wilson, D. B.,** Production of phosphoinositide-derived messengers, *Cell*, 37, 701, 1984.
288. **Exton, J. H.,** Role of calcium and phosphoinositides in the actions of certain hormones and neurotransmitters, *J. Clin. Invest.*, 75, 1753, 1985.
289. **Boss, W. F. and Massel, M. O.,** Polyphosphoinositides are present in plant tissue culture cells, *Biochem. Biophys. Res. Comm.*, 132, 1018, 1985.
290. **Nakanishi, H., Nomura, H., Kikkawa, U., Kishimoto, A., and Nishikura, Y.,** Rat brain and liver soluble phospholipase C: resolution of two forms with different requirements for calcium, *Biochem. Biophys. Res. Comm.*, 132, 582, 1985.
291. **Storey, D. J., Shears, S. B., Kirk, C. J., and Michell, R. H.,** Stepwise enzymatic dephosphorylation of inositol 1,4,5-trisphosphate to inositol in liver, *Nature (London)*, 312, 374, 1984.
292. **Cockroft, S. and Gomperts, B. D.,** Role of guanine nucleotide binding protein in the activation of polyphosphoinositide phosphodiesterase, *Nature (London)*, 314, 534, 1985.
293. **Wilson, D. B., Bross, T. E., Sherman, W. R., Berger, R. A., and Majerus, P. W.,** Inositol cyclic phosphates are produced by cleavage of phosphatidylphosphoinositoids (polyphosphoinositides) with purified sheep seminal vesicle phospholipase C enzymes, *Proc. Natl. Acad. Sci. U.S.A.*, 82, 4013, 1985.
294. **Ashendel, C. L.,** The phorbol ester receptor: a phospholipid-regulated protein kinase, *Biochim. Biophys. Acta*, 822, 219, 1985.
295. **Takai, Y., Kaibuchi, K., Tsuda, T., and Hoshijima, M.,** Role of protein kinase C in transmembrane signaling, *J. Cell. Biochem.*, 29, 156, 1985.
296. **Schwantki, N., Le Bouffant, F., Dorée, M., and Le Peuch, C. J.,** Protein kinase C: properties and possible role in cellular division and differentiation, *Biochimie*, 67, 1103, 1985.
297. **Horn, F., Gschwendt, M., and Marks, F.,** Partial purification and characterization of the calcium-dependent and phospholipid-dependent protein kinase C from chick oviduct, *Eur. J. Biochem.*, 148, 533, 1985.
298. **Schäfer, A., Bygrave, F., Matzenauer, S., and Marmé,** Identification of a calcium- and phospholipid-dependent protein kinase in plant tissue, *FEBS Lett.*, 187, 25, 1985.
299. **Cochet, C., Gill, G. N., Meisenhelder, J., Cooper, J. A., and Hunter, T.,** C-kinase phosphorylates the epidermal growth factor receptor and reduces its epidermal growth factor-stimulated tyrosine protein kinase activity, *J. Biol. Chem.*, 259, 2553, 1984.
300. **Witters, L. A., Vater, C. A., and Lienhard, G. E.,** Phosphorylation of the glucose transporter *in vitro* and *in vivo* by protein kinase C, *Nature (London)*, 315, 777, 1985.
301. **Wolf, M., Cuatrecasas, P., and Sahyoun, N.,** Interaction of protein kinase C with membranes is regulated by Ca^{2+}, phorbol esters, and ATP, *J. Biol. Chem.*, 260, 15718, 1985.

302. **McDonald, J. R. and Walsh, M. P.**, Inhibition of the Ca^{2+}- and phospholipid-dependent protein kinase by a novel M_r 17,000 Ca^{2+}-binding protein, *Biochem. Biophys. Res. Comm.*, 129, 603, 1985.
303. **Murakami, K. and Routtenberg, A.**, Direct activation of purified protein kinase C by unsaturated fatty acids (oleate and arachidonate) in the absence of phospholipids and Ca^{2+}, *FEBS Lett.*, 192, 189, 1985.
304. **Girard, P. R., Mazzei, G. J., and Kuo, J. F.**, Immunological quantitation of phospholipid/Ca^{2+}-dependent protein kinase and its fragments: tissue levels, subcellular distribution, and ontogenic changes in brain and heart, *J. Biol. Chem.*, 261, 370, 1986.
305. **Rozengurt, E., Rodriguez-Pena, A., Coombs, M., and Sinnett-Smith, J.**, Diacylglycerol stimulates DNA synthesis and cell division in mouse 3T3 cells: role of Ca^{2+}-sensitive phospholipid-dependent protein kinase, *Proc. Natl. Acad. Sci. U.S.A.*, 81, 5748, 1984.
306. **Donnelly, T. E., Sittler, R., and Scholar, E. M.**, Relationship between membrane-bound protein kinase C activity and calcium-dependent proliferation of BALB/c 3T3 cells, *Biochem. Biophys. Res. Comm.*, 126, 741, 1985.
307. **Nabika, T., Velletri, P. A., Lovenberg, W., and Beaven, M. A.**, Increase in cytosolic calcium and phosphoinositide metabolism induced by angiotensin II and (Arg)vasopresin in vascular smooth muscle cells, *J. Biol. Chem.*, 260, 4661, 1985.
308. **Wolf, M., Le Vine, H., III, May, W. S., Jr., Cuatrecasas, P., and Sahyoun, N.**, A model for intracellular translocation of protein kinase C involving synergism between Ca^{2+} and phorbol esters, *Nature (London)*, 317, 546, 1985.
309. **May, W. S., Jr., Sahyoun, N., Wolf, M., and Cuatrecasas, P.**, Role of intracellular calcium mobilization in the regulation of protein kinase C-mediated membrane processes, *Nature (London)*, 317, 549, 1985.
310. **Boynton, A. L., Klein, L. P., Whitfield, J. F., and Bossi, D.**, Involvement of the Ca^{2+}/phospholipid-dependent protein kinase in the G1 transit of T51B rat liver epithelial cells, *Exp. Cell Res.*, 160, 197, 1985.
311. **Whitaker, M. and Irvine, R. F.**, Inositol 1,4,5-triphosphate microinjection activates sea urchin eggs, *Nature (London)*, 312, 636, 1984.
312. **Aberdam, E. and Dekel, N.**, Activators of protein kinase C stimulate meiotic maturation of rat oocytes, *Biochem. Biophys. Res. Comm.*, 132, 570, 1985.
313. **Kuo, J. F., Andersson, R. G. G., Wise, B. G., Mackerlova, L., Salomonsson, I., Brackett, N. L., Katoh, N., Shoji, M., and Wrenn, R. W.**, Calcium-dependent protein kinase: widespread occurrence in various tissues and phyla of the animal kingdom and comparison of the effects of phospholipid, calmodulin, and trifluoperizine, *Proc. Natl. Acad. Sci. U.S.A.*, 77, 7039, 1980.
314. **Leffert, H. L.**, Monovalent cations, cell proliferation and cancer: an overview, in *Ions, Cell Proliferation, and Cancer*, Boynton, A. L., McKeehan, W. L., and Whitfield, J. F., Eds., Academic Press, New York, 1982, 93.
315. **Rozengurt, E.**, Monovalent ion fluxes, cyclic nucleotides and the stimulation of DNA synthesis in quiescent cells, in *Ions, Cell Proliferation, and Cancer*, Boynton, A. L., McKeehan, W. L., and Whitfield, J. F., Eds., Academic Press, New York, 1982, 259.
316. **Rossoff, P. M. and Cantley, L. C.**, Lipopolysaccharide and phorbol esters induce differentiation but have opposite effects on phosphatidylinositol turnover and Ca^{2+} mobilization in 70Z/3 pre-B lymphocytes, *J. Biol. Chem.*, 260, 9209, 1985.
317. **Vara, F., Schneider, J. A., and Rozengurt, E.**, Ionic responses rapidly elicited by activation of protein kinase C in quiescent Swiss 3T3 cells, *Proc. Natl. Acad. Sci. U.S.A.*, 82, 2384, 1985.
318. **Vicentini, L. M. and Villereal, M. L.**, Activation of Na^+/H^+ exchange in cultured fibroblasts: synergism and antagonism between phorbol ester, Ca^{2+} ionophore, and growth factors, *Proc. Natl. Acad. Sci. U.S.A.*, 82, 8053, 1985.
319. **Parker, P. J., Katan, M., Waterfield, M. D., and Leader, D. P.**, The phosphorylation of eukaryotic ribosomal protein S6 by protein kinase C, *Eur. J. Biochem.*, 148, 579, 1985.
320. **Werth, D. K., Wiedel, J. E., and Pastan, I.**, Vinculin, a cytoskeletal substrate of protein kinase C, *J. Biol. Chem.*, 258, 11423, 1983.
321. **Werth, D. K. and Pastan, I.**, Vinculin phosphorylation in response to Ca^{2+} and phorbol esters in intact cells, *J. Biol. Chem.*, 259, 5264, 1984.
322. **Kimura, K., Katoh, N., Sakurada, K., and Kubo, S.**, Phosphorylation of high mobility group 1 protein by phospholipid-sensitive Ca^{2+}-dependent protein kinase from pig testis, *Biochem. J.*, 227, 271, 1985.
323. **Carney, D. H., Scott, D. L., Gordon, E. A., and LaBelle, E. F.**, Phosphoinositides in mitogenesis: neomycin inhibits thrombin-stimulated phosphoinositide turnover and initiation of cell proliferation, *Cell*, 42, 479, 1985.
324. **Fisher, G. J., Bakshian, S., and Baldassare, J. J.**, Activation of human platelets by ADP causes a rapid rise in cytosolic free calcium without hydrolysis of phosphatidylinositol-4,5-bisphosphate, *Biochem. Biophys. Res. Comm.*, 129, 958, 1985.
325. **Sasaki, T. and Hasegawa-Sasaki, H.**, Breakdown of phosphatidylinositol 4,5-bisphosphate in a T-cell leukaemia line stimulated by phytohaemagglutinin is not dependent on Ca^{2+} mobilization, *Biochem. J.*, 227, 971, 1985.

326. **Leeb-Lundberg, L. M. F., Cotecchia, S., Lomasney, J. W., DeBernardis, J. F., Lefkowitz, R. J., and Caron, M. G.,** Phorbol esters promote alpha$_1$-adrenergic receptor phosphorylation and receptor uncoupling from inositol phospholipid metabolism, *Proc. Natl. Acad. Sci. U.S.A.*, 82, 5651, 1985.
327. **Habenicht, A. J. R., Glomset, J. A., King, W. C., Nist, C., Mitchell, C. D., and Ross, R.,** Early changes in phosphatidylinositol and arachidonic acid metabolism in quiescent Swiss 3T3 cells stimulated to divide by platelet-derived growth factor, *J. Biol. Chem.*, 256, 12329, 1981.
328. **Sawyer, S. T. and Cohen, S.,** Enhancement of calcium uptake and phosphatidylinositol turnover by epidermal growth factor in A431 cells, *Biochemistry*, 20, 6280, 1981.
329. **Sugden, D., Vanecek, J., Klein, D. C., Thomas, T. P., and Anderson, W. B.,** Activation of protein kinase C potentiates isoprenaline-induced cyclic AMP accumulation in rat pinealocytes, *Nature (London)*, 314, 359, 1985.
330. **Farese, R. V., Larson, R. E., and Sabir, M. A.,** Insulin acutely increases phospholipids in the phosphatidate-inositide cycle in rat adipose tissue, *J. Biol. Chem.*, 257, 4042, 1982.
331. **Honeyman, T. W., Strohsnitter, W., Scheid, C. R., and Schimmel, R. J.,** Phosphatidic acid and phosphatidylinositol labelling in adipose tissue: relationship to the metabolic effects of insulin-like agents, *Biochem. J.*, 212, 489, 1983.
332. **Besterman, J. M., Watson, S. P., and Cuatrecasas, P.,** Lack of association of epidermal growth factor-, insulin-, and serum-induced mitogenesis with stimulation of phosphoinositide degradation in BALB/c 3T3 fibroblasts, *J. Biol. Chem.*, 261, 723, 1986.
333. **Taylor, D., Uhing, R. J., Blackmore, P. F., Prpić, V., and Exton, J. H.,** Insulin and epidermal growth factor do not affect phosphoinositide metabolism in rat liver plasma membranes and hepatocytes, *J. Biol. Chem.*, 260, 2011, 1985.
334. **Thompson, D. M., Cochet, C., Chambaz, E. M., and Gill, G. N.,** Separation and characterization of a phosphatidylinositol kinase activity that co-purifies with the epidermal growth factor receptor, *J. Biol. Chem.*, 260, 8824, 1985.
335. **Warden, C. H. and Friedkin, M.,** Regulation of phosphatidylcholine biosynthesis by mitogenic growth factors, *Biochim. Biophys. Acta*, 792, 270, 1984.
336. **Warden, C. H. and Friedkin, M.,** Regulation of choline kinase activity and phosphatidylcholine biosynthesis by mitogenic growth factors in 3T3 fibroblasts, *J. Biol. Chem.*, 260, 6006, 1985.
337. **Olson, J. W.,** Enhanced phosphatidylinositol kinase activity is associated with early stages of hepatocarcinogenesis and hepatocellular carcinoma, *Biochem. Biophys. Res. Comm.*, 132, 969, 1985.
338. **Montesano, R., Orci, L., and Vassalli, P.,** Human endothelial cell cultures: phenotypic modulation by leukocyte interleukin, *J. Cell. Physiol.*, 122, 424, 1985.
339. **Folkman, J.,** Tumor angiogenesis, *Adv. Cancer Res.*, 43, 175, 1985.
340. **Castagna, M., Takai, Y., Kaibuchi, K., Sano, K., Kiddawa, U., and Nishikura, Y.,** Direct activation of calcium-activated, phospholipid-dependent protein kinase by tumor-promoting phorbol esters, *J. Biol. Chem.*, 257, 7847, 1982.
341. **Vandenbark, G. R. and Niedel, J. E.,** Phorbol diesters and cellular differentiation, *J. Natl. Cancer Inst.*, 73, 1013, 1984.
342. **Gschwendt, M., Horn, F., Kittstein, W., Fürstenberger, G., Besemfelder, E., and Marks, F.,** Calcium and phospholipid-dependent protein kinase activity in mouse epidermis cytosol: stimulation by complete and incomplete tumor promoters and inhibition by various compounds, *Biochem. Biophys. Res. Comm.*, 124, 63, 1984.
343. **Parker, P. J., Stabel, S., and Waterfield, M. D.,** Purification of protein kinase C from bovine brain: identity with the phorbol ester receptor, *EMBO J.*, 3, 953, 1984.
344. **Cabot, M. C.,** Tumor promoting phorbol diesters: substrates for diacylglycerol lipase, *Biochem. Biophys. Res. Comm.*, 123, 170, 1984.
345. **de Chaffoy, de Courcelles, D., Roevens, P., and van Belle, H.,** 12-O-tetradecanoylphorbol 13-acetate stimulates inositol lipid phosphorylation in intact human platelets, *FEBS Lett.*, 173, 389, 1984.
346. **Taylor, M. V., Metcalfe, J. C., Hesketh, T. R., Smith, G. A., and Moore, J. P.,** Mitogens increase phosphorylation of phosphoinositides in thymocytes, *Nature (London)*, 312, 462, 1984.
347. **Tapley, P. M. and Murray, A. W.,** Modulation of Ca^{2+}-activated, phospholipid-dependent protein kinase in platelets treated with a tumor-promoting phorbol ester, *Biochem. Biophys. Res. Comm.*, 122, 158, 1984.
348. **Wickremasinghe, R. G., Piga, A., Campana, D., Yaxley, J. C., and Hoffbrand, A. V.,** Rapid downregulation of protein kinase C and membrane association in phorbol ester-treated leukemia cells, *FEBS Lett.*, 190, 50, 1985.
349. **Orellana, S. A., Solski, P. A., and Brown, J. H.,** Phorbol ester inhibits phosphoinositide hydrolysis and calcium mobilization in cultured astrocytoma cells, *J. Biol. Chem.*, 260, 5236, 1985.
350. **Welch, W. J.,** Phorbol ester, calcium ionophore, or serum added to quiescent rat embryo fibroblast cells all result in the elevated phosphorylation of two 28,000-dalton mammalian stress proteins, *J. Biol. Chem.*, 260, 3058, 1985.

351. **Voellmy, R.**, The heat shock genes: a family of highly conserved genes with a superbly complex expression pattern, *BioEssays*, 1, 213, 1984.
352. **Bienz, M.**, Transient and developmental activation of heat-shock genes, *Trends Biochem. Sci.*, 10, 157, 1985.
353. **Kingston, R. E., Baldwin, A. S., and Sharp, P. A.**, Regulation of heat shock protein 70 gene expression by c-*myc*, *Nature (London)*, 312, 280, 1984.
354. **Oppermann, H., Levinson, W., and Bishop, J. M.**, A cellular protein that associates with the transforming protein of Rous sarcoma virus is also a heat-shock protein, *Proc. Natl. Acad. Sci. U.S.A.*, 78, 1067, 1981.
355. **Fujiki, H., Tanaka, Y., Miyake, R., Kikkawa, U., Nishizuka, T., and Sugimura, T.**, Activation of calcium-activated, phospholipid-dependent protein kinase (protein kinase C) by new classes of tumor promoters: teleocidin and debromoaplysiatoxin, *Biochem. Biophys. Res. Comm.*, 120, 339, 1984.
356. **Nishizuka, Y.**, Protein kinases in signal transduction, *Trends Biochem. Sci.*, 9, 163, 1984.
357. **Malkinson, A. M., Conway, K., Bartlett, S., Butley, M. S., and Conroy, C.**, Strain differences among inbred mice in protein kinase C activity, *Biochem. Biophys. Res. Comm.*, 122, 492, 1984.
358. **Kiss, Z. and Steinberg, R. A.**, Interactions between cyclic AMP- and phorbol ester-dependent phosphorylation systems in S49 mouse lymphoma cells, *J. Cell. Physiol.*, 125, 200, 1985.
359. **Koeffler, H. P.**, Induction of differentiation of human acute myelogenous leukemia cells: therapeutic implications, *Blood*, 62, 709, 1983.
360. **Ebeling, J. G., Vandenbark, G. R., Kuhn, J., L. J., Ganong, B. R., Bell, R. M., and Niedel, J. E.**, Diacylglycerols mimic phorbol ester induction of leukemic cell differentiation, *Proc. Natl. Acad. Sci. U.S.A.*, 82, 815, 1985.
361. **Pincus, S. M., Beckman, B. S., and George, W. J.**, Inhibition of dimethylsulfoxide-induced differentiation in Friend erythroleukemic cells by diacylglycerols and phospholipase C, *Biochem. Biophys. Res. Comm.*, 125, 491, 1984.
362. **Durham, J. P., Emler, C. A., Butcher, F. R., and Fontana, J. A.**, Calcium-activated, phospholipid-dependent protein kinase activity and protein phosphorylation in HL60 cells induced to differentiate by retinoic acid, *FEBS Lett.*, 185, 157, 1985.
363. **Anderson, N. L., Gemmell, M. A., Coussens, P. M., Murao, S., and Huberman, E.**, Specific protein phosphorylation in human promyelocytic HL-60 leukemia cells susceptible or resistant to induction of cell differentiation by phorbol-12-myristate-13-acetate, *Cancer Res.*, 45, 4955, 1985.
364. **Rance, A. J., Thönnes, M., and Issinger, O.-G.**, Ribosomal protein S6 phosphorylation and morphological changes in response to the tumour promoter 12-O-tetradecanoylphorbol 13-acetate in primary human tumour cells, established and transformed cell lines, *Biochim. Biophys. Acta*, 847, 128, 1985.
365. **Zylber-Katz, E. and Glazer, R. I.**, Phospholipid- and Ca^{2+}-dependent protein kinase activity and protein phosphorylation patterns in the differentiation of human promyelocytic leukemia cell line HL-60, *Cancer Res.*, 45, 5159, 1985.
366. **Mita, S., Nakaki, T., Yamamoto, S., and Kato, R.**, Phosphorylation and dephosphorylation of human promyelocytic leukemia cell (HL-60) proteins by tumor promoter, *Exp. Cell Res.*, 154, 492, 1984.
367. **Kreutter, D., Caldwell, A. B., and Morin, M. J.**, Dissociation of protein kinase C activation from phorbol ester-induced maturation of HL-60 leukemia cells, *J. Biol. Chem.*, 260, 5979, 1985.
368. **Schwizer, R. W.**, The phorbol ester, 12-O-tetradecanoylphorbol 13-acetate enhances the long-term maintenance of pancreatic islet B cell differentiated function in tissue culture, *Biochem. Biophys. Res. Comm.*, 128, 1315, 1985.
369. **Lin, T.**, The role of calcium/phospholipid-dependent protein kinase in Leydig cell steroidogenesis, *Endocrinology*, 117, 119, 1985.
370. **Beguinot, L., Hanover, J. A., Ito, S., Richert, N. D., Willingham, M. C., and Pastan, I.**, Phorbol esters induce transient internalization without degradation of unoccupied growth factor receptors, *Proc. Nat. Acad. Sci. U.S.A.*, 82, 2774, 1985.
371. **Kohno, H., Taketani, S., and Tokunaga, R.**, Tumor-promoting, phorbol ester-induced phosphorylation of cell-surface transferrin receptors in human erythroleukemia cells, *Cell Struct. Function*, 10, 95, 1985.
372. **May, W. S., Sahyoun, N., Jacobs, S., Wolf, M., and Cuatrecasas, P.**, Mechanism of phorbol diester-induced regulation of surface transferrin receptor involves the action of activated protein kinase C and an intact cytoskeleton, *J. Biol. Chem.*, 260, 9419, 1985.
373. **Costa, S.d., Fabbro, D., Regazzi, R., Küng, W., and Eppenberger, U.**, The cytosolic phorboid receptor correlates with hormone dependency in six mammary carcinoma cell lines, *Biochem. Biophys. Res. Comm.*, 133, 814, 1985.
374. **Garte, S. J.**, Differential effects of phorbol ester on the beta-adrenergic response of normal and *ras*-transformed NIH 3T3 cells, *Biochem. Biophys. Res. Comm.*, 133, 702, 1985.
375. **Bishop, R., Martinez, R., Weberg, M. J., Blackshear, P. J., Beatty, S., Lim, R., and Herschman, H. R.**, Protein phosphorylation in tetradecanoyl phorbol acetate-nonproliferative variant of 3T3 cells, *Mol. Cell. Biol.*, 5, 2231, 1985.

376. **Patskan, G. J. and Baxter, C. S.**, Specific stimulation of histone H2B and H4 phosphorylation in some mouse lymphocytes by 12-O-tetradecanoylphorbol 13-acetate, *J. Biol. Chem.*, 260, 12899, 1985.
377. **Soma, G.-I., Murata, M., Kitahara, N., Gatanaga, T., Shibai, H., Morioka, H., and Andoh, T.**, Detection of a countertranscript in promyelocytic leukemia cells HL-60 during early differentiation by TPA, *Biochem. Biophys. Res. Comm.*, 132, 100, 1985.
378. **Jetten, A. M., Ganong, B. R., Vandenbark, G. R., Shirley, J. E., and Bell, R. M.**, Role of protein kinase C in diacylglycerol-mediated induction of ornithine decarboxylase and reduction of epidermal growth factor binding, *Proc. Natl. Acad. Sci. U.S.A.*, 82, 1941, 1985.
379. **Buckley, A. R., Putnam, C. W., and Russell, D. H.**, In vivo induction of rat hepatic ornithine decarboxylase and plasminogen activator by 12-O-tetradecanoylphorbol 13-acetate, *Biochim. Biophys. Acta*, 841, 127, 1985.
380. **Rinehart, C. A., Jr. and Canellakis, E. S.**, Induction of ornithine decarboxylase activity by insulin and growth factors is mediated by amino acids, *Proc. Natl. Acad. Sci. U.S.A.*, 82, 4365, 1985.
381. **Gelfand, E. W., Cheung, R. K., Mills, G. B., and Grinstein, S.**, Mitogens trigger a calcium-independent signal for proliferation in phorbol-ester-treated lymphocytes, *Nature (London)*, 315, 419, 1985.
382. **Moolenaar, W. H., Tertoolen, L. G. J., and de Laat, S. W.**, Phorbol ester and diacylglycerol mimic growth factors in raising cytoplasmic pH, *Nature (London)*, 312, 371, 1984.
383. **Diringer, H. and Friis, R. R.**, Changes in phosphatidylinositol metabolism correlated to growth state and Rous sarcoma virus-transformed Japanese quail cells, *Cancer Res.*, 37, 2979, 1977.
384. **Richert, N. D., Blithe, D. L., and Pastan, I.**, Properties of the src kinase purified from Rous sarcoma virus-induced rat tumors, *J. Biol. Chem.*, 257, 7143, 1982.
385. **Graziani, Y., Erikson, E., and Erikson, R. L.**, Evidence that the Rous sarcoma virus transforming gene product is associated with glycerol kinase activity, *J. Biol. Chem.*, 258, 2126, 1983.
386. **Sugimoto, Y., Whitman, M., Cantley, L. C., and Erikson, R. L.**, Evidence that the Rous sarcoma virus transforming gene product phosphorylates phosphatidylinositol and diacylglycerol, *Proc. Natl. Acad. Sci. U.S.A.*, 81, 2117, 1984.
387. **Whitman, M., Kaplan, D. R., Schaffahausen, B., Cantley, L., and Roberts, T. M.**, Association of phosphatidylinositol kinase activity with polyoma middle-T competent for transformation, *Nature (London)*, 315, 239, 1985.
388. **Macara, I. G., Marinetti, G. V., and Balduzzi, P. C.**, Transforming protein of avian sarcoma virus UR2 is associated with phosphatidylinositol kinase activity: possible role in tumorigenesis, *Proc. Natl. Acad. Sci. U.S.A.*, 81, 2728, 1984.
389. **MacDonald, M. L., Kuenzel, E. A., Glomset, J. A., and Krebs, E. G.**, Evidence from two transformed cell lines that the phosphorylation of peptide tyrosine and phosphatidylinositol are catalyzed by different proteins, *Proc. Natl. Acad. Sci. U.S.A.*, 82, 3993, 1985.
390. **Sugano, S. and Hanafusa, H.**, Phosphatidylinositol kinase activity in virus-transformed and nontransformed cells, *Mol. Cell. Biol.*, 5, 2399, 1985.
391. **Fischer, S., Fagard, R., Comoglio, P., and Gacon, G.**, Phosphoinositides are not phosphorylated by the very active tyrosine protein kinase from the murine lymphoma LSTRA, *Biochem. Biophys. Res. Comm.*, 132, 481, 1985.
392. **Purchio, A. F., Shoyab, M., and Gentry, L. E.**, Site-specific increased phosphorylation of $pp60^{v-src}$ after treatment of RSV-transformed cells with a tumor promoter, *Science*, 229, 1393, 1985.
393. **Gould, K. L., Woodgett, J. R., Cooper, J. A., Buss, J. E., Shalloway, D., and Hunter, T.**, Protein kinase C phosphorylates $pp60^{src}$ at a novel site, *Cell*, 42, 849, 1985.
394. **Persson, H., Hennighausen, L., Taub, R., De Grado, W., and Leder, P.**, Antibodies to human c-*myc* oncogene product: evidence of an evolutionarily conserved protein induced during cell proliferation, *Science*, 225, 687, 1984.
395. **Kelly, K. and Siebenlist, U.**, The role of c-myc in the proliferation of normal and neoplastic cells, *J. Clin. Immunol.*, 5, 65, 1985.
396. **Kelly, K., Cochran, B. H., Stiles, C. D., and Leder, P.**, Cell-specific regulation of the c-*myc* gene by lymphocyte mitogens and platelet-derived growth factors, *Cell*, 35, 603, 1983.
397. **Makino, R., Hayashi, K., and Sugimura, T.**, c-*myc* transcript is induced in rat liver at a very early stage of regeneration or by cycloheximide treatment, *Nature (London)*, 310, 697, 1984.
398. **Reed, J. C., Nowell, P. C., and Hoover, R. G.**, Regulation of c-*myc* mRNA levels in normal human lymphocytes by modulators of cell proliferation, *Proc. Natl. Acad. Sci. U.S.A.*, 82, 4221, 1985.
399. **Goustin, A. S., Betsholtz, C., Pfeifer-Ohlsson, S., Persson, H., Rydnert, J., Bywater, M., Holmgren, G., Heldin, C.-H., Westermark, B., and Ohlsson, R.**, Coexpression of the *sis* and *myc* proto-oncogenes in developing human placenta suggests autocrine control of trophoblastic growth, *Cell*, 41, 301, 1985.
400. **Donner, P., Greiser-Wilke, I., and Moelling, K.**, Nuclear localization and DNA binding of the transforming gene product of avian myelocytomatosis virus, *Nature (London)*, 296, 262, 1982.

401. **Hann, S. R., Abrams, H. D., Rohrschneider, L. R., and Eisenman, R. N.,** Proteins encoded by v-*myc* and c-*myc* oncogenes: identification and localization in acute leukemia virus transformants and bursal lymphoma cell lines, *Cell,* 34, 789, 1983.
402. **Persson, H. and Leder, P.,** Nuclear localization and DNA binding properties of a protein expressed by human c-*myc* oncogene, *Science,* 225, 718, 1984.
403. **Eisenman, R. N., Tachibana, C. Y., Abrams, H. D., and Hann, S. R.,** v-*myc* and c-*myc*-encoded proteins are associated with the nuclear matrix, *Mol. Cell. Biol.,* 5, 114, 1985.
404. **Eisenman, R. N. and Hann, S. R.,** Proteins expressed by the c-*myc* oncogene in lymphomas of human and avian origin, *Proc. R. Soc. London Ser. B,* 226, 73, 1985.
405. **Evan, G. I., Lewis, G. K., Ramsay, G., and Bishop, J. M.,** Isolation of monoclonal antibodies specific for the human c-*myc* proto-oncogene product, *Mol. Cell. Biol.,* 5, 3610, 1985.
406. **Beimling, P., Benter, T., Sander, T., and Moelling, K.,** Isolation and characterization of the human cellular *myc* gene product, *Biochemistry,* 24, 6349, 1985.
407. **Reed, J. C., Sabath, D. E., Hoover, R. G., and Prystowsky, M. B.,** Recombinant interleukin 2 regulates levels of c-*myc* mRNA in a cloned murine T lymphocyte, *Mol. Cell. Biol.,* 5, 3361, 1985.
408. **Coughlin, S. R., Lee, W. M. F., Williams, P. W., Giels, G. M., and Williams, L. T.,** c-*myc* gene expression is stimulated by agents that activate protein kinase C and does not account for the mitogenic effect of PDGF, *Cell,* 43, 243, 1985.
409. **Grosso, L. E. and Pitot, H. C.,** Modulation of c-myc expression in the HL-60 cell line, *Biochem. Biophys. Res. Comm.,* 119, 473, 1984.
410. **Lachman, H. M. and Skoultchi, A. I.,** Expression of c-*myc* changes during differentiation of mouse erythroleukaemia cells, *Nature (London),* 310, 592, 1984.
411. **Filmus, J. and Buick, R. N.,** Relationship of c-*myc* expression to differentiation and proliferation of HL-60 cells, *Cancer Res.,* 45, 822, 1985.
412. **Sejersen, T., Sümegi, J., and Ringertz, N. R.,** Density-dependent arrest of DNA replication is accompanied by decreased levels of c-*myc* mRNA in myogenic but not in differentiation-defective myoblasts, *J. Cell. Physiol.,* 125, 465, 1985.
413. **Fleming, W. H., Murphy, P. R., Murphy, L. J., Hatton, T. W., Matusik, R. J., and Friesen, H. G.,** Human growth hormone induces and maintains c-*myc* gene expression in Nb2 lymphoma cells, *Endocrinology,* 117, 2547, 1985.
414. **Vandenbark, G. R., Kuhn, L. J., and Niedel, J. E.,** Possible mechanism of phorbol diester-induced maturation of human promyelocytic leukemia cells, *J. Clin. Invest.,* 73, 448, 1984.
415. **Grosso, L. E. and Pitot, H. C.,** Transcriptional regulation of c-*myc* during chemically induced differentiation of HL-60 cultures, *Cancer Res.,* 45, 847, 1985.
416. **Grosso, L. E. and Pitot, H. C.,** Chromatin structure of the c-*myc* gene in HL-60 cells during alterations of transcriptional activity, *Cancer Res.,* 45, 5035, 1985.
417. **Faletto, D. L., Arrow, A. S., and Macara, I. G.,** An early decrease in phosphatidylinositol turnover occurs on induction of Friend cell differentiation and precedes the decrease in c-*myc* expression, *Cell,* 43, 315, 1985.
418. **de Bustros, A., Baylin, S. B., Berger, C. L., Roos, B. A., Leong, S. S., and Nelkin, B. D.,** Phorbol esters increase calcitonin gene transcription and decrease c-*myc* mRNA levels in cultured human medullary thyroid carcinoma, *J. Biol. Chem.,* 260, 98, 1985.
419. **Thiele, C. J., Reynolds, C. P., and Israel, M. A.,** Decreased expression of N-*myc* precedes retinoic acid-induced morphological differentiation of human neuroblastoma, *Nature (London),* 313, 404, 1985.
420. **Amatruda, T. T., III, Sidell, N., Ranyard, J., and Koeffler, H. P.,** Retinoic acid treatment of human neuroblastoma cells is associated with decreased N-*myc* expression, *Biochem. Biophys. Res. Comm.,* 126, 1189, 1985.
421. **Einat, M., Resnitzky, D., and Kimchi, A.,** Close link between reduction of c-*myc* expression by interferon and G_0/G_1 arrest, *Nature (London),* 313, 597, 1985.
422. **Tsuda, T., Kaibuchi, K., West, B., and Takai, Y.,** Involvement of Ca^{2+} in platelet-derived growth factor-induced expression of c-*myc* oncogene in Swiss 3T3 fibroblasts, *FEBS Lett.,* 187, 43, 1985.
423. **Sefton, B. M., Hunter, T., and Raschke, W. C.,** Evidence that the Abelson virus protein functions *in vivo* as a protein kinase that phosphorylates tyrosine, *Proc. Natl. Acad. Sci. U.S.A.,* 78, 1552, 1981.
424. **Fry, M. J., Gebhardt, A., Parker, P. J., and Foulkes, J. G.,** Phosphatidylinositol turnover and transformation of cells by Abelson murine leukaemia virus, *EMBO J.,* 4, 3173, 1985.
425. **Blackshear, P. J., Witters, L. A., Girard, P. R., Kuo, J. F., and Quamo, S. N.,** Growth factor-stimulated protein phosphorylation in 3T3-L1 cells: evidence for protein kinase C-dependent and -independent pathways, *J. Biol. Chem.,* 260, 13304, 1985.
426. **Hunter, T. and Sefton, B. M.,** Transforming gene product of Rous sarcoma virus phosphorylates tyrosine, *Proc. Nat. Acad. Sci. U.S.A.,* 77, 1311, 1980.
427. **Malkinson, A. M. and McSwigan, C. E.,** Protein phosphorylation in normal and neoplastic development: phosphorylation of proteins endogenous to foetal tissues and tumours, *Biochem. J.,* 172, 423, 1978.

428. **Senger, D. R., Wirth, D. F., and Hynes, R. O.**, Transformed mammalian cells secrete specific proteins and phosphoproteins, *Cell*, 16, 885, 1979.
429. **Senger, D. R. and Perruzzi, C. A.**, Secreted phosphoprotein markers for neoplastic transformation of human epithelial and fibroblastic cells, *Cancer Res.*, 45, 5818, 1985.
430. **Phan-Dinh-Tuy, F., Henry, J., Boucheix, C., Perrot, J. Y., Rosenfeld, C., and Kahn, A.**, Protein kinases in human leukemic cells, *Am. J. Hematol.*, 19, 209, 1985.
431. **Collett, M. S. and Erikson, R. L.**, Protein kinase activity associated with the avian sarcoma virus *src* gene product, *Proc. Natl. Acad. Sci. U.S.A.*, 75, 2021, 1978.
432. **Erikson, E., Cook, R., Miller, G. J., and Erikson, R. L.**, The same normal cell protein is phosphorylated after transformation by avian sarcoma viruses with unrelated transforming genes, *Mol. Cell. Biol.*, 1, 43, 1981.
433. **Cooper, J. A. and Hunter, T.**, Four different classes of retroviruses induce phosphorylation of tyrosines present in similar cellular proteins, *Mol. Cell. Biol.*, 1, 394, 1981.
434. **Patschinsky, T. and Sefton, B. M.**, Evidence that there exist four classes of RNA tumor viruses which encode proteins associated with tyrosine protein kinase activity, *J. Virol.*, 39, 104, 1981.
435. **Cooper, J. A. and Hunter, T.**, Regulation of cell growth and transformation by tyrosine-specific protein kinases: the search for important cellular substrate proteins, *Curr. Top. Microbiol. Immunol.*, 107, 125, 1983.
436. **Heldin, C-H. and Westermark, B.**, Growth factors: mechanism of action and relation to oncogenes, *Cell*, 37, 9, 1984.
437. **Gilmore, T., DeClue, J. E., and Martin, G. S.**, Protein phosphorylation at tyrosine is induced by the v-*erbB* gene product in vivo and in vitro, *Cell*, 40, 609, 1985.
438. **Sefton, B. M.**, Oncogenes encoding protein kinases, *Trends Genet.*, 1, 306, 1985.
439. **Snyder, M. A., Bishop, J. M., McGrath, J. P., and Levinson, A. D.**, A mutation at the ATP-binding site of $pp60^{v-src}$ abolishes kinase activity, transformation, and tumorigenicity, *Mol. Cell. Biol.*, 5, 1772, 1985.
440. **Uehara, Y., Hori, M., Takeuchi, T., and Umezawa, H.**, Screening of agents which convert ''transformed morphology'' of Rous sarcoma virus-infected rat kidney cells to ''normal morphology'': identification of an active agent as herbimycin and its inhibition of intracellular *src* kinase, *Jpn. J. Cancer Res.*, 76, 675, 1985.
441. **Bolen, J. B., Rosen, N., and Israel, M. A.**, Increased $pp60^{c-src}$ tyrosyl kinase activity in human neuroblastomas is associated with amino-terminal tyrosine phosphorylation of the *src* gene product, *Proc. Natl. Acad. Sci. U.S.A.*, 82, 7275, 1985.
442. **Wang, J. Y. J.**, Isolation of antibodies for phosphotyrosine by immunization with a v-*abl* oncogene-encoded protein, *Mol. Cell. Biol.*, 5, 3640, 1985.
443. **Konopka, J. B., Watanabe, S. M., and Witte, O. N.**, An alteration of the human c-*abl* protein in K562 leukemia cells unmasks associated tyrosine kinase activity, *Cell*, 37, 1035, 1984.
444. **Davis, R. L., Konopka, J. B., and Witte, O. N.**, Activation of the c-*abl* oncogene by viral transduction or chromosomal translocation generates altered c-*abl* proteins with similar in vitro kinase properties, *Mol. Cell. Biol.*, 5, 204, 1985.
445. **Kloetzer, M., Kurzrock, R., Smith, L., Talpaz, M., Spiller, M., Gutterman, J., and Arlinghaus, R.**, The human cellular *abl* gene product in the chronic myelogenous leukemia cell line K562 has an associated tyrosine protein kinase activity, *Virology*, 140, 230, 1985.
446. **Konopka, J. B., Watanabe, S. M., Singer, J. W., Collins, S. J., and Witte, O. N.**, Cell lines and clinical isolates derived from Ph¹-positive chronic myelogenous leukemia patients express c-*abl* proteins with a common structural alteration, *Proc. Natl. Acad. Sci. U.S.A.*, 82, 1810, 1985.
447. **Konopka, J. B. and Witte, O. N.**, Detection of c-*abl* tyrosine kinase activity in vitro permits direct comparison of normal and altered *abl* gene products, *Mol. Cell. Biol.*, 5, 3116, 1985.
448. **Konopka, J. B. and Witte, O. N.**, Activation of the *abl* oncogene in murine and human leukemias, *Biochim. Biophys. Acta*, 823, 1, 1985.
449. **Stam, K., Heisterkamp, N., Grosveld, G., de Klein, A., Verma, R. S., Coleman, M., Dosik, H., and Groffen, J.**, Evidence of a new chimeric bcr/c-*abl* mRNA in patients with chronic myelocytic leukemia and the Philadelphia chromosome, *N. Engl. J. Med.*, 313, 1429, 1985.
450. **Dean, M., Park, M., Le Beau, M. M., Robins, T. S., Diaz, M. O., Rowley, J. D., Blair, D. G., and Vande Woude, G. F.**, The human *met* oncogene is related to the tyrosine kinase oncogenes, *Nature (London)*, 318, 385, 1985.
451. **Kloetzer, W. S., Maxwell, S. A., and Arlinghaus, R. B.**, Further characterization of the $p85^{gag-mos}$-associated protein kinase activity, *Virology*, 138, 143, 1984.
452. **Moelling, K., Heimann, B., Beimling, P., Rapp, U. R., and Sander, T.**, Serine- and threonine-specific protein kinase activities of purified gag-mil and gag-raf proteins, *Nature (London)*, 312, 558, 1984.
453. **Maxwell, S. A. and Arlinghaus, R. B.**, Serine kinase activity associated with Moloney murine sarcoma virus-124-encoded $p37^{mos}$, *Virology*, 143, 321, 1985.

454. **Maxwell, S. A. and Arlinghaus, R. B.**, cAMP-independent serine threonine kinase activity is associated with the *mos* sequences of ts 110 Moloney murine sarcoma virus-encoded p85 *gag-mos*, *J. Gen. Virol.*, 66, 2135, 1985.
455. **Kamps, M. P., Taylor, S. S., and Sefton, B. M.**, Direct evidence that oncogenic tyrosine kinases and cyclic AMP-dependent protein kinase have homologous ATP-binding sites, *Nature (London)*, 310, 589, 1984.
456. **Wadsworth, S. C., Madhavan, K., and Bilodeau-Wentworth, D.**, Maternal inheritance of transcripts from three *Drosophila src*-related genes, *Nucleic Acids Res.*, 13, 2153, 1985.
457. **Simon, M. A., Drees, B., Kronberg, T., and Bishop, J. M.**, The nucleotide sequence and tissue-specific expression of Drosophila c-*src*, *Cell*, 42, 841, 1985.
458. **Kasuga, M., Zick, Y., Blithe, D. L., Crettaz, M., and Kahn, C. R.**, Insulin stimulates tyrosine phosphorylation of the insulin receptor in a cell-free system, *Nature (London)*, 298, 667, 1982.
459. **Kasuga, M., Fujita-Yamaguchi, T., Blithe, D. L., and Kahn, C. R.**, Tyrosine-specific protein kinase activity is associated with the purified insulin receptor, *Proc. Natl. Acad. Sci. U.S.A.*, 80, 2137, 1983.
460. **Rubin, J. B., Shia, M. A., and Pilch, P. F.**, Stimulation of tyrosine-specific phosphorylation *in vitro* by insulin-like growth factor I, *Nature (London)*, 305, 438, 1983.
461. **Hunter, T. and Cooper, J. A.**, Epidermal growth factor induces rapid tyrosine phosphorylation of proteins in A431 tumor cells, *Cell*, 24, 741, 1981.
462. **Reynolds, F. H., Todaro, G. J., Fryling, C., and Stephenson, J. R.**, Human transforming growth factors induce tyrosine phosphorylation of EGF receptors, *Nature (London)*, 292, 259, 1981.
463. **Pike, L. J., Marquardt, H., Todaro, G. J., Gallis, B., Casnellie, J. E., Bornstein, P., and Krebs, E. G.**, Transforming growth factor and epidermal growth factor stimulate the phosphorylation of a synthetic, tyrosine-containing peptide in a similar manner, *J. Biol. Chem.*, 257, 14628, 1982.
464. **Ek, B. and Heldin, C-H.**, Characterization of a tyrosine-specific kinase activity in human fibroblast membranes stimulated by platelet-derived growth factor, *J. Biol. Chem.*, 257, 10486, 1982.
465. **Migliaccio, A., Rotondi, A., and Auricchio, F.**, Calmodulin-stimulated phosphorylation of 17 beta-estradiol receptor on tyrosine, *Proc. Natl. Acad. Sci. U.S.A.*, 81, 5921, 1984.
466. **Ganguly, C. L., Dale, J. B., Courtney, H. S., and Beachey, E. H.**, Tyrosine phosphorylation of a 94-kDa protein of human fibroblasts stimulated by streptococcal lipoteichoic acid, *J. Biol. Chem.*, 260, 13342, 1985.
467. **Olson, J. W.**, Liver tyrosine kinase activation during early stages of chemical hepatocarcinogenesis, *J. Cell. Biochem.*, 27, 175, 1985.
468. **Bishop, R., Martinez, R., Nakamura, K. D., and Weber, M. J.**, A tumor promoter stimulates phosphorylation on tyrosine, *Biochem. Biophys. Res. Comm.*, 115, 536, 1983.
469. **Grunberger, G., Zick, Y., Taylor, S. I., and Gorden, P.**, Tumor-promoting phorbol ester stimulates tyrosine phosphorylation in U-937 monocytes, *Proc. Natl. Acad. Sci. U.S.A.*, 81, 2762, 1984.
470. **Srivastava, A. K.**, Stimulation of tyrosine protein kinase activity by dimethyl sulfoxide, *Biochem. Biophys. Res. Comm.*, 126, 1042, 1985.
471. **Michiel, D. F. and Wang, J. H.**, Stimulation of tyrosine phosphorylation of rat brain membrane proteins by calmodulin, *FEBS Lett.*, 190, 11, 1985.
472. **Klein, H. H., Freidenberg, G. R., Cordera, R., and Olefsky, J. M.**, Substrate specificities of insulin and epidermal growth factor receptor kinases, *Biochem. Biophys. Res. Comm.*, 127, 254, 1985.
473. **Pike, L. J., Kuenzel, E. A., Casnellie, J. E., and Krebs, E. G.**, A comparison of the insulin- and epidermal growth factor-stimulated protein kinases from human placenta, *J. Biol. Chem.*, 259, 9913, 1984.
474. **Kuzumaki, N., Yamagiwa, S., and Oikawa, T.**, No expression of a Rous sarcoma virus-induced tumor antigen in mammalian cells infected with retroviruses transducing other oncogenes of the *src* gene family, *J. Natl. Cancer Inst.*, 74, 889, 1985.
475. **Maher, P. A., Pasquale, E. B., Wang, J. Y. J., and Singer, S. J.**, Phosphotyrosine-containing proteins are concentrated in focal adhesions and intercellular junctions in normal cells, *Proc. Natl. Acad. Sci. U.S.A.*, 82, 6576, 1985.
476. **Frackelton, A. R. Jr., Tremble, P. M., and Williams, L. T.**, Evidence for the platelet-derived growth factor-stimulated tyrosine phosphorylation of the platelet-derived growth factor receptor *in vivo*: immunopurification using a monoclonal antibody to phosphotyrosine, *J. Biol. Chem.*, 259, 7909, 1984.
477. **Cooper, J. A. and Hunter, T.**, Major substrate for growth factor-activated protein-tyrosine kinases is a low-abundance protein, *Mol. Cell. Biol.*, 5, 3304, 1985.
478. **Wong, T. W. and Goldberg, A. R.**, Tyrosyl protein kinases in normal rat liver: identification and partial characterization, *Proc. Natl. Acad. Sci. U.S.A.*, 80, 2529, 1983.
479. **Wong, T. W. and Goldberg, A. R.**, Purification and characterization of the major species of tyrosine protein kinase in rat liver, *J. Biol. Chem.*, 259, 8505, 1984.
480. **Yoshikawa, K., Usui, H., Imazu, M., Tsukamoto, H., and Takeda, M.**, Comparison of tyrosine protein kinases in membrane fractions from mouse liver and Ehrlich ascites tumor, *J. Biol. Chem.*, 260, 15091, 1985.

481. **Galski, H., De Groot, N., and Hochberg, A. A.**, Phosphorylation of tyrosine in cultured human placenta, *Biochim. Biophys. Acta,* 761, 284, 1983.
482. **Dangott, L. J., Puett, D., Garbers, D. L., and Melner, M. H.**, Tyrosine protein kinase activity in purified rat Leydig cells, *Biochem. Biophys. Res. Comm.,* 116, 400, 1983.
483. **Swarup, G., Dasgupta, J. D., and Garbers, D. L.**, Tyrosine protein kinase activity of rat spleen and other tissues, *J. Biol. Chem.,* 258, 10341, 1983.
484. **Nakamura, S., Takeuchi, F., Kondo, H., and Yamamura, H.**, High tyrosine protein kinase activities in soluble and particulate fractions in bone marrow cells, *FEBS Lett.,* 170, 139, 1984.
485. **Gacon, G., Piau, J-P., Blaineau, C., Fagard, R., Genetet, N., and Fischer, S.**, Tyrosine phosphorylation in human T lymphoma cells, *Biochem. Biophys. Res. Comm.,* 117, 843, 1983.
486. **Harrison, M. L., Low, P. S., and Geahlen, R. L.**, T and B lymphocytes express distinct tyrosine protein kinases, *J. Biol. Chem.,* 259, 9348, 1984.
487. **Earp, H. S., Austin, K. S., Gillespie, G. Y., Buesson, S. C., Davies, A. A., and Parker, P. J.**, Characterization of distinct tyrosine-specific protein kinases in B and T lymphocytes, *J. Biol. Chem.,* 260, 4351, 1985.
488. **Tuy, F. P. D., Henry, J., Rosenfeld, C., and Kahn, A.**, High tyrosine kinase activity in normal nonproliferating cells, *Nature (London),* 305, 435, 1983.
489. **Kuratsune, H., Owada, K., Machii, H., Nishimori, Y., Ueda, E., Tokumine, Y., Tagawa, S., Taniguchi, N., Fujio, H., and Kitani, T.**, Association of specific tyrosine phosphorylation with stages of B-cell differentiation in human lymphoid leukemias, *Biochem. Biophys. Res. Comm.,* 133, 598, 1985.
490. **Levy, B. T., Sorge, L. K., Meymandi, A., and Maness, P. F.**, pp60^{c-src} kinase is in chick and human embryonic tissues, *Devel. Biol.,* 104, 9, 1984.
491. **Sorge, L. K., Levy, B. T., and Maness, P. F.**, pp60^{c-src} is developmentally regulated in the neural retina, *Cell,* 36, 249, 1984.
492. **Barnekow, A. and Bauer, H.**, The differential expression of the cellular *src*-gene product pp60^{src} and its phosphokinase activity in normal chicken cells and tissues, *Biochim. Biophys. Acta,* 782, 94, 1984.
493. **Fults, D. W., Towle, A. C., Lauder, J. M., and Maness, P. F.**, pp60^{c-src} in the developing cerebellum, *Mol. Cell. Biol.,* 5, 27, 1985.
494. **Neer, E. J. and Lok, J. M.**, Partial purification and characterization of a pp60^{v-src}-related tyrosine kinase from bovine brain, *Proc. Natl. Acad. Sci. U.S.A.,* 82, 6025, 1985.
495. **Brugge, J. S., Cotton, P. C., Queral, A. E., Barrett, J. N., Nonner, D., and Keane, R. W.**, Neurones express high levels of a structurally modified, activated form of pp60^{c-src}, *Nature (London),* 316, 554, 1985.
496. **Alemà, S., Casalbore, P., Agostini, E., and Tatò, F.**, Differentiation of PC12 phaeochromocytoma cells induced by v-*src* oncogene, *Nature (London),* 316, 557, 1985.
497. **Sefton, B. M., Hunter, T., Beemon, K., and Eckhart, W.**, Evidence that the phosphorylation of tyrosine is essential for cellular transformation by Rous sarcoma virus, *Cell,* 20, 807, 1980.
498. **Castellanos, R. M. P., and Mazon, M. J.**, Identification of phosphotyrosine in yeast proteins and of a protein tyrosine kinase associated with the plasma membrane, *J. Biol. Chem.,* 260, 8240, 1985.
499. **Schieven, G., Thorner, J., and Martin, G. S.**, Protein tyrosine kinase activity in *Saccharomyces cerevisiae, Science,* 231, 390, 1986.
500. **Vallejos, R. H., Holuigue, L., Lucero, H. A., and Torruella, M.**, Evidence of tyrosine kinase activity in the photosynthetic bacterium *Rhodospirillium rubrum, Biochem. Biophys. Res. Comm.,* 126, 678, 1985.
501. **Montagnier, L., Chamaret, S., and Dauguet, C.**, Augmentation, dans des extraits de cellules cancéreuses ou transformées, de l'activité d'une protéine-kinase phosphorylant la tyrosine, *C. R. Acad. Sci. Paris,* 295, 375, 1982.
502. **Wickremasinghe, R. G., Piga, A., Mire, A. R., Reza Taheri, M., Yaxley, J. C., and Hoffbrand, A. V.**, Tyrosine protein kinases and their substrates in human leukemia cells, *Leukemia Res.,* 9, 1443, 1985.
503. **Marth, J. D., Peet, R., Krebs, E. G., and Perlmutter, R. M.**, A lymphocyte-specific protein-tyrosine kinase gene is rearranged and overexpressed in the murine T cell lymphoma LSTRA, *Cell,* 43, 393, 1985.
504. **Fischer, S., Flamand, A., Fagard, R., Chich, F., Piau, J.-P., Reibel, L., and Gacon, G.**, Vesicular stomatitis virus produced from infected LSTRA lymphoma cells bear tyrosine protein kinase activity (p56), *J. Biol. Chem.,* 260, 14406, 1985.
505. **Casnellie, J. E., Harrison, M. L., Hellström, K. E, and Krebs, E. G.**, A lymphoma cell line expressing elevated levels of tyrosine protein kinase activity, *J. Biol. Chem.,* 258, 10738, 1983.
506. **Marchildon, G. A., Casnellie, J. E., Walsh, K. A., and Krebs, E. G.**, Covalently bound myristate in a lymphoma tyrosine kinase, *Proc. Natl. Acad. Sci. U.S.A.,* 81, 7679, 1984.
507. **Gacon, G., Piau, J. P., Blaineau, C., Fagard, Genetet, N., and S.**, Tyrosine phosphorylation in human T lymphoma cells, *Biochem. Biophys. Res. Comm.,* 117, 843, 1983.
508. **Ogawa, R., Ohtsuka, M., Watanabe, Y., Noguchi, K., Arimori, S., and Sasadaira, H.**, Immunofluorescent staining of human leukemic cells with monoclonal antibody to phosphotyrosine, *Jpn. J. Cancer Res.,* 76, 567, 1985.

509. **Piga, A., Taheri, M. R., Yaxley, J. C., Wickremasinghe, R. G., and Hoffbrand, A. V.,** High tyrosine protein kinase activity in resting lymphocytes than in proliferating normal or leukaemic blood cells, *Biochem. Biophys. Res. Comm.,* 124, 766, 1984.
510. **Swarup, G. and Subrahmanyam, G.,** Activation of a cellular tyrosine-specific protein kinase by phosphorylation, *FEBS Lett.,* 188, 131, 1985.
511. **Voronova, A. F., Buss, J. E., Patschinsky, T., Hunter, T., and Sefton, B. M.,** Characterization of the protein apparently responsible for the elevated tyrosine protein kinase activity in LSTRA cells, *Mol. Cell. Biol.,* 4, 2705, 1984.
512. **Pavloff, N., Biquard, J-M., Hanania, N., and Semmel, M.,** Isolation of proteins with kinase activity and related to pp60 src from human cells, *Biochem. Biophys. Res. Comm.,* 121, 779, 1984.
513. **Buss, J. E. and Sefton, B. M.,** Myristic acid, a rare fatty acid, is the lipid attached to the transforming protein of Rous sarcoma virus and its cellular homolog, *J. Virol.,* 53, 7, 1985.
514. **Parker, R. C., Varmus, H. E., and Bishop, J. M.,** Expression of v-src and chicken c-*src* in rat cells demonstrates qualitative differences between pp60^{v-src} and pp60^{c-src}, *Cell,* 37, 131, 1984.
515. **Iba, H., Takeya, Cross, F. R., Hanafusa, T., and Hanafusa, H.,** Rous sarcoma virus variants that carry the cellular *src* gene instead of the viral *src* gene cannot transform chicken embryo fibroblasts, *Proc. Natl. Acad. Sci. U.S.A.,* 81, 4424, 1984.
516. **Shalloway, D., Coussens, P. M., and Yaciuk, P.,** Overexpression of the c-*src* protein does not induce transformation of NIH 3T3 cells, *Proc. Natl. Acad. Sci. U.S.A.,* 81, 7071, 1984.
517. **Iba, H., Cross, F. R., Garber, E. A., and Hanafusa, H.,** Low level of cellular protein phosphorylation by nontransforming overproduced p60^{c-src}, *Mol. Cell. Biol.,* 5, 1058, 1985.
518. **Dasgupta, J. D. and Garbers, D. L.,** Tyrosine protein kinase activity during embryogenesis, *J. Biol. Chem.,* 258, 6174, 1983.
519. **Nakamura, K. D., Martinez, R., and Weber, M. J.,** Tyrosine phosphorylation of specific proteins after mitogen stimulation of chicken embryo fibroblasts, *Mol. Cell. Endocrinol.,* 3, 380, 1983.
520. **Gentleman, S., Russell, P., Martensen, T. M., and Chader, G. J.,** Characteristics of protein tyrosine kinase activities of Y-79 retinoblastoma cells and retina, *Arch. Biochem. Biophys.,* 239, 130, 1985.
521. **Lee, W-H., Murphree, A. L., and Benedict, W. F.,** Expression and amplification of the N-*myc* gene in primary retinoblastoma, *Nature (London),* 309, 458, 1984.
522. **Swarup, G., Cohen, S., and Garbers, D. L.,** Selective dephosphorylation of proteins containing phosphotyrosine by alkaline phosphatases, *J. Biol. Chem.,* 256, 8197, 1981.
523. **Swarup, G., Speeg, K. V., Jr., Cohen, S., and Garbers, D. L.,** Phosphotyrosyl-protein phosphatase of TCRC-2 cells, *J. Biol. Chem.,* 257, 7298, 1982.
524. **Leis, J. F. and Kaplan, N. O.,** An acid phosphatase in the plasma membrane of human astrocytoma showing marked specificity toward phosphotyrosine protein, *Proc. Natl. Acad. Sci. U.S.A.,* 79, 6507, 1982.
525. **Foulkes, J. G., Erikson, E., and Erikson, R. L.,** Separation of multiple phosphotyrosyl- and phosphoseryl-protein phosphatases from chicken brain, *J. Biol. Chem.,* 258, 431, 1983.
526. **Shriner, C. L. and Brautigan, D. L.,** Cytosolic protein phosphotyrosine phosphatases from rabbit kidney: purification of two distinct enzymes that bind to Zn^{2+}-iminodiacetate agarose, *J. Biol. Chem.,* 259, 11383, 1984.
527. **Nelson, R. L. and Branton, P. E.,** Identification, purification, and characterization of phosphotyrosine-specific protein phosphatases from cultured chicken embryo fibroblasts, *Mol. Cell. Biol.,* 4, 1003, 1984.
528. **Klarlund, J. K.,** Transformation of cells by an inhibitor of phosphatases acting on phosphotyrosine in proteins, *Cell,* 41, 707, 1985.
529. **Martin, G. S., Radke, K., Carter, C., Moss, P., Dehazya, P., and Gilmore, T.,** The role of protein phosphorylation at tyrosine in transformation and mitogenesis, *J. Cell. Physiol.,* 3, 139, 1984.
530. **Zajac, J. D., Livesey, S. A., and Martin, T. J.,** Selective activation of cyclic AMP dependent protein kinase by calcitonin in a calcitonin secreting lung cancer cell line, *Biochem. Biophys. Res. Comm.,* 122, 1040, 1984.
531. **Perisic, O. and Traugh, J. A.,** Protease-activated kinase II as the mediator of epidermal growth factor-stimulated phosphorylation of ribosomal protein S6, *FEBS Lett.,* 183, 215, 1985.
532. **Fontana, J. A., Emler, C., Ku, K., McClung, J. K., Butcher, F. R., and Durham, J. P.,** Cyclic AMP-dependent and -independent protein kinases and protein phosphorylation in human promyelocytic leukemia (HL60) cells induced to differentiate by retinoic acid, *J. Cell. Physiol.,* 120, 49, 1984.
533. **Yasui, W., Sumiyoshi, H., Ochiai, A., Yamahara, M., and Tahara, E.,** Type I and II cyclic adenosine 3′:5′-monophosphate-dependent protein kinase in human gastric mucosa and carcinomas, *Cancer Res.,* 45, 1565, 1985.
534. **Papkoff, J., Nigg, E. A., and Hunter, T.,** The transforming protein of Moloney murine sarcoma virus is a soluble cytoplasmic protein, *Cell,* 33, 161, 1983.
535. **Barker, W. C. and Dayhoff, M. O.,** Viral src gene products are related to the catalytic chain of mammalian cAMP-dependent protein kinase, *Proc. Natl. Acad. Sci. U.S.A.,* 79, 2836, 1982.

536. **Baldwin, G. S.**, Epidermal growth factor precursor is related to the translation product of the Moloney sarcoma virus oncogene *mos*, *Proc. Natl. Acad. Sci. U.S.A.*, 82, 1921, 1985.
537. **Propst, F. and Vande Woude, G. F.**, Expression of c-*mos* proto-oncogene transcripts in mouse tissues, *Nature (London)*, 315, 516, 1985.
538. **Maller, J. L., Foulkes, J. G., Erikson, E., and Baltimore, D.**, Phosphorylation of ribosomal protein S6 on serine after microinjection of the Abelson murine leukemia virus tyrosine-specific protein kinase into *Xenopus* oocytes, *Proc. Natl. Acad. Sci. U.S.A.*, 82, 272, 1985.
539. **Blenis, J., Spivack, J. G., and Erikson, R. L.**, Phorbol ester, serum, and Rous sarcoma virus transforming gene product induce similar phosphorylations of ribosomal protein S6, *Proc. Natl. Acad. Sci. U.S.A.*, 81, 6408, 1984.
540. **Blenis, J. and Erikson, R. L.**, Regulation of a ribosomal protein S6 kinase activity by the Rous sarcoma virus transforming protein, serum, or phorbol ester, *Proc. Natl. Acad. Sci. U.S.A.*, 82, 7621, 1985.
541. **Kruse, C., Johnson, S. P., and Warner, J. R.**, Phosphorylation of the yeast equivalent of ribosomal protein S6 is not essential for growth, *Proc. Natl. Acad. Sci. U.S.A.*, 82, 7515, 1985.
542. **Talha, S. and Harel, L.**, Early stimulation of ATP turnover induced by growth factors: synergistic effect of EGF and insulin and correlation with DNA synthesis, *Exp. Cell Res.*, 158, 311, 1985.
543. **Durban, E., Goodenough, M., Mills, J., and Busch, H.**, Topoisomerase I phosphorylation *in vitro* and in rapidly growing Novikoff hepatoma cells, *EMBO J.*, 4, 2921, 1985.
544. **Durban, E., Mills, J. S., Roll, D., Busch, H.**, Phosphorylation of purified Novikoff hepatoma topoisomerase I, *Biochem. Biophys. Res. Comm.*, 11, 897, 1983.
545. **Miskimins, R., Miskimins, W. K., Bernstein, H., and Shimizu, N.**, Epidermal growth factor-induced topoisomerase(s): intracellular translocation and relation to DNA synthesis, *Exp. Cell Res.*, 146, 53, 1983.
546. **Mroczkowski, B., Mosig, G., and Cohen, S.**, ATP-stimulated interaction between epidermal growth factor receptor and supercoiled DNA, *Nature (London)*, 309, 270, 1984.
547. **Tse-Dinh, Y-C., Wong, T. W., and Goldberg, A. R.**, Virus- and cell-encoded tyrosine protein kinases inactivate DNA topoisomerases *in vitro*, *Nature (London)*, 312, 785, 1984.
548. **Basu, M., Frick, K., Sen-Majumdar, A., Scher, C. D., and Das, M.**, EGF receptor-associated DNA-nicking activity is due to a M_r-100,000 dissociable protein, *Nature (London)*, 316, 640, 1985.
549. **Yang, L., Rowe, T. C., and Liu, L. F.**, Identification of DNA topoisomerase II as an intracellular target of antitumor epipodophyllotoxins in simian virus 40-infected monkey cells, *Cancer Res.*, 45, 5872, 1985.
550. **Priel, E., Aboud, M., Feigelman, H., and Segal, S.**, Topoisomerase-II activity in human leukemic and lymphoblastoid cells, *Biochem. Biophys. Res. Comm.*, 130, 325, 1985.
551. **Maytin, E. V., Balduzzi, P. C., Notter, M. F. D., and Young, D. A.**, Changes in the synthesis and phosphorylation of cellular proteins in chick fibroblasts transformed by two avian sarcoma viruses, *J. Biol. Chem.*, 259, 12135, 1984.

Chapter 3

INSULIN

I. INTRODUCTION

Insulin is an anabolic hormone with a wide range of effects on metabolic functions, including stimulation of the synthesis of DNA, RNA, proteins, glycogen, and lipids. In addition, insulin has important actions on the control mechanisms of cell proliferation and cell differentiation both in vivo and in vitro.[1-3] The molecular structure of human insulin is well characterized.[4] More recently, there have been important advances in the characterization of cellular insulin receptors.[5,6]

II. INSULIN STRUCTURE AND THE INSULIN GENE

Insulin is a polypeptide hormone consisting of 51 amino acids arranged in two chains (A and B) linked by disulfide bonds. Insulin is normally synthesized in the beta cells of the pancreatic islets of Langerhans and is secreted into the blood in variable amounts according to the action of a diversity of endogeneous and exogenous stimuli.

A. The Human Insulin Gene and Insulin Biosynthesis

The structural gene for insulin (INS) is located on the short arm of human chromosome 11, region chromosome 11p14.1, which corresponds to the same region where the human c-H-*ras*-1 oncogene has been located.[7] Human chromosome 11 also contains in its short arm the genes coding for parathormone (PTH) and beta-globin (HBG). The linear order of genes in human chromosome 11p is probably cen-PTH-HBG-(H-*ras*-1-INS)-pter.[7,8] The distance between c-*ras*-1 and INS genes is approximately 8 cmorgans but it is not known which of these two genes is more proximal to the centromere. Deletions of chromosome 11 at location 11p13-14 have been detected in the tumor cells of children with the aniridia-Wilm's tumor association (AWTA anomaly).[9,10] The deletion may resulting either hemizygosity or homozygosity of the corresponding allele, the latter being produced by duplication of the respective DNA segment remaining in the unaffected chromosome.[11-14] In some patients with Wilms' tumor one allele of the insulin gene, as well as one allele of the c-H-*ras* proto-oncogene, are deleted on the tumor cells.[13,15] Deletion of the same chromosome region (11p13-14) has also been detected in the tumor cells of a human hepatocellular carcinoma in relation to the integration of an hepatitis B-virus (HBV) genome copy.[16] A similar deletion may occur in bladder cancer.[17] The biological significance of such partial deletions of human chromosome 11 are not understood.

1. Structure of the Insulin Gene

The complete nucleotide sequence of the human insulin gene has been determined.[18-22] The human insulin gene is composed of two exons and two introns (intervening sequences) and is associated with elements of the *Alu* family of short (<400 bp) interspersed repeated DNA sequences. Tandemly repeating DNA sequences have been identified in the 5' flanking region of the human insulin gene, with specific allelic sequences showing different frequencies among the human population.[23,24] It has been suggested that particular types of polymorphisms in the 5' flanking region of the human insulin gene may be associated with increased incidence of noninsulin-dependent diabetes mellitus,[25-28] but this suggestion has not been confirmed.[29,30] Particular types of mutation of the human insulin coding DNA sequences, resulting in amino acid substitutions in the insulin molecule, may be associated with rare cases of diabetes mellitus.[31] Such cases have been described under the generic name of insulinopathies.

While the human genome contains only one functional insulin gene, two structural genes encoding for insulin are present in the genome of rat, mouse, and three fish species.[32] The rat preproinsulin gene I contains a single intron and is a functional semiprocessed gene generated by reinsertion into the genome (retroposition) of a cDNA copy of preproinsulin gene II transcripts.[33] In other words, the rat preproinsulin I gene is a functional retroposon. Insulin-I and insulin-II genes are asyntenic in the mouse but are syntenic in the rat, both being located on chromosome 1.[34] Point mutations close to the AUG initiator codon may affect the efficiency of translation of rat preproinsulin II mRNA in vivo.[35] This finding provided the first evidence that sequence context influences the ability of an AUG triplet to be recognized as an initiator codon by eukaryotic ribosomes.

Rat insulin-I locus show polymorphism due to the presence or absence of a 2.7-kb repeated DNA element in a region located 2.2 kb upstream of the coding sequence.[36] The biological significance of DNA polymorphisms due to the insertion of short or long repeated nucleotide sequences is unknown but the presence of such elements could potentially exert profound effects on the contiguous DNA sequences, including effects on gene transcriptional activity and the occurrence of structural changes such as deletion, duplication, and other types of DNA rearrangements.

2. Insulin Biosynthesis in Metazoans and Microbes

Insulin is synthesized in the form of a high-molecular weight precursor, preproinsulin, which is processed via an intermediate precursor, proinsulin, to the mature insulin molecule. In the human and other mammals insulin is normally synthesized in, and secreted by, the beta cells of the islets of Langerhans in the pancreas. There is, however, evidence that insulin may be synthesized in a diversity of mammalian organs and tissues, including the lung, the intestine, and the central nervous system, although these extra-beta cell sites would lack the capacity to secrete the hormone.[37-39]

Insulin is present in insects such as *Drosophila melanogaster* and annelids such as the earthworm *Annelida oligocheta*.[40] The amino-terminal amino acid sequence of the prothoracicotropic hormone of the adult silkworm *Bombyx mori* exhibit significant homology with insulin and IGFs.[41] Moreover, there is evidence that insulin, or a molecule closely related to insulin, is produced by microbes such as *E. coli* and *Tetrahymena*.[42] The possible role of the insulin-like material produced by microbes is unknown. An insulin-related gene has not been characterized as yet in microbes.

3. Pancreatic Beta-Cell Tumors in Transgenic Mice Expressing Recombinant Insulin/SV40 Genes

Recombinant genes composed of the upstream region of the rat preproinsulin gene II linked to sequences coding for SV40 large-T antigen have been transferred into fertilized mouse eggs.[43] SV40 large-T antigen was detected exclusively in the beta-cells of the endocrine pancreas of transgenic mice, whereas the alpha and delta cells normally found in the islets of Langerhans were rare and disordered. The transgenic animals developed hyperplasia of the islets followed by the formation of well vascularized beta-cell tumors. Expression of the hybrid insulin/SV40 gene is tissue-specific to the pancreas, being restricted to the beta-cells of the islets where the transferred insulin sequences specify correct expression. Apparently, the hybrid insulin/SV40 genes are expressed in all beta-cells, thereby inducing their proliferation, which leads to hyperplasia followed by oncogenic transformation.[43]

III. INSULIN RECEPTORS

The insulin receptor is a protein kinase located at the cell surface. The structural gene for human insulin receptor has been localized to chromosome 19, bands p13.2-13.3, a region which is frequently involved in pre-B-cell acute leukemia.[44] However, the physiological significance of this association, if any, is not understood.

A. Structure of Insulin Receptors

Insulin receptors on the cell surface are dimers with an estimated molecular weight of 350,000 daltons and are composed of two alpha subunits of 135,000 daltons and two beta subunits of 95,000 daltons linked by sulfhydryl groups as a (beta-s-s-alpha)-s-s-(alpha-s-s-beta) complex.[5,6] In addition to this alpha$_2$beta$_2$ form, the native insulin receptor, as characterized in rat hepatoma cells, exists under other different forms, including free-alpha and -beta subunits and the following combinations of disulfide-linked oligomers: alpha-beta, alpha$_2$, and alpha$_2$-beta.[45] Two hydrodynamic forms of the insulin receptor, termed R_I and R_{II}, correspond to larger and smaller oligomeric forms of the receptor, respectively, and the R_{II} receptor form contains a large proportion of the free receptor beta-subunit.[46] The possible physiological role of the different molecular forms of the insulin receptor in hepatoma cells is not known but the alpha and beta subunits maybe in close physical association in the plasma membrane when they are not linked by disulfide bonds. Since only the alpha chains of the receptor bind insulin and each receptor molecule has two alpha chains, it has been assumed that insulin receptor is bivalent for the ligand. Homogeneous bivalent insulin receptor has been purified using insulin coupled to 1,1'-carbonyldiimidazole-activated agarose.[47] However, there is evidence that the purified insulin receptor from human placenta binds only one molecule of insulin with high affinity.[48]

Disulfide bonds and sulfhydryl exchange reactions are involved in interconversion of different molecular forms of the insulin receptors as well as in the formation of groups of insulin receptors on the cell surface, which maybe important for the biological actions of insulin.[49-52] The insulin receptor is a high mannose glycoprotein but the exact arrangement of the carbohydrate residues of in the receptor molecule is unknown.[53]

B. Insulin Receptor Synthesis

Insulin receptor biosynthesis occurs from a precursor polypeptide.[54] A cDNA clone of approximately 5 kb corresponding to the coding sequences of the insulin receptor gene from human placenta has been constructed,[55] which will allow a better characterization of the structural basis for the transmembrane signaling activated by the hormone. The nucleotide sequence of the cDNA clone predicts a 1382-amino acid precursor of the human insulin receptor. The alpha subunit (735 amino acids) comprises the amino-terminal portion of the precursor and contains a cysteine-rich cross-linking domain. The beta subunit (620 amino acids), corresponding to the carboxy-terminal portion of the precursor, contains a transmembrane domain and a cytoplasmic domain with the structural elements for a tyrosine-specific protein kinase. The molecular weight of the polypeptide coded by the cDNA of the human insulin receptor is 153,917, whereas the estimated molecular weight of the receptor solubilized from human placenta is 220,000.[55] Thus a substantial fraction of the mass of the insulin receptor molecule must be carbohydrate. The cloned human insulin receptor cDNA has been expressed as a functional insulin receptor protein in Chinese hamster ovary cells.[56]

C. Formation and Phosphorylation of the Insulin Receptor Complex

The cellular actions of insulin are initiated by its binding to the specific receptors on the cell surface. Binding of insulin to its receptor is pH- and temperature-dependent but the reaction does not conform to a simple reversible bimolecular model, which is due to the existence of negative cooperativity in the hormone-receptor interaction as well as to internalization and degradation of the hormone-receptor complex. Furthermore, occupancy of binding sites is not stoichiometrically related to the biological effects of insulin and in most tissues there is an excess of insulin receptors, called "spare receptors".[57]

1. Insulin Receptor Phosphorylation

Insulin binds to the alpha subunit of the receptor and after this binding it stimulates the phosphorylation of the beta subunit of the receptor.[58-60] Insulin-induced receptor autophosphorylation occurs partially on tyrosine residues and the receptor acquires protein kinase activity with specificity for tyrosine residues of different cellular proteins.[61-66] In addition to tyrosine, at least another six sites of receptor phosphorylation are present, including serine and threonine residues.[67] While in cell-free systems insulin-dependent receptor phosphorylation is only observed on tyrosine residues, in intact cells receptor phosphorylation takes place essentially on serine residues and, in a lower extent, on tyrosine residues.[68] However, phosphorylation on tyrosine residues is one of the earliest molecular events occurring after insulin binding, preceding serine phosphorylation of the beta subunit,[69] and tyrosine autophosphorylation occurs in receptors containing little or no phosphoserine and phosphothreonine which suggests that receptor activity maybe regulated intracellularly by phosphorylation processes.[70]

Although it is generally believed that phosphorylation of the insulin receptor is associated in some way with its functional activation, the particular roles of phosphorylations occurring on specific tyrosine and nontyrosine residues of the receptor are not understood in molecular terms. Treatment of the purified insulin receptor with 1 mM dithiothreitol completely reduces disulfide linkages between the receptor subunits and it has been found that the monomeric alpha-beta form of the receptor exhibits much higher insulin-dependent kinase activity that the intact receptor in the alpha$_2$beta$_2$ form.[71] Removal of sialic acid from the purified insulin receptor results in enhanced binding and kinase activities.[72] The production of an antipeptide antibody that specifically inhibits insulin receptor autophosphorylation and protein kinase activity has been reported recently.[73]

2. Defective Insulin Receptors

The importance of insulin receptor phosphorylation for insulin action is suggested by the recognition that insulin resistance in a diabetic patient was associated with normal insulin binding but defective phosphorylation of the beta subunit of the receptor.[74] However, a selective defect in phosphorylation of the insulin receptor beta subunit is apparently not responsible for most cases of insulin resistance.[75] Insulin receptor-associated tyrosine kinase is also defective in insulin-resistant obese mice.[76] This defect in the enzyme function of the receptor was observed for both receptor autophosphorylation and the ability of the receptor to catalyze phosphorylation of a synthetic substrate. Impaired insulin-induced effects on RNA synthesis has been observed in a disorder (the Alström syndrome) associated with a genetically defective insulin receptor.[77] In the cultured fibroblasts of these patients, insulin-receptor binding and insulin-stimulated glucose uptake are in the normal range. The molecular basis of the insulin resistance in patients with the Alström syndrome is unknown but the results suggest that, after binding of insulin to its receptor, the mechanisms involved in the early cellular action of insulin (glucose uptake) may be mediated by mechanisms which are different from those involved in the late effects (RNA synthesis). The possible role of insulin receptor phosphorylation at different sites in relation to early and late insulin effects is not understood.

D. Relationship between the Insulin Receptor and Oncogene Protein Products

A structural homology between the insulin receptor and oncogene protein products was initially suggested by the fact that the insulin receptors present in cultured human lymphocytes (IM-9 cells) are specifically immunoprecipitated by antibodies to pp60^{src}.[78] The precipitation is competitively inhibited by purified pp60^{v-src} but not by the product of the v-*raf* oncogene. It is unlikely that pp60^{src} represents a truncated form of the insulin receptor because three other antibodies to pp60^{v-src}, including a monoclonal antibody, are unable to induce precipitation of the receptor.[78] Most probably, there would be a similarity between certain epitopes of both proteins. The solubilized insulin receptor is phosphorylated and activated by the pp60^{v-src} kinase.[79,80] However, the biological significance of this fact is not understood. pp60^{v-src} can also induce the phosphorylation of exogenous proteins like angiotensin and calmodulin.[81,82]

As deduced from its entire amino acid sequence, the human insulin receptor is structurally related to the tyrosine kinase family of oncogenes, i.e., to the *src* gene family.[83] In particular, there is a striking homology between the insulin receptor and the predicted sequence of the protein product of the v-*ros* oncogene.[55] Moreover, the overall structure of the insulin receptor is reminiscent of the EGF receptor and this receptor is highly homologous to the v-*erb*-B oncogene protein product.

E. Regulation of Insulin Receptor Expression

Numerous endogenous and exogenous factors are capable of regulating the expression of insulin receptors, including age, menstrual cycle, pregnancy, diet, physical exercise, and a diversity of pathological conditions.[6,57] In general, the expression of insulin receptors on the cell surface is inversely correlated with the concentrations of circulating insulin. Obese individuals (which are frequently hyperinsulinemic) are characterized by a decreased number of insulin receptors and are resistant to the physiological effects of insulin.

The number of hormone receptors expressed on the surface of different types of cells are partially regulated by the environmental concentration of hormones. Chronic exposure to high concentrations of insulin induce a loss of insulin receptors expression on the cell surface, which is called receptor down regulation.[84-87] The mechanisms involved in the down-regulation of insulin receptors are not totally clear. In MCF-7 human breast cancer cells cultured in a medium containing fetal calf serum (FCS), preincubation of the cells with insulin (50 µg/mℓ for 24 hr) causes a decrease in insulin receptor binding by 47% but no insulin-induced receptor loss is observed when the cells are cultured in serum-free medium.[88] The absence of down-regulation of insulin receptors in serum-free medium suggests that some factor(s) present in FCS play an important role in the regulation of insulin receptors. In contrast, EGF receptors are down-regulated in both FCS and SF media.[88]

As stated above, insulin stimulates the phosphorylation of its own receptor on the cell surface. Insulin receptor phosphorylation is probably regulated by humoral and intracellular factors acting in a complex way. Thyroid hormone has been identified as a factor involved in the regulation of the insulin receptor beta subunit autophosphorylation.[89] Thyroidectomy increases the autophosphorylation of this subunit without changing the number or affinity of the insulin receptor. The mechanisms responsible for insulin receptor regulation by thyroid hormone are unknown. Phorbol esters stimulate phosphorylation of insulin receptors in intact hepatoma cells at serine and threonine residues, the phosphorylation occurring at as many as nine sites in the receptor beta subunit.[90] The action of phorbol ester on insulin receptor phosphorylation is probably mediated by protein kinase C activity.

F. Insulin Receptor Abnormalities in Neoplastic Cells

Abnormalities of insulin binding and receptor phosphorylation have been described in some malignant cells, for example, in the insulin-resistant mouse melanoma cell line Cloudman S91.[91] In contrast to the almost universal proliferation-inducing action of insulin in cultured cells, insulin acts as a potent, reversible inhibitor of proliferation in Cloudman S91 mouse melanoma cells.[92] This cell line is defective both in its affinity for insulin and its autophosphorylation properties, one site of autophosphorylation being absent in the altered receptor, which would make it unable to stimulate cell growth. An abnormally large insulin receptor has been detected in a cultured human monocyte cell line, U-937, derived from a patient with generalized histiocytic lymphoma, but the receptor appeared as functionally normal.[93] The difference in molecular size of the insulin receptor in U-937 cells could be due to either differences in the core protein or to an alteration in the receptor glycosylation pattern. The changes of insulin receptors in transformed cells are discussed further with those referring to insulin-like growth factors receptors.

G. Functional Role of the Insulin Receptor

The role of insulin cell surface receptors in the cellular responses to insulin is not totally clear. The solely presence or emergence of insulin receptors does not ensure insulin responsiveness because insulin receptors are present in many types of cells and tissues where a specific action of insulin has not been characterized.[6] Cells lacking insulin receptors, like the Madin-Darby canine kidney cells (MDCK cells), do not respond to insulin by increasing the incorporation of radiolabeled glucose into glycogen or the uptake of radiolabeled alpha-aminobutyrate. However, MDCK cells may display some responses considered as characteristic of insulin action, including stimulation of glycogen synthesis, when they are exposed to insulin mimickers such as insulin-ricin B hybrid molecules, which suggests that other plasma membrane receptors may act as functional alternates for insulin receptors as initial sites for hormone action.[94] Moreover, lectins and hydrogen peroxyde can elicit insulin-like responses in MDCK cells, which indicates that it is possible to bypass plasma membrane insulin binding sites but still elicit cellular responses characteristically mediated by insulin.[95]

The physiological role of insulin receptor phosphorylation is also not completely understood. After binding to its cellular receptor, insulin stimulates phosphorylation not only on the beta-subunit of its own receptor but also on exogenously added substrates such as casein, histone H2b, and a synthetic peptide.[96] Insulin receptor phosphorylation may not be a prerequisite for acute insulin action,[97] and the effects of insulin on its receptor at the cell surface are apparently not absolutely specific because treatment of rat adipocytes or human placenta with trypsin also stimulates phosphorylation of the insulin receptor on tyrosine residues.[98]

H. Internalization of the Insulin-Insulin Receptor Complex

After its formation, the insulin-receptor complex is internalized and processed intracellularly.[84,87,99] Internalized insulin receptors maybe recycled back to the cell surface.[100] Intracellular insulin "receptors" have been described, even at the level of nuclear membrane, where they would be potentially related to mRNA metabolism in the effector cells.[101-103] It has also been suggested that the internalized insulin-receptor complex may retain a weak, but significant, capacity to stimulate both glucose transport and phosphodiesterase activities.[104] However, the physiological role of these phenomena in intact cells is not understood. After their internalization insulin-receptor complexes are degraded, mainly at the level of lysosomes.[85]

IV. TRANSDUCTIONAL MECHANISMS OF INSULIN ACTION

The possible action of classical mediators of hormone action, such as cyclic nucleotides and calcium ions, in the cellular mechanisms of action of insulin is a subject of high controversy.[105-112] A critical role for Ca^{2+} in the cellular mechanisms of action of insulin is suggested by the fact that the insulin receptor contains a calmodulin-binding domain.[113] Elevated levels of calmodulin have been detected in the heart and kidney of two types of diabetic mice, and in muscle and abdominal fat of streptozotocin-induced diabetes in mice.[114]

A. Insulin Mediators

It has been postulated that after binding of insulin to its cellular receptor several insulin-specific mediators are formed, possibly by limited proteolytic processes.[112,115,116] These mediators would be peptides of 1000 to 3000 mol wt acting intracellularly at a number of sites, controlling the activity of different enzymes. However, all the efforts oriented to the purification and characterization of such peptides have been so far unsuccessful. If such insulin mediators exist, they could be responsible for the insulin-like effects produced by treatment of cells with trypsin.

B. Phosphoinositides as Transducers of Insulin Action

Insulin, as other hormones, may provoke rapid changes in phospholipid metabolism. Insulin enhances the incorporation of radiolabeled orthophosphate into different subcellular fractions from rat adipose tissue in vitro.[117,118] It has been demonstrated that the hormone increases the incorporation of ^{32}P-labeled orthophosphate into phosphatidate, phosphatidylinositol, and diacylglycerol in the same tissue.[119] Administration of insulin in vivo provokes rapid increase in the concentration of phosphatidic acid, phosphatidylinositol and phosphoinositides in rat adipose tissue and it also increases phosphatidylinositol levels in vitro.[120] The results suggest that stimulation of phosphatidic acid synthesis may serve as an effector mechanism for insulin, at least in adipose tissue.

Enzymatic methylation of phospholipids plays a role in the transduction of receptor-mediated signals through the membranes of a variety of cells, and it has been suggested that this process may also play an important role in the transductional mechanisms of insulin action.[121] Plasma membranes prepared from rat adipocytes contain a phosphatidylethanolamine methyltransferase system and insulin stimulates this enzyme in a concentration-dependent manner, the effects being observed as early as 15 sec after the addition of the hormone.

Phorbol esters stimulate phosphorylation of the insulin receptor on serine and threonine residues, altering insulin receptor autophosphorylation on tyrosine residues and insulin activation of cellular enzymes (glycogen synthetase and tyrosine aminotransferase).[90] Since phorbol esters induce a direct activation of the calcium-activated and phospholipid-dependent enzyme protein kinase C, these results suggest a physiological relation between insulin action and protein kinase C. However, activation of phosphoinositide metabolism is apparently not a universal pathway of insulin action. In rat liver plasma membranes or isolated hepatocytes neither insulin nor EGF stimulate the synthesis of phosphatidylinositol, phosphatidylinositol 4-phosphate, or phosphatidylinositol 4,5-bisphosphate, which suggests that the insulin- and EGF-stimulated receptor kinases do not act on phosphoinositides in the liver.[122] Moreover, it has been clearly demonstrated that in isolated fat cells insulin does not stimulate phosphoinositide breakdown but that it only increases the *de novo* synthesis of phosphatidylinositol and phosphatidylinositol 4,5-bisphosphate.[123] The precise role of stimulated phosphoinositide synthesis in the mechanisms of insulin action remains unclear.

V. POST-TRANSDUCTIONAL MECHANISMS OF INSULIN ACTION

In insulin-responsive cells the formation of an insulin-receptor complex induces an array of functional changes at the level of the plasma membrane, including stimulation of ion transport and the uptake of essential nutrients, especially glucose and amino acids.[124,125] Subsequently, changes in the activities of different enzymes may occur. Activation of cellular enzymes by insulin action can occur by phosphorylation processes depending on the tyrosine-specific protein kinase activity of the insulin-receptor complex as well as by interaction of this complex with serine/threonine kinases.[126]

Under the action of insulin, the synthesis of specific proteins and/or other compounds maybe increased or decreased, depending on the type of cell, its stage of differentiation and other physiological conditions. By the action of insulin, RNA synthesis can also be either stimulated or inhibited.[117,118,127] In some cases, after prolonged insulin action, stimulation of DNA synthesis and mitogenic effects may occur.[1,2] In density-inhibited chick embryo cell cultures the following sequence of events was characteristically observed after stimulation with microgram quantities of insulin: an early increase in sugar uptake and decrease in leucine uptake, increase in cell volume, stimulation of RNA and protein synthesis, increase in thymidine uptake, DNA synthesis, mitosis, and cell division.[128] However, very little effect of insulin was observed in the absence of serum, which strongly suggests that some other factors present in serum are required for the stimulation. The precise sequence of the complex insulin-induced cellular changes, the relationships existing between them, and the interaction of insulin with other hormones and peptide growth factors in the origin and development of such changes are not understood. In the whole animal, insulin is an important anabolic agent. Insulin treatment can reverse cachexia in tumor-bearing rats, with suppression of anorexia and preservation of host weight and without stimulation of tumor growth.[129] Despite these beneficial effects, long-term insulin treatment slightly, but significantly, shortens survival in tumor-bearing animals.

A. Insulin Action on Protein Phosphorylation

Protein phosphorylation/dephosphorylation processes are importantly involved in the action of insulin in insulin-responsive cells. Different cellular proteins are phosphorylated on serine, threonine and/or tyrosine residues as a consequence of the interaction of insulin with its receptor. One of them is the 40S ribosomal protein S6,[130] which maybe involved in the biochemical processes leading to initiation of protein synthesis. Protein S6 is phosphorylated on serine by the action of serum, EGF, prostaglandins, and phorbol esters.[131-134] S6 is also phosphorylated by oncogene products with protein kinase activity, including pp60$^{v\text{-}src}$.[135] It has been suggested that the enzyme mediating insulin-stimulated phosphorylation of protein S6 is a soluble kinase, termed S6 kinase, which is neither cyclic AMP-dependent nor phospholipid- and Ca^{2+}-dependent.[136] S6 kinase is activated within minutes of exposure of differentiated 3T3-L1 adipocytes to nanomolar concentrations of insulin or phorbol ester. Insulin-mediated phosphorylation of the ribosomal protein S6 may be related to induction of cell proliferation,[137] but a direct regulation of DNA synthesis through S6 phosphorylation seems unlikely. Under conditions in which EGF maximally stimulates S6 phosphorylation and DNA synthesis, insulin further stimulates DNA synthesis but not S6 phosphorylation.[138] Probably, insulin and EGF use different pathways of action for stimulating DNA synthesis.

Other cellular protein whose phosphorylation is stimulated by insulin is a 96,000-dalton (96 k) cytosol protein, which may be involved in some step leading to initiation of DNA synthesis.[139] Cellular 110,000-, 120,000-, and 185,000-dalton proteins may be substrates for the insulin receptor kinase,[140-142] but the functions of these proteins are unknown. A 90,000-dalton phosphoprotein (pp90), an associated protein kinase, and a specific phosphatase are involved in the regulation of the Cloudman melanoma cell line proliferation by insulin.[143] The proliferation of wild-type Cloudman melanoma cell line is inhibited by insulin and the evidence suggests that phosphorylation and dephosphorylation of pp90 may be one step in the insulin-mediated control of proliferation of this cell line.

The activated insulin receptor kinase may catalyze the phosphorylation of protein kinase C and calmodulin, which may result in phosphorylation of different cellular proteins.[126] Phosphorylation of a 40-kdalton protein is stimulated by insulin in a Ca^{2+}-dependent manner.[144]

Ca^{2+}-stimulated phosphorylation of the same protein was also enhanced in membranes incubated with phospholipase C, which catalyzes the production of endogenous 1,2-diacylglycerol.

Microtubules are involved in the mechanisms of action of insulin, and tubulin and microtubule-associated proteins are phosphorylated by the action of the insulin receptor associated protein kinase activity.[140,145] In general, it is not known which cellular proteins are primary targets of the insulin-receptor complex kinase activity or if they are phosphorylated by other, secondarily activated kinases.

1. Insulin and Phosphorylation of the v-ras Protein

Insulin, as EGF, is apparently capable of inducing an increase in phosphorylation and guanine nucleotide binding capacity of the viral oncogene product p21$^{v\text{-}ras}$ in membranes isolated from NRK cells transformed by Harvey MuSV,[146] but the possible biological significance of this phenomenon is unknown.

B. Postreceptor Defects of Insulin Action in Neoplastic Cells

Little information is available at present about possible postreceptor defects of insulin action occurring in neoplastic cells. In spontaneously transformed cloned rat hepatocytes lack of a serum requirement for growth is accompanied by several defects in insulin action, including loss of insulin stimulation of protein synthesis and reduction in insulin's ability to activate glycogen synthesis from glucose.[147] Postreceptor mechanisms are the most important mechanisms involved in the mediation of insulin-induced desensitization phenomena occurring in a rat hepatoma cell line (h-35) exposed to the action of insulin.[148] Although insulin resistance observed in this hepatoma cell line is accompanied by a 50 to 60% decrease in insulin receptor number, this decrease cannot account for the marked impairment observed in the responses of these cells to insulin action, which should be attributed to postreceptor phenomena functioning abnormally in the tumor cells. In cultured rat hepatoma cells exhibiting different degrees of differentiation it was observed that the number of insulin receptors at the cell surface is higher in the well differentiated cells than in dedifferentiated cells, and that some postreceptor insulin effects (e.g., induction of tyrosine aminotransferase (TAT) activity) are lost in the more differentiated tumor cells, whereas other effects occurring at the same level (e.g., conversion of the D-form to the I-form of glycogen synthase activity) maybe preserved.[149]

C. Effects of Insulin on DNA Synthesis, Cell Proliferation, and Cell Differentiation

Insulin and insulin-like growth factors, as well as other factors contained in serum, are importantly involved in the processes leading to DNA synthesis, cell proliferation, and cell differentiation. It has been shown, for example, that insulin initiates DNA synthesis in the epithelial cells of mammary gland explants derived from 3-month-old virgin mice and that the effects are elicited, at least in part, by an insulin-dependent emergence of DNA polymerase activity.[150] Insulin and insulin-like growth factors (IGFs) stimulate the incorporation of thymidine into DNA in cultured human skin fibroblasts.[151] In the latter system, IGFs are active at concentrations lower than those required for insulin. The combined action of insulin and IGFs fails to give additive results, which suggests the existence of a common mechanism for both polypeptides, whereas combined action of insulin and serum, or IGFs and serum, give additive results.[151]

Insulin has intrinsic mitogenic activity.[152] The hormone is important in the control of processes related to cell growth and proliferation, both in vivo and in vitro.[1-3] Insulin and IGF-I have highly synergistic effects with growth factors such as EGF and PDGF in cell cycle control and stimulation of DNA synthesis in cultured cells.[153,154] Each growth factor (insulin, IGF-I, EGF, PDGF) may control discrete cellular events within the cell cycle.[153]

The presence of insulin and other growth factors are also required for the optimal growth and development of tissues in vivo, especially fetal tissues. Insulin is important in the control of differentiation processes occurring at particular stages of ontogeny. Insulin, glucagon, and epidermal growth factor play important roles in the control of various endodermally derived organs.[155] Insulin and insulin-like growth factors maybe required for the action of other cellular growth factors, including, for example, the action of nerve growth factor (NGF) for the induction of neurite formation in cultured human neuroblastoma cells.[156] Insulin acts synergistically with progesterone to induce meiosis of Rana pipiens oocytes in vitro.[157]

The mechanisms involved in the growth-promoting action of insulin are little known, but the growth and metabolic activities of the insulin molecule have probably evolved in a nonparallel fashion and may involve separate functional domains of the molecule.[158] Insulin, as well as several growth factors with mitogenic properties (EGF, PDGF, NGF), induce the expression ornithine decarboxylase activity in the presence of amino acids like asparagine, and this activity is closely related to the synthesis of polyamines and the induction of cell proliferation.[159] Inhibition of ornithine decarboxylase without the concurrent administration of polyamines leads to cessation of DNA synthesis and cell proliferation. It seems thus likely that polyamines, acting synergistically with certain amino acids, and perhaps also with other agents, are involved in the regulation of cell proliferation stimulated by hormones and growth factors.

VI. INSULIN REQUIREMENT BY NEOPLASTIC CELLS

On the basis of experiments performed in vivo and in vitro it has been postulated that most types of tumor cells are dependent on insulin for their proliferation.[1,2] For example, mammary carcinomas induced in rats by 7,12-dimethylbenz(a)anthracene (DMBA) or methylnitrosourea (MNU) are dependent on insulin and most of these tumors regress, or grow more slowly, after the animals are made diabetic with administration of alloxan, streptozotocin, or diazoxide.[160-163] Prolonged insulin treatment may have remarkable effects on tumor growth in mice. Administration of insulin in pharmacological doses exerts a cocarcinogenic effect on a nonhormonally dependent tumor, squamous cell carcinoma induced in mice by 3-methylcholanthrene (MCA).[164] The hormone changes tumor evolution, increasing DNA synthesis and inducing morphological alterations in the cancer cells, affecting the expression of differentiated characteristics. In contrast with these experimental observations, little is known about the relationship between diabetes and neoplastic diseases in humans and other animal species. In a recent study diabetes was shown to be a risk factor for cancer of the uterus, vulva, and vagina, as well as kidney and skin, but not pancreas.[165] No association between type of treatment, including insulin, and cancer risk was noted.

Malignant cells may require less insulin than normal cells for their growth in culture,[161] but the results of many studies indicate that most tumor cell lines require insulin as an essential component of the medium.[1,2] The growth of a human breast cancer cell line (MCF-7) is sensitive to physiological concentrations of insulin, which produces a shortening of the G_1 transit time and a marked increase in the fraction of S-phase cells by effects that are probably mediated via the insulin receptor.[167] Anchorage-independent growth of murine melanoma cells in serum-free medium is dependent or insulin on MSH.[168] Insulin has also been recognized as a potent, specific growth factor in hepatoma cell lines.[169] The molecular events involved in such insulin actions are not understood but there is evidence that phospholipids may play a role in mediating the effects of insulin on growth of experimental tumors.[170]

Insulin may be required for the expression of a transformed phenotype. The addition of insulin to cultured cells (BALB/3T3 cells) markedly enhances the yield of transformation induced by x-irradiation.[171] The latter effect was dependent on the concentration of insulin added to the medium and was consistently observed in cell cultures irradiated at various dosage levels. Tumoral cells producing insulin may show resistance to insulin.[172]

A. Insulin Action and the Induction of Differentiation in Neoplastic Cells

Insulin and its receptors may have an important role in the processes of differentiation induced in vitro by several types of compounds. Differentiation of the U-937 human monocyte-like cell line by calcitriol or by retinoic acid is accompanied by altered, although opposite, effects on insulin receptors.[173] Cytosolic Ca^{2+} is able to regulate the function of insulin receptors in U-937 cells induced to differentiation by calcitriol, while it remains without effect in the uninduced cells.[174] This phenomenon is synergistic with the action of phorbol esters on the binding of insulin. Retinoids are effective inhibitors of mammary gland carcinogenesis in the rat, and the addition of retinoic acid to media of organ cultures of rat mammary gland carcinomas blocks the stimulatory effect of insulin on DNA synthesis.[175]

VII. SUMMARY

The precise signal by which insulin regulates intracellular metabolic and functional processes remains to be identified. Moreover, it is not known whether all of the cellular actions of insulin are mediated similarly or whether each effect, either acute (glucose and amino acid uptake, ion transport) or chronic (protein, RNA and DNA synthesis), is evoked by a unique signal.

Very little is known at present about the biochemical mechanisms involved in the promotion of cell proliferation and cell differentiation by insulin, and this is especially true with reference to tumor tissues. The existence of at least some pathways of insulin action shared with the mechanisms of action of certain oncogene protein products is suggested by the presence of tyrosine-specific protein kinase activity in the insulin-receptor complex, the same or a similar activity being present in the protein products of several cellular and viral oncogenes, especially oncogenes from the *src* family. Moreover, oncogene protein products with tyrosine kinase activity, such as $pp60^{v\text{-}src}$, may induce phosphorylation of the solubilized insulin receptor, and the insulin receptors present in cultured human lymphocytes are immunoprecipitated by antibodies to $pp60^{src}$. Changes in phosphoinositide metabolism and Ca^{2+} compartmentalization, with activation of protein kinase C and phosphorylation of different cellular proteins on serine residues, are also shared by insulin, several growth factors, and certain oncogene protein products. A structural and functional relationship has been established between the human insulin receptor and the tyrosine kinase family of oncogenes, which is typified by the *src* oncogene.

REFERENCES

1. **Straus, D. S.**, Effects of insulin on cellular growth and proliferation, *Life Sci.*, 29, 2131, 1981.
2. **Straus, D. S.**, Growth-stimulatory actions of insulin *in vitro* and *in vivo*, *Endocr. Rev.*, 5, 356, 1984.
3. **Hill, D. J. and Milner, R. D. G.**, Insulin as a growth factor, *Pediatr. Res.*, 19, 879, 1985.
4. **Saunders, D. J., Brandenburg, D., Shang-chuan, C., and Chih-chen, W.**, Insulin chemistry — pathway to understanding insulin action, *Endeavour*, 6, 146, 1982.
5. **Czech, M. P.**, Insulin action, *Am. J. Med.*, 70, 142, 1981.
6. **Kaplan, S. A.**, The insulin receptor, *J. Pediatr.*, 104, 327, 1984.
7. **Chaganti, R. S. K., Jhanwar, S. C., Antonarakis, S. E., and Hayward, W. S.**, Germ-line chromosomal localization in chromosome 11p linkage: parathyroid hormone, beta-globin, c-Ha-ras-1, and insulin, *Somat. Cell Mol. Genet.*, 11, 197, 1985.

8. **Zabel, B. U., Kronenberg, H. M., Bell, G. I., and Shows, T. B.,** Chromosome mapping of genes on the short arm of human chromosome 11: parathyroid hormone gene is at 11p15 together with the genes for insulin, c-Harvey-*ras* 1, and beta-hemoglobin, *Cytogenet. Cell Genet.*, 39, 200, 1985.
9. **Yunis, J. J. and Ramsay, N. K. C.,** Familial occurrence of the aniridia-Wilms' tumor syndrome with deletion 11p13-14.1, *J. Pediatr.*, 96, 1027, 1980.
10. **Kaneko, Y., Egues, M. C., and Rowley, J. D.,** Interstitial deletion of short arm of chromosome 11 limited to Wilms' tumor cells in a patient without aniridia, *Cancer Res.*, 41, 4577, 1981.
11. **Koufos, A., Hansen, M. F., Lampkin, B. C., Workman, M. L., Copeland, N. G., Jenkins, N. A., and Cavenee, W. K.,** Loss of alleles at loci on human chromosome 11 during genesis of Wilms' tumour, *Nature (London)*, 309, 170, 1984.
12. **Orkin, S. H., Goldman, D. S., and Sallan, S. E.,** Development of homozygosity for chromosome 11p markers in Wilms' tumour, *Nature (London)*, 309, 172, 1984.
13. **Fearon, E. R., Vogelstein, B., and Feinberg, A. P.,** Somatic deletion and duplication of genes on chromosome 11 in Wilms' tumours, *Nature (London)*, 309, 176, 1984.
14. **Solomon, E.,** Recessive mutation in aetiology of Wilms' tumour, *Nature (London)*, 309, 111, 1984.
15. **Reeve, A. E., Housiaux, P. J., Gardner, R. J. M., Chewings, W. E., Grindley, R. M., and Millow, L. J.,** Loss of a Harvey *ras* allele in sporadic Wilms' tumour, *Nature (London)*, 309, 174, 1984.
16. **Rogler, C. E., Sherman, M., Su, C. Y., Shafritz, D. A., Summers, J., Shows, T. B., Henderson, A., and Kew, M.,** Deletion in chromosome 11p associated with a hepatitis B integration site in hepatocellular carcinoma, *Science*, 230, 319, 1985.
17. **Fearon, E. R., Feinberg, A. P., Hamilton, S. H., and Vogelstein, B.,** Loss of genes on the short arm of chromosome 11 in bladder cancer, *Nature (London)*, 318, 377, 1985.
18. **Bell, G. I., Swain, W. F., Pictet, R., Cordell, B., Goodman, H. M., and Rutter, W. J.,** Nucleotide sequence of a cDNA clone encoding human preproinsulin, *Nature (London)*, 282, 525, 1979.
19. **Bell, G. I., Pictet, R. L., Rutter, W. J., Cordell, B., Tischer, E., and Goodman, H. M.,** Sequence of the human insulin gene, *Nature (London)*, 284, 26, 1980.
20. **Sures, I., Goeddel, D. V., Gray, A., and Ullrich, A.,** Nucleotide sequence of human preproinsulin complementary DNA, *Science*, 208, 57, 1980.
21. **Bell, G. I., Pictet, R., and Rutter, W. J.,** Analysis of the regions flanking the human insulin gene and sequence of an Alu family member, *Nucleic Acids Res.*, 8, 4091, 1980.
22. **Ullrich, A., Dull, T. J., Gray, A., Brosius, J., and Sures, I.,** Genetic variation in the human insulin gene, *Science*, 209, 612, 1980.
23. **Bell, G. I., Selby, M. J., and Rutter, W. J.,** The highly polymorphic region near the human insulin gene is composed of simple tandemly repeating sequences, *Nature (London)*, 295, 31, 1982.
24. **Ullrich, A., Dull, T. J., Gray, A., Philips, J. A., III, and Peter, S.,** Variation in the sequence and modification state of the human insulin gene flanking regions, *Nucleic Acids Res.*, 10, 2225, 1982.
25. **Rotwein, P., Chyn, R., Chirgwin, J., Cordell, B., Goodman, H. M., and Permutt, M. A.,** Polymorphism in the 5'-flanking region of the human insulin gene and its possible relation to type 2 diabetes, *Science*, 213, 117, 1981.
26. **Owerbach, D. and Nerup, J.,** Restriction fragment length polymorphism of the insulin gene in diabetes mellitus, *Diabetes*, 31, 275, 1982.
27. **Owerbach, D., Poulsen, S., Billesbolle, P., and Nerup, J.,** DNA insertion sequences near the insulin gene affect glucose regulation, *Lancet*, i, 880, 1982.
28. **Rotwein, P. S., Chirgwin, J., Province, M., Knowler, W. C., Pettitt, D. J., Cordell, B., Goodman, H. M., and Permutt, M. A.,** Polymorphism in the 5' flanking region of the human insulin gene: a genetic marker for non-insulin-dependent diabetes, *N. Engl. J. Med.*, 308, 65, 1983.
29. **Yokoyama, S.,** Polymorphism in the 5'-flanking region of the human insulin gene and the incidence of diabetes, *Am. J. Human Genet.*, 35, 193, 1983.
30. **Permutt, M. A., Rotwein, P., Andreone, T., Ward, W. K., and Porte, D., Jr.,** Istet beta-cell function and polymorphism in the 5'-flanking region of the human insulin gene, *Diabetes*, 34, 311, 1985.
31. **Tager, H. S.,** Abnormal products of the human insulin gene, *Diabetes*, 33, 693, 1984.
32. **Lomedico, P., Rosenthal, N., Efstratiadis, A., Gilbert, W., Kolodner, R., and Tizard, R.,** The structure and evolution of the two nonallelic rat preproinsulin genes, *Cell*, 18, 545, 1979.
33. **Soares, M. B., Schon, E., Henderson, A., Karathanasis, S. K., Cate, R., Zeitlin, S., Chirgwin, J., and Efstratiadis, A.,** RNA-mediated gene duplication: the rat preproinsulin I gene is a functional retroposon, *Mol. Cell. Biol.*, 5, 2090, 1985.
34. **Todd, S., Yoshida, M. C., Fang, X. E., McDonald, L., Jacobs, J., Heinrich, G., Bell, G. I., Naylor, S. L., and Sakaguchi, A. Y.,** Genes for insulin I and II, parathyroid hormone, and calcitonin are on rat chromosome 1, *Biochem. Biophys. Res. Comm.*, 131, 1175, 1985.
35. **Kozak, M.,** Point mutations close to the AUG initiator codon affect the efficiency of translation of rat preproinsulin *in vivo*, *Nature (London)*, 308, 241, 1984.

36. **Lakshmikumaran, M. S., D'Ambrosio, E., Laimins, L. A., Lin, D. T., and Furano, A. V.,** Long interspersed repeated DNA (LINE) causes polymorphism at the rat insulin 1 locus, *Mol. Cell. Biol.*, 5, 2197, 1985.
37. **Rosenzweig, J. L., Havrankova, J., Lesniak, M. A., Brownstein, M., and Roth, J.,** Insulin is ubiquitous in extrapancreatic tissues of rats and humans, *Proc. Natl. Acad. Sci. U.S.A.*, 77, 572, 1980.
38. **Murakami, K., Taniguchi, H., and Baba, S.,** Presence of insulin-like immunoreactivity and its biosynthesis in rat and human parotid gland, *Diabetologia*, 22, 358, 1982.
39. **Raizada, M. K.,** Localization of insulin-like immunoreactivity in the neurons from primary cultures of rat brain, *Exp. Cell Res.*, 143, 351, 1983.
40. **LeRoith, D., Lesniak, M. A., and Roth, J.,** Insulin in insects and annelids, *Diabetes*, 30, 70, 1981.
41. **Nagasawa, H., Kataoka, H., Isogai, A., Tamura, S., Suzuki, A., Ishizaki, H., Mizoguchi, A., Fujiwara, Y., and Susuki, A.,** Amino-terminal amino acid sequence of the silkworm prothoracicotrophic hormone: homology with insulin, *Science*, 226, 1344, 1984.
42. **LeRoith, D., Shiloach, J., Heffron, R., Rubinowitz, C., Tanenbaum, R., and Roth, J.,** Insulin-related material in microbes: similarities and differences from mammalian insulins, *Can. J. Biochem. Cell Biol.*, 63, 839, 1985.
43. **Hanahan, D.,** Heritable formation of pancreatic beta-cell tumours in transgenic mice expressing recombinant insulin/simian virus 40 oncogenes, *Science*, 315, 115, 1985.
44. **Yang-Feng, T. L., Francke, U., and Ullrich, A.,** Gene for human insulin receptor: localization to site on chromosome 19 involved in pre-B-cell leukemia, *Science*, 228, 728, 1985.
45. **Chvatchko, Y., Gazzano, H., Van Obberghen, E., and Fehlmann, M.,** Subunit arrangement of insulin receptors in hepatoma cells, *Mol. Cell. Endocrinol.*, 36, 59, 1984.
46. **Maturo, J. M., III and Hollenberg, M. D.,** Distinct hydrodynamic forms of the insulin receptor: electrophoretic analysis of the R_I and R_{II} species, *Can. J. Physiol. Pharmacol.*, 63, 987, 1985.
47. **Newman, J. D. and Harrison, L. C.,** Homogeneous bivalent insulin receptor: purification using insulin coupled to 1,1'-carbonyldiimidazole-activated agarose, *Biochem. Biophys. Res. Comm.*, 132, 1059, 1985.
48. **Pang, D. T. and Shafer, J. A.,** Evidence that insulin receptor from human placenta has a high affinity for only one molecule of insulin, *J. Biol. Chem.*, 259, 8589, 1984.
49. **Massagué, J. and Czech, M. P.,** Role of disulfides in the subunit structure of the insulin receptor: reduction of class I disulfides does not impair transmembrane signalling, *J. Biol. Chem.*, 257, 6729, 1982.
50. **Jarett, L. and Smith, R. M.,** Partial disruption of naturally occurring groups of insulin receptors on adipocyte plasma membranes by dithiothreitol and N-ethylmaleimide: the role of disulfide bonds, *Proc. Natl. Acad. Sci. U.S.A.*, 80, 1023, 1983.
51. **Maturo, J. M., III, Hollenberg, M. D., and Aglio, L. S.,** Insulin receptor: insulin-modulated interconversion between distinct molecular forms involving disulfide-sulfydryl exchange, *Biochemistry*, 22, 2579, 1983.
52. **Aglio, L. S., Maturo, J. M., III, and Hollenberg, M. D.,** Receptors for insulin and epidermal growth factor: interaction with organomercurial agarose, *J. Cell. Biochem.*, 28, 143, 1985.
53. **Hedo, J. A., Kahn, C. R., Hayashi, M., Yamada, K. M., and Kasuga, M.,** Biosynthesis and glycosylation of the insulin receptor: evidence for a single polypeptide precursor of the two major subunits, *J. Biol. Chem.*, 258, 10020, 1983.
54. **Hedo, J. A. and Gorden, P.,** Biosynthesis of the insulin receptor, *Horm. Metab. Res.*, 17, 487, 1985.
55. **Ebina, Y., Ellis, L., Jarnagin, K., Edery, M., Graf, L., Clauser, E., Ou, J., Masiarz, F., Kan, Y. W., Goldfine, I. D., Roth, R. A., and Rutter, W. J.,** The human insulin receptor cDNA: the structural basis for hormone-activated transmembrane signalling, *Cell*, 40, 747, 1985.
56. **Ebina, Y., Edery, M., Ellis, L., Standring, D., Beaudoin, J., Roth, R. A., and Rutter, W. J.,** Expression of a functional human insulin receptor from a cloned cDNA in Chinese hamster ovary cells, *Proc. Natl. Acad. Sci. U.S.A.*, 82, 8014, 1985.
57. **Thomopoulos, P.,** Les récepteurs insuliniques des cellules sanguines circulantes, *Diabete Metab.*, 7, 207, 1981.
58. **Kasuga, M., Karlsson, F. A., and Kahn, C. R.,** Insulin stimulates the phosphorylation of the 95,000-dalton subunit of its own receptor, *Science*, 215, 185, 1982.
59. **Häring, H.-U., Kasuga, M. and Kahn, C. R.,** Insulin receptor phosphorylation in intact adipocytes, *Biochem. Biophys. Res. Comm.*, 108, 1538, 1982.
60. **Van Obberghen, E., Rossi, B., Kowalski, A., Gazzano, H., and Ponzio, G.,** Receptor-mediated phosphorylation of the hepatic insulin receptor: evidence that the M_r 95,000 receptor subunit is its own kinase, *Proc. Natl. Acad. Sci. U.S.A.*, 80, 945, 1983.
61. **Kasuga, M., Zick, Y., Blithe, D. L., Crettaz, M., and Kahn, C. R.,** Insulin stimulates tyrosine phosphorylation of the insulin receptor in a cell-free system, *Nature (London)*, 298, 667, 1982.
62. **Zick, Y., Whittaker, J., and Roth, J.,** Insulin stimulated phosphorylation of its own receptor: activation of a tyrosine-specific protein kinase that is tightly associated with the receptor, *J. Biol. Chem.*, 258, 3431, 1983.

63. **Kasuga, M., Fujita-Yamaguchi, Y., Blithe, D. L., and Kahn, C. R.**, Tyrosine-specific protein kinase activity is associated with the purified insulin receptor, *Proc. Natl. Acad. Sci. U.S.A.*, 80, 2137, 1983.
64. **Cobb, M. H. and Rosen, O. M.**, The insulin receptor and tyrosine protein kinase activity, *Biochim. Biophys. Acta*, 738, 1, 1983.
65. **Rees-Jones, R. W., Hendricks, S. A., Quarum, M., and Roth, J.**, The insulin receptor of rat brain is coupled to tyrosine kinase activity, *J. Biol. Chem.*, 259, 3470, 1984.
66. **Petruzzelli, L., Herrera, R., and Rosen, O. M.**, Insulin receptor is an insulin-dependent tyrosine protein kinase: copurification of insulin-binding activity and protein kinase activity to homogeneity from human placenta, *Proc. Natl. Acad. Sci. U.S.A.*, 81, 3327, 1984.
67. **Yu, K.-T. and Czech, M. P.**, Tyrosine phosphorylation of the insulin receptor beta subunit activates the receptor-associated tyrosine kinase activity, *J. Biol. Chem.*, 259, 5277, 1984.
68. **Kasuga, M., Zick, Y., Blithe, D. L., Karlsson, F. A., Häring, H. U., and Kahn, C. R.**, Insulin stimulation of phosphorylation of the beta subunit of the insulin receptor: formation of both phosphoserine and phosphotyrosine, *J. Biol. Chem.*, 257, 9891, 1982.
69. **White, M. F., Takayama, S., and Kahn, C. R.**, Differences in the sites of phosphorylation of the insulin receptor *in vivo* and *in vitro*, *J. Biol. Chem.*, 260, 9470, 1985.
70. **Pang, D. T., Sharma, B. R., Shafer, J. A., White, M. F., and Kahn, C. R.**, Predominance of tyrosine phosphorylation of insulin receptors during the initial response of intact cells to insulin, *J. Biol. Chem.*, 260, 7131, 1985.
71. **Fujita-Yamaguchi, Y. and Kathuria, S.**, The monomeric alpha beta form of the insulin receptor exhibits much higher insulin-dependent tyrosine-specific protein kinase activity than the intact $alpha_2 beta_2$ form of the receptor, *Proc. Natl. Acad. Sci. U.S.A.*, 82, 6095, 1985.
72. **Fujita-Yamaguchi, Y., Sato, Y., and Kathuria, S.**, Removal of sialic acids from the purified insulin receptor results in enhanced insulin-binding and kinase activities, *Biochem. Biophys. Res. Comm.*, 129, 739, 1985.
73. **Herrera, R., Petruzzelli, L., Thomas, N., Bramson, H. N., Kaiser, E. T., and Rosen, O. M.**, An antipeptide antibody that specifically inhibits insulin receptor autophosphorylation and protein kinase activity, *Proc. Natl. Acad. Sci. U.S.A.*, 82, 7899, 1985.
74. **Grunberger, G., Zick, Y., and Gorden, P.**, Defect in phosphorylation of insulin receptors in cells from an insulin-resistant patient with normal insulin binding, *Science*, 223, 932, 1984.
75. **Grunberger, G., Comi, R. J., Taylor, S. I., and Gorden, P.**, Tyrosine kinase activity of the insulin receptor of patients with type A extreme insulin resistance: studies with circulating mononuclear cells and cultured lymphocytes, *J. Clin. Endocrinol. Metab.*, 59, 1152, 1984.
76. **Le Marchand-Brustel, Y., Grémeaux, T., Ballotti, R., and Van Obberghen, E.**, Insulin receptor tyrosine kinase is defective in skeletal muscle of insulin-resistant obese mice, *Nature (London)*, 315, 676, 1985.
77. **Rüdiger, H. W., Ahrens, P., Dreyer, M., Frorath, B., Löffel, C., and Schmidt-Preuss, U.**, Impaired insulin-induced RNA synthesis secondary to a genetically defective insulin receptor, *Human Genet.*, 69, 76, 1985.
78. **Perrotti, N., Taylor, S. I., Richert, N. D., Rapp, U. R., Pastan, I. H., and Roth, J.**, Immunoprecipitation of insulin receptors from cultured human lymphocytes (IM-9 cells) by antibodies to $pp60^{src}$, *Science*, 227, 761, 1985.
79. **White, M. F., Werth, D. K., Pastan, I., and Kahn, C. R.**, Phosphorylation of the solubilized insulin receptor by the gene product of the Rous sarcoma virus, $pp60^{src}$, *J. Cell. Biochem.*, 26, 169, 1984.
80. **Yu, K-T., Werth, D. K., Pastan, I. H., and Czech, M. P.**, src kinase catalyzes the phosphorylation and activation of the insulin receptor kinase, *J. Biol. Chem.*, 260, 5838, 1985.
81. **Wong, T. W. and Goldberg, A. R.**, Kinetics and mechanism of angiotensin phosphorylation by the transforming gene product of Rous sarcoma virus, *J. Biol. Chem.*, 259, 3127, 1984.
82. **Fukami, Y. and Lipmann, F.**, Purification of the Rous sarcoma virus src kinase by casein-agarose affinity chromatography, *Proc. Natl. Acad. Sci. U.S.A.*, 82, 321, 1985.
83. **Ullrich, A., Bell, J. R., Chen, E. Y., Herrera, R., Petruzzelli, L. M., Dull, T. J., Gray, A., Coussens, L., Liao, Y.-C., Tsubokawa, M., Mason, A., Seeburg, P. H., Grunfeld, C., Rosen, O. M., and Ramachandran, J.**, Human insulin receptor and its relationship to the tyrosine kinase family of oncogenes, *Nature (London)*, 313, 756, 1985.
84. **Marshall, S. and Olefsky, J. M.**, Characterization of insulin-induced receptor loss and evidence for internalization of the insulin receptor, *Diabetes*, 30, 746, 1981.
85. **Carpentier, J.-L., Van Obberghen, E., Gorden, P., and Orci, L.**, Binding, membrane redistribution, internalization and lysosomal association of (^{125}I)anti-insulin receptor antibody in IM-9-cultured human lymphocyte, *Exp. Cell Res.*, 134, 81, 1981.
86. **Lane, M. D.**, The regulation of insulin receptor level and activity, *Nutr. Rev.*, 39, 417, 1981.
87. **Berhanu, P., Olefsky, J. M., Tsai, P., Thamm, P., Saunders, D., and Brandenburg, D.**, Internalization and molecular processing of insulin receptors in isolated rat adipocytes, *Proc. Natl. Acad. Sci. U.S.A.*, 79, 4069, 1982.

88. Hwang, D. L., Papolan, T., Barseghian, G., Josefsberg, Z., and Lev-Ran, A., Absence of down-regulation of insulin receptors in human breast cancer cells (MCF-7) cultured in serum-free medium: comparison with epidermal growth factor receptor, *J. Receptor Res.*, 5, 27, 1985.
89. Correze, C., Pierre, M., Thibout, H., and Toru-delbauffe, D., Autophosphorylation of the insulin receptor in rat adipocytes is modulated by thyroid hormone status, *Biochem. Biophys. Res. Comm.*, 126, 1061, 1985.
90. Takayama, S., White, M. F., Lauris, V., and Kahn, C. R., Phorbol esters modulate insulin receptor phosphorylation and insulin action in cultured hepatoma cells, *Proc. Natl. Acad. Sci. U.S.A.*, 81, 7797, 1984.
91. Haring, H. U., White, M. F., Kahn, C. R., Kasuga, M., Lauris, V., Fleischmann, R. F., Murray, M., and Pawelek, J., Abnormality of insulin binding and receptor phosphorylation in an insulin-resistant melanoma cell culture, *J. Cell Biol.*, 99, 900, 1984.
92. Kahn, R., Murray, M., and Pawelek, J., Inhibition of proliferation of Cloudman S91 melanoma cells by insulin and characterization of some insulin-resistant variants, *J. Cell. Physiol.*, 103, 109, 1980.
93. McElduff, A., Grunberger, G., and Gorden, P., An alteration in apparent molecular weight of the insulin receptor from the human monocyte cell line U-937, *Diabetes*, 34, 686, 1985.
94. Hofmann, C. A., Lotan, R. M., Ku, W. W., and Oeltmann, T. N., Insulin-ricin B hybrid molecules mediate an insulin-associated effect on cells which do not bind insulin, *J. Biol. Chem.*, 258, 11774, 1983.
95. Hofmann, C. A., Crettaz, M., Bruns, P., Hessel, P., and Hadawi, G., Cellular responses elicited by insulin mimickers in cells lacking detectable plasma membrane insulin receptors, *J. Cell. Biochem.*, 27, 401, 1985.
96. Grunberger, G., Zick, Y., Roth, J., and Gorden, P., Protein kinase activity of the insulin receptor in human circulating and cultured mononuclear cells, *Biochem. Biophys. Res. Comm.*, 115, 560, 1983.
97. Simpson, I. A. and Hedo, J. A., Insulin receptor phosphorylation may not be a prerequisite for acute insulin action, *Science*, 223, 1301, 1984.
98. Tamura, S., Fujita-Yamaguchi, Y., and Larner, J., Insulin-like effect of trypsin on the phosphorylation of rat adipocyte insulin receptor, *J. Biol. Chem.*, 258, 14749, 1983.
99. Jarett, L., Insulin uptake and receptor reutilization (Symposium Summary), *Fed. Proc.*, 42, 2553, 1983.
100. Fehlmann, M., Carpentier, J.-L., Van Obberghen, E., Freychet, P., Thamm, P., Saunders, D., Brandenburg, D., and Orci, L., Internalized insulin receptors are recycled to the cell surface in rat hepatocytes, *Proc. Natl. Acad. Sci. U.S.A.*, 79, 5921, 1982.
101. Goldfine, I. D., Clawson, G. A., Smuckler, E. A., Purrello, F., and Vigneri, R., Action of insulin at the nuclear envelope, *Mol. Cell. Biochem.*, 48, 3, 1982.
102. Purrello, F., Vigneri, R., Clawson, G. A., and Goldfine, I. D., Insulin stimulation of nucleoside triphosphate activity in isolated nuclear envelopes, *Science*, 216, 1005, 1982.
103. Purrello, F., Burnham, D. B., and Goldfine, I. D., Insulin regulation of protein phosphorylation in isolated rat liver nuclear envelopes: potential relationship to mRNA metabolism, *Proc. Natl. Acad. Sci. U.S.A.*, 80, 1189, 1983.
104. Ueda, M., Robinson, F. W., Smith, M. M., and Kono, T., Effects of monensin on insulin processing in adipocytes: evidence that the internalized insulin-receptor complex has some physiological activities, *J. Biol. Chem.*, 260, 3941, 1985.
105. Planchart, A. and Barros-Pita, J. C., Role of divalent cations in cold and ouabain sensitive glucose uptake of adipose tissue stimulated by insulin, *Acta Cient. Venez.*, 28, 385, 1977.
106. McDonald, J. M., Bruns, D. E., and Jarett, L., Ability of insulin to increase calcium uptake by adipocyte endoplasmic reticulum, *J. Biol. Chem.*, 253, 3504, 1978.
107. Hobson, C. H., Upton, J. D., Loten, E. G., and Rennie, P. I. C., Is extracellular calcium required for insulin action?, *J. Cyclic Nucleotide Res.*, 6, 179, 1980.
108. Malchoff, D. M. and Bruns, D. E., Dissociation of insulin's effects on cell metabolism and on subcellular calcium transport systems of 3T3-L1 adipocytes, *Biochem. Biophys. Res. Comm.*, 100, 501, 1981.
109. Walaas, O. and Horn, R. S., The controversial problem of insulin action, *Trends Pharmacol. Sci.*, 2, 196, 1981.
110. Chan, K-M. and McDonald, J. M., Identification of an insulin-sensitive calcium-stimulated phosphoprotein in rat adipocyte plasma membranes, *J. Biol. Chem.*, 257, 7443, 1982.
111. Pershadsingh, H. A. and McDonald, J. M., Hormone-receptor coupling and the molecular mechanism of insulin action in the adipocyte: a paradigm for Ca^{2+} homeostasis in the initiation of the insulin-induced metabolic cascade, *Cell Calcium*, 5, 111, 1984.
112. Cheng, K., Thompson, M., Craig, J., Schwartz, C., Locher, E., and Larner, J., Cell membrane signals in the mechanism of insulin action, *Ann. Clin. Lab. Sci.*, 14, 78, 1984.
113. Graves, C. B., Goewert, R. R., and McDonald, J. M., The insulin receptor contains a calmodulin-binding domain, *Science*, 230, 827, 1985.

114. **Morley, J. E., Levine, A. S., Brown, D. M., and Handwerger, B. S.,** Calmodulin levels in diabetic mice, *Biochem. Biophys. Res. Comm.*, 108, 1418, 1982.
115. **Larner, J., Galasko, G., Cheng, K., DePaoli-Roach, A. A., Huang, L., Daggy, P., and Kellogg, J.,** Generation by insulin of a chemical mediator that controls protein phosphorylation and dephosphorylation, *Science*, 206, 1408, 1979.
116. **Stevens, E. V. J. and Husbands, D. R.,** Insulin-dependent production of low-molecular-weight compounds that modify key enzymes in metabolism, *Comp. Biochem. Physiol.*, 81B, 1, 1985.
117. **Pimentel, E., Gonzalez, C. A., and Gonzalez, F.,** Biochemical effects of insulin on subcellular fractions from rat adipose tissue, *Acta Endocrinol. (Kbh)*, Suppl. 173, 121, 1973.
118. **Pimentel, E., Gonzalez, C. A., and Gonzalez-Mujica, F.,** Effects of insulin and glucose on subcellular fractions from rat adipose tissue, *Acta Diabetol. Lat.*, 11, 206, 1974.
119. **Honeyman, T. W., Strohsnitter, W., Scheid, C. R., and Schimmel, R. J.,** Phosphatidic acid and phosphatidylinositol labelling in adipose tissue, *Biochem. J.*, 212, 489, 1983.
120. **Farese, R. V., Larson, R. E., and Sabir, M. A.,** Insulin acutely increases phospholipids in the phosphatidate-inositide cycle in rat adipose tissue, *J. Biol. Chem.*, 257, 4042, 1982.
121. **Kelly, K. L., Kiechle, F. L., and Jarett, L.,** Insulin stimulation of phospholipid methylation in isolated rat adipocyte plasma membranes, *Proc. Natl. Acad. Sci. U.S.A.*, 81, 1089, 1984.
122. **Taylor, D., Ushing, R. J., Blackmore, P. F., Prpić, V., and Exton, J. H.,** Insulin and epidermal growth factor do not affect phosphoinositide metabolism in rat liver plasma membranes and hepatocytes, *J. Biol. Chem.*, 260, 2011, 1985.
123. **Pennington, S. R. and Martin, B. R.,** Insulin-stimulated phosphoinositide metabolism in isolated fat cells, *J. Biol. Chem.*, 260, 11039, 1985.
124. **Czech, M. P.,** Insulin action and the regulation of hexose transport, *Diabetes*, 29, 399, 1980.
125. **Moore, R. D.,** Effects of insulin upon ion transport, *Biochim. Biophys. Acta*, 737, 1, 1983.
126. **Haring, H. U., White, M. F., Kahn, C. R., Ahmad, Z., DePaoli-Roach, A. A., and Roach, P. J.,** Interaction of the insulin receptor kinase with serine/threonine kinases in vitro, *J. Cell. Biochem.*, 28, 171, 1985.
127. **Pimentel, E.,** Transcriptional and translational effects of insulin, Manuscript in preparation, 1987.
128. **Vaheri, A., Ruoslahti, E., Hovi, T., and Norling, S.,** Stimulation of density-inhibited cell cultures by insulin, *J. Cell. Physiol.*, 81, 355, 1973.
129. **Moley, J. F., Morrison, S. D., and Norton, J. A.,** Insulin reversal of cancer cachexia in rats, *Cancer Res.*, 45, 4925, 1985.
130. **Rosen, O. M., Rubin, C. S., Cobb, M. H., and Smith, C. J.,** Insulin stimulates the phosphorylation of ribosomal protein S6 in a cell-free system derived from 3T3-L1 adipocytes, *J. Biol. Chem.*, 256, 3630, 1981.
131. **Thomas, G., Martin-Pérez, J., Siegmann, M., and Otto, A. M.,** The effect of serum, EGF, $PGF_{2\,alpha}$, and insulin on S6 phosphorylation and the initiation of protein and DNA synthesis, *Cell*, 30, 235, 1982.
132. **Martin-Pérez, J., Siegmann, M., and Thomas, G.,** EGF, $PGF_{2\,alpha}$ and insulin induce the phosphorylation of identical S6 peptides in Swiss mouse 3T3 cells: effect of cAMP on early sites of phosphorylation, *Cell*, 36, 287, 1984.
133. **Blenis, J., Spivack, J. G., and Erikson, R. L.,** Phorbol ester, serum, and Rous sarcoma virus transforming gene product induce similar phosphorylations of ribosomal protein S6, *Proc. Natl. Acad. Sci. U.S.A.*, 81, 6408, 1984.
134. **Trevillyan, J. M., Perisic, O., Traugh, J. A., and Byus, C. V.,** Insulin- and phorbol ester-stimulated phosphorylation of ribosomal protein S6, *J. Biol. Chem.*, 260, 3041, 1985.
135. **Decker, S.,** Phosphorylation of ribosomal protein S6 in avian sarcoma virus-transformed chicken embryo fibroblasts, *Proc. Natl. Acad. Sci. U.S.A.*, 78, 4112, 1981.
136. **Tabarini, D., Heinrich, J., and Rosen, O. M.,** Activation of S6 kinase activity in 3T3-L1 cells by insulin and phorbol ester, *Proc. Natl. Acad. Sci. U.S.A.*, 82, 4369, 1985.
137. **Kulkarni, R. K. and Straus, D. S.,** Insulin-mediated phosphorylation of ribosomal protein S6 in mouse melanoma cells and melanoma x fibroblast hybrid cells in relation to cell proliferation, *Biochim. Biophys. Acta*, 762, 542, 1983.
138. **Nilsen-Hamilton, M., Hamilton, R. T., Allen, W. R., and Potter-Perigo, S.,** Synergistic stimulation of S6 ribosomal protein phosphorylation and DNA synthesis by epidermal growth factor and insulin in quiescent 3T3 cells, *Cell*, 31, 237, 1982.
139. **Kletzien, R. F. and Day, P.,** Modulation of the G_0 to S phase transit time by insulin: potential involvement of protein phosphorylation, *J. Cell. Physiol.*, 105, 533, 1980.
140. **Rees-Jones, R. W. and Taylor, S. I.,** An endogenous substrate for the insulin receptor-associated tyrosine kinase, *J. Biol. Chem.*, 260, 4461, 1985.
141. **Sadoul, J.-L., Peyron, J.-F., Ballotti, R., Debant, A., Fehlmann, M., and Van Obberghen, E.,** Identification of a cellular 110,000-Da protein substrate for the insulin-receptor kinase, *Biochem. J.*, 227, 887, 1985.

142. **White, M. F., Maron, R., and Kahn, C. R.,** Insulin rapidly stimulates tyrosine phosphorylation of a M_r-185,000 protein in intact cells, *Nature (London)*, 318, 183, 1985.
143. **Fleischmann, R. D. and Pawelek, J. M.,** Evidence that a 90-kDa phosphoprotein, an associated kinase, and a specific phosphatase are involved in the regulation of Cloudman melanoma cell proliferation by insulin, *Proc. Natl. Acad. Sci. U.S.A.*, 82, 1007, 1985.
144. **Graves, C. B. and McDonald, J. M.,** Insulin and phorbol ester stimulate phosphorylation of a 40-kDa protein in adipocyte plasma membranes, *J. Biol. Chem.*, 260, 11286, 1985.
145. **Kadowaki, T., Fujita-Yamaguchi, Y., Nishida, E., Takaku, F., Akiyama, T., Kathuria, S., Akanuma, Y., and Kasuga, M.,** Phosphorylation of tubulin and microtubule-associated proteins by the insulin receptor kinase, *J. Biol. Chem.*, 260, 4016, 1985.
146. **Kamata, T. and Feramisco, J. R.,** Epidermal growth factor stimulates guanine nucleotide binding activity and phosphorylation of *ras* oncogene proteins, *Nature (London)*, 310, 147, 1984.
147. **Petersen, B. and Blecher, M.,** Insulin receptors and functions in normal and spontaneously transformed cloned rat hepatocytes, *Exp. Cell Res.*, 120, 119, 1979.
148. **Krett, N. L., Heaton, J. H., and Gelehrter, T. D.,** Insulin resistance in H-35 rat hepatoma cells is mediated by post-receptor mechanisms, *Mol. Cell. Endocrinol.*, 32, 91, 1983.
149. **Crettaz, M. and Kahn, C. R.,** Analysis of insulin action using differentiated and dedifferentiated hepatoma cells, *Endocrinology*, 113, 1201, 1983.
150. **Lockwood, D. H., Voytovich, A. E., Stockdale, F. E., and Topper, Y. J.,** Insulin-dependent DNA polymerase and DNA synthesis in mammary epithelial cells in vitro, *Proc. Natl. Acad. Sci. U.S.A.*, 58, 658, 1967.
151. **Rechler, M. M., Podskalny, J. M., Goldfine, I. D., and Wells, C. A.,** DNA synthesis in human fibroblasts: stimulation by insulin and by nonsuppressible insulin-like activity (NSILA-S), *J. Clin. Endocrinol. Metab.*, 39, 512, 1974.
152. **Petrides, P. E. and Böhlen, P.,** The mitogenic activity of insulin: an intrinsic property of the molecule, *Biochem. Biophys. Res. Comm.*, 95, 1138, 1980.
153. **O'Keefe, E. J. and Pledger, W. J.,** A model of cell cycle control: sequential events regulated by growth factors, *Mol. Cell. Endocrinol.*, 31, 167, 1983.
154. **Shipley, G. D., Childs, C. B., Volkenant, M. E., and Moses, H. L.,** Differential effects of epidermal growth factor, transforming growth factor, and insulin on DNA and protein synthesis and morphology in serum-free cultures of AKR-2B cells, *Cancer Res.*, 44, 710, 1984.
155. **Scheving, L. A., Scheving, L. E., Tsai, T. H., and Pauly, J. E.,** Circadian stage-dependent effects of insulin and glucagon on incorporation of (^{3}H)thymidine into deoxyribonucleic acid in the esophagus, stomach, duodenum, jejunum, ileum, caecum, colon, rectum, and spleen of the adult female mouse, *Endocrinology*, 111, 308, 1982.
156. **Recio-Pinto, E., Lang, F. F., and Ishii, D. N.,** Insulin and insulin-like growth factor II permit nerve growth factor binding and the neurite formation response in cultured neuroblastoma cells, *Proc. Natl. Acad. Sci. U.S.A.*, 81, 2562, 1984.
157. **Lessman, C. A. and Schuetz, A. W.,** Insulin induction of meiosis of *Rana pipiens* oocytes' relation to endogenous progesterone, *Gamete Res.*, 6, 95, 1982.
158. **King, G. L. and Kahn, C. R.,** Non-parallel evolution of metabolic and growth-promoting functions of insulin, *Nature (London)*, 292, 644, 1981.
159. **Rinehart, C. A., Jr. and Canellakis, E. S.,** Induction of ornithine decarboxylase activity by insulin and growth factors is mediated by amino acids, *Proc. Natl. Acad. Sci. U.S.A.*, 82, 4365, 1985.
160. **Heuson, J. C. and Legros, N.,** Influence of insulin deprivation on growth of the 7,12-dimethylbenz(*a*)anthracene-induced mammary carcinoma in rats subjected to alloxan diabetes and food restriction, *Cancer Res.*, 31, 226, 1972.
161. **Cohen, N. D. and Hilf, R.,** Influence of insulin on growth and metabolism of 7,12-dimethylbenz(*a*)anthracene-induced mammary tumors, *Cancer Res.*, 34, 3245, 1974.
162. **Shafie, S. M. and Hilf, R.,** Insulin receptor levels and magnitude of insulin-induced responses in 7,12-dimethylbenz(*a*)anthracene-induced mammary tumors in rats, *Cancer Res.*, 41, 826, 1981.
163. **Berger, M. R., Fink, M., Feichter, G. E., and Janetschek, P.,** Effects of diazoxide-induced reversible diabetes on chemically induced autochthonous mammary carcinomas in Sprague-Dawley rats, *Int. J. Cancer*, 35, 395, 1985.
164. **Lupulescu, A. P.,** Effect of prolonged insulin treatment on carcinoma formation in mice, *Cancer Res.*, 45, 3288, 1985.
165. **O'Mara, B. A., Byers, T., and Schenfeld, E.,** Diabetes mellitus and cancer risk: a multisite case-control study, *J. Chronic Dis.*, 38, 435, 1985.
166. **Powers, S., Fisher, P. B., and Pollack, R.,** Analysis of the reduced growth factor dependency of simian virus 40-transformed 3T3 cells, *Mol. Cell. Biol.*, 4, 1572, 1984.
167. **Gross, G. E., Boldt, D. H., and Osborne, C. K.,** Perturbation by insulin of human breast cancer cell cycle kinetics, *Cancer Res.*, 44, 3570, 1984.

168. **Bregman, M. D., Abdel Malek, Z. A., and Meyskens, F. L., Jr.,** Anchorage-independent growth of murine melanoma in serum-less media is dependent on insulin or melanocyte-stimulating hormone, *Exp. Cell Res.*, 157, 419, 1985.
169. **Koontz, J. W. and Iwahashi, M.,** Insulin as a potent, specific growth factor in a rat hepatoma cell line, *Science*, 211, 947, 1981.
170. **Narayanan, U., Ribes, J. A., and Hilf, R.,** Effects of streptozotocin-induced diabetes and insulin on phospholipid content of R3230AC mammary tumor cells, *Cancer Res.*, 45, 4833, 1985.
171. **Umeda, M., Tanaka, K., and Ono, T.,** Effect of insulin on the transformation of BALB/3T3 cells by X-ray irradiation, *Gann*, 74, 864, 1983.
172. **Sener, A. and Malaisse, W. J.,** Resistance to insulin of tumoral insulin-producing cells, *FEBS Lett.*, 193, 150, 1985.
173. **Rouis, M., Thomopoulos, P., Louache, F., Testa, U., Hervy, C., and Titeaux, M.,** Differentiation of U-937 human monocyte-like cell line by 1 alpha,25-dihydroxyvitamin D_3 or by retinoic acid, *Exp. Cell Res.*, 157, 539, 1985.
174. **Rouis, M., Thomopoulos, P., Cherier, C., and Testa, U.,** Inhibition of insulin receptor binding by A23187: synergy with phorbol esters, *Biochem. Biophys. Res. Comm.*, 130, 9, 1985.
175. **Welsch, C. W., DeHoog, J. V., Scieszka, K. M., and Aylsworth, C. F.,** Retinoid feeding, hormone inhibition, and/or immune stimulation and the progression of *N*-methyl-*N*-nitrosourea-induced rat mammary carcinoma: suppression by retinoids of peptide hormone-induced tumor cell proliferation *in vivo* and *in vitro*, *Cancer Res.*, 44, 166, 1984.

Chapter 4

INSULIN-LIKE GROWTH FACTORS

I. INTRODUCTION

Insulin-like growth factors (IGFs), also called somatomedins, are polypeptides with marked homologies to insulin, possessing potent anabolic and mitogenic effects both in vivo and in vitro.[1,2] IGFs are constituents of a complex of compounds present in serum and designated as nonsuppressible insulin-like activity (NSILA).[3] Insulin-like growth factor I (IGF-I or somatomedin C) and insulin-like growth factor II (IGF-II or somatomedin A) are members of a family of peptide hormones that mediate many, but not all, of the growth-promoting actions of growth hormone (GH).[4] These factors are required for normal fetal and postnatal growth and development as well as for the growth of cultured cells, especially for the control of progression through the G_1 phase of the cell cycle. IGF-I (somatomedin C) may have a critical role for the entry of cells into the S phase of the cell cycle, apparently acting by a post-transcriptional mechanism.[5]

IGF-I and IGF-II are synthesized in many, if not all tissues, including liver, heart, lung, kidney, pancreas, spleen, small intestine, colon, brain, and pituitary gland, although the abundance of IGF-I and IGF-II synthesis in each tissue varies.[6] There is evidence that insulin-related mRNA sequences, possibly including both insulin and IGFs, are transcribed in the normal human placenta, where they may represent from 0.03 to 0.1% of the total polyadenylated RNA.[7] Moreover, placentas from diabetic women express much more of these sequences which may be involved in stimulating the growth of the human fetus. IGFs may be involved not only in growth stimulation but also in enhancing differentiation processes occurring in specific cell types.[8] The physiological properties of IGF-I are different from those of IGF-II. IGF-I is more GH-dependent and more mitogenic than the other somatomedin, IGF-II, which is more insulin-like in its actions and is present in the blood at levels three times greater than those of IGF-I. A monoclonal antibody to IGF-I is capable of blocking the stimulation of DNA synthesis by human plasma or calf serum.[9]

A factor called multiplication-stimulating activity (MSA) has been isolated from fetal rat liver and has been found at high concentrations in fetal rat serum.[10] The cellular receptors for growth and metabolic activities of MSA are separate from the insulin receptors.[11] Rat MSA is probably identical to human IGF-II.[12]

II. THE IGFs GENES

The gene coding for IGF-II consists of four exons spanning a total region of 14 kb.[13] A cDNA sequence of the human IGF-II gene with capability of coding for the IGF-II polypeptide precursor molecule has been constructed.[14] The IGF-II gene resides on human chromosome 11p15, in close proximity to the loci for insulin and the c-H-*ras*-1 proto-oncogene, and the gene for IGF-I is localized on human chromosome 12q22-q24.1.[15-17] The gene order on human chromosome 11p is 5'-INS-IGF2-3'. These genes have the same polarity and the 5' end of the IGF-II gene (IGF2) is within 12.6 kbp of the 3' end of the insulin gene (INS), the genes being separated by an *Alu* sequence.[6] The possible biological significance of these syntenies is unknown.

More precisely, the gene coding for IGF-II has been mapped to human chromosome band 11p14.1.[18] This region is close to 11p13, a band which is deleted in Wilms' tumor (nephroblastoma), an embryonal neoplasm occurring in children in either hereditary or spontaneous forms. Interestingly, high level of expression of IGF-II gene transcripts have been detected in the cells of Wilms' tumors.[18,19] The possible role of this alteration in the origin and/or development of Wilms' tumor is unknown. IGF-II transcripts are elevated in the cells of this tumor when compared with adult tissues but the levels are similar to those found in several fetal tissues including kidney, liver, adrenals, and striated muscle. It is thus not clear whether the high level of IGF-II gene expression in Wilms' tumor reflects only the stage of tumor differentiation or whether IGF-II contributes to the malignant process by mitogenic properties through an autostimulatory action.

III. STRUCTURE AND SYNTHESIS OF IGFs

Determination of the primary structures of IGF-I and IGF-II demonstrated that they are different from the insulin molecule.[20,21] Human IGF-I consists of a single chain of 70 amino acids with 3 disulfide bridges which is synthesized as a 130-amino acid precursor. A synthetic gene coding for an IGF-I analogue replacing the only methionine of the protein at position 59 has been constructed.[22] The artificial protein was found to be 60% as active as native IGF-I in a radioimmunoassay and 50% as potent as native IGF-I in a radioreceptor assay, and displayed mitogenic activity for BALB/c 3T3 cells.

Human IGF-II consists of a single chain of 67 amino acids with three disulfide bridges and is synthesized as a 180-amino acid precursor. Human proIGF-II has an 89-amino acid carboxy-terminal extension called the E-domain, whose function is unknown.[6,23] A similar structure is contained in the rat IGF II precursor polypeptide. The artificial synthesis of biologically active human IGF-II by the solid-phase method has been reported.[24] IGF-II is synthesized by rat liver cells as a polypeptide precursor of approximately 20 kdaltons which is processed to intermediate and mature IGF-II species at the time of secretion or shortly thereafter.[23,25,26] The mature 7884-dalton IGF-II molecule corresponds to the amino-terminus of the 20-kdalton IGF-II precursor.[27]

IV. FUNCTIONS OF IGFs

Insulin and IGFs are importantly involved in both fetal and postnatal growth. IGF-I and IGF-II immunoreactivity has been detected in the tissues or body fluids of fetuses from different mammalian species, and in some cases the activity may be markedly elevated in comparison to postnatal levels.[10] IGFs may also be produced by cultured cells derived from embryos and fetuses. IGF-II activity is produced, for example, by endoderm-like cells derived from embryonal carcinoma cells, which suggests that IGF-II produced by endoderm cells, particularly visceral endoderm, may serve as an early embryonic growth factor.[28] IGF-I is produced by cultured porcine aortic smooth muscle cells, and the synergistical action of IGF-I and PDGF induces in these cells an increase in DNA synthesis that exceeds the sum of the individual effects obtained when either growth factor is added alone.[29]

IGF-II, but not IGF-I, is present in the human cerebrospinal fluid and is apparently synthesized in the brain.[30] Five distinct size classes of IGF-II have been separated from the human brain on the basis of their immunoreactivity, the smallest component having a molecular weight of 7.5 kdaltons, identical to that of purified IGF-II from human serum. The highest concentrations of IGF-II occur in the anterior pituitary gland. The structure of the larger forms of IGF-II-like immunoreactive material and the function of IGF-II in the brain are still unknown.

Many different types of endogenous and exogenous factors contribute to regulating the synthesis and physiological actions of IGFs. EGF, insulin, and GH contribute to the regulation of IGF-I production by human and mouse embryonic tissues as well as by human adult tumor-derived and transformed cells.[31,32]

V. CELLULAR RECEPTORS FOR IGFs

The cellular receptors for IGF-I and IGF-II are structurally different from the insulin receptor.

A. IGF-I Receptor

Insulin may interact on the cell surface with its own receptor or with the receptor for IGF-I, whose structure is very similar to that of the insulin receptor. The IGF-I receptor is composed of two alpha 125,000-dalton subunits and two beta 90,000-dalton subunits constituting a heterotetrameric disulfide-linked complex.[33] After binding with its receptor, IGF-I stimulates phosphorylation of tyrosine residues on the beta-subunits of both its own receptor and the insulin receptor.[34-36] The possible physiological role of this phenomenon is not understood. The human placenta contains two distinct binding and immunoreactive species of IGF-I receptors.[37] A monoclonal antibody against IGF-I is capable of inhibiting the mitogenic effect of IGF-I but not insulin, and also blocks the stimulation of DNA synthesis by human plasma or calf serum.[9] The growth-promoting actions of insulin in most tissues seem to be solely mediated by the binding of the hormone to its high-affinity receptor,[38,39] although in cells such as cultured human skin fibroblasts and breast cells, IGF-I receptors may mediate the growth-stimulating action of insulin.[11,40]

B. IGF-II Receptor

In contrast to the IGF-I receptor, the IGF-II receptor does not appear to have a significant affinity for insulin, although insulin may increase the expression of IGF-II receptors on the cell surface.[41] Upon IGF-II binding, IGF-II receptors are internalized and are subsequently rapidly recycled back to the cell surface.[42] Insulin modulates this IGF-II receptor recycling process such that it mediates an increased rate of cellular IGF-II uptake and degradation. This insulin effect primarily results from the insulin-mediated increase in cell surface IGF-II receptor numbers leading to an increasing number of ligand-bound receptors internalized per unit of time.[42] The IGF-II receptor is constituted by a monomeric structure with an approximate 250,000 mol wt and has been purified to homogeneity from rat placenta and human chondrosarcoma cells.[43,44] Glycosylation of the IGF-II receptor protein is required for the acquisition of IGF-II binding activity.[45]

The physiological role of the IGF-II receptor is not clear at present. The IGF-II receptor present in rat hepatoma H35 cells (which are cells that express abundant IGF-II receptors but have no detectable IGF-I receptors) does not directly mediate the action of IGF-II on DNA synthesis.[46] Furthermore, the IGF-II receptor is devoid of protein kinase activity and does not appear to be subjected to down regulation phenomena. IGF-II receptor is abundantly present in many cell types, including human T-lymphocytes,[47] but its precise physiological role is not understood. In any case, the IGF-II receptor would not act as a direct mediator of at least some of the biological actions of IGF-II. Insulin and IGF-II enhance erythroid colony formation by human bone marrow cells and this stimulation maybe explained by activation of a common receptor or postreceptor system.[48] Further studies are required to characterize the receptor and postreceptor mechanisms of action of insulin, IGF-I and IGF-II.

VI. INSULIN AND IGFs RECEPTORS IN TRANSFORMED CELLS

Receptors for insulin and IGFs are present in many types of tumor cells.[49,50] Similar or identical receptors are present in cell lines derived from human or nonhuman tumors.

A. Insulin Receptors in Primary Tumors

Comparison of insulin binding in tissue samples from a diversity of human tumors (breast carcinoma, colonic adenocarcinoma, gastric carcinoma, adrenocortical carcinoma, lymphoma, and pheochromocytoma) with normal insulin-responsive tissues revealed marked similarity in binding site concentration and affinity.[49] The results suggest that similar mechanisms operate to influence insulin binding in human neoplastic and non-neoplastic tissue and that insulin receptors are preserved during malignant transformation.

Resting normal human lymphocytes do not have insulin receptors, but the receptors emerge on the cell surface upon stimulation with mitogens and other substances.[51,52] In human leukemia, insulin receptors are only present on immature cells, although some cases of poorly differentiated malignant lymphoma show no insulin receptors in the tumor cells.[53] Lymphoblasts from acute lymphoblastic leukemias possess insulin-binding sites, the number of receptors per cell being higher in the null-cell type than in the T-cell type, whereas in 6 from 14 patients with chronic lymphoid leukemias the tumor cells lacked insulin receptors and in the remaining 8 patients the number of insulin binding sites per cell were low.[54] In a study of insulin binding to lymphoblasts in 46 children with leukemia, including 35 children with acute lymphoblastic leukemia (ALL), variable levels of insulin receptors were detected but no correlation with several different clinical parameters was found.[55] The authors of the latter study concluded that "although study of insulin binding by malignant lymphoid cells may be important in understanding the biology of leukemic cells, it does not appear to have any obvious clinical utility".[55]

Insulin (and glucagon) receptors are present in the plasma membranes of rat Morris hepatomas of varying growth rates.[56] The binding of insulin, however, is diminished in Morris hepatomas when compared to plasma membranes from rat normal liver, which is apparently due to either a decrease in binding affinity or a change in site-site interactions. A significant correlation exists between the binding of radiolabeled insulin and the growth rate of different Morris hepatomas.[56]

B. Insulin and IGFs Receptors in Neoplastic Cell Lines

Insulin and IGFs receptors are present at variable levels in most of the cultured neoplastic cell lines examined. The capacity of insulin binding maybe either decreased, normal, or increased in the neoplastic cell lines when compared with the respective nontransformed counterparts.[57-59]

Several cultured human breast cancer cell lines contain IGF-I and IGF-II receptors, and the mitogenic action of insulin on some of these cells (T47D cells) appears to be mediated at least partially by IGF receptors.[40,60] However, insulin is mitogenic to T47D cells only when it is used in supraphysiological concentrations, probably due to its limited binding to the IGF receptors, and it seems likely that IGFs are directly involved in the growth regulation of this tumor cell line, and other similar cell lines, through interaction with their own cellular receptors. Absence of down-regulation of insulin receptors is observed in the human breast cancer cell line MCF-7 cultured in serum-free medium.[61]

High-affinity and low-affinity insulin receptors are abundantly present in cell lines derived from some human hematopoietic neoplasms, such as the IM-9 line, which are B-type lymphoblasts derived from a patient with multiple myeloma.[62] In contrast, almost undetectable levels of insulin receptors are present in other hematopoietic cell lines, for example, in some murine lymphoid tumor cell lines.[63]

In most neoplastic cell lines the insulin receptors are of apparently normal structure but abnormalities in receptor affinity,[64] suggesting possible changes in the structure of insulin receptors, have been detected in some cell lines. An alteration in apparent molecular weight of the insulin receptor has been detected in the human monocytic cell line U-937.[65] Abnormalities of insulin binding and receptor phosphorylation have been described in the insulin-resistant mouse melanoma cell line Cloudman S91.[66] This cell line is defective in both its affinity for insulin and its autophosphorylation properties, one site of autophosphorylation being absent in the altered receptor, which would make it unable to stimulate cell growth.

C. Insulin and IGFs Receptors in Induction of Differentiation of Neoplastic Cells

In the human promyelocytic leukemia cell line HL-60, induction to differentiation by means of different substances (calcitriol, retinoic acid, dimethylsulfoxide, phorbol ester) is accompanied by an increase in the number of plasma membrane insulin receptors.[67] This cell line produces a peptide, or peptides, with insulin-like activity which is distinct from insulin or IGFs and which may possibly play a role in the growth of HL-60 cells.[68] In the U-937 human monocyte-like cell line, induction of differentiation with calcitriol is accompanied by increased insulin binding, whereas induction of differentiation by retinoic acid decreases the hormone binding.[69] Abnormal insulin binding and altered plasma membrane physical properties have been described in a Friend erythroleukemia cell clone resistant to differentiation induced by DMSO.[70] Phorbol esters, which are potent tumor promoters, enhance the phosphorylation of both insulin and IGF-I receptors in a human lymphocyte cell line (IM-9).[71] Phorbol esters also modulate insulin receptor phosphorylation and insulin action in cultured hepatoma cells.[72]

D. Conclusion

In conclusion, receptors for insulin and insulin-like growth factors are preserved in normal amounts and with normal affinity in most tumors and it seems likely that insulin and IGFs may help maintain or stimulate the growth of human and nonhuman tumors. However, the precise role of insulin and IGFs in neoplastic transformation and tumor growth remains undetermined. Expression of insulin and IGFs receptors in both normal and neoplastic cells could be associated with different types of stimulation for cell proliferation. An increase in IGF-II receptors is observed, for example, in rat kidney during compensatory growth.[73] An increase in IGF-II binding to kidney microsomal membranes appears in heminephrectomized rats 4 days after operation, thus preceding compensatory kidney growth for a few days, which suggests that this increase may represent a stimulus for growth in normal tissues.

VII. SUMMARY

Insulin and insulin-like growth factors are very important for the regulation of growth and differentiation in most, if not all, tissues. Functional receptors for insulin and IGFs are present in most types of tumor cells from both primary tumors and neoplastic cell lines. The possible interconnections between the mechanisms of action of insulin and IGFs and oncogene protein products are still little understood. The structural and functional relationship between the insulin receptor kinase and the protein products of the *src* oncogene family, which also possess tyrosine-specific protein kinase activity, makes likely the existence of such interconnections. Further studies are needed in order to characterize the action of carcinogens and tumor promoters on insulin and IGFs receptors present in normal cells, as well as the action of differentiation-inducing compounds on the same receptors present in transformed cells. The postreceptor mechanisms of insulin action in malignant cells have been little studied and possible alterations of these mechanisms remain essentially unknown.

REFERENCES

1. **Herington, A. C., Cornell, H. J., and Kuffer, A. D.**, Recent advances in the biochemistry and physiology of the insulin-like growth factor/somatomedin family, *Int. J. Biochem.*, 15, 1201, 1983.
2. **Vetter, U. and Teller, W. M.**, Somatomedine und ihre Bedeutung für die Pädiatrie, *Klin. Padiat.*, 197, 378, 1985.
3. **Zapf, J., Rinderknecht, E., Humbel, R. E., and Froesch, E. R.**, Nonsuppressible insulin-like activity (NSILA) from human serum: recent accomplishments and their physiological implications, *Metabolism*, 27, 1803, 1978.
4. **Phillips, L. S. and Vassilopoulou-Sellin, R.**, Somatomedins, *N. Engl. J. Med.*, 302, 371, 1980.
5. **Campisi, J. and Pardee, A. B.**, Post-transcriptional control of the onset of DNA synthesis by an insulin-like growth factor, *Mol. Cell. Biol.*, 4, 1807, 1984.
6. **Bell, G. I., Gerhard, D. S., Fong, N. M., Sanchez-Pescador, R., and Rall, L. B.**, Isolation of the human insulin-like growth factor genes: insulin-like growth factor II and insulin genes are contiguous, *Proc. Natl. Acad. Sci. U.S.A.*, 82, 6450, 1985.
7. **Liu, K-S., Wang, C-Y., Mills, N., Gyves, M., and Ilan, J.**, Insulin-related genes expressed in human placenta from normal and diabetic pregnancies, *Proc. Natl. Acad. Sci. U.S.A.*, 82, 3668, 1985.
8. **Schmid, C., Steiner, T., and Froesch, E. R.**, Insulin-like growth factor I supports differentiation of cultured osteoblast-like cells, *FEBS Lett.*, 173, 48, 1984.
9. **Russell, W. E., Van Wyk, J. J., and Pledger, W. J.**, Inhibition of the mitogenic effects of plasma by a monoclonal antibody to somatomedin C, *Proc. Natl. Acad. Sci. U.S.A.*, 81, 2389, 1984.
10. **Moses, A. C., Nissley, S. P., Short, P. A., Rechler, M. M., White, R. M., Knight, A. B., and Higa, O. Z.**, Increased levels of multiplication-stimulatory activity, an insulin-like growth factor, in fetal rat serum, *Proc. Natl. Acad. Sci. U.S.A.*, 77, 3649, 1980.
11. **King, G. L., Kahn, C. R., Rechler, M. M., and Nissley, S. P.**, Direct demonstration of separate receptors for growth and metabolic activities of insulin and multiplication-stimulating activity (an insulin-like growth factor) using antibodies to the insulin receptor, *J. Clin. Invest.*, 66, 130, 1980.
12. **Marquardt, H., Todaro, G. J., Henderson, L. E., and Oroszlan, S.**, Purification and primary structure of a polypeptide with multiplication stimulating activity from rat liver cell cultures: homology with human insulin-like growth factor II, *J. Biol. Chem.*, 256, 6859, 1981.
13. **de Pagter-Holthuizen, P., Höppener, J. W. M., Jansen, M., Geurts van Kessel, A. H. M., van Ommen, G. J. B., and Sussenbach, J. S.**, Chromosomal localization and preliminary characterization of the human gene encoding insulin-like growth factor II, *Human Genet.*, 69, 170, 1985.
14. **Bell, G. I., Merryweather, J. P., Sanchez-Pescador, R., Stempien, M. M., Priestley, L., Scott, J., and Rall, L. B.**, Sequence of a cDNA clone encoding human preproinsulin-like growth factor II, *Nature (London)*, 310, 775, 1984.
15. **Brissenden, J. E., Ullrich, A., and Francke, U.**, Human chromosomal mapping of genes for insulin-like growth factors I and II and epidermal growth factor, *Nature (London)*, 310, 781, 1984.
16. **Tricoli, J. V., Rall, L. B., Scott, J., Bell, G. I., and Shows, T. B.**, Localization of insulin-like growth factor genes to human chromosomes 11 and 12, *Nature (London)*, 310, 784, 1984.
17. **Morton, C. C., Byers, M. G., Nakai, H., Bell, G. I., and Shows, T. B.**, Human genes for insulin-like growth factors I and II and epidesmal growth factor are located on 12q22→q24.1, 11p15, and 4q25-7 q27, respectively, *Cytogenet. Cell Genet.*, 41, 245, 1986.
18. **Reeve, A. E., Eccles, M. R., Wilkins, R. J., Bell, G. I., and Millow, L. J.**, Expression of insulin-like growth factor-II transcripts in Wilms' tumour, *Nature (London)*, 317, 258, 1985.
19. **Scott, J., Cowell, J., Robertson, M. E., Priestley, L. M., Wadey, R., Hopkins, B., Pritchard, J., Bell, G. I., Rall, L. B., Graham, C. F., and Knott, T. J.**, Insulin-like growth factor-II gene expression in Wilms' tumour and embryonic tissue, *Nature (London)*, 317, 260, 1985.
20. **Rinderknecht, E. and Humbel, R. E.**, The amino acid sequence of human insulin-like growth factor I and its structural homology with proinsulin, *J. Biol. Chem.*, 253, 2769, 1978.
21. **Rinderknecht, E. and Humbel, R. E.**, Primary structure of human insulin-like growth factor II, *FEBS Lett.*, 89, 283, 1978.
22. **Peters, M. A., Lau, E. P., Snitman, D. L., Van Wyk, J. J., Underwood, L. E., Russell, W. E., and Svoboda, M. E.**, Expression of a biologically active analogue of somatomedin-C/insulin-like growth factor I, *Gene*, 35, 83, 1985.
23. **Dull, T. J., Gray, A., Hayflick, J. S., and Ullrich, A.**, Insulin-like growth factor II precursor gene organization in relation to insulin gene family, *Nature (London)*, 310, 777, 1984.
24. **Li, C. H., Yamashiro, D., Hammonds, R. G., Jr., and Westphal, M.**, Synthetic insulin-like growth factor II, *Biochem. Biophys. Res. Comm.*, 127, 420, 1985.
25. **Whitfield, H. J., Bruni, C. B., Frunzio, R., Terrell, J. E., Nissley, S. P., and Rechler, M. M.**, Isolation of a cDNA clone encoding rat insulin-like growth factor-II precursor, *Nature (London)*, 312, 277, 1984.

26. **Yang, Y. W.-H., Romanus, J. A., Liu, T-Y., Nissley, S. P., and Rechler, M. M.,** Biosynthesis of rat insulin-like growth factor II. I. Immunochemical demonstration of a 20-kilodalton biosynthetic precursor of rat insulin-like growth factor II in metabolically labeled BRL-3A rat liver cells, *J. Biol. Chem.,* 260, 2570, 1985.
27. **Yang, Y. W.-H., Rechler, M. M., Nissley, S. P., and Coligan, J. E.,** Biosynthesis of rat insulin-like growth factor II. II. Localization of mature rat insulin-like growth factor II (7484 daltons) to the amino terminus of the 20-kilodalton biosynthetic precursor by radiosequence analysis, *J. Biol. Chem.,* 260, 2578, 1985.
28. **Nagarajan, L., Anderson, W. B., Nissley, S. P., Rechler, M. M., and Jetten, A. M.,** Production of insulin-like growth factor II (MSA) by endoderm-like cells derived from embryonal carcinoma cells: possible mediator of embryonic cell growth, *J. Cell. Physiol.,* 124, 199, 1985.
29. **Clemmons, D. R.,** Exposure to platelet-derived growth factor modulates the porcine aortic smooth muscle cell response to somatomedin C, *Endocrinology,* 117, 77, 1985.
30. **Haselbacher, G. K., Schwab, M. E., Pasi, A., and Humbel, R. E.,** Insulin-like growth factor II (IGF II) in human brain: regional distribution of IGF II and of higher molecular mass forms, *Proc. Natl. Acad. Sci. U.S.A.,* 82, 2153, 1985.
31. **Atkison, P. R., Bala, R. M., and Hollenberg, M. D.,** Somatomedin-like activity from cultured embryo-derived cells: partial characterization and stimulation of production by epidermal growth factor (urogastrone), *Can. J. Biochem. Cell Biol.,* 62, 1335, 1984.
32. **Atkison, P. R., Hayden, L. J., Bala, R. M., and Hollenberg, M. D.,** Production of somatomedin-like activity by human adult tumor-derived, transformed, and normal cell cultures and by cultured rat hepatocytes: effects of culture conditions and of epidermal growth factor (urogastrone), *Can. J. Biochem. Cell Biol.,* 62, 1343, 1984.
33. **Massagué, J. and Czech, M. P.,** The subunit structures of two distinct receptors for insulin-like growth factors I and II and their relationship to the insulin receptor, *J. Biol. Chem.,* 257, 5038, 1982.
34. **Rubin, J. B., Shia, M. A., and Pilch, P. F.,** Stimulation of tyrosine-specific phosphorylation *in vitro* by insulin-like growth factor I, *Nature (London),* 305, 438, 1983.
35. **Zick, Y., Sasaki, N., Rees-Jones, R. W., Grunberger, G., Nissley, S. P., and Rechler, M. M.,** Insulin-like growth factor-I (IGF-I) stimulates tyrosine kinase activity in purified receptors from a rat liver cell line, *Biochem. Biophys. Res. Comm.,* 119, 6, 1984.
36. **Sasaki, N., Rees-Jones, R. W., Zick, Y., Nissley, S. P., and Rechler, M. M.,** Characterization of insulin-like growth factor I-stimulated tyrosine kinase activity associated with the beta-subunit of type I insulin-like growth factor receptors of rat liver cells, *J. Biol. Chem.,* 260, 9793, 1985.
37. **Jonas, H. A. and Harrison, L. C.,** The human placenta contains two distinct binding and immunoreactive species of insulin-like growth factor-I receptors, *J. Biol. Chem.,* 260, 2288, 1985.
38. **Massagué, J., Blinderman, L. A., and Czech, M. P.,** The high affinity insulin receptor mediates growth stimulation in rat hepatoma cells, *J. Biol. Chem.,* 257, 13958, 1982.
39. **Czech, M. P., Oppenheimer, C. L., and Massagué, J.,** Interrelationships among receptor structures for insulin and peptide growth factors, *Fed. Proc.,* 42, 2598, 1983.
40. **Furnaletto, R. W. and DiCarlo, J. N.,** Somatomedin-C receptors and growth effects in human breast cells maintained in long-term tissue culture, *Cancer Res.,* 44, 2122, 1984.
41. **Oka, Y., Mottola, C., Oppenheimer, C. L., and Czech, M. P.,** Insulin activates the appearance of insulin-like growth factor II receptors on the adipocyte cell surface, *Proc. Natl. Acad. Sci. U.S.A.,* 81, 4028, 1984.
42. **Oka, Y., Rozek, L. M., and Czech, M. P.,** Direct demonstration of rapid insulin-like growth factor II receptor internalization and recycling in rat adipocytes: insulin stimulates ^{125}I-insulin-like growth factor II degradation by modulating the IGF-II receptor recycling process, *J. Biol. Chem.,* 260, 9435, 1985.
43. **Oppenheimer, C. L. and Czech, M. P.,** Purification of type II insulin-like growth factor receptor from rat placenta, *J. Biol. Chem.,* 258, 8539, 1983.
44. **Cooper, J. L. and Smith, G. L.,** Insulin-like growth factor II binding to cultured human chondrosarcoma cells, *Proc. Soc. Exp. Biol. Med.,* 179, 68, 1985.
45. **MacDonald, R. G. and Czech, M. P.,** Biosynthesis and processing of the type II insulin-like growth factor receptor in H-35 hepatoma cells, *J. Biol. Chem.,* 260, 11357, 1985.
46. **Mottola, C. and Czech, M. P.,** The type II insulin-like growth factor receptor does not mediate increased DNA synthesis in H-35 hepatoma cells, *J. Biol. Chem.,* 259, 12705, 1984.
47. **Brown, T. J., Ercolani, L., and Ginsberg, B. H.,** Demonstration of receptors for insulin-like growth factor-II on human T lymphocytes, *J. Receptor Res.,* 5, 297, 1985.
48. **Dainiak, N. and Kreczko, S.,** Interactions of insulin, insulinlike growth factor II, and platelet-derived growth factor in erythropoietic culture, *J. Clin. Invest.,* 76, 1237, 1985.
49. **Benson, A. E., and Holdaway, I. M.,** Insulin receptors in human cancer, *Br. J. Cancer,* 44, 917, 1981.
50. **Wong, M. and Holdaway, I. M.,** Insulin binding by normal and neoplastic colon tissue, *Int. J. Cancer,* 35, 335, 1985.

51. Helderman, J. H., Reynolds, T. C., and Strom, T. B., The insulin receptors as a universal marker of activated lymphocytes, *Eur. J. Immunol.*, 8, 589, 1978.
52. Helderman, J. H., Role of insulin in the intermediary metabolism of the activated thymic derived lymphocytes, *J. Clin. Invest.*, 67, 1636, 1981.
53. Chen, P., Kwan, S., Hwang, T., Chiang, B. N., and Chou, C-K., Insulin receptors on leukemia and lymphoma cells, *Blood*, 62, 251, 1983.
54. Thomopoulos, P. and Marie, J. P., Insulin receptors in acute and chronic lymphoid leukaemias, *Eur. J. Clin. Invest.*, 10, 387, 1980.
55. Pui, C-H. and Costlow, M. E., Clinical and biologic correlates of insulin binding by leukemia lymphoblastas, *Leuk. Res.*, 9, 843, 1985.
56. Pezzino, V., Vigneri, R., Siperstein, M. D., and Goldfine, I. D., Insulin and glucagon receptors in Morris hepatomas of varying growth rates, *Cancer Res.*, 39, 1443, 1979.
57. Hoffmann, S. S. and Kolodny, G. M., Insulin receptors in 3T3 fibroblasts: relationship to growth phase, transformation and differentiation into new cell types, *Exp. Cell Res.*, 107, 293, 1977.
58. Petersen, B. and Blecher, M., Insulin receptors and functions in normal and spontaneously transformed cloned rat hepatocytes, *Exp. Cell Res.*, 120, 119, 1979.
59. Hofmann, C., Marsh, J. W., Miller, B., and Steiner, D. F., Cultured hepatoma cells as a model system for studying insulin processing and biological responsiveness, *Diabetes*, 29, 865, 1980.
60. Myal, Y., Shiu, R. P. C., Bhaumick, B., and Bala, M., Receptor binding and growth-promoting activity of insulin-like growth factors in human breast cancer cells (T-47-D) in culture, *Cancer Res.*, 44, 5486, 1984.
61. Hwang, D. L., Papoian, T., Barseghian, G., Josefsberg, Z., and Lev-Ran, A., Absence of down-regulation of insulin receptors in human breast cancer cells (MCF-7) cultured in serum-free medium: comparison with epidermal growth factor, *J. Receptor Res.*, 5, 27, 1985.
62. Kaplan, S. A., The insulin receptor, *J. Pediatr.*, 104, 327, 1984.
63. Straus, D. S. and Pang, K. J., Insulin receptors on cultured murine lymphoid tumor cell lines, *Mol. Cell Biochem.*, 47, 161, 1982.
64. Maturo, J. M., III and Hollenberg, M. D., Insulin receptors in transformed fibroblasts and in adipocytes: a comparative study, *Can. J. Biochem.*, 57, 497, 1979.
65. McElduff, A. M., Grunberger, G., and Gorden, P., An alteration in apparent molecular weight of the insulin receptors from the human monocyte cell line U-937, *Diabetes*, 34, 686, 1985.
66. Haring, H. U., White, M. F., Kahn, C. R., Kasuga, M., Lauris, V., Fleischmann, R., Murray, M., and Pawelek, J., Abnormality of insulin binding and receptor phosphorylation in an insulin-resistant melanoma cell line, *J. Cell Biol.*, 99, 900, 1984.
67. Yamanouchi, T., Tsushima, T., Murakami, H., Sato, Y., Shizume, K., Oshimi, K., and Mizoguchi, H., Differentiation of human promyelocytic leukemia cells is accompanied by an increase in insulin receptors, *Biochem. Biophys. Res. Comm.*, 108, 414, 1982.
68. Yamanouchi, T., Tsushima, T., Kasuga, M., and Takaku, F., Variables that regulate the production of insulin-like peptide(s) in human leukemia cell line (HL-60), *Biochem. Biophys. Res. Comm.*, 129, 293, 1985.
69. Rouis, M., Thomopoulos, P., Louache, F., Testa, U., Hervy, C., and Titeux, M., Differentiation of U-937 human monocyte-like cell line by 1 alpha,25-dihycroxyvitamin D_3 or by retinoic acid: opposite effects on insulin receptors, *Exp. Cell Res.*, 157, 539, 1985.
70. Simon, I., Brown, T. J., and Ginsberg, B. H., Abnormal insulin binding and membrane physical properties of a Friend erythroleukemia clone resistant to dimethylsulfoxide-induced differentiation, *Biochim. Biophys. Acta*, 803, 39, 1984.
71. Jacobs, S., Sahyoun, N. E., Saltiel, A. R., and Cuatrecasas, P., Phorbol esters stimulate the phosphorylation of receptors for insulin and somatomedin C, *Proc. Natl. Acad. Sci. U.S.A.*, 80, 6211, 1983.
72. Takayama, S., White, M. F., Lauris, V., and Kahn, C. R., Phorbol esters modulate insulin receptor phosphorylation and insulin action in cultured hepatoma cells, *Proc. Natl. Acad. Sci. U.S.A.*, 81, 7797, 1984.
73. Polychronakos, C., Guyda, H. J., and Posner, B. I., Increase in the type 2 insulin-like growth factor receptors in the rat kidney during compensatory growth, *Biochem. Biophys. Res. Comm.*, 132, 418, 1985.

Chapter 5

EPIDERMAL GROWTH FACTOR

I. INTRODUCTION

Epidermal growth factor (EGF) is a polypeptide with potent mitogenic activity in a diversity of cell types in vivo and a wide variety of cultured cells in vitro.[1-4] The factor has been isolated from the submaxillary glands of mice, from human urine and other sources. EGF is also synthesized and secreted by cultured cells such as cultured human fibroblasts.[5,6] It has been suggested that EGF maybe considered, at least in some aspects, as a polypeptide hormone because it circulates with the blood and can act at far distant sites.[7] In the mouse, the normal serum levels of EGF are of approximately 1 ng/mℓ (1.7×10^{-10} M) and these levels may be regulated by hormones like thyroxine and testosterone, and may show variation according to sex and stage of development.[8,9]

II. STRUCTURE OF EGF

Mouse EGF is a 6045-daltons single-chain polypeptide of 53 amino acids with three disulfide bonds.[10] For an unknown reason it is found in high concentration in the submaxillary glands of adult male mice.[11,12] Human EGF, also named urogastrone (URO) because it is a potent inhibitor of gastric acid secretion, has been isolated from human urine and its amino acid sequence is similar to that of mouse EGF, differing in 16 positions.[13,14] In spite of these differences, mouse EGF and human URO are biologically equipotent and both bind to human fibroblast receptors with similar affinities.[15,16] A polypeptide isolated from human milk is identical to EGF as well as to urogastrone from human urine.[17] Two forms of mouse EGF have been detected by reversed-phase high performance liquid chromatography, alpha-EGF and beta-EGF, and it has been shown that beta-EGF represents the des-asparaginyl-1 form of the polypeptide and that both forms are essentially equipotent as mitogens.[18] Mouse EGF and human EGF/URO seem to derive from a primitive protein related to certain blood coagulation factors.[19]

A. EGF and the Vaccinia Virus Growth Factor

A 140-amino acid polypeptide encoded by one of the early genes of vaccinia virus is related closely to EGF and TGF-α and might be able to bind the EGF cellular receptor and act as a growth factor.[20,21] The factor, called vaccinia virus growth factor (VVGF), is found in the culture medium of vaccinia virus-infected cells, is recognized by antibodies to mouse EGF, and is capable of stimulating autophosphorylation of the EGF receptor.[22] In this manner, the vaccinia virus protein could mediate binding of the virus to the cell surface of cells expressing EGF receptors and could also stimulate growth of the infected cells and/or the neighboring cells. Interestingly, EGF receptor occupancy inhibits vaccinia virus infection.[23] In any case, the detected homology suggests that the capture of cellular genes by viruses, especially genes related to peptide growth factors, is not limited to retroviruses, where they are transduced as viral oncogenes, but may also occur with DNA viruses and may be more common than has been suspected until now.

III. BIOSYNTHESIS OF EGF

According to the results obtained with mouse x Chinese hamster cell hybrids, the EGF gene has been assigned to mouse chromosome 3.[24] In the human, the gene coding for EGF is located on chromosome 4, region 4q25-q27.[24,25] The human EGF gene has been cloned and expressed in *Escherichia coli*, which results in the synthesis and secretion of human EGF by the manipulated bacteria.[26] The human EGF gene has also been chemically synthesized by phosphite-coupling procedures and its expression in yeast yielded a single-chain polypeptide of 53 amino acids that was shown to induce biological effects characteristics of EGF action, like promotion of epithelial cell proliferation and inhibition of gastric acid secretion.[27] Studies with synthetic peptides corresponding to fragments of the amino acid EGF sequence demonstrate that residues 20 to 31 define a primary receptor binding site of EGF, whereas the amino- and carboxyl-terminal regions of the EGF molecule would provide the necessary conformational stability for binding of the middle region.[28]

A. The EGF Precursor

EGF is synthesized in many mammalian tissues from an unexpectedly large precursor.[29-31] The mRNA for EGF contains about 4750 nucleotides and predicts the synthesis of an EGF precursor (prepro-EGF) of 1217 amino acids (130,000 daltons). The amino-terminal segment of the EGF precursor contains seven peptides with sequences similar but not identical to EGF.[29] Whether EGF-like peptides have distinct biological roles and whether they act through their own receptors is not yet known. The polypeptides p788 and p789, which are considered as reliable markers for neoplastic transformation of human fibroblasts, show an amino acid composition which is remarkably similar to that of residues 630 to 680 of the EGF precursor.[32]

The levels of the EGF precursor protein in the mouse kidney are unexpectedly high, being only twofold less than in the mouse submaxillary gland in spite that the kidney contains approximately 2000 times less mature EGF than the submaxillary gland.[31] The relatively low levels of EGF in the kidney may thus reflect differences between the processing of the precursor in this tissue and the submaxillary gland. In the kidney, the highest amounts of EGF precursor are found in the distal tubules, which are involved in the fine regulation of urine composition. The function of the EGF precursor at this site is unknown. EGF precursor maybe a membrane-bound protein as it contains an internal hydrophobic domain and maybe anchored in the basal or luminal surface of the cell membrane. It could act as a receptor in the cells of the distal tubules regulating membrane transport events or may be processed by enzymes present in serum or urine.[31] The latter possibility is supported by the fact that several EGF-related growth factors are present in the urine of normal and human-tumor-bearing athymic mice.[33]

1. Structural Homologies of the EGF Precursor

Homology has been detected between a portion of prepro-EGF (amino acid residues 63 to 880) and a polypeptide, p788 (26,000 mol wt), considered as a reliable marker for neoplastic transformation in human fibroblasts.[32] After carcinogen-induced transformation, the rate of p788 synthesis is elevated about 30-fold, which constitutes the most dramatic example of quantitatively enhanced gene expression among the 700 most abundant polypeptides in the human KD diploid fibroblast strain. However, elevated synthesis of p788 and a related polypeptide, p789, appears to be unrelated to the degree of tumorigenicity of transformed cell lines.[32]

An unexpected sequence homology between the mouse EGF precursor and a region of the bovine low-density lipoprotein (LDL) receptor has been uncovered, which suggests a common ancestor gene for both proteins.[34] Another unexpected homology has been observed between the EGF precursor and a peptide sequence encoded by the second exon of atrial natriuretic factor (ANF) precursor.[35] ANF is known to play an important role in regulating blood pressure as well as the extracellular fluid volume. The evolutionary and biological significance of these structural homologies remain to be determined.

2. Homology between the EGF Precursor and the c-mos Protein

The murine EGF precursor has three regions of sequence homology with the murine c-*mos* proto-oncogene and the v-*mos* oncogene of Moloney MuSV.[36] The three regions together comprise 17% of the EGF precursor and 58% of the c-*mos* sequence. Similarity is greatest between the carboxy-terminal region of the v-*mos* protein (residues 317 to 360) and part of the cytoplasmic domain of the EGF precursor (residues 1127 to 1174). Similarities are also observed between two regions of the murine c-*mos* protein sequence (residues 48 to 134 and 196 to 275) and parts of the extracellular domain of the EGF precursor (residues 565 to 651 and 741 to 817, respectively).[36] These structural relationships suggest that the gene of the EGF precursor and the c-*mos* proto-oncogene may have evolved from a common ancestor.

IV. PRODUCTION AND BIOLOGICAL ACTIONS OF EGF AND EGF-RELATED GROWTH FACTORS

EGF-like activity is present in the developing chick embryo.[37] The biological activity of mouse and human EGF is almost identical. Under certain conditions, the polypeptide acts as a potent mitogen and anabolic agent for a variety of tissues of ectodermal and endodermal origin, being involved in wound repair.[38] Both human and murine EGF, as well as TGF-alpha, are active in promoting eyelid opening in newborn mice.[39] In the mouse, EGF is involved in the control of secondary palate formation by a developmentally regulated process which includes a quantitative modulation of glycosaminglycan synthesis.[40] EGF immunoreactive material has been detected in the central nervous system of the rat, especially in the forebrain and midbrain structures of pallidal areas of the brain, where it could act as either a neurotransmitter or neuromodulator.[41] It is also possible that EGF may act as a mitogenic agent during development of the central nervous system.

EGF-like mitogens are produced and secreted by cultured cells from bovine pituitary glands.[42] The importance of EGF in the growth and function of the mammary gland is indicated by the fact that sialoadenectomy of mice decreases milk production and increases offspring mortality during the lactation period.[43] Administration of EGF to sialoadenectomized pregnant mice increases survival rate of offspring to almost a normal level. Since the mouse submaxillary gland is a rich source of EGF, these results are more easily explained by assuming that the gland has an endocrine function associated with EGF secretion.

In certain organs and tissues the mitogenic action of EGF maybe supplemented, or perhaps replaced, by other mitogenic substances. For example, gastrin-releasing peptide (GRP), which is an analogous of bombesin (a tetradecapeptide discovered in amphibian skin), is particularly abundant in the human fetal lung where it is produced in neuroendocrine cells and may have important mitogenic effects which are independent on the presence of EGF.[44]

In addition to its mitogenic effects, EGF displays a diversity of important physiological activities, including stimulation of ion transport, enhancement of endogenous protein phosphorylation, alterations in cell morphology, and stimulation of DNA synthesis. EGF maybe important in regulating glycogen synthase, a key enzyme involved in the control of glycogen metabolism, through phosphorylation and dephosphorylation mechanisms.[45] EGF has striking effects on the control of vascular function, especially on arterial contractility, and these effects are mediated by prostaglandins, being abolished by indomethacin.[46] EGF also stimulates the synthesis of prostaglandins in kidney cells and modulates hormone production by pituitary cells.[47,48] EGF has been shown to stimulate bone resorption in neonatal mouse calvaria in organ culture via a prostaglandin-mediated mechanism.[49] In addition to these stimulating effects, EGF regulates protein breakdown in A431 human cells.[50] EGF is a potent inhibitor of gastric acid secretion.[1,2]

Intact EGF, as well as EGF fragments, can be isolated from human urine.[51] Urine from nude mice contains EGF as well as a minor acid-stable component with an apparent molecular weight of 20,000, which competes with EGF for binding to EGF membrane receptors and which promotes colony formation by NRK cells in soft agar.[33] The urinary levels of this EGF-related factor are increased in nude mice bearing tumors following subcutaneous injection of cultured human tumor cells and the elevated levels of this growth factor appear to be dependent on tumor burden. The 20,000 mol wt urinary component is separable into four EGF competing activities and the major species is immunologically related to mouse submaxillary gland EGF.

V. THE EGF RECEPTOR

The plasma membrane receptor for EGF has been purified from different sources, including human placenta,[52-54] mouse liver, and the human epidermoid carcinoma cell line A431, the latter expressing a very high number of EGF receptors ($1\text{-}3 \times 10^6$ receptors per cell).[55,56] Paradoxically, the growth of A431 cells is inhibited by EGF.[57] Monoclonal antibodies to EGF receptor have been produced and used for its isolation, characterization, and purification.[58-63] The majority of these monoclonal antibodies recognize oligosaccharide determinants in the EGF receptor molecule. One antibody of the IgM type inhibits EGF binding to its receptor and mimics the biological effects of EGF.[58] Only one EGF-binding site is contained in each EGF receptor molecule,[64] but local aggregation of hormone-receptor complexes is required for activation by EGF.[63,65]

A. The EGF Receptor Gene

The EGF receptor in human cells is encoded by a gene located on chromosome 7p13-q22,[66] which is expressed in different types of human cell lines.[67] Amplification, rearrangement and enhanced expression of the EGF receptor gene occurs in A431 cells.[68] These cells may contain 15 to 25 copies of the EGF receptor gene.[69] cDNA copies of the EGF receptor gene have been obtained from the same cells.[70,71] This cDNA is homologous to a variety of RNAs overproduced in A431 cells,[72] but it is not known whether these RNAs are translated, and whether overproduction of some particular species of EGF receptor-related RNA and/or protein products may be associated with the appearance of a malignant phenotype in A431 cells and perhaps also in other cells. The mechanisms responsible for generation of EGF receptor-related transcripts are unknown but could be associated with structural genomic changes or with alterations in RNA transcription or RNA processing phenomena. Aberrant EGF receptor-coding RNAs may be created in A431 cells by gene rearrangement within chromosome 7, resulting in a fusion of the 5' portion of the EGF receptor gene to an unidentified region of genomic DNA.[73]

Amplification of the EGF receptor gene is also present in the MDA-468 human breast cancer cell line.[74] This cell line expresses a high number of EGF receptors on the cell surface and its growth is inhibited by exogenous EGF. A similar gene amplification is observed in squamous cell carcinomas where the EGF receptor protein is produced in excessive amounts.[75] A 6- to 60-fold amplification of the EGF receptor gene was detected in 4 of 10 primary human brain tumors of glial origin but not in other types of human brain tumors.[69]

The promoter region of the human EGF receptor gene has some unusual characteristics.[76] In contrast to most other eukaryotic genes, the promoter region of the EGF receptor gene contains neither a "TATA box" nor a "CAAT box", but has an extremely high G + C content (88%) and contains five CCGCCC repeats and four (TCC)TCCTCCTCC repeats. This promoter region is situated close to, or within, a DNase I-hypersensitive site in A431 cells.[76] The promoter region of the EGF receptor shows a striking similarity with the promoter region of the human c-H-*ras* proto-oncogene, and this similarity may be relevant to the molecular mechanisms by which the expression of such growth control genes is regulated.[77]

Variant Swiss-Webster 3T3 cell lines unable to mount a mitogenic response to EGF are characterized by the absence of functionally active EGF receptors.[78] Attempts to complement by co-cultivation the EGF-nonresponsive phenotype of three independently isolated variants of these cell lines were unsuccessful.

B. Structure of the EGF Receptor

The mature EGF receptor is a transmembrane glycoprotein of 175,000 daltons consisting of a single polypeptide chain of 1186 amino acids and N-linked carbohydrates. As deduced from a cDNA coding for the EGF receptor precursor, the receptor can be divided into two functional domains: an extracellular domain of 621 amino acids containing the EGF receptor binding site and a cytoplasmic domain of 542 amino acids containing the sequence responsible for tyrosine-specific protein kinase activity.[71] Both domains are linked by a transmembrane region constituted by a stretch of 23 predominantly hydrophobic amino acids. A striking feature of the EGF receptor is that the cytoplasmic domain contains 9 cysteine residues, a value well within the range of most cytoplasmic proteins, but the extracellular domain contains 51 cysteine residues.[71] Most of these extracellular cysteine residues are concentrated within two regions of approximately 170 amino acids and there is evidence indicating the presence of sulfhydryl groups in the EGF receptor.[79] The purified EGF receptor can be reconstituted into artificial phospholipid bilayers by using a detergent-dialysis method.[80] In this system the receptor is uniformly oriented within the bilayer, with the EGF binding domain facing outside the liposomes, and the incorporated receptor is functional in binding EGF and a monoclonal antireceptor antibody.

A model for the ATP-binding site fo the EGF receptor and other structurally related proteins (the v-*erb*-B and c-*src* oncogene products, mammalian cyclic AMP-dependent protein kinase, and the cell division control protein CDC28) has been proposed on the basis of the conservation of certain key amino acid residues.[81] The amino acid residues corresponding to the nucleotide binding site of the EGF receptor have been identified.[82] Monoclonal antibodies to the EGF receptor present in A431 cells act as inhibitors of EGF binding to its receptor on the cell surface and are antagonists of EGF-stimulated protein tyrosine kinase activity.[83] Another monoclonal antibody to the EGF receptor of human A431 cells may function as noncompetitive agonist of EGF action.[63]

The EGF receptor structure is highly conserved in different vertebrate species (human, baboon, dog, rat, mouse, frog, and chicken), which indicates an essential function in different cell types. EGF receptor-related nucleotide sequences are also present in invertebrates, including *Drosophila melanogaster*.[84] Such sequences show varying degrees of homology to members of the *src* oncogene family, including the oncogenes v-*src*, v-*abl*, v-*fes*, v-*fps*, v-*yes*, and v-*fms*. The complete nucleotide sequence of the *Drosophila* EGF receptor homolog gene has been determined.[85] The protein encoded by this gene has, as the human EGF receptor, three distinct domains: an extracellular putative EGF binding domain, a hydrophobic transmembrane region, and a cytoplasmic kinase domain. The overall amino acid homology between the human and the *Drosophila* receptor is 41% in the extracellular domain and 55% in the kinase domain. Two cysteine regions, a hallmark of the human ligand-binding domain, have also been conserved and it is clear from the sequence that the extracellular and the cytoplasmic domains of the receptor have been part of the same molecule for over 800 millions of years.[85] However, the physiological ligand of the *Drosophila* EGF receptor has not been characterized and a specific EGF clone has not been isolated from the *Drosophila* genome.

C. Biosynthesis of the EGF Receptor

The biosynthesis of EGF receptor in A431 cells is initiated in form of a 70,000-dalton protein which is subjected to different post-translational modifications,[86] including the addition of seven or more N-linked high-mannose oligosaccharide chains onto a polypeptide of 138,000 daltons to form a 160,000-dalton intermediate. These oligosaccharide chains are subsequently modified by addition of terminal sugars, including fucose and sialic acid, to give the mature 175,000-daltons form of the receptor.[87] The external EGF-binding domain of the receptor contains about 30% of carbohydrate,[64] and blood group-related antigens are expressed on the carbohydrate chains of the EGF receptor in A431 cells.[88] In addition to high-mannose oligosaccharides, the EGF receptor contains N-linked backbone structures of oligosaccharide chains and peripheral monosaccharides conferring blood group and other polymorphic antigen properties. Glycosylation is necessary for the acquisition of ligand-binding capacity of the receptor.[89] It is not known if glycosylation *per se* is required for ligand binding or if glycosylation must precede a later processing step where binding activity is acquired but the acquisition of binding activity occurs relatively late in the processing pathway of the EGF receptor precursor. The biosynthesis of the EGF receptor in normal human fibroblasts is similar to that found in the neoplastic human A431 cells.[90,91]

D. Expression of EGF Receptors in Different Types of Cells

The EGF receptor is present in a variety of fetal and adult tissues, including human fetal membranes,[92] and its number per cell increases during embryo development.[93] All of the cell types, but not noncellular elements, found in human amnion, chorion, decidua, and placenta contain EGF receptors and the number of EGF receptors per cell is significantly higher at midpregnancy compared to term pregnancy.[94] Functional EGF receptors are present in the rat uterus,[95] but it is not known if all uterine cell types contain the receptors and the possible role of EGF in uterine physiology is unknown. A similar ignorance exists in relation to many other organs and tissues where EGF receptors have been identified and in which the precise role of EGF in regulatory phenomena is not understood.

1. Functional Heterogeneity of EGF Receptors in Different Types of Cells

Most cells contain about 10^5 EGF receptors per cell but the human epidermoid vulval carcinoma cell line A431 contains as many as 2×10^6 EGF receptors per cell and has been used as an excellent source for the purification and characterization of these receptors.[64] The EGF receptor gene is amplified 15- to 20-fold in A431 cells, yet amplification cannot explain the high level of expression of EGF receptors in these cells because it does not lead to concomitant increase of 5.8- and 10.5-kb mRNA transcripts which are responsible for the synthesis of the receptor.[71] In contrast to normal cells, A431 cells synthesize a 2.8-kb transcript at levels 100-fold greater than that of the larger mRNA species. This mRNA encodes a 70,000-mol wt secreted polypeptide representing almost the entire extracellular domain of the receptor. Since this truncated receptor is secreted, it cannot account for the over-expression of functionally intact EGF receptors in A431 cells.

The human breast cancer cell line MDA-468 is also characterized by a very high number of EGF receptors associated with amplification and over-expression of the EGF receptor gene.[74] From 22 grade III or IV human astrocytoma cell lines examined, one cell line (SK-MG-3) exhibited an unusually high number of specific EGF binding sites associated with amplification and rearrangement of the EGF receptor gene.[96] Double minute (DM) chromosomes were detected in SK-MG-3 cells but no homogeneously staining regions (HSRs) were observed. No abnormal EGF receptor-related mRNA species were detected in the same line.[96] Of ten primary human glioblastomas graded as glioblastoma multiforme (astrocytoma grade III or IV) four exhibited an elevated EGF receptor level associated with amplification of the EGF receptor gene.[69] The rarity of EGF receptor gene amplification in astrocytoma cell lines selected and propagated in monolayer suggests that the amplification does not represent an advantage and may even lead to counter-selection, at least under certain conditions of in vitro culture.[96]

The growth of both A-431 and MDA-468 cells maybe stimulated in vitro by low concentrations of EGF (between 0.1 and 10 pM) but is inhibited at higher concentrations (0.1 to 10 nM) which stimulate most other cells. An MDA-468 clone selected for resistance to EGF-induced growth inhibition shows a number of receptors within the normal range, which suggests that a correlation between EGF receptor number and EGF-induced proliferative response may be a general phenomenon.[74] Sustained high levels of EGF achieved in vivo by the administration of testosterone to female athymic mice do not affect the growth of solid A431 tumors in the animals, whereas low levels of EGF stimulate growth of the tumor.[97] These data suggest that the mechanism(s) involved in the inhibition of A431 cell growth in vitro does not operate in vivo and that the effect of EGF in vivo is associated with growth promotion in both normal and tumor cells.

EGF receptors constitute a functionally heterogeneous population and occur in both high and low affinity states in A431 cells and other types of cells. Two classes of EGF receptors are present in rat pheochromocytoma cells (clone PC12), one high-affinity class with 7600 sites per cell and another low-affinity class with 62,000 sites per cell.[98] The two subpopulations of EGF receptors may respond in different fashion to stimulatory agents like TGF-β.[99] High-affinity receptors mediate growth effects of EGF and can be distinguished from the major population of low-affinity receptors by their considerably reduced rate of lateral diffusion.[100] The molecular basis of this heterogeneity is not understood. In particular, it is not known whether a small proportion of a heterogeneous population of EGF receptors associate with an effector molecule to give multiple classes in equilibrium or whether a heterogeneous population of EGF receptors pre-exists but responds independently and in different ways to EGF.

In addition to the mature EGF receptor located on the cell membrane, A431 cells synthesize and secrete a 105-kdalton EGF receptor-related polypeptide (ERRP) which is not derived from the mature receptor but is separately produced by the cell.[101] ERRP synthesis may be from a distinct mRNA generated via alternate splicing.

E. Phosphorylation and Processing of the EGF Receptor

There are three functional sites distinguished in the EGF receptor: the EGF binding site, the protein kinase catalytic site, and the major autophosphorylation site.[102] The functional properties of the EGF receptor on the cell surface could depend on its state of phosphorylation at different amino acid residues.[103] This assumption is not supported by studies with antibodies to the autophosphorylation sites of the receptor.[104] In cells untreated with EGF the EGF receptor is already phosphorylated at several sites on tyrosine, threonine, and serine residues located in the carboxyl-terminal region of the molecule, and phosphorylation at these sites increases following treatment with EGF, which also induces phosphorylation at threonine in the amino-terminal, EGF-binding domain of the molecule.[105] After binding of EGF to its cellular receptor, the receptor is autophosphorylated on tyrosine residues,[55,106,107] which may result in the acquisition of tyrosine kinase activity with affinity for different protein substrates.[108-111] Three major in vitro autophosphorylation sites near the carboxy-terminus of the EGF receptor have been assigned as follows: tyrosine 1,068 (site P3), tyrosine 1,148 (site P2), and tyrosine 1,173 (site P1).[112] Each of these residues is immediately preceded by an amino acid residue with an acidic side chain. P1 is the site selectively phosphorylated in vivo while sites P2 and P3 are autophosphorylation sites used to a lesser extent in vivo. In contrast to the insulin receptor and the viral oncogene products possessing tyrosine-specific protein kinase activity, the EGF receptor region that is active as a protein kinase lacks the major site(s) of tyrosine phosphorylation.[113] This fact indicates that tyrosine phosphorylation within the catalytic domain of the EGF receptor molecule is not important for the kinase activity associated with the EGF receptor.

In addition to tyrosine, the EGF receptor is phosphorylated on serine and threonine residues by the catalytic subunit of cyclic AMP-dependent protein kinase,[114] and analysis of the in vivo phosphorylated EGF receptor reveals the presence of phosphotyrosine, phosphoserine and phosphothreonine.[107] EGF receptor affinity is modulated, via altered receptor phosphorylation, by intracellular Ca^{2+} and protein kinase C activity.[115] A functional consequence of C-kinase-catalyzed phosphorylation at specific threonine residues in the EGF receptor kinase may be the inhibition of EGF-stimulated tyrosine-specific protein kinase activity.[116] Protein kinase C phosphorylates the EGF receptor at threonine-654, which is in a very basic sequence of nine amino acid residues close to the cytoplasmic face of the plasma membrane, in the region before the protein kinase domain of the molecule.[117] In the EGF receptor, threonine-654 is thus located at a position where it can be involved in the modulation of signaling between the internal domain and external EGF-binding domains.

At least some phosphorylation of the receptor occurs during its process of maturation, before the addition of high-mannose oligosaccharide which determines the formation of a 160,000-daltons form of receptor precursor.[90] EGF-stimulated protein kinase shows similar activity in the liver of adult and senescent mouse.[118] Calcineurin purified from bovine brain catalyzes the complete dephosphorylation of the phosphotyrosine and phosphoserine residues in the human placental EGF receptor.[119]

F. Internalization and Degradation of the EGF-Receptor Complex

The EGF-receptor complex is internalized into the cell and is, at least partially, fused with lysosomes, which results in the production of proteolytic fragments and simultaneous down-regulation of the receptors located at the cell surface.[120-124] At least three intermediates in the degradation of EGF have been detected in KB human epidermoid carcinoma cells. Some proteolysis of EGF may already occur in the endocytic vesicles while the final proteolytic products are formed after arrival of EGF in the lysosomal compartment.[125] On the other hand, internalized EGF is processed to a number of high molecular weight products in PANC-1 human pancreatic carcinoma cells. In these cells EGF undergoes only limited processing, being able to bypass the cellular degradative pathways and being rebound to the cell after its slow release into the culture medium.[126] Little is known about the physiological relevance of these processes. Recycling or inactivation of the EGF receptor may not occur during down-regulation and it is likely that EGF receptors are rapidly degraded in the presence of EGF.[127] Binding of EGF induces a decrease in the population of mature EGF receptors but has no discernible effect on the synthesis or stability, which suggests that the observed reduction in EGF binding capacity observed during down-regulation is produced solely by a change in the rate of degradation of receptors.[91] The comparison of different cell lines which are either responsive or nonresponsive to EGF indicates that the presence of specific saturable EGF receptors does not correlate with cell growth but differences in the processing of EGF after binding to the receptor were clearly apparent in the same cell lines.[128]

In contrast to EGF-induced EGF receptor internalization, phorbol ester-induced EGF receptor internalization do not cause delivery of the EGF receptors to lysosomes but rather, the receptors reappear after about 1 hr on the cell surface, even in the continuous presence of phorbol ester and probably as a consequence of receptor recycling.[129] Since phorbol ester induces phosphorylation of the EGF receptor on a threonine residue that lies close to the cytosolic face of the plasma membrane,[117] the data suggest that phorbol ester-induced phosphorylation of a specific threonine residue by protein kinase C (which is activated by phorbol ester) is directly involved in inducing EGF receptor internalization.[129]

Studies with monoclonal antibodies directed against the EGF receptor suggest that the internalization of ligand-receptor complexes is necessary for EGF to exert its mitogenic effects.[58] Monensin and methylamine prevent degradation of ^{125}I-labeled EGF and cause intracellular accumulation of EGF receptors, blocking the EGF-induced mitogenic response.[130] These results are consistent with the hypothesis that sequestration of the EGF-receptor complexes in a cytosolic compartment maybe a primary signal required for correct intracellular transport of EGF and for events that might lead to generation of a mitogenic response. The EGF receptor of A431 cells is associated with the cytoskeleton both at the cell surface and at intracellular sites, and this structural association of the receptor may contribute to the modulation of the activity and substrate specificity of the EGF receptor kinase.[131]

G. The erb-B Oncogene Protein Product and the EGF Receptor

A particular type of acute leukemia virus (ALV) is the avian erythroblastosis virus (AEV), which induces primarily erythroblastosis when injected into susceptible chickens but can also occasionally induce sarcomas or carcinomas.[132] AEV also induces transformation of immature erythroid cells and fibroblasts in culture. Differentiation can occur in AEV-transformed embryonic erythroid cells.[133] The transforming sequences of AEV correspond to the v-erb oncogene, whose cellular counterpart, c-erb, is present in the genome of all of the vertebrate species examined, including humans and fishes.[134] AEV transduces two different oncogenes, v-erb-A and v-erb-B, derived from separate, unlinked DNA sequences in the cellular genome.[135] An AEV variant, called AEV-H, transduces only the erb-B oncogene and is also capable of inducing both erythroblastosis and sarcomas in chickens.[136] The erb-B gene is frequently transduced by another virus, the Rous-associated virus type 1 (RAV-1), which is also capable of inducing rapid-onset erythroblastosis when inoculated into 1-week-old chickens.[137]

A striking structural homology has been observed between the transforming protein of the v-erb-B oncogene of AEV and the EGF receptor protein purified from A431 cells and normal human placenta.[138] Furthermore, both the c-erb-B proto-oncogene and the EGF receptor gene are located in the same region of human chromosome 7,[66,139] which suggests their possible identity. Moreover, the 3' coding domain of the human EGF receptor gene cloned from A431 cells shows striking homology to the v-erb-B oncogene.[70] The c-erb-B/EGF receptor locus has been mapped to mouse chromosome 11, which also contains the loci for alpha-globin and for two hematopoietic growth factors, CSF-2 and IL-3.[140] However, the possible biological significance of these associations is not understood.

The product of the v-erb-B oncogene is a 68,000-dalton glycoprotein, gp68$^{v\text{-}erb\text{-}B}$, which is modified further to a 74,000-dalton protein, gp74$^{v\text{-}erb\text{-}B}$, the latter being located at the cell surface.[141,142] The use of three different glycoprotein processing inhibitors has demonstrated that incorrectly glycosylated v-erb-B protein is inserted normally into the plasma membrane and is still capable of exerting its oncogenic activity on susceptible cells.[143]

The v-erb-B oncogene product corresponds to a truncated form of the EGF receptor, with deletion of the amino-terminal end, which may explain the absence of cross-reactions between antisera against these two proteins. However, site-specific antibodies to the v-erb-B protein product precipitate EGF receptor.[144] The v-erb-B protein contains only the transmembrane and tyrosine-specific protein kinase domains of the EGF receptor, and lacks most of the extracellular domain responsible for EGF binding.[71,138] Thus, the v-erb-B protein corresponds to a growth factor receptor which has lost the regulatory binding domain but has retained the membrane locating and biochemical activating domains. The v-erb-B protein represents a kind of unregulated receptor which may be expressed constitutively in the activated state.

In contrast to the truncated intracytoplasmic version of the receptor corresponding to the expression of the v-erb-B oncogene, A431 carcinoma cells synthesize high amounts of 2,8-kb mRNA transcripts which encode a secreted polypeptide representing almost the entire extracellular EGF binding domain of the receptor.[71] Since this abnormal version of the EGF receptor is secreted, it cannot account for the very high number of functionally intact EGF receptors that are expressed in A431 cells.

The v-erb-B 68,000-dalton protein product is phorphorylated primarily on serine and threonine residues but it contains minor amounts of phosphotyrosine and protein phosphorylation at tyrosine residues is induced by the v-*erb*-B protein in vivo and in vitro.[145,146] Both the EGF receptor and the v-*erb*-B proteins are tyrosine-specific protein kinases. AEV-transformed chicken embryo fibroblasts show enhanced tyrosine phosphorylation of a number of cellular polypeptides, including 36- and 42-kdalton proteins.[146] The major in vivo tyrosine autophosphorylation site is tyrosine 1,173 (site P1), which is located 14 residues from the carboxy-terminus of the receptor molecule and is not found in the v-*erb*-B protein.[112] The latter protein is fused at the carboxy-terminus to four residues of the viral *env* protein. Therefore, v-*erb*-B represents an oncogene which is terminated prematurely with respect to the cellular EGF receptor sequences from which it is derived.[112] Antibodies to a synthetic oligopeptide have been used as a probe for the kinase activity of the avian EGF receptor and v-*erb*-B protein.[147]

H. The c-*neu* Oncogene Protein Product and the EGF Receptor

A proto-oncogene, termed c-*neu*, has been detected by DNA transfection experiments, using the NIH/3T3 test system, in the genome of rat neuroglioblastoma cell lines derived from tumors induced by ethylnitrosourea (ENU).[148,149] The c-*neu* gene is associated with the expression of a common tumor antigen, which is a 185,000-dalton protein, termed p185, that is glycosylated and possesses intrinsic tyrosine-specific protein kinase activity. This protein has antigenic determinants in common with the EGF receptor but its size is somewhat larger. The c-*neu* protein is encoded by a distinct proto-oncogene which is located on human chromosome 17, region 17q21, and which is not coamplified with the c-*erb*-B/EGF receptor gene in A431 cells.[150] There is nucleotide sequence homology between the *neu* and *erb*-B proto-oncogenes and p185 is antigenically related to the EGF receptor protein.[149] The c-*neu* gene, like c-*erb*-B, is a member of the *src* gene family and at least one region of the c-*neu* sequence, that which encodes the tyrosine-specific protein kinase domain, is closely related to the c-*erb*-B/EGF receptor sequence.[150] These similarities suggest that the p185 c-*neu* protein product may function in normal cells as a growth factor receptor. A receptor with tyrosine protein kinase activity and with extensive homology to EGF receptor shares chromosomal location with the c-*neu* proto-oncogene.[151] Monoclonal antibodies against the p185 c-*neu* gene product can induce reversion of the transformed phenotype in NIH/3T3 cells transformed by transfection of a c-*neu* gene.[152,153]

A novel proto-oncogene of the tyrosine kinase family, termed c-*erb*-B-2 has been found to be amplified in a human salivary gland adenocarcinoma and a human mammary carcinoma.[154,155] The predicted amino acid sequence of c-*erb*-B-2 protein is highly homologous to the human c-*erb*-B/EGF receptor proto-oncogene product and shows 42 to 52% homology with the predicted amino acid sequences of other tyrosine kinase-encoding genes, including *src*, *abl*, *fms*, and the human insulin receptor. The tyrosine-specific protein kinase domain of the putative c-*erb*-B-2 protein shows 82% homology with that of the EGF receptor. The proto-oncogenes C-*neu* and C-*erb*-B-2 are probably identical.

I. The v-*src* Oncogene Protein Product and the EGF Receptor

Antibodies generated against two synthetic peptides corresponding to two defined regions of the RSV oncogene protein product, pp60^{v-src}, interact specifically with the EGF receptor protein.[156] The results indicate that the carboxy-terminal (cytoplasmic) portion of the EGF receptor, which contains the tyrosine-specific protein kinase and autophosphorylating functional domains of the molecule, is antigenically related to pp60^{v-src}. A similar antigenic and structural homology exists between the cytoplasmic portion of the EGF receptor, the *erb*-B oncogene, and other members of the *src* oncogene family, including the oncogenes *fes*, *mos*, *yes*, *fps*, and *abl*.[157] This region would also have homology with hormone and peptide growth factor receptors possessing tyrosine-specific protein kinase activity, including the receptors for insulin, IGF-I and PDGF. Indeed, the human insulin receptor is precipitated by antisera against pp60$^{v\text{-}src}$.[158]

As deduced from the entire 1,370 amino acid sequence of the human insulin receptor precursor, the insulin receptor is structurally related to both the EGF receptor and the members of the *src* gene family of proteins.[159] The cytoplasmic domain of the EGF receptor has the stretch of amino acid sequence (approximately, residues 690 to 940) that is shared with proteins of the *src* family.[71] The ATP-binding site of the EGF receptor/kinase has structural features in common with the homologous sites in pp60$^{v\text{-}src}$ as well as with cyclic AMP-dependent protein kinase.[82] Furthermore, these kinases share structural homology with a variety of nucleotide-binding proteins, including the p21 protein products of the c-*ras* proto-oncogenes.[160]

Both the activated EGF receptor/kinase and the oncogene protein pp60$^{v\text{-}src}$ are able to interact with and nick supercoiled double-stranded DNA in an ATP-stimulated manner.[161] This phenomenon could account for some of the functional changes produced by the EGF receptor kinase and the pp60$^{v\text{-}src}$ protein at the nuclear level.

J. EGF Receptors in Human Tumors

The possible role of EGF receptors in the origin and/or maintenance of a transformed phenotype remains little understood. The amount of EGF receptors is highly variable among different human tumor cell lines, in some lines the receptors being readily detectable whereas in other lines the number of these receptors are below the limits of detection.[67] As stated above, very high amounts of EGF receptors (30 to 100 times more than normal) associated with amplification and over-expression of the EGF receptor gene are exceptionally found in the human epidermoid carcinoma cell line A431 and the human breast cancer cell line MDA-468, whose growths are inhibited by EGF at concentrations that stimulate most other cell lines.[56,74]

EGF receptors are present in some human breast cancer cell lines but in other similar lines they are apparently absent. Moreover, there is no clearcut correlation between EGF binding and EGF-induced mitogenic activity in cell lines with EGF receptors,[162] and even cell lines expressing exceptionally large number of EGF receptors per cell may not respond to EGF.[163] Changes in EGF receptors have also been observed in human primary breast cancer and have been compared to changes observed in estrogen receptors.[164] Significant levels of EGF receptors were detected in approximately half of 137 biopsies from unselected primary and metastatic human breast cancer tumors.[165] In the latter study, no significant association was found between the magnitude of EGF binding in individual tumors and either estrogen or progesterone receptor levels.

EGF receptors were also present in cell lines derived from different types of other human tumors (pulmonary, gastric, bone, oral, laryngeal, and cervical cancers) as well as in normal human fibroblasts.[166,167] A wide variation in the number of EGF receptor sites was found in the latter study among different types of tumor cell lines, in some of them the receptor levels being much lower and in other lines higher than in normal human fibroblasts. Increased levels of EGF receptors have been detected in some human sarcomas.[168] EGF receptors have been detected in tumor cell lines established from human osteosarcoma and giant tumors of the bone but their characteristics are comparable or identical to those of normal cells, and the same receptors are present in normal bone, which suggests that EGF maybe involved in normal bone metabolism.[167] A human choriocarcinoma cell line (JEG-3) contains EGF receptors and the cells respond to addition of EGF (or phorbol ester) by a marked increase in the secretion of human chorionic gonadotropin into the medium.[169] EGF receptors are present in normal human liver and may be decreased in human primary hepatoma.[170,171] Variation in EGF receptor levels have been observed in human bladder cancer when invasive and superficial tumors have been compared.[172]

High levels of EGF receptors have been found in human brain tumors, especially in tumors of non-neuronal origin such as glioblastomas and meningiomas, whereas the receptor levels present in neuroblastomas were similar to those present in the brain of patients who died from diseases not related to the central nervous system.[69,173] The EGF receptor levels in human glioblastoma multiforme tumors were highly variable, which could be attributed to the cell type diversity within these tumors. Qualitative differences may exist between EGF receptors present in different human tumor cell lines. EGF receptors in the human glioblastoma cell line SF268 differ from those in epidermoid carcinoma cell line A431.[174]

K. Factors Involved in Changes of EGF Receptors

Many factors can produce changes in EGF receptor number or affinity as well as in EGF receptor phosphorylation and activity.[175] EGF receptor levels in the liver, the uterus and the mammary gland may show significant changes according to various physiological states, including hormonal changes.[176-178] Estrogens can regulate acutely the levels of EGF receptors in the uterus, which raises the possibility that events coupled to this receptor may play a role in estrogen-stimulated growth.[178] In the thyroid gland, EGF receptor expression, as well as responsiveness to the mitogenic action of EGF, is modulated in vitro by TSH through a cAMP-dependent process.[179] Mouse embryo cells arrested in G1 due to nutrient deficiency show a reduction in EGF binding to 10 to 20% of that observed under other conditions.[180] This effect appears to be due to decreased receptor number and could represent the mechanism by which cells are able to enter a quiescent state when faced with the possibility of becoming deficient in essential low molecular weight nutrients. No changes in the number of kinase activity of the EGF receptor has been observed in the liver of senescent mice when compared with the liver of adult mice.[118]

Transforming growth factor beta (TGF-β) controls EGF receptor levels in cultured NRK fibroblasts.[181] The expression of EGF and growth hormone receptors in a human breast cancer cell line (T-47D) is regulated by progestins, but not by testosterone, estradiol or hydrocortisone.[182] Thyroid hormone is involved in the regulation of EGF receptor levels in vivo, the levels being markedly reduced in hypothyroid animals.[176] It is possible that at least some effects of thyroid hormone on cell growth and differentiation are mediated through alterations in EGF receptor levels. Dimethyl sulfoxide (DMSO), which may produce important effects on the growth and differentiation of cells in vitro, stimulates tyrosine kinase activity of the EGF receptor.[183]

L. Expression of EGF Receptors in Malignant Cells

Besides the A431 human epidermoid carcinoma cell line, the MDA-468 human breast cancer cell line, and the SK-MG-3 human astrocytoma cell line, which are characterized by amplification of the EGF receptor gene, some other tumor cell lines, mainly derived from tumors of the female urogenital system, contain increased amounts of the EGF receptor protein although they do not possess amplified EGF receptor gene sequences.[67] EGF receptor is expressed selectively by human melanoma cells which show the presence of an extra copy of chromosome 7 (which contains the locus for the EGF receptor).[184] Cells of benign pigmented lesions (nevi) or radial growth phase (nonmetastatic) primary melanoma do not express EGF receptors and do not have extra copies of chromosome 7. Low or undetectable levels of EGF receptors are contained in cultured normal human fibroblasts as well as in some cell lines derived from normal or malignant blood cells. Presence of EGF receptors in the tumor cells may be an indicator of poor prognosis in patients with breast cancer.[185]

Oncogenic viruses, carcinogenic agents, and tumor promoters may produce important changes in EGF receptors. Infection of cells with SV40 may produce important quantitative and qualitative changes in EGF receptors, which maybe related to changes in the oligosaccharide portion of the receptor molecule.[186] Acute transforming retroviruses induce a rapid and profound decrease in EGF receptor levels.[187] The same phenomenon is observed in vivo when renal tumors are induced with Kirsten MuSV.[188] Decreased levels of EGF receptors are also observed in chemically transformed cells at saturation density in vitro.[189] Gradual decrease in EGF receptor expression is observed in the course of spontaneous neoplastic progression of whole Chinese hamster embryo (CHE) lineages during serial passage in culture.[190]

1. Biological Significance of EGF Receptor Alterations in Malignant Cells

The mechanisms related to decrease of EGF receptors in transformed cells are unknown. It has been suggested that the decreased levels of EGF receptors observed in these cells could result either from an indirect effect of transforming genes on the synthesis, degradation and/or location of functional receptors, or from production in transformed cells of endogenous EGF or a related substance that binds to EGF receptors.[187] Rat kidney cells transformed by K-MuSV lose the ability to bind labeled EGF.[188] However, the presence of EGF receptors in K-MuSV-transformed cells can be demonstrated by immunoprecipitation with monoclonal antibody to the receptor, and furthermore, two-dimensional peptide mapping indicates that the EGF receptor present in the transformed cells has apparently the same structure as its normal counterpart.[191] A possible explanation for such discrepancy is the presence of conformational changes in membrane components surrounding the EGF receptors in virus-transformed cells. On the other hand, there is evidence that activation of c-*erb*-B in ALV-induced erythroblastosis leads to the expression of a truncated EGF receptor kinase.[192]

Four different calmodulin antagonists induce decreased binding of ^{125}I-labeled EGF to SV40-transformed human fibroblasts (WI38) in a dose-dependent manner, but the same phenomenon is not observed in cultured normal human cells.[193] The effect observed in transformed cells is apparently due to a decrease in the affinity of the plasma membrane EGF receptor for the EGF molecule when calmodulin activity is inhibited.

It is possible that a decrease of EGF receptors is associated with activation of the EGF-dependent cellular growth regulatory system, and that certain oncogene products maybe involved in the induction of this phenomenon. The protein products of two oncogenes, v-*src* and v-*fps*, which have tyrosine kinase activity, lead to a rapid decrease in cellular EGF receptors, probably through activation of the EGF-dependent growth regulatory system.[194]

However, a direct relationship between the EGF receptor reduction and cellular tumorigenic properties appears to be nonexistent,[195] and a decrease in EGF receptors is probably not a general phenomenon of neoplastic transformation since it is not observed in cells transformed by DNA viruses or by most types of chemical carcinogens.[196] TGF-β induces not a decrease but a rapid increase in the number of EGF receptors on the plasma membrane.[181] Although the growth of human tumor cells in athymic mice may be inhibited by monoclonal antibodies against EGF receptor,[197] a monoclonal antibody to the EGF receptor is unable to block "autocrine" growth stimulation in TGF-secreting melanoma cells.[198] The disparate variations in the number of cellular EGF receptors under different conditions make rather unlikely the possibility that the process of transformation should always be mediated by an activation of the EGF-dependent cellular proliferation regulatory system.

M. Tumor Promoters and EGF Receptors

A marked decrease in EGF receptor levels can occur under the influence of potent tumor promoters, such as the phorbol ester TPA.[199] Phorbol diesters as well as other types of tumor promoters (indole alkaloids and polyacetates) inhibit EGF-stimulated phosphorylation on tyrosine residues, which correlates with loss of EGF binding to the high-affinity EGF receptor.[200] Phorbol diesters like TPA and PMA induce increased serine and threonine phosphorylation within the binding domain of the EGF receptor.[90,105,201,202] The major site of phorbol ester-induced EGF receptor phosphorylation is not a tyrosine residue but a threonine residue located on the amino-terminal, EGF-binding domain of the molecule.[105] Phorbol esters cause the phosphorylation of EGF receptors in normal human fibroblasts at a unique site, threonine in position 654, which is located nine amino acids away from the predicted transmembrane domain of the EGF receptor on the cytoplasmic side of the plasma membrane.[203] Threonine-654 maybe involved in regulating the tyrosine kinase activity of the EGF receptor as well as the binding of EGF.

Tumor promoters can exert their effects on the phosphorylation and affinity of EGF receptors by substituting for diacylglycerol at the level of activation of protein kinase C. A stringent correlation has been observed between the potency of 11 diacylglycerol analogues to modulate protein kinase C activity in vitro and their capacity to mimic the action of phorbol ester in vivo.[204] The results suggest that tumor-promoting phorbol esters regulate the EGF receptor by a mechanism that can be accounted for by their effects on the activity of protein kinase C. TPA itself shares several biological activities with EGF, stimulating the proliferation of mammary gland epithelium and preventing DNA fragmentation induced by serum deprivation in cultured fibroblasts.[205,206]

The functional properties of EGF receptors may be modified by tumor promoters. Phorbol esters may inhibit the binding of labeled EGF to its receptor on the cell surface of different types of tumor cell lines.[166,169] This inhibition is accompanied by either potentiation or blockage of tyrosine phosphorylation of EGF receptors in A431 cells.[116,200,201] However, the possible role of protein kinase C in these effects is not clear. Protein kinase C itself is a phorbol ester receptor,[207] but phorbol receptors other than protein kinase C may exist,[208] and they could mediate the effect of phorbol esters on EGF receptor phosphorylation. In general, the relationship between phosphorylation of the EGF receptor on tyrosine residues and the physiological actions of EGF is not clear. There is no temporal correlation between EGF-enhanced tyrosine phosphorylation and the growth response, and EGF does not elicit a mitogenic response in A431 cells at concentrations that stimulate EGF-dependent tyrosine kinase activity in membranes.[57]

It is clear that further studies are required for a better understanding of the role of the EGF receptor in normal and transformed cells and for a better characterization of its functional relationship with the protein product of the c-erb-B proto-oncogene.

VI. POSTRECEPTOR MECHANISMS OF ACTION OF EGF

The postreceptor mechanisms involved in EGF action are very complex and, at present, there is an almost complete ignorance about their sequential operation in either normal or transformed cells. In particular, the nature of the possible intracellular chemical mediators or second messengers responsible for the pleiotropic EGF-induced physiological effects is unknown.[209]

A. Redistribution of Cell Membrane Components
EGF induces a rapidly initiated series of structural and functional changes on the surface of responsive cells,[210] but the mechanisms accounting for these changes are not understood. An important action of EGF may consist in induction of a redistribution of cell membrane components, including the transferrin receptor. EGF induces a rapid and transient increase in the number of transferrin receptors at the cell surface with no change in total cellular receptor content, which suggests that the newly appearing transferrin receptors are temporarily unable to be internalized.[211] This type of action may be associated with the pleiotropic responses that cells are known to undergo as a consequence of EGF treatment.

B. EGF-Stimulated Phosphorylation of Cellular Proteins
Phosphorylation of different cellular proteins is stimulated after the formation of a complex

between EGF and its cellular receptor but it is difficult to determine which of these proteins are direct substrates of the EGF-stimulated kinase activity of the receptor and which are phosphorylated by other, secondarily activated kinases.[55,109,212] EGF receptor kinase participates in the phosphorylation of a 34,000 to 39,000-dalton protein, usually termed 34K or 36p protein, which is located on the inner aspect of the plasma membrane.[213] The 34K protein is also a cellular substrate for the protein product of the v-*src* oncogene,[214] and both protein kinases, the EGF receptor kinase and the pp60$^{v\text{-}src}$ kinase, phosphorylate the protein exclusively on tyrosine residues.[215] The 34K protein is detectable in fibroblastic cells of connective tissue and in endothelial cells in all of the mouse and rat tissues examined but is present at very low or undetectable levels in skeletal and smooth muscle cells, erythrocytes and lymphocytes.[216] In vivo phosphorylation of the 34K protein in normal cells treated with EGF has not been demonstrated and the function of 34K protein is unknown. A 35,000-dalton calcium substrate protein is phosphorylated by EGF action in intact A431 cells.[217]

Other substrate for EGF-stimulated phosphorylation is the 40S ribosomal protein S6,[218] and S6 kinase is most likely activated by phosphorylation.[216] Phosphorylation of protein S6 is stimulated by the action of other hormones and growth factors in different tissues, including insulin in liver cells and human chorionic gonadotropin (hCG) in Leydig cells.[220,221] S6 phosphorylation is also stimulated by the protein products of oncogenes like v-*src* and v-*fps*.[222] The mechanisms of S6 phosphorylation are little understood. A small amount of S6 phosphorylation by action of EGF and other mitogens may depend on the activity of a cAMP-dependent protein kinase.[223] A potential mediator of EGF-induced S6 phosphorylation is the cAMP-independent, Ca^{2+}-independent protein kinase named protease-activated kinase II (PAK-II).[224] The mechanisms involved in regulation of PAK-II activity in vivo are unknown. Phosphorylation of S6 could contribute to stimulate cell proliferation under particular physiological conditions,[225] although a direct regulation of DNA synthesis through S6 phosphorylation seems unlikely.

In other cell types the major increase in tyrosine phosphorylation under the action of EGF (or other growth factors and mitogens) occurs on a 42K protein, and to a lesser extent on a 40K protein.[226] The physiological role of the phosphorylation of different cellular proteins by EGF action is not understood. Exogenous substrates can also be phosphorylated to different extents by EGF-stimulated kinase activity, including proteins such as histones, protamines and tubulin, as well as small synthetic peptides.[55] Hormones such as gastrin-17 and growth hormone, and hormone receptors such as progesterone receptor subunits can also be phosphorylated by EGF-stimulated kinase activity.[227-229] The possible physiological role of these modifications is not understood. The differences between stimulation and inhibition of growth induced by EGF and other growth factors, or by oncogene protein products, may not lie in unique tyrosine phosphorylations in one growth state, but in the balance between phosphorylated and nonphosphorylated forms of many cellular proteins.[230] Monoclonal antibodies to phosphotyrosyl proteins from growth factor-stimulated cells and viral oncogene-transformed cells may contribute to the isolation and characterization of specific protein substrates involved in the action mechanisms of growth factors and oncogene protein products.[231]

C. EGF and Phosphorylation of the v-*ras* Protein Product

EGF receptor kinase could also participate in the phosphorylation of oncogene protein products. A synthetic peptide containing the autophosphorylation site of the v-H-*ras* p21 product is phosphorylated solely on tyrosine by the EGF-stimulated receptor kinase of A431 cell membranes.[232]

Membranes isolated from NRK cells transformed by H-MuSV show a three- to fivefold increase in the phosphorylation of the v-H-*ras* p21 oncogene product when EGF is added to the incutation medium.[233] This increase in p21$^{v\text{-}ras}$ phosphorylation may reflect the activation of a kinase or inhibition of a phosphatase by EGF. Apparently, insulin may have a similar effect on v-*ras* p21 phosphorylation. The biological significance of these findings is unknown.

D. EGF Action and Phosphoinositide Metabolism

Phosphatidylinositol kinase activity is not present in the activated EGF receptor molecule and the mechanisms of EGF action on phosphoinositide metabolism are little understood.[234] There is evidence, however, that at least in some cellular systems EGF can stimulate phosphoinositide turnover with activation of protein kinase C and diacylglycerol kinase, which results in 1,2-diacylglycerol breakdown to phosphatidic acid.[235] In turn, 1,2-diacylglycerol can modulate binding and phosphorylation of the EGF receptor, probably through activation of protein kinase C.[115,236,237] Treatment of 3T3 fibroblasts with diacylglycerol results in a rapid decrease in the affinity of EGF receptors for the specific ligand.[238] Redistribution of Ca^{2+} maybe associated with changes in phosphoinositide metabolism. The calmodulin antagonist trifluoperazine (TFP) is a potent inhibitor of EGF-induced tyrosine phosphorylation of the EGF receptor, although this effect is apparently not due to calmodulin antagonism since other calmodulin antagonist, W7, has only a slight effect on EGF receptor phosphorylation.[239]

Neither insulin nor EGF stimulate the synthesis of phosphatidylinositol, phosphatidylinositol 4-phosphate, or phosphatidylinositol 4,5-bisphosphate in rat liver plasma membranes or intact hepatocytes, which suggests that the insulin- and EGF-stimulated receptor tyrosine kinases do not act on phosphoinositides in the liver.[240,241] Further studies are required for a better characterization of the complex interrelationships between EGF action and changes in phosphoinositide metabolism.

E. EGF Actions at the Nuclear Level

Very low concentrations of EGF can stimulate both RNA and DNA synthesis in human fibroblasts and these effects can be markedly affected by minute amounts of cholera toxin, which acts by raising the intracellular levels of cAMP.[242] There is evidence of the existence in monkey cells of a 48,000-dalton growth inhibitor protein capable of antagonizing EGF-mediated effects on stimulation of DNA synthesis through mechanisms depending on RNA synthesis.[243] These results lend support to a model for the control of cell proliferation based on opposed actions of growth factors and growth inhibitors. Moreover, certain growth factors may act as growth inhibitors in specific cell systems, as demonstrated by the inhibition of proliferation of A431 human carcinoma cells by EGF in spite of the presence of an extremely high number of EGF receptors in these cells. In AKR-2B mouse embryo cells, EGF induces the synthesis of particular species of mRNAs (VL30 sequence elements) closely related to sequences present in integrated retroviruses and certain classes of transposable genetic elements.[244,245]

The mRNAs induced by EGF remain poorly characterized. Stimulation of quiescent rat fibroblasts by either EGF or serum results in increase of different molecular species of mRNAs, the most abundant being those coding for proteins with extensive homology to the glycolytic enzymes lactate dehydrogenase, enolase, and triose phosphate isomerase.[246] In addition, another EGF-induced mRNA encodes for actin. It is not clear, however, if the elevation in the levels of these mRNAs, and presumably of the corresponding proteins, is directly related to the triggering of the particular mitogenic signal or if it is a consequence of the general growth response. A particular species of mRNA, termed pTR1 RNA, is present in high amounts in rat fibroblasts transformed by polyoma virus, the v-*src* oncogene, and the EJ mutant human c-H-*ras* oncogene, and the same RNA is rapidly induced by the addition of EGF (5 ng/mℓ) to the culture medium of normal rat fibroblasts.[246] Both the growth factor and the oncogenes control expression of the corresponding gene at the transcriptional level. pTR1 RNA is not present in serum-stimulated cells and is apparently not a participant of the "normal" or general cellular growth response. The function of the pTR1-derived protein has not been established but the results indicate that growth factors and oncogene protein products may act by controlling the transcriptional activity of similar or identical genes.

Cyclin maybe involved in the control of cell proliferation by EGF. EGF inhibits the proliferation of A431 cells, and the inhibition is paralleled by a decreased synthesis of cyclin, both phenomena being absent in A431 cell variants resistant to the growth inhibitory effect of EGF.[247]

EGF may have inhibitory effects on the synthesis of certain RNAs and cellular proteins. In a primary culture system of the dog thyroid gland, EGF decreases thyroglobulin mRNA and protein synthesis to undetectable levels, affecting profoundly the gland's morphology and the capacity for iodide trapping.[248] EGF inhibits casein production and the accumulation of casein mRNA induced by insulin, cortisol and prolactin in primary cultures of mammary epithelial cells from pregnant mice.[249] Various cAMP derivatives counteract the inhibitory effect of EGF, which suggests a possible modulatory function of the cyclic nucleotide in regulation of casein production at either the transcriptional or the translational level. In contrast, cAMP do not reverse the stimulatory effect of EGF on mammary cell proliferation, which indicates that cAMP selectively counteracts the effect of EGF on mammary gland differentiation.[249] An interesting feature is that the action of cAMP on casein production is observed in the mammary gland organ culture system but not in mammary cell cultures. These results suggest that EGF may not act directly on mammary epithelial cells but that its effects are mediated through the action on nonepithelial cells of the gland, such as fat or mesenchymal cells, which are present in tissue explants but not in cell culture systems.[249] The possibility should be considered that cAMP acts on the epithelial cells of the mammary gland through the production of growth factors synthesized by the nonepithelial cells of the same gland.

EGF induces in mouse embryo cells a rapid increase in actin mRNA levels, which could involve modulation of a specific, labile repressor of actin gene transcription.[250] This fact is interesting because sequences related to the cytoskeletal actin gene have been found in the v-*onc* gene of the Gardner-Rasheed feline sarcoma virus, the *fes/fgr* oncogene.[251] Actin or its higher ordered derivatives are the principal components of microfilaments, structures believed to be involved in important cellular processes, including morphogenesis, motility, and mitosis. Microfilaments may play a necessary role in the initiation of DNA synthesis in response to EGF,[252] and an altered microfilament organization is characteristically observed in cells transformed by diverse biological and chemical agents.

F. EGF-Mediated Induction of Proto-Oncogene Expression

EGF can mediate the expression of proto-oncogenes in particular types of EGF-responsive cells. In NIH/3T3 cells, EGF induces a rapid increase of c-*myc* messenger RNA levels.[253] In addition, EGF stimulates guanine nucleotide binding activity and phosphorylation of the *ras* oncogene protein products.[233] Stimulation and inhibition of growth by EGF in different A431 cell clones is accompanied by the rapid induction of c-*fos* and c-*myc* proto-oncogenes.[254] In contrast to fibroblasts, A431 cells, which are characterized by the presence of a very high number of EGF receptors, respond to EGF with a decreased growth rate but, in spite of this fact, EGF induces in A431 cells transient expression of c-*fos* and c-*myc* messenger RNA and proteins. These observations suggest that the induction of both proto-oncogenes by EGF is not strictly correlated with proliferative activity but should be attributed to the primary interaction of EGF with its receptor. Moreover, c-*fos* and c-*myc* expression is also induced by cyanide bromide-cleaved EGF (CNBR-EGF), a molecule with no mitogenic activity but still capable of triggering some of the early cellular actions of EGF.[254]

Strong induction of c-*fos* and, to a lesser extent, of c-*myc* is produced by the phorbol ester TPA and by the calcium ionophore A23187, which suggests that protein kinase C may be involved in proto-oncogene activation by growth factors.[254] EGF and NGF induce rapid transient changes in proto-oncogene expression in PC12 rat pheochromocytoma cells.[255]

G. EGF Action on Cell Proliferation

EGF is an important mitogenic factor involved in regulation of the proliferation of different types of cells, especially epithelial cells. In primary cultures of adult rat hepatocytes, EGF stimulates DNA synthesis and this effect is greatly enhanced by norepinephrine which reduces binding of EGF to its receptor at the cell surface.[256] Unfortunately, the mechanisms responsible for EGF-induced cell proliferation are not understood. There is evidence that calcium ions may play a central role in such mechanisms, perhaps by acting as an intracellular mediator or second messenger because EGF induces by over 50-fold the extracellular Ca^{2+} requirement for multiplication of cultured normal skin fibroblasts.[257,258] Pretreatment of human fibroblasts with interferon abolishes the mitogenic effect of EGF without affecting either the receptor binding of EGF or the down-regulation of the EGF receptor and without blocking early events (increased amino acid transport) in the course of EGF-stimulated thymidine incorporation.[259]

EGF receptor phosphorylation may be necessary, but insufficient in itself, to trigger mitogenesis. The TNR9 variant line of Swiss 3T3 cells does not respond mitogenically to phorbol esters like TPA but does respond mitogenically to EGF. TPA binding, protein kinase C activation, and EGF receptor phosphorylation, however, proceed normally in the non-mitogenic variant.[260] The results clearly show that EGF and other growth factors do not depend solely on the protein kinase C pathway to elicit their mitogenic effect.

A possible mechanism for EGF-induced modification of genomic functions is the regulation of DNA topoisomerases and DNA nicking activity. EGF binding to mouse and human cells has been shown to enhance topoisomerase activity, and this activity in the nucleus corresponds with DNA synthesis in the cells.[261] Purified EGF receptors of both human and murine origin can nick supercoiled double-stranded DNA in an ATP-dependent fashion.[161] However, DNA-nicking activity is not intrinsic to the EGF receptor but is apparently mediated by a 100,000-mol wt cellular protein.[262]

Oncogene protein products may either diminish or abolish the requirement of EGF for cellular proliferation. Clonal BALB/c mouse epidermal keratinocyte (BALB/MK) cells have an absolute requirement for nanomolar concentrations of EGF for their proliferation. Infection of these cells with BALB- or K-MuSV induces a complete abrogation of their requirement for EGF,[263] which indicates that v-*ras* protein products confer to epidermal cells the rapid acquisition of EGF-independent growth. Members of both the *ras* and *src* oncogene families may supplant the EGF requirement of BALB/MK-2 keratinocytes and may induce alterations in the terminal differentiation processes of these cells.[264]

H. EGF Influence on Carcinogenic Processes

It is difficult to evaluate the exact role of EGF in carcinogenic processes. EGF may even display opposite effects according to different experimental protocols. For example, EGF promotes radiation-induced cell transformation in the C3H10T1/2 system,[265] but the same factor is capable of suppressing transformation induced by the complete chemical carcinogen, methylcholanthrene (MCA), in the same system.[266] When applied to mouse skin following a carcinogenic dose of MCA, EGF shorts the latent period prior to tumor appearance and increases the frequency of tumors.[267]

The submaxillary gland is a rich source of EGF in mice (virgin female mice of C3H/HeN strain) and its surgical removal (sialoadenectomy) reduces the incidence of mammary tumors in the animals.[268] Long-term treatment of the sialoadenectomized mice with EGF increases the incidence of mammary tumors. Moreover, sialoadenectomy of mammary tumor-bearing mice causes a rapid and sustained cessation of tumor growth, but EGF administration quickly restores the rate of tumor growth to the usual level.[268] These results suggest that EGF produced by the submaxillary glands plays a crucial role in mouse mammary tumorigenesis.

VII. SUMMARY

EGF is an important factor involved in the regulation of growth and metabolism of normal cells. The EGF precursor protein (prepro-EGF) is a large protein of 1217 amino acids which shows three regions of sequence homology with the c-*mos* proto-oncogene protein product. EGF receptor has close structural and functional homology with the protein product of the v-*erb*-B oncogene. The v-*erb*-B protein represents a truncated version of the EGF receptor protein, containing only the transmembrane attaching part and the cytoplasmic tyrosine-specific protein kinase domains of the EGF receptor and lacking most of the extracellular domain responsible for EGF binding, which suggests that abnormal expression of EGF receptors could contribute to uncontrolled cell proliferation and tumorigenesis. The cytoplasmic (carboxy-terminal) portion of the EGF receptor, containing the kinase and autophosphorylating domains of the molecule, is antigenically related to the protein products of the *src* gene family. Moreover, some cellular protein substrates of the EGF receptor-associated kinase activity maybe shared with similar activities that are present in the protein products of the *src* oncogene family. EGF enhances guanylate cyclase activity in vivo and in vitro and stimulates guanine nucleotide binding activity and phosphorylation of *ras* oncogene protein products. In addition, EGF could be involved in regulating the expression of several proto-oncogenes, including c-*fos* and c-*myc*, in particular types of cells, which may influence cell proliferation and/or cell differentiation. Further studies are required for a proper evaluation of the exact role of EGF in the regulation of growth in normal tissues as well as in experimental and spontaneous carcinogenic processes.

REFERENCES

1. **Carpenter, G. and Cohen, S.,** Epidermal growth factor, *Ann. Rev. Biochem.,* 48, 193, 1979.
2. **Hollenberg, M. D.,** Epidermal growth factor-urogastrone, a polypeptide acquiring hormonal status, *Vitam. Horm.,* 37, 69, 1979.
3. **Schlessinger, J., Schreiber, A. B., Levi, A., Lax, I., Libermann, T., and Yarden, Y.,** Regulation of cell proliferation by epidermal growth factor, *CRC Crit. Rev. Biochem.,* 14, 93, 1982.
4. **Hollenberg, M. D. and Armstrong, G. D.,** Epidermal growth factor-urogastrone and its receptor, in *Polypeptide Hormone Receptors,* Posner, B. I., Ed., Marcel Dekker, New York, 1985, 201.
5. **Kurobe, M., Furukawa, S., and Hayashi, K.,** Synthesis and secretion of an epidermal growth factor (EGF) by human fibroblast cells in culture, *Biochem. Biophys. Res. Comm.,* 131, 1080, 1985.
6. **Sato, M., Yoshida, H., Hayashi, Y., Miyakami, K., Bando, T., Yanagawa, T., Yura, Y., Azuma, M., and Ueno, A.,** Expression of epidermal growth factor and transforming growth factor-beta in a human salivary gland adenocarcinoma cell line, *Cancer Res.,* 45, 6160, 1985.
7. **Hollenberg, M. D.,** Epidermal growth factor: a polypeptide acquiring hormonal status, *PAABS Rev.,* 5, 265, 1976.
8. **Perheentupa, J., Lakshmanan, J., Hoath, S. B., and Fisher, D. A.,** Hormonal modulation of mouse plasma concentration of epidermal growth factor, *Acta Endocrinol.,* 107, 571, 1984.
9. **Perheentupa, J., Lakshmanan, J., Hoath, S. B., Beri, U., Kim, H., Macaso, T., and Fisher, D. A.,** Epidermal growth factor measurements in mouse plasma: method, ontogeny, and sex difference, *Am. J. Physiol.,* 248, E391, 1985.
10. **Savage, C. R., Jr., Inagami, T., and Cohen, S.,** The primary structure of epidermal growth factor, *J. Biol. Chem.,* 247, 7612, 1972.
11. **Cohen, S.,** Isolation of a mouse submaxillary gland protein accelerating incisor eruption and eyelid opening in the newborn animal, *J. Biol. Chem.,* 237, 1555, 1962.

12. **Turkington, R. W., Males, J. L., and Cohen, S.**, Synthesis and storage of epithelial-epidermal growth factor in submaxillary gland, *Cancer Res.*, 31, 252, 1971.
13. **Gregory, H.**, Isolation and structure of urogastrone and its relationship to epidermal growth factor, *Nature (London)*, 257, 325, 1975.
14. **Gregory, H. and Preston, B. M.**, The primary structure of human urogastrone, *Int. J. Pept. Protein Res.*, 9, 107, 1977.
15. **Hollenberg, M. D. and Gregory, H.**, Human urogastrone and mouse epidermal growth factor share a common receptor site in cultured human fibroblasts, *Life Sci.*, 20, 267, 1976.
16. **Hollenberg, M. D. and Gregory, H.**, Epidermal growth factor-urogastrone: biological activity and receptor binding derivatives, *Mol. Pharmacol.*, 17, 314, 1980.
17. **Petrides, P. E., Hosang, M., Shooter, E., Esch, F. S., and Böhlen, P.**, Isolation and characterization of epidermal growth factor from human milk, *FEBS Lett.*, 187, 89, 1985.
18. **DiAugustine, R. P., Walker, M. P., Klapper, D. G., Grove, R. I., Willis, W. D., Harvan, D. J., and Hernandez, O.**, beta-Epidermal growth factor is the des-asparaginyl[1] form of the polypeptide, *J. Biol. Chem.*, 260, 2807, 1985.
19. **Doolittle, R. F., Feng, D. F., and Johnson, M. S.**, Computer-based characterization of epidermal growth factor precursor, *Nature (London)*, 307, 558, 1984.
20. **Brown, J. P., Twardzik, D. R., Marquardt, H., and Todaro, G. J.**, Vaccinia virus encodes a polypeptide homologous to epidermal growth factor and transforming growth factor, *Nature (London)*, 313, 491, 1985.
21. **Twardzik, D. R., Brown, J. P., Ranchalis, J. E., Todaro, G. J., and Moss, B.**, Vaccinia virus-infected cells release a novel polypeptide functionally related to transforming and epidermal growth factors, *Proc. Natl. Acad. Sci. U.S.A.*, 82, 5300, 1985.
22. **Stroobant, P., Rice, A. P., Gullick, W. J., Cheng, D. J., Kerr, I. M., and Waterfield, M. D.**, Purification and characterization of vaccinia virus growth factor, *Cell*, 42, 383, 1985.
23. **Eppstein, D. A., Marsh, Y. V., Schreiber, A. B., Newman, S. R., Todaro, G. J., and Nestor, J. J., Jr.**, Epidermal growth factor receptor occupancy inhibits vaccinia virus infection, *Nature (London)*, 318, 663, 1985.
24. **Zabel, B. U., Eddy, R. L., Lalley, P. A., Scott, J., Bell, G. I., and Shows, T. B.**, Chromosomal locations of the human and mouse genes for precursors of epidermal growth factor and the beta subunit of nerve growth factor, *Proc. Natl. Acad. Sci. U.S.A.*, 82, 469, 1985.
25. **Morton, C. C., Byers, M. G., Nakai, H., Bell, G. I., and Shows, T. B.**, Human genes for insulin-like growth factors I and II and epidermal growth factors are located on 12q22→q24.1, 11p15, and 4q25→q27, respectively, *Cytogenet. Cell Genet.*, 41, 245, 1986.
26. **Oka, T., Sakamoto, S., Miyoshi, K.-I., Fuwa T., Yoda, K., Yamasaki, M., Tamura, G., and Miyake, T.**, Synthesis and secretion of human epidermal growth factor by *Escherichia coli*, *Proc. Natl. Acad. Sci. U.S.A.*, 82, 7212, 1985.
27. **Urdea, M. S., Merryweather, J. P., Mullenbach, G. T., Coit, D., Heberlein, U., Valenzuela, P., and Barrk, P. J.**, Chemical synthesis of a gene for human epidermal growth factor urogastrone and its expression in yeast, *Proc. Natl. Acad. Sci. U.S.A.*, 80, 7461, 1983.
28. **Komoriya, A., Hortsch, M., Meyers, C., Smith, M., Kanety, H., and Schlessinger, J.**, Biologically active synthetic fragments of epidermal growth factor: localization of a major receptor-binding region, *Proc. Natl. Acad. Sci. U.S.A.*, 81, 1351, 1984.
29. **Scott, J., Urdea, M., Quiroga, M., Sanchez-Pescador, R., Fong, N., Selby, M., Rutter, W. J., and Bell, G. I.**, Structure of a mouse submaxillary messenger RNA encoding epidermal growth factor and seven related proteins, *Science*, 221, 236, 1983.
30. **Gray, A., Dull, T. J., and Ullrich, A.**, Nucleotide sequence of epidermal growth factor cDNA predicts a 128,000-molecular weight protein precursor, *Nature (London)*, 303, 722, 1983.
31. **Rall, L. B., Scott, J., Bell, G. I., Crawford, R. J., Penschow, J. D., Niall, H. D., and Coghlan, J. P.**, Mouse prepro-epidermal growth factor synthesis by the kidney and other tissues, *Nature (London)*, 313, 228, 1985.
32. **Burbeck, S., Latter, G., Metz, E., and Leavitt, J.**, Neoplastic human fibroblast proteins are related to epidermal growth factor precursor, *Proc. Natl. Acad. Sci. U.S.A.*, 81, 5360, 1984.
33. **Twardzik, D. R., Kimball, E. S., Sherwin, S. A., Ranchalis, J. E., and Todaro, G. J.**, Comparison of growth factors functionally related to epidermal growth factor in the urine of normal and human tumor-bearing athymic mice, *Cancer Res.*, 45, 1934, 1985.
34. **Russell, D. W., Schneider, W. J., Yamamoto, T., Luskey, K. L., Brown, M. S., and Goldstein, J. L.**, Domain map of the LDL receptor: sequence homology with the epidermal growth factor precursor, *Cell*, 37, 577, 1984.
35. **Hayashida, H. and Miyata, T.**, Sequence similarity between epidermal growth factor precursor and atrial natriuretic factor precursor, *FEBS Lett.*, 185, 125, 1985.
36. **Baldwin, G. S.**, Epidermal growth factor precursor is related to the translation product of the Moloney sarcoma virus oncogene *mos*, *Proc. Natl. Acad. Sci. U.S.A.*, 82, 1921, 1985.

37. **Mesiano, S., Browne, C. A., and Thorburn, G. D.,** Detection of endogenous epidermal growth factor-like activity in the developing chick embryo, *Dev. Biol.,* 110, 23, 1985.
38. **Buckley, A., Davidson, J. M., Kamerath, C. D., Wolt, T. B., and Woodward, S. C.,** Sustained release of epidermal growth factor accelerates wound repair, *Proc. Natl. Acad. Sci. U.S.A.,* 82, 7340, 1985.
39. **Smith, J. M., Sporn, M. B., Roberts, A. B., Derynk, R., Winkler, M. E., and Gregory, H.,** Human transforming growth factor-alpha causes precocious eyelid opening in newborn mice, *Nature (London),* 315, 515, 1985.
40. **Turley, E. A., Hollenberg, M. D., and Pratt, R. M.,** Effect of epidermal growth factor/urogastrone on glycosaminoglycan synthesis and accumulation in vitro in the developing mouse palate, *Differentiation,* 28, 279, 1985.
41. **Fallon, J. H., Seroogy, K. B., Loughlin, S. E., Morrison, R. S., Bradshaw, R. A., Knauer, D. J., and Cunningham, D. D.,** Epidermal growth factor immunoreactive material in the central nervous system: location and development, *Science,* 224, 1107, 1984.
42. **Kudlow, J. E. and Kobrin, M. S.,** Secretion of epidermal growth factor-like mitogens by cultured cells from bovine anterior pituitary glands, *Endocrinology,* 115, 911, 1984.
43. **Okamoto, S. and Oka, T.,** Evidence for physiological function of epidermal growth factor: pregestational sialoadenectomy of mice decreases milk production and increases offspring mortality during lactation period, *Proc. Natl. Acad. Sci. U.S.A.,* 81, 6059, 1984.
44. **Willey, J. C., Lechner, J. F., and Harris, C. C.,** Bombesin and the C-terminal tetradecapeptide of gastrin-releasing peptide are growth factors for normal human bronchial epithelial cells, *Exp. Cell Res.,* 153, 245, 1984.
45. **Chan, C. P. and Krebs, E. G.,** Epidermal growth factor stimulates glycogen synthase activity in cultured cells, *Proc. Natl. Acad. Sci. U.S.A.,* 82, 4563, 1985.
46. **Muramatsu, I., Hollenberg, M. D., and Lederis, K.,** Vascular actions of epidermal growth factor-urogastrone: possible relationship to prostaglandin production, *Can. J. Physiol. Pharmacol.,* 63, 994, 1985.
47. **Levine, L. and Hassid, A.,** Epidermal growth factor stimulates prostaglandin biosynthesis by canine kidney (MDCK) cells, *Biochem. Biophys. Res. Comm.,* 76, 1181, 1977.
48. **Schonbrunn, A., Kransoff, M., Westendorf, J. M., and Tashjian, A. H.,** Epidermal growth factor and thyrotropin-releasing hormone act similarly on a clonal pituitary cell strain: modulation of hormone production and inhibition of cell proliferation, *J. Cell. Physiol.,* 85, 786, 1980.
49. **Tashjian, A. H., Jr. and Levine, L.,** Epidermal growth factor stimulates prostaglandin production and bone resorption in cultured mouse calvaria, *Biochem. Biophys. Res. Comm.,* 85, 966, 1978.
50. **Ballard, F. J.,** Regulation of protein breakdown by epidermal growth factor in A431 cells, *Exp. Cell Res.,* 157, 172, 1985.
51. **Mounts, C. D., Lukas, T. J., and Orth, D. N.,** Purification and characterization of epidermal growth factor (beta-urogastrone) and epidermal growth factor fragments from large volumes of human urine, *Arch. Biochem. Biophys.,* 240, 33, 1985.
52. **Hock, R. A., Nexo, E., and Hollenberg, M. D.,** Isolation of the human placenta receptor for epidermal growth factor-urogastrone, *Nature (London),* 277, 403, 1979.
53. **Hock, R. A. and Hollenberg, M. D.,** Characterization of the receptor for epidermal growth factor-urogastrone in human placenta membranes, *J. Biol. Chem.,* 255, 10731, 1980.
54. **Hock, R. A., Nexo, E., and Hollenberg, M. D.,** Solubilization and isolation of the human placenta receptor for epidermal growth factor-urogastrone, *J. Biol. Chem.,* 255, 10737, 1980.
55. **Carpenter, G.,** The biochemistry and physiology of the receptor-kinase for epidermal growth factor, *Mol. Cell. Endocrinol.,* 31, 1, 1983.
56. **Carpenter, G.,** Properties of the receptor for epidermal growth factor, *Cell,* 37, 357, 1984.
57. **Gill, G. N. and Lazar, C. S.,** Increased phosphotyrosine content and inhibition of proliferation in epidermal growth factor-treated A431 cells, *Nature (London),* 293, 305, 1981.
58. **Schreiber, A. B., Libermann, T. A., Lax, I., Yarden, Y., and Schlessinger, J.,** Biological role of epidermal growth factor receptor clustering, *J. Biol. Chem.,* 258, 846, 1983.
59. **Richert, N. D., Willingham, M. C., and Pastan, I.,** Epidermal growth factor receptor: characterization of a monoclonal antibody specific for the receptor of A431 cells, *J. Biol. Chem.,* 258, 8902, 1983.
60. **Gooi, H. C., Schlessinger, J., Lax, L., Yarden, Y., Libermann, T. A., and Feizi, T.,** Monoclonal antibody reactive with the human epidermal-growth-factor receptor recognizes the blood-group-A antigen, *Biosci. Rep.,* 3, 1045, 1983.
61. **Parker, P. J., Young, S., Gullick, W. J., Mayes, E. L. V., Bennett, P., and Waterfield, M. D.,** Monoclonal antibodies against the human epidermal growth factor receptor from A431 cells: isolation, characterization, and use in the purification of active epidermal growth factor receptor, *J. Biol. Chem.,* 259, 9906, 1984.
62. **Yarden, Y., Harari, I., and Schlessinger, J.,** Purification of an active EGF receptor kinase with monoclonal antireceptor antibodies, *J. Biol. Chem.,* 260, 315, 1985.

63. **Fernandez-Pol, J. A.**, Epidermal growth factor receptor of A431 cells: characterization of a monoclonal anti-receptor antibody noncompetitive agonist of epidermal growth factor action, *J. Biol. Chem.*, 260, 5003, 1985.
64. **Weber, W., Bertics, P. J., and Gill, G. N.**, Immunoaffinity purification of the epidermal growth factor receptor: stoichiometry of binding and kinetics of self-phosphorylation, *J. Biol. Chem.*, 259, 14631, 1984.
65. **Schechter, Y., Hernaez, L., Schlessinger, J., and Cuatrecasas, P.**, Local aggregation of hormone-receptor complexes is required for activation by epidermal growth factor, *Nature (London)*, 278, 835, 1979.
66. **Kondo, I. and Shimizu, N.**, Mapping of the human gene for epidermal growth factor receptor (EGFR) on the p13-q22 region of chromosome 7, *Cytogenet. Cell Genet.*, 35, 9, 1983.
67. **Xu, Y-H., Richert, N., Ito, S., Merlino, G. T., and Pastan, I.**, Characterization of epidermal growth factor receptor gene expression in malignant and normal human cell lines, *Proc. Natl. Acad. Sci. U.S.A.*, 81, 7308, 1984.
68. **Merlino, G. T., Xu, Y-H., Ishii, S., Clark, A. J. L., Semba, K., Toyoshima, K., Yamamoto, T., and Pastan, I.**, Amplification and enhanced expression of the epidermal growth factor receptor gene in A431 human carcinoma cells, *Science*, 224, 417, 1984.
69. **Libermann, T. A., Nusbaum, H. R., Razon, N., Kris, R., Lax, I., Soreq, H., Whittle, N., Waterfield, M. D., Ullrich, A., and Schlessinger, J.**, Amplification, enhanced expression and possible rearrangement of EGF receptor gene in primary human brain tumours of glial origin, *Nature (London)*, 313, 144, 1985.
70. **Lin, C. R., Chen, W. S., Kruiger, W., Stolarsky, L. S., Weber, W., Evans, R. M., Verma, I. M., Gill, G. N., and Rosenfeld, M. G.**, Expression cloning of human EGF receptor complementary DNA: gene amplification and three related messenger RNA products in A431 cells, *Science*, 224, 843, 1984.
71. **Ullrich, A., Coussens, L., Hayflick, J. S., Dull, T. J., Gray, A., Tam, A. W., Lee, J., Yarden, Y., Libermann, T. A., Schlessinger, J., Downward, J., Mayes, E. L. V., Whittle, N., Waterfield, M. D., and Seeburg, P. H.**, Human epidermal growth factor cDNA sequence and aberrant expression of the amplified gene in A431 epidermoid carcinoma cells, *Nature (London)*, 309, 418, 1984.
72. **Xu, Y-H., Ishii, S., Clark, A. J. L., Sullivan, M., Wilson, R. K., Ma, D. P., Roe, B. A., Merlino, G. T., and Pastan, I.**, Human epidermal growth factor receptor cDNA is homologous to a variety of RNAs overproduced in A431 carcinoma cells, *Nature (London)*, 309, 806, 1984
73. **Merlino, G. T., Ishii, S., Whang-Peng, J., Knutsen, T., Xu, Y-H., Clark, A. J. L., Stratton, R. H., Wilson, R. K., Ma, D. P., Roe, B. A., Hunts, J. H., Shimizu, N., and Pastan, I.**, Structure and localization of genes encoding aberrant and normal epidermal growth factor receptor RNAs from A431 human carcinoma cells, *Mol. Cell. Biol.*, 5, 1722, 1985.
74. **Filmus, J., Pollak, M. N., Cailleau, R., and Buick, R. N.**, MDA-468, a human breast cancer cell line with a high number of epidermal growth factor (EGF) receptors, has an amplified EGF receptor gene and is growth inhibited by EGF, *Biochem. Biophys. Res. Comm.*, 128, 898, 1985.
75. **Hunts, J., Ueda, M., Ozawa, S., Abe, O., Pastan, I., and Shimizu, N.**, Hypermethylation and gene amplification of the epidermal growth factor receptor in squamous cell carcinomas, *Jpn. J. Cancer Res.*, 76, 663, 1985.
76. **Ishii, S., Xu, Y-H., Stratton, R. H., Roe, B. A., Merlino, G. T., and Pastan, I.**, Characterization and sequence of the promoter region of the human epidermal growth factor receptor gene, *Proc. Natl. Acad. Sci. U.S.A.*, 82, 4920, 1985.
77. **Ishii, S., Merlino, G. T., and Pastan, I.**, Promoter region of receptor proto-oncogene promoter, *Science*, 230, 1378, 1985.
78. **Terwilliger, E. and Herschman, H. R.**, 3T3 variants unable to bind epidermal growth factor cannot complement in coculture, *Biochem. Biophys. Res. Comm.*, 118, 60, 1984.
79. **Aglio, L. S., Maturo, J. M., III, and Hollenberg, M. D.**, Receptors for insulin and epidermal growth factor: interaction with organomercurial agarose, *J. Cell. Biochem.*, 28, 143, 1985.
80. **Panayotou, G. N., Magee, A. I., and Geisow, M. J.**, Reconstitution of the epidermal growth factor receptor in artificial lipid bilayers, *FEBS Lett.*, 183, 321, 1985.
81. **Sternberg, M. J. E. and Taylor, W. R.**, Modelling the ATP-binding site of oncogene products, the epidermal growth factor receptor and related proteins, *FEBS Lett.*, 175, 387, 1984.
82. **Russo, M. W., Lukas, T. J., Cohen, S., and Staros, J. V.**, Identification of residues in the nucleotide binding site of the epidermal growth factor receptor/kinase, *J. Biol. Chem.*, 260, 5205, 1985.
83. **Gill, G. N., Kawamoto, T., Cochet, C., Le, A., Sato, J. D., Masui, H., McLeod, C., and Mendelsohn, J.**, Monoclonal anti-epidermal growth factor receptor antibodies which are inhibitors of epidermal growth factor binding and antagonists of epidermal growth factor-stimulated tyrosine protein kinase activity, *J. Biol. Chem.*, 259, 7755, 1984.
84. **Wadsworth, S. C., Vincent, W. S., III, and Bilodeau-Wentworth, D.**, A *Drosophila* genomic sequence with homology to human epidermal growth factor receptor, *Nature (London)*, 314, 178, 1985.
85. **Livneh, E., Glazer, L., Segal, D., Schlessinger, J., and Shilo, B-Z.**, The Drosophila EGF receptor gene homolog: conservation of both hormone binding and kinase domains, *Cell*, 40, 599, 1985.

86. **Carlin, C. R. and Knowles, B. B.,** Biosynthesis of the epidermal growth factor receptor in human epidermoid carcinoma-derived A431 cells, *J. Biol. Chem.,* 259, 7902, 1984.
87. **Mayes, E. L. V. and Waterfield, M. D.,** Biosynthesis of the epidermal growth factor receptor in A431 cells, *EMBO J.,* 3, 531, 1984.
88. **Childs, R. A., Gregoriou, M., Scudder, P., Thorpe, S. J., Rees, A. R., and Feizi, T.,** Blood group-active chains on the receptor for epidermal growth factor of A431 cells, *EMBO J.,* 3, 2227, 1984.
89. **Slieker, L. J. and Lane, M. D.,** Post-translational processing of the epidermal growth factor receptor: glycosylation-dependent acquisition of ligand-binding capacity, *J. Biol. Chem.,* 260, 687, 1985.
90. **Decker, S. J.,** Effects of epidermal growth factor and 12-O-tetradecanoylphorbol-13-acetate on metabolism of the epidermal growth factor receptor in normal human fibroblasts, *Mol. Cell. Biol.,* 4, 1718, 1984.
91. **Stoscheck, C. M., Soderquist, A. M., and Carpenter, G.,** Biosynthesis of the epidermal growth factor receptor in cultured human cells, *Endocrinology,* 116, 528, 1985.
92. **Rao, C. V., Carman, F. R., Chegini, N., and Schultz, G. S.,** Binding sites for epidermal growth factor in human fetal membranes, *J. Clin. Endocrinol. Metab.,* 58, 1034, 1984.
93. **Nexo, E., Hollenberg, M. D., Figueroa, A., and Pratt, R. M.,** Detection of epidermal growth factor-urogastrone and its receptor during fetal mouse development, *Proc. Natl. Acad. Sci. U.S.A.,* 77, 2782, 1980.
94. **Chegini, N. and Rao, C. V.,** Epidermal growth factor binding to human amnion, chorion, decidua, and placenta from mid- and term pregnancy: quantitative light microscopic autoradiographic studies, *J. Clin. Endocrinol. Metab.,* 61, 529, 1985.
95. **Mukku, V. R. and Stancel, G. M.,** Receptors for epidermal growth factor in the rat uterus, *Endocrinology,* 117, 149, 1985.
96. **Filmus, J., Pollak, M. N., Cairncross, J. G., and Buick, R. N.,** Amplified, overexpressed and rearranged epidermal growth factor receptor gene in ahuman astrocytoma cell line, *Biochem. Biophys. Res. Comm.,* 131, 207, 1985.
97. **Ginsburg, E. and Vonderhaar, B. K.,** Epidermal growth factor stimulates the growth of A431 tumors in athymic mice, *Cancer Lett.,* 28, 143, 1985.
98. **Boonstra, J., Mummery, C. L., van der Saag, P. T., and de Laat, S. W.,** Two receptor classes for epidermal growth factor on pheochromocytoma cells, distinguishable by temperature, lectins, and tumor promoters, *J. Cell. Physiol.,* 123, 347, 1985.
99. **Assoian, R. K.,** Biphasic effects of type beta transforming growth factor on epidermal growth factor receptors in NRK fibroblasts: functional consequences for epidermal growth factor stimulated mitosis, *J. Biol. Chem.,* 260, 9613, 1985.
100. **Rees, A. R., Gregoriou, M., Johnson, P., and Garland, P. B.,** High affinity epidermal growth factor receptors on the surface of A431 cells have restricted lateral diffusion, *EMBO J.,* 3, 1843, 1984.
101. **Weber, W., Gill, G. N., and Spiess, J.,** Production of an epidermal growth factor receptor-related protein, *Science,* 224, 294, 1984.
102. **Basu, M., Biswas, R., and Das, M.,** 42,000-molecular weight EGF receptor has protein kinase activity, *Nature (London),* 311, 477, 1984.
103. **Bertics, P. J., Weber, W., Cochet, C., and Gill, G. N.,** Regulation of the epidermal growth factor receptor by phosphorylation, *J. Cell. Biochem.,* 29, 195, 1985.
104. **Gullick, W. J., Downward, J., and Waterfield, M. D.,** Antibodies to the autophosphorylation sites of the epidermal growth factor receptor protein-tyrosine kinase as probes of structure and function, *EMBO J.,* 4, 2869, 1985.
105. **Chinkers, M. and Garbers, D. L.,** Phorbol ester-induced threonine phosphorylation of the human epidermal growth factor receptor occurs within the EGF binding domain, *Biochem. Biophys. Res. Comm.,* 123, 618, 1984.
106. **Ushiro, H. and Cohen, S.,** Identification of phosphotyrosine as a product of epidermal growth factor-activated protein kinase in A-431 cell membranes, *J. Biol. Chem.,* 255, 8363, 1980.
107. **Hunter, T. and Cooper, J. A.,** Epidermal growth factor induces rapid tyrosine phosphorylation of proteins in A431 human tumor cells, *Cell,* 24, 741, 1981.
108. **Soderquist, A. M. and Carpenter, G.,** Developments in the mechanism of growth factor action: activation of protein kinase by epidermal growth factor, *Fed. Proc.,* 42, 2615, 1983.
109. **Zendegui, J. G. and Carpenter, G.,** Substrates of the epidermal growth factor receptor-kinase, *Cell Biol. Int. Rep.,* 8, 619, 1984.
110. **Downward, J., Waterfield, M. D., and Parker, P. J.,** Autophosphorylation and protein kinase C phosphorylation of the epidermal growth factor receptor. Effect on tyrosine kinase activity and ligand binding affinity, *J. Biol. Chem.,* 260, 14538, 1985.
111. **Bertics, P. J. and Gill, G. N.,** Self-phosphorylation enhances the protein-tyrosine kinase activity of the epidermal growth factor receptor, *J. Biol. Chem.,* 260, 14642, 1985.
112. **Downward, J., Parker, P., and Waterfield, M. D.,** Autophosphorylation sites on the epidermal growth factor receptor, *Nature (London),* 311, 483, 1984.

113. **Chinkers, M. and Brugge, J. S.**, Characterization of structural domains of the human epidermal growth factor receptor obtained by partial proteolysis, *J. Biol. Chem.*, 259, 11534, 1984.
114. **Rackoff, W. R., Rubin, R. A., and Earp, H. S.**, Phosphorylation of the hepatic EGF receptor with cAMP-dependent protein kinase, *Mol. Cell. Endocrinol.*, 34, 113, 1984.
115. **Fearn, J. C. and King, A. C.**, EGF receptor affinity is regulated by intracellular calcium and protein kinase C, *Cell*, 40, 991, 1985.
116. **Cochet, C., Gill, G. N., Meisenhelder, J., Cooper, J. A., and Hunter, T.**, C-kinase phosphorylates the epidermal growth factor receptor and reduces its epidermal growth factor-stimulated tyrosine protein kinase activity, *J. Biol. Chem.*, 259, 2553, 1984.
117. **Hunter, T., Ling, N., and Cooper, J. A.**, Protein kinase C phosphorylation of the EGF receptor at a threonine residue close to the cytoplasmic face of the plasma membrane, *Nature (London)*, 311, 480, 1984.
118. **Finocchiaro, L., Komano, O., and Loeb, J.**, Epidermal growth factor stimulated protein kinase shows similar activity in liver of senescent and adult mice, *FEBS Lett.*, 187, 96, 1985.
119. **Pallen, C. J., Valentine, K. A., Wang, J. H., and Hollenberg, M. D.**, Calcineurin-mediated dephosphorylation of the human placental membrane receptor for epidermal growth factor urogastrone, *Biochemistry*, 24, 4727, 1985.
120. **O'Connor-McCourt, M. and Hollenberg, M. D.**, Receptors, acceptors, and the action of polypeptide hormones: illustrative studies with epidermal growth factor (urogastrone), *Can. J. Biochem. Cell Biol.*, 61, 670, 1983.
121. **Beguinot, L., Lyal, R. M., Willingham, M. C., and Pastan, I.**, Down-regulation of the epidermal growth factor receptor in KB cells is due to receptor internalization and subsequent degradation in lysosomes, *Proc. Natl. Acad. Sci. U.S.A.*, 81, 2384, 1984.
122. **Matrisian, L. M., Planck, S. R., and Magun, B. E.**, Intracellular processing of epidermal growth factor, I. Acidification of ^{125}I-epidermal growth factor in intracellular organelles, *J. Biol. Chem.*, 259, 3047, 1984.
123. **Planck, S. R., Finch, J. S., and Magun, B. E.**, Intracellular processing of epidermal growth factor, II. Intracellular cleavage of the COOH-terminal region of ^{125}I-epidermal growth factor, *J. Biol. Chem.*, 259, 3053, 1984.
124. **Wiley, H. S., Van Nostrand, W., McKinney, D. N., and Cunningham, D. D.**, Intracellular processing of epidermal growth factor and its effect on ligand-receptor interactions, *J. Biol. Chem.*, 260, 5290, 1985.
125. **Haigler, H. T., Wiley, H. S., Moehring, J. M., and Moehring, T. J.**, Altered degradation of epidermal growth factor in a diphtheria toxin-resistant clone of KB cells, *J. Cell. Physiol.*, 124, 322, 1985.
126. **Korc, M. and Magun, B. E.**, Recycling of epidermal growth factor in a human pancreatic carcinoma cell line, *Proc. Natl. Acad. Sci. U.S.A.*, 82, 6172, 1985.
127. **Stoscheck, C. M. and Carpenter, G.**, Characterization of the metabolic turnover of epidermal growth factor receptor protein in A-431 cells, *J. Cell. Physiol.*, 120, 296, 1984.
128. **Schaudies, R. P., Harper, R. A., and Savage, C. R., Jr.**, ^{125}I-EGF binding to responsive and nonresponsive cells in culture: loss of cell-associated radioactivity relates to growth induction, *J. Cell. Physiol.*, 124, 493, 1985.
129. **Beguinot, L., Hanover, J. A., Ito, S., Richert, N. D., Willingham, M. C., and Pastan, I.**, Phorbol esters induce transient internalization without degradation of unoccupied epidermal growth factor receptors, *Proc. Natl. Acad. Sci. U.S.A.*, 82, 2774, 1985.
130. **King, A. C.**, Monensin, like methylamine, prevents degradation of ^{125}I-epidermal growth factor, causes intracellular accumulation of receptors and blocks the mitogenic response, *Biochem. Biophys. Res. Comm.*, 124, 585, 1984.
131. **Landreth, G. E., Williams, L. K., and Rieser, G. D.**, Association of the epidermal growth factor receptor kinase with the detergent-insoluble cytoskeleton of A431 cells, *J. Cell Biol.*, 101, 1341, 1985.
132. **Graf, T. and Beug, H.**, Avian leukemia viruses: interaction with their target cells in vivo and in vitro, *Biochim. Biophys. Acta*, 516, 269, 1978.
133. **Jurdic, P., Bouabdelli, M., Moscovici, M. G., and Moscovici, C.**, Embryonic erythroid cells transformed by avian erythroblastosis virus may proliferate and differentiate, *Virology*, 144, 73, 1985.
134. **Saule, S., Roussel, M., Lagrou, C., and Stehelin, D.**, Characterization of the oncogene (*erb*) of avian erythroblastosis virus and its cellular progenitor, *J. Virol.*, 38, 409, 191.
135. **Graf, T. and Beug, H.**, Role of the v-*erbA* and v-*erbB* oncogenes of avian erythroblastosis virus in erythroid cell transformation, *Cell*, 34, 7, 1983.
136. **Yamamoto, T., Hihara, H., Nishida, T., Kawai, S., and Toyoshima, K.**, A new avian erythroblastosis virus, AEV-H, carries *erbB* gene responsible for the induction of both erythroblastosis and sarcomas, *Cell*, 34, 225, 1983.
137. **Miles, B. D. and Robinson, H. L.**, High-frequency transduction of c-*erb*B in avian leukosis virus-induced erythroblastosis, *J. Virol.*, 54, 295, 1985.
138. **Downward, J., Yarden, Y., Mayes, E., Scrace, G., Totty, N., Stockwell, P., Ullrich, A., Schlessinger, J., and Waterfield, M. D.**, Close similarity of epidermal growth factor receptor and v-*erb*-B oncogene protein sequences, *Nature (London)*, 307, 521, 1984.

139. **Spurr, N. K., Solomon, E., Jansson, M., Sheer, D., Goodfellow, P. N., Bodmer, W. F., and Vennstrom, B.**, Chromosomal localisation of the human homologues to the oncogenes *erb*A and B, *EMBO J.*, 3, 159, 1984.
140. **Silver, J., Whitney, J. B., III, Kozak, C., Hollis, G., and Kirsch, I.**, *ErbB* is linked to the alpha-globin locus on mouse chromosome 11, *Mol. Cell. Biol.*, 5, 1784, 1985.
141. **Hayman, M. J., Ramsey, G., Savin, K., Kitchener, G., Graf, T., and Beug, H.**, Identification and characterization of the avian erythroblastosis virus *erb*B gene product as a membrane glycoprotein, *Cell*, 32, 579, 1983.
142. **Hayman, M. J. and Beug, H.**, Identification of a form of the avian erythroblastosis virus *erb-B* gene product at the cell surface, *Nature (London)*, 309, 460, 1984.
143. **Schmidt, J. A., Beug, H., and Hayman, M. J.**, Effects of inhibitors of glycoprotein processing on the synthesis and biological activity of the *erb B* oncogene, *EMBO J.*, 4, 105, 1985.
144. **Akiyama, T., Yamada, Y., Ogawara, H., Richert, N., Pastan, I., Yamamoto, T., and Kasuga, M.**, Site-specific antibodies to the *erb*B oncogene product immunoprecipitate epidermal growth factor receptor, *Biochem. Biophys. Res. Comm.*, 123, 797, 1984.
145. **Decker, S. J.**, Phosphorylation of the erbB gene product from an avian erythroblastosis virus-transformed chick fibroblast cell line, *J. Biol. Chem.*, 260, 2003, 1985.
146. **Gilmore, T., DeClue, J. E., and Martin, G. S.**, Protein phosphorylation at tyrosine is induced by the v-*erb*B gene product in vivo and in vitro, *Cell*, 40, 609, 1985.
147. **Kris, R. M., Lax, I., Gullick, W., Waterfield, M. D., Ullrich, A., Fridkin, M., and Schlessinger, J.**, Antibodies against a synthetic peptide as a probe for the kinase activity of the avian EGF receptor and v-erbB protein, *Cell*, 40, 619, 1985.
148. **Shih, C., Padhy, L. C., Murray, M., and Weinberg, R. A.**, Transforming genes of carcinomas and neuroblastomas introduced into mouse fibroblasts, *Nature (London)*, 290, 261, 1981.
149. **Schechter, A. L., Stern, D. F., Vaidyanathan, L., Decker, S. J., Drebin, J. A., Greene, M. I., and Weinberg, R. A.**, The *neu* oncogene: an *erb-B*-related gene encoding a 185,000-M_r tumour antigen, *Nature (London)*, 312, 513, 1984.
150. **Schechter, A. L., Hung, M-C., Vaidyanathan, L., Weinberg, R. A., Yang-Feng, T. L., Francke, U., Ullrich, A., and Coussens, L.**, The *neu* gene: an *erb*B-homologous gene distinct from and unlinked to the gene encoding the EGF receptor, *Science*, 229, 976, 1985.
151. **Coussens, L., Yang-Feng, T. L., Liao, Y-C., Chen, E., Gray, A., McGrath, J., Seeburg, P. H., Libermann, T. A., Schlessinger, J., Francke, U., Levinson, A., and Ullrich, A.**, Tyrosine kinase receptor with extensive homology to EGF receptor shares chromosomal location with *neu* oncogene, *Science*, 230, 1132, 1985.
152. **Drebin, J. A., Stern, D. F., Link, V. C., Weinberg, R. A., and Greene, M. I.**, Monoclonal antibodies identify a cell-surface antigen associated with an activated cellular oncogene, *Nature (London)*, 226, 545, 1984.
153. **Drebin, J. A., Link, V. C., Stern, D. F., Weinberg, R. A., and Greene, M. I.**, Down-modulation of an oncogene protein product and reversion of the transformed phenotype by monoclonal antibodies, *Cell*, 41, 695, 1985.
154. **King, C. R., Kraus, M. H., and Aaronson, S. A.**, Amplification of a novel v-*erb*B-related gene in a human mammary carcinoma, *Science*, 229, 974, 1985.
155. **Semba, K., Kamata, N., Toyoshima, K., and Yamamoto, T.**, A v-*erbB*-related proto-oncogene, c-*erbB-1*/epidermal growth factor-receptor gene and is amplified in a human salivary gland adenocarcinoma, *Proc. Natl. Acad. Sci. U.S.A.*, 82, 6497, 1985.
156. **Lax, I., Bar-Eli, M., Yarden, Y., Liberman, T. A., and Schlessinger, J.**, Antibodies to two defined regions of the transforming protein pp60^{src} interact specifically with the epidermal growth factor receptor kinase system, *Proc. Natl. Acad. Sci. U.S.A.*, 81, 5911, 1984.
157. **Privalsky, M. L., Ralston, R., and Bishop, J. M.**, The membrane glycoprotein encoded by the retroviral oncogene v-*erb*-B is structurally related to tyrosine-specific protein kinase, *Proc. Natl. Acad. Sci. U.S.A.*, 81, 704, 1984.
158. **Perrotti, N., Taylor, S. I., Richert, N. D., Rapp, U. R., Pastan, I., and Roth, J.**, Immunoprecipitation of insulin receptors from cultured human lymphocytes (IM-9 cells) by antibodies to pp60^{src}, *Science*, 227, 761, 1985.
159. **Ullrich, A., Bell, J. R., Chen, E. Y., Herrera, R., Petruzzelli, L. M., Dull, T. J., Gray, A., Coussens, L., Liao, Y. C., Tsubokawa, M., Mason, A., Seeburg, P. H., Grunfeld, C., Rosen, O. M., and Ramachandran, J.**, Human insulin receptor and its relationship to the tyrosine kinase family of oncogenes, *Nature (London)*, 313, 756, 1985.
160. **Wierenga, R. K. and Hol, W. G. J.**, Predicted nucleotide-binding properties of p21 protein and its cancer-associated variant, *Nature (London)*, 302, 842, 1983.
161. **Mroczkowski, B., Mosig, G., and Cohen, S.**, ATP-stimulated interaction between epidermal growth factor receptor and supercoiled DNA, *Nature (London)*, 309, 270, 1984.

162. **Imai, Y., Leung, C. K. H., Friesen, H. G., and Shiu, R. P. C.**, Epidermal growth factor receptors and effect of epidermal growth factor on growth of human breast cancer cells in long-term tissue culture, *Cancer Res.*, 42, 4394, 1982.
163. **Fitzpatrick, S. L., LaChance, M. P., and Schultz, G. S.**, Characterization of epidermal growth factor receptor and action on human breast cancer cells in culture, *Cancer Res.*, 44, 3442, 1984.
164. **Sainsbury, J. R. C., Farndon, J. R., Sherbet, G. V., and Harris, A. L.**, Epidermal growth factor receptors and oestrogen receptors in human breast cancer, *Lancet*, i, 364, 1985.
165. **Fitzpatrick, S. L., Brightwell, J., Wittliff, J. L., Barrows, G. H., and Schultz, G. S.**, Epidermal growth factor binding by breast tumor biopsies and relationship to estrogen receptor and progestin receptor levels, *Cancer Res.*, 44, 3448, 1984.
166. **Hirata, Y., Uchihashi, M., Fujita, T., Matsukura, S., Motoyama, T., Kaku, M., and Koshimizu, K.**, Characteristics of specific binding of epidermal growth factor (EGF) on human tumor cell lines, *Endocrinol. Jpn.*, 30, 601, 1983.
167. **Hirata, Y., Uchihashi, M., Nakashima, H., Fujita, T., Matsukura, S., and Matsui, K.**, Specific receptors for epidermal growth factor in human bone tumour cells and its effect on synthesis of prostaglandin E_2 by cultured osteosarcoma cell line, *Acta Endocrinol.*, 107, 125, 1984.
168. **Gusterson, B., Cowley, G., McIlhinney, J., Ozanne, B., Fisher, C., and Reeves, B.**, Evidence for increased epidermal growth factor receptors in human sarcomas, *Int. J. Cancer*, 36, 689, 1985.
169. **Ilekis, J. and Benveniste, R.**, Effects of epidermal growth factor, phorbol myristate acetate, and arachidonic acid on chroriogonadotropin secretion by cultured human choriocarcinoma cells, *Endocrinology*, 116, 2400, 1985.
170. **Costrini, N. V. and Beck, R.**, Epidermal growth factor-urogastrone receptors in normal human liver and primary hepatoma, *Cancer*, 51, 2191, 1983.
171. **Lev-Ran, A., Hwang, D., Josefsberg, Z., Barseghian, G., Kemeny, M., Meguid, M., and Beatty, D.**, Binding of epidermal growth factor (EGF) and insulin to human liver microsomes and Golgi fractions, *Biochem. Biophys. Res. Comm.*, 119, 1181, 1984.
172. **Neal, D. E., Marsh, C., Bennett, M. K., Abel, P. D., Hall, R. R., Sainsbury, J. R. C., and Harris, A. L.**, Epidermal growth factor receptors in human bladder cancer: comparison of invasive and superficial tumors, *Lancet*, i, 366, 1985.
173. **Libermann, T. A., Razon, N., Bartal, A. D., Yarden, Y., Schlessinger, J., and Soreq, H.**, Expression of epidermal growth factor receptors in human brain tumors, *Cancer Res.*, 44, 753, 1984.
174. **Westphal, M., Harsh, G. R., IV, Rosenblum, M. L., and Hammonds, R. G., Jr.**, Epidermal growth factor receptors in the human glioblastoma cell line SF268 differ from those in epidermoid carcinoma cell line A431, *Biochem. Biophys. Res. Comm.*, 132, 284, 1985.
175. **Adamson, E. D. and Rees, A. R.**, Epidermal growth factor receptors, *Mol. Cell. Biochem.*, 34, 129, 1981.
176. **Mukku, V. R.**, Regulation of epidermal growth factor receptor levels by thyroid hormone, *J. Biol. Chem.*, 259, 6543, 1984.
177. **Edery, M., Pang, K., Larson, L., Colosi, T., and Nandi, S.**, Epidermal growth factor receptor levels in mouse mammary glands in various physiological states, *Endocrinology*, 117, 405, 1985.
178. **Mukku, V. R. and Stancel, G. M.**, Regulation of epidermal growth factor receptor by estrogen, *J. Biol. Chem.*, 260, 9820, 1985.
179. **Westermark, K., Karlson, F. A., and Westermark, B.**, Thyrotropin modulates EGF receptor function in porcine thyroid follicle cells, *Mol. Cell. Endocrinol.*, 40, 17, 1985.
180. **Robinson, R. A., Vokenant, M. E., Ryan, R. J., and Moses, H. L.**, Decreased epidermal growth factor binding in cells growth arrested in G_1 by nutrient deficiency, *J. Cell. Physiol.*, 109, 517, 1981.
181. **Assoian, R. K., Frolik, C. A., Roberts, A. B., Miller, D., and Sporn, M. B.**, Transforming growth factor-beta controls receptor levels for epidermal growth factor in NRK fibroblasts, *Cell*, 36, 35, 1984.
182. **Murphy, L. J., Sutherland, R. L., and Lazarus, L.**, Regulation of growth hormone and epidermal growth factor receptors by progestins in breast cancer cells, *Biochem. Biophys. Res. Comm.*, 131, 767, 1985.
183. **Rubin, R. A. and Earp, H. S.**, Dimethyl sulfoxide stimulates tyrosine residue phosphorylation of rat liver epidermal growth factor receptor, *Science*, 219, 60, 1983.
184. **Koprowski, H., Herlyn, M., Balaban, G., Parmiter, A., Ross, A., and Nowell, P.**, Expression of the receptor for epidermal growth factor correlates with increased dosage of chromosome 7 in malignant melanoma, *Somat. Cell Mol. Genet.*, 11, 297, 1985.
185. **Sainsbury, J. R. C., Malcolm, A. J., Appleton, D. R., Farndon, J. R., and Harris, A. L.**, Presence of epidermal growth factor receptor as an indicator of poor prognosis in patients with breast cancer, *J. Clin. Pathol.*, 38, 1225, 1985.
186. **Berhanu, P. and Hollenberg, M. D.**, Epidermal growth factor-urogastrone receptor: selective alteration in simian virus 40 transformed mouse fibroblasts, *Arch. Biochem. Biophys.*, 203, 134, 1980.

187. **Todaro, G. J., De Larco, J. E., and Cohen, S.**, Transformation by murine and feline sarcoma viruses specifically blocks binding of epidermal growth factor to cells, *Nature (London)*, 264, 26, 1976.
188. **Usui, T., Moriyama, N., Ishibe, T., and Nakatsu, H.**, Loss of epidermal growth factor receptor on the renal neoplasm induced in vivo with xenotropic pseudotype Kirsten murine sarcoma virus, *Biochem. Biophys. Res. Comm.*, 120, 879, 1984.
189. **Robinson, R. A., Branum, E. L., Volkenant, M. E., and Moses, H. L.**, Cell cycle variation in ^{125}I-labeled epidermal growth factor binding in chemically transformed cells, *Cancer Res.*, 42, 2633, 1982.
190. **Wakshull, E., Kraemer, P. M., and Wharton, W.**, Multistep change in epidermal growth factor receptors during spontaneous neoplastic progression in Chinese hamster embryo fibroblasts, *Cancer Res.*, 45, 2070, 1985.
191. **Chua, C. C., Geiman, D. E., Schreiber, A. B., and Ladda, R. L.**, Nonfunctional epidermal growth factor receptor in cells transformed by Kirsten sarcoma virus, *Biochem. Biophys. Res. Comm.*, 118, 538, 1984.
192. **Lax, I., Kris, R., Sasson, I., Ullrich, A., Hayman, M. J., Beug, H., and Schlessinger, J.**, Activation of c-*erb*-B in avian leukosis virus-induced erythroblastosis leads to the expression of a truncated EGF receptor kinase, *EMBO J.*, 4, 3179, 1985.
193. **Bodine, P. V. and Tupper, J. T.**, Calmodulin antagonists decrease binding of epidermal growth factor to transformed, but not to normal human fibroblasts, *Biochem. J.*, 218, 629, 1984.
194. **Decker, S.**, Reduced binding of epidermal growth factor by avian sarcoma virus-transformed rat cells, *Biochem. Biophys. Res. Comm.*, 113, 678, 1983.
195. **Hollenberg, M. D., Barrett, J. C., Ts'o, P. O. P., and Berhanu, P.**, Selective reduction in receptors for epidermal growth factor-urogastrone in chemically transformed tumorigenic Syrian hamster embryo fibroblasts, *Cancer Res.*, 39, 4166, 1979.
196. **Todaro, G. J., De Larco, J. E., and Fryling, C. M.**, Sarcoma growth factor and other transforming peptides produced by human cells: interactions with membrane receptors, *Fed. Proc.*, 41, 2996, 1982.
197. **Masui, H., Kawamoto, T., Sato, J. D., Wolf, B., Sato, G., and Mendelsohn, J.**, Growth inhibition of human tumor cells in athymic mice by anti-epidermal growth factor receptor monoclonal antibodies, *Cancer Res.*, 44, 1002, 1984.
198. **Kudlow, J. E., Khosravi, M. J., Kobrin, M. S., and Mak, W. W.**, Inability of anti-epidermal growth factor receptor monoclonal antibody to block "autocrine" growth stimulation in transforming growth factor-secreting melanoma cells, *J. Biol. Chem.*, 259, 11895, 1984.
199. **Hollenberg, M. D., Nexo, E., Berhanu, P., and Hock, R.**, Phorbol ester and the selective modulation of receptors for epidermal growth factor-urogastrone, in *Receptor-Mediated Binding and Internalization of Toxins and Hormones,* Academic Press, New York, 1981, 181.
200. **Friedman, B. A., Frackelton, A. R., Jr., Ross, A. H., Connors, J. M., Fujiki, H., Sugimura, T., and Rosner, M. R.**, Tumor promoters block tyrosine-specific phosphorylation of the epidermal growth factor receptor, *Proc. Natl. Acad. Sci. U.S.A.*, 81, 3034, 1984.
201. **Moon, S. O., Palfrey, H. C., and King, A. C.**, Phorbol esters potentiate tyrosine phosphorylation of epidermal growth factor receptors in A431 membranes by a calcium-independent mechanism, *Proc. Natl. Acad. Sci. U.S.A.*, 81, 2298, 1984.
202. **Davis, R. J. and Czech, M. P.**, Tumor-promoting phorbol diesters mediate phosphorylation of the epidermal growth factor receptor, *J. Biol. Chem.*, 259, 8545, 1984.
203. **Davis, R. J. and Czech, M. P.**, Tumor-promoting phorbol diesters cause the phosphorylation of epidermal growth factor receptors in normal human fibroblasts at threonine-654, *Proc. Natl. Acad. Sci. U.S.A.*, 82, 1974, 1985.
204. **Davis, R. J., Ganong, B. R., Bell, R. M., and Czech, M. P.**, Structural requirements for diacylglycerols to mimic tumor-promoting phorbol diester action on the epidermal growth factor receptor, *J. Biol. Chem.*, 260, 5315, 1985.
205. **Taketani, Y. and Oka, T.**, Tumor promoter 12-*O*-tetradecanoylphorbol 13-acetate, like epidermal growth factor, stimulates cell proliferation and inhibits differentiation of mouse mammary epithelial cells in culture, *Proc. Natl. Acad. Sci. U.S.A.*, 80, 1646, 1983.
206. **Kanter, P., Leister, K. J., Tomei, L. D., Wenner, P. A., and Wenner, C. E.**, Epidermal growth factor and tumor promoters prevent DNA fragmentation by different mechanisms, *Biochem. Biophys. Res. Comm.*, 118, 392, 1984.
207. **Parker, P. J., Stabel, S., and Waterfield, M. D.**, Purification to homogeneity of protein kinase C from bovine brain — identity with the phorbol ester receptor, *EMBO J.*, 3, 953, 1984.
208. **Ashendel, C. L.**, The phorbol ester receptor: a phospholipid-regulated protein kinase, *Biochim. Biophys. Acta,* 822, 219, 1985.
209. **Fox, C. F., Linsley, P. S., and Wrann, M.**, Receptor remodeling and regulation in the action of epidermal growth factor, *Fed. Proc.*, 41, 2988, 1982.

210. **Connolly, J. L., Green, S. A., and Greene, L. A.,** Comparison of rapid changes in surface morphology and coated pit formation of PC12 cells in response to nerve growth factor, epidermal growth factor, and dibutiryl cyclic AMP, *J. Cell Biol.,* 98, 457, 1984.
211. **Wiley, H. S. and Kaplan, J.,** Epidermal growth factor rapidly induces a redistribution of transferrin receptor pools in human fibroblasts, *Proc. Natl. Acad. Sci. U.S.A.,* 81, 7456, 1984.
212. **Giugni, T. D., James, L. C., and Haigler, H. T.,** Epidermal growth factor stimulates tyrosine phosphorylation of specific proteins in permeabilized human fibroblast, *J. Biol. Chem.,* 260, 15081, 1985.
213. **Greenberg, M. E. and Edelman, G. M.,** The 34 kd pp60^{src} substrate is located at the inner face of the plasma membrane, *Cell,* 33, 767, 1983.
214. **Erikson, E. and Erikson, R. L.,** Identification of a cellular protein substrate phosphorylated by the avian sarcoma virus-transforming gene product, *Cell,* 21, 829, 1980.
215. **Ghosh-Dastidar, P. and Fox, C. F.,** Epidermal growth factor and epidermal growth factor receptor-dependent phosphorylation of a $M_r = 34,000$ protein substrate for pp60^{src}, *J. Biol. Chem.,* 258, 2041, 1983.
216. **Gould, K. L., Cooper, J. A., and Hunter, T.,** The 46,000-dalton tyrosine protein kinase substrate is widespread, whereas the 36,000-dalton substrate is only expressed at high levels in certain rodent tissues, *J. Cell Biol.,* 98, 487, 1984.
217. **Sawyer, S. T. and Cohen, S.,** Epidermal growth factor stimulates the phosphorylation of the calcium-dependent 35,000-dalton substrate in intact A431 cells, *J. Biol. Chem.,* 260, 8233, 1985.
218. **Thomas, G., Martin-Pérez, J., Siegmann, M., and Otto, A. M.,** The effect of serum, EGF, PGF$_{2\,alpha}$ and insulin on S6 phosphorylation and the initiation of protein and DNA synthesis, *Cell,* 30, 235, 1982.
219. **Novak-Hofer, I. and Thomas, G.,** Epidermal growth factor-mediated activation of an S6 kinase in Swiss mouse 3T3 cells, *J. Biol. Chem.,* 260, 10314, 1985.
220. **Rosen, O. M., Rubin, C. S., Cobb, M. H., and Smith, C. J.,** Insulin stimulates the phosphorylation of ribosomal protein S6 in a cell-free system derived from 3T3-L1 adipocytes, *J. Biol. Chem.,* 256, 3630, 1981.
221. **Dazord, A., Genot, A., Langlois-Gallet, D., Mombrial, C., Haour, F., and Saez, J. M.,** hCG-increased phosphorylation of proteins in primary culture of Leydig cells: further characterization, *Biochem. Biophys. Res. Comm.,* 118, 8, 1984.
222. **Blenis, J. and Erikson, R. L.,** Phosphorylation of the ribosomal protein S6 is elevated in cells transformed by a variety of tumor viruses, *J. Virol.,* 50, 966, 1984.
223. **Martin-Pérez, J., Siegmann, M., and Thomas, G.,** EGF, PGF$_{2\,alpha}$ and insulin induce the phosphorylation of identical S6 peptides in Swiss mouse 3T3 cells: effect of cAMP on early sites of phosphorylation, *Cell,* 36, 287, 1984.
224. **Perisic, O. and Traugh, J. A.,** Protease-activated kinase II as the mediator of epidermal growth factor-stimulated phosphorylation of ribosomal protein S6, *FEBS Lett.,* 183, 215, 1985.
225. **Kulkarni, R. K. and Straus, D. S.,** Insulin-mediated phosphorylation of ribosomal protein S6 in mouse melanoma cells and melanoma x fibroblas hybrid cells in relation to cell proliferation, *Biochim. Biophys. Acta,* 762, 542, 1983.
226. **Nakamura, K. D., Martinez, R., and Weber, M. J.,** Tyrosine phosphorylation of specific proteins after mitogen stimulation of chicken embryo fibroblasts, *Mol. Cell. Biol.,* 3, 380, 1983.
227. **Baldwin, G. S., Knesel, J., and Monckton, J. M.,** Phosphorylation of gastrin-17 by epidermal growth factor-stimulated tyrosine kinase, *Nature (London),* 301, 435, 1983.
228. **Baldwin, G. S., Grego, B., Hearn, M. T. W., Knesel, J. A., Morgan, F. J., and Simpson, R. J.,** Phosphorylation of human growth hormone by the epidermal growth factor-stimulated tyrosine kinase, *Proc. Natl. Acad. Sci. U.S.A.,* 80, 5276, 1983.
229. **Ghosh-Dastidar, P. and Fox, C. F.,** c-AMP-dependent protein kinase stimulates epidermal growth factor-dependent phosphorylation of epidermal growth factor receptors, *J. Biol. Chem.,* 259, 3864, 1984.
230. **Buss, J. E., Chouvet, C., and Gill, G. N.,** Comparison of protein phosphorylations in variant A431 cells with different growth responses to epidermal growth factor, *J. Cell. Physiol.,* 119, 296, 1984.
231. **Frackelton, A. R., Jr., Ross, A. H., and Eisen, H. N.,** Characterization and use of monoclonal antibodies for isolation of phosphoryrosyl proteins from retrovirus-transformed cells and growth factor-stimulated cells, *Mol. Cell. Biol.,* 3, 1343, 1983.
232. **Baldwin, G. S., Stanley, I. J., and Nice, E. C.,** A synthetic peptide containing the autophosphorylation site of the transforming protein of Harvey sarcoma virus is phosphorylated by the EGF-stimulated tyrosine kinase, *FEBS Lett.,* 153, 257, 1983.
233. **Kamata, T. and Feramisco, J. R.,** Epidermal growth factor stimulates guanine nucleotide binding activity and phosphorylation of *ras* oncogene proteins, *Nature (London),* 310, 147, 1984.
234. **Thompson, D. M., Cochet, C., Chambaz, E. M., and Gill, G. N.,** Separation and characterization of a phosphatidylinositol kinase activity that co-purifies with the epidermal growth factor receptor, *J. Biol. Chem.,* 260, 8824, 1985.

235. **Kato, M., Homma, Y., Nagai, Y., and Takenawa, T.,** Epidermal growth factor stimulates diacylglycerol kinase in isolated plasma membrane vesicles from A431 cells, *Biochem. Biophys. Res. Comm.,* 129, 375, 1985.
236. **Brown, K. D., Blay, J., Irvine, R. F., Heslop, J. P., and Berridge, M. J.,** Reduction of epidermal growth factor receptor affinity by heterologous ligands: evidence for a mechanism involving the break down of phosphoinositides and the activation of protein kinase C, *Biochem. Biophys. Res. Comm.,* 123, 377, 1984.
237. **McCaffrey, P. G., Friedman, B-A., and Rosner, M. R.,** Diacylglycerol modulates binding and phosphorylation of the epidermal growth factor receptor, *J. Biol. Chem.,* 259, 12501, 1984.
238. **Sinnett-Smith, J. W. and Rozengurt, E.,** Diacylglycerol treatment rapidly decreases the affinity of the epidermal growth factor receptors of Swiss 3T3 cells, *J. Cell. Physiol.,* 124, 81, 1985.
239. **Ross, A. H., Damsky, C., Phillips, P. D., Hwang, F., and Vance, P.,** Inhibition of epidermal growth factor-induced phosphorylation by trifluoperazine, *J. Cell. Physiol.,* 124, 499, 1985.
240. **Taylor, D., Uhing, R. J., Blackmore, P. F., Prpić, V., and Exton, J. H.,** Insulin and epidermal growth factor do not affect phosphoinositide metabolism in rat liver plasma membranes and hepatocytes, *J. Biol. Chem.,* 260, 2011, 1985.
241. **Raben, D. M. and Cunningham, D. D.,** Effects of EGF and thrombin on inositol-containing phospholipids of cultured fibroblasts: stimulation of phosphatidylinositol synthesis by thrombin but not EGF, *J. Cell. Physiol.,* 125, 582, 1985.
242. **Hollenberg, M. D. and Cuatrecasas, P.,** Epidermal growth factor: receptors in human fibroblasts and modulation of action by cholera toxin, *Proc. Natl. Acad. Sci. U.S.A.,* 70, 2964, 1973.
243. **Nilsen-Hamilton, M. and Holley, R. W.,** Rapid selective effects by a growth inhibitor and epidermal growth factor on the incorporation of (^{35}S)methionine into protein secreted by African green monkey (BSC-1) cells, *Proc. Natl. Acad. Sci. U.S.A.,* 80, 5636, 1983.
244. **Courtney, M. G., Schmidt, L. J., and Getz, M. J.,** Organization and expression of endogenous viruslike (VL-30) DNA sequences in nontransformed and chemically transformed mouse embryo cells in culture, *Cancer Res.,* 42, 569, 1982.
245. **Foster, D. N., Schmidt, L. J., Hodgson, C. P., Moses, H. L., and Getz, M. J.,** Polyadenylated RNA complementary to a mouse retrovirus-like multigene family is rapidly and specifically induced by epidermal growth factor stimulation of quiescent cells, *Proc. Natl. Acad. Sci. U.S.A.,* 79, 7317, 1982.
246. **Matrisian, L. M., Rautmann, G., Magun, B. E., and Breatnach, R.,** Epidermal growth factor or serum stimulation of rat fibroblasts induces an elevation in mRNA levels for lactate dehydrogenese and other glycolytic enzymes, *Nucleic Acids Res.,* 13, 711, 1985.
247. **Bravo, R.,** Epidermal growth factor inhibits the synthesis of the nuclear protein cyclin in A431 human carcinoma cells, *Proc. Natl. Acad. Sci. U.S.A.,* 81, 4848, 1984.
248. **Roger, P. P., Van Heuverswyn, B., Lambert, C., Reuse, S., Vassart, G., and Dumont, J. E.,** Antagonistic effects of thyrotropin and epidermal growth factor on thyroglobulin mRNA level in cultured thyroid cells, *Eur. J. Biochem.,* 152, 239, 1985.
249. **Arakawa, M., Perry, J. W., Cossu, M. F., and Oka, T.,** Further characterization of the inhibition of casein production in a primary mouse mammary epithelial cell culture by epidermal growth factor, *Exp. Cell Res.,* 158, 111, 1985.
250. **Elder, P. K., Schmidt, L. J., Ono, T., and Getz, M. J.,** Specific stimulation of actin gene transcription by epidermal growth factor and cycloheximide, *Proc. Natl. Acad. Sci. U.S.A.,* 81, 7476, 1984.
251. **Naharro, G., Robbins, K. C., and Reddy, E. P.,** Gene product of v-*fgr onc*: hybrid protein containing a portion of actin and a tyrosine-specific protein kinase, *Science,* 223, 63, 1984.
252. **Maness, P. F. and Walsh, R. C., Jr.,** Dihydrocytochalasin B disorganizes actin cytoarchiteture and inhibits initiation of DNA synthesis in 3T3 cells, *Cell,* 30, 253, 1982.
253. **Müller, R., Bravo, R., Burckhardt, J., and Curran, T.,** Induction of c-*fos* gene and protein by growth factors precedes activation of c-*myc*, *Nature (London),* 312, 716, 1984.
254. **Bravo, R., Burckhardt, J., Curran, T., and Müller, R.,** Stimulation and inhibition of growth by EGF in different A431 cell clones is accompanied by the rapid induction of c-*fos* and c-*myc* proto-oncogenes, *EMBO J.,* 4, 1193, 1985.
255. **Greenberg, M. E., Greene, L. A., and Ziff, E. B.,** Nerve growth factor and epidermal growth factor induce rapid transient changes in proto-oncogene transcription in PC12 cells, *J. Biol. Chem.,* 260, 14101, 1985.
256. **Cruise, J. L. and Michalopoulos, G.,** Norepinephrine and epidermal growth factors: dynamics of their interaction in the stimulation of hepatocyte DNA synthesis, *J. Cell. Physiol.,* 125, 45, 1985.
257. **McKeehan, W. L. and McKeehan, K. A.,** Epidermal growth factor modulates extracellular Ca^{2+} requirement for multiplication of normal human skin fibroblasts, *Exp. Cell Res.,* 123, 397, 1979.
258. **Lechner, J. F.,** Interdependent regulation of epithelial cell replication by nutrients, hormones, growth factors, and cell density, *Fed. Proc.,* 43, 116, 1984.

259. **Lin, S. L., Ts'o, P. O. O., and Hollenberg, M. D.,** The effects of interferon on epidermal growth factor action, *Biochem. Biophys. Res. Comm.,* 96, 168, 1980.
260. **Bishop, R., Martinez, R., Weber, M. J., Blackshear, P. J., Beatty, S., Lim, R., and Herschman, H. R.,** Protein phosphorylation in a tetradecanoyl phorbol acetate-nonproliferative variant of 3T3 cells, *Mol. Cell. Biol.,* 5, 2231, 1985.
261. **Miskimins, R., Miskimins, W. K., Bernstein, H., and Shimizu, N.,** Epidermal growth factor-induced topoisomerase(s): intracellular translocation and relation to DNA synthesis, *Exp. Cell Res.,* 146, 53, 1983.
262. **Basu, M., Frick, K., Sen-Majumdar, A., Scher, C. D., and Das, M.,** EGF receptor-associated DNA-nicking activity is due to a M_r-100,000 dissociable protein, *Nature (London),* 316, 640, 1985.
263. **Weissman, B. E. and Aaronson, S. A.,** BALB and Kirsten murine sarcoma viruses alter growth and differentiation of EGF-dependent BALB/c mouse epidermal keratinocyte lines, *Cell,* 32, 599, 1983.
264. **Weissman, B. and Aaronson, S. A.,** Members of the *src* and *ras* oncogene families supplant the epidermal growth factor requirement of BALB/MK-2 keratinocytes and induce distinct alterations in their terminal differentiation process, *Mol. Cell. Biol.,* 5, 3386, 1985.
265. **Fisher, P. B., Mufson, R. A., Weinstein, I. B., and Little, J. B.,** Epidermal growth factor, like tumor promoters, enhances viral and radiation-induced cell transformation, *Carcinogenesis,* 2, 183, 1981.
266. **Herschman, H. R. and Brankow, D.,** Interaction of epidermal growth factor with initiators and complete carcinogens in the C3H10T1/2 cell culture system, *J. Cell. Biochem.,* 28, 1, 1985.
267. **Rose, S. P., Stahn, R., Passovoy, D. S., and Herschman, H. R.,** Epidermal growth factor enhancement of skin tumor induction in mice, *Experientia,* 32, 913, 1976.
268. **Kurachi, H., Okamoto, S., and Oka, T.,** Evidence for the involvement of the submandibular gland epidermal growth factor in mouse mammary tumorigenesis, *Proc. Natl. Acad. Sci. U.S.A.,* 82, 6940, 1985.

Chapter 6

FIBROBLAST, MELANOCYTE, AND NERVE GROWTH FACTORS

I. INTRODUCTION

Fibroblast, melanocyte, and nerve growth factors are involved in the regulation of growth and function of fibroblasts, melanocytes, and neural cells, respectively. The relationships between these factors and the structure and function of proto-oncogenes are still little understood.

II. FIBROBLAST GROWTH FACTOR

Fibroblast growth factor (FGF), also called fibroblast-derived growth factor (FDGF), is a polypeptide growth factor of 14,000 to 16,000 mol wt with mitogenic properties for fibroblasts as well as for a variety of neuroectoderm- and mesoderm-derived cell types. FGF has been isolated from bovine pituitary and adrenal gland tissue as well as from bovine brain tissue.[1-6]

A. Types of FGFs

Two types of FGF have been distinguished, basic FGF and acidic FGF. Although these mitogens resemble each other with respect to their in vitro mitogenic activity on cell types such as fibroblasts and endothelial cells and with respect to certain physicochemical characteristics, acidic FGF is structurally different from basic FGF as judged by molecular weight, amino acid composition, and sequence.[7] Acidic FGF is 30 to 100 times less potent than basic FGF, depending on the cell type. Bovine acidic FGF shows sequence homology with human IL-1 and is a potent mitogen for vascular endothelial cells in culture.[8] In the presence of heparin, acidic FGF induces blood vessel growth in vivo and the evidence suggests that it may be similar or identical to the tumor angiogenesis factor (TAF). Moreover, acidic FGF may be similar or identical to both a factor described as endothelial cell growth factor (ECGF) and other factor called eye-derived growth factor II (EDGF-II).[9] The complete amino acid sequence of bovine brain-derived acidic FGF has been reported recently.[10]

FDF secreted by an SV40-transformed baby hamster kidney cell line (SV28) has been purified close to homogeneity.[11] The coincidence of the physical, biological, and immunological characteristics of this factor and PDGF strongly suggests that they are closely related in structure and may be due to activation of the c-*sis* proto-oncogene.

B. Functions of FGF

In addition to the pituitary, adrenal gland, and brain, FGF is also present in a diversity of other tissues, including ovary, kidney, and retina. A factor detected in macrophages, called macrophage-derived growth factor (MDGF), which is a potent mitogen for nonlymphoid mesenchymal cells, including fibroblasts, smooth muscle cells, and endothelial cells, is probably identical to FGF.[12] Moreover, in the pituitary FGF may participate not in cell growth regulation but in an intrapituitary mechanism regulating normal prolactin and TSH secretion.[13] This finding also indicates that substances classically considered as growth factors may participate in regulatory functions other than cell growth regulation.

FGF induces cell growth in responsive cells such as differentiated muscle cells in culture (the clonal mouse muscle cell line BC3H1), and both pituitary-derived FGF or brain-derived FGF are as effective as serum in repressing the synthesis of creatine phosphokinase in these cells.[14]

C. Mechanisms of Action of FGF

The mechanisms of action of FGF are little understood, but generation of 1,2-diacylglycerol, mobilization of Ca^{2+}, and activation of protein kinase C are observed in quiescent cultures of Swiss 3T3 fibroblasts after addition of FGF to the medium.[15]

III. MELANOCYTE GROWTH FACTOR

Fibroblasts may participate in regulating the growth of melanocytes. A growth factor of approximate 40,000 mol wt extracted from the fibroblast cell line WI-38, derived from human embryonic lung, supports continued proliferation of cultured melanocytes in the absence of a tumor promoter (PMA).[16] Moreover, exposure of melanocytes to the combined action of TPA and active WI-38 extracts results in a synergistic effect on cell growth. In contrast to normal melanocytes, human melanoma cells generally grow vigorously in vitro in the absence of TPA, suggesting that their independent growth might be associated with the endogenous production of autostimulatory growth factors which may include TGFs and/or a melanocyte growth factor(s).[16,17]

IV. NERVE GROWTH FACTOR

Many proteins present in soluble tissue extracts and in the intercellular matrix influence the survival and development of cultured neurons.[18] Such proteins, termed neurotrophic factors, may have an important role in maintaining the structure and function of the nervous system in vivo. For example, a 43,000-mol wt protein called neurite promoting factor (NPF) is released by brain cells and glioma cells in culture and is capable of inducing neurite outgrowth in neuroblastoma cells.[19] An NPF has been purified and characterized from chicken gizzard smooth muscle cells.[20] So far, however, only one trophic factor has been shown to be responsible for the epigenetic determination of neurone survival: the protein called nerve growth factor (NGF) was demonstrated to be required for the survival of developing peripheral sympathetic and sensory neurons by showing that neutralization of endogenous NGF by antibodies to NGF results in the death of these neurons.[18,21-23]

NGF was discovered when mouse sarcoma tissue was transplanted into chick embryos and it was observed that the transplants caused a marked increase in the size of spinal sensory and sympathetic ganglia. These effects were then attributed to the release from the sarcomatous tissue of a humoral factor into the chick's circulation. Subsequently, two particularly rich sources of NGF were discovered, namely, snake venoms, and mouse submaxillary glands.[23]

A. The NGF Gene and the Structure of NGF

NGF is a multimeric protein consisting in three different types of subunits (alpha, beta, and gamma). The biological activity of NGF is specifically associated with the beta subunit which is a 118-amino acid polypeptide synthesized from a 307-amino acid precursor. The gene coding for the beta subunit of NGF is located on human chromosome 1, region 1p21-p22.1; in the mouse, the gene is located on chromosome 3.[24]

B. Mechanisms of Action of NGF

The cellular mechanisms of action of NGF are little understood. NGF modulates the differentiation of several neuronal cell types in vivo and in vitro.[25] In the rat pheochromocytoma cell line PC12, NGF acts via a specific receptor to induce a diversity of morphological and physiological changes that include neurite growth, modification of the cytoskeleton, and changes in neurotransmitter synthesis. PC12 cells respond to NGF by shifting a chromaffin-cell-like phenotype to a neurite-bearing sympathetic neurone-like phenotype. The enzyme ornithine decarboxylase (ODC) is induced by NGF in the same cell system. ODC catalyzes the formation of putrescine from ornithine, which is the first and rate-limiting step in polyamine biosynthesis.[26] Benzodiazepines can interact with NGF to modify NGF action both on neurite outgrowth and on the induction of ODC.[27]

The actions of NGF on PC12 cells maybe divided in two categories.[27] The first category consists of early effects that do not require activation of transcription and include the phosphorylation of endogenous substrates. The second category correspond to delayed actions that depend on RNA synthesis. The molecular phenomena involved in the regulation of transcriptional processes by NGF have not been characterized. NGF does not activate Na^+/H^+ exchange in PC12 cells.[28]

1. NGF-Induced Protein Phosphorylation

Treatment of PC12 cells with NGF induces increased phosphorylation of a set of cellular proteins including tyrosine hydroxylase, ribosomal protein S6, and several histone and nonhistone nuclear proteins.[29] At least some of these phosphorylations are mediated by activation of cyclic AMP-dependent protein kinases. NGF stimulates the phosphorylation of a specific nonhistone nuclear protein of approximate 30,000 mol wt through activation of a kinase which is not cAMP-dependent, nor is it similar to protein kinase C or casein kinase.[30] The increased phosphorylation produced by NGF is not transient, the stimulation being persistent for at least 3 days in the continuous presence of NGF.

2. NGF and c-fos Proto-Oncogene Expression

NGF and EGF induce rapid transient changes in proto-oncogene transcription in PC12 cells.[31] Of the transcriptional changes induced by NGF, induction of c-*fos* expression is the most rapid described so far, being detected 30 min after treatment.[27,32] This induction is enhanced more than 100-fold in the presence of peripherally active benzodiazepines. Moreover, the effect is specific as very little change is observed in the levels of transcripts of other oncogenes, including c-H-*ras*, c-K-*ras*, c-*myc*, and c-N-*myc*. This specific effect could be mediated by a transcriptional enhancer element which is located upstream from the c-*fos* proto-oncogene.[33] Under the conditions used in these experiments, NGF treatment ultimately results in neurite growth, with a reduction or cessation of cell division. Thus, NGF-induced expression of the c-*fos* proto-oncogene in PC12 cells is associated with cellular differentiation and not with cellular proliferation.

3. Abrogation of NGF Requirement by Oncogene Protein Products

Microinjection of the T24 mutant, oncogenic form of the c-*ras* protein, but not the normal c-*ras* protein, into PC12 cells results in the induction of a morphologically differentiated phenotype with the outgrowth of neuron-like processes.[34] The differentiation occurs in the absence of NGF, which suggests that the mutant c-*ras* protein has an ability to eliminate the exogenous signal represented by the specific growth factor.

Expression of a viral *src* oncogene has also an inductive effect that resembles the physiological action of NGF.[35] However, it is not known whether the c-*src* proto-oncogene is able to display a similar effect and, in case of positive answer whether the c-*src* protein and NGF share a common pathway for their action on neural tissue.

REFERENCES

1. **Holley, R. W. and Kiernan, J. A.,** "Contact inhibition" of cell division in 3T3 cells, *Proc. Natl. Acad. Sci. U.S.A.,* 60, 300, 1968.
2. **Böhlen, P., Baird, A., Esch, F., Ling, N., and Gospodarowicz, D.,** Isolation and partial molecular characterization of pituitary fibroblast growth factor, *Proc. Natl. Acad. Sci. U.S.A.,* 81, 5364, 1984.
3. **Gospodarowicz, D., Cheng, J., Lui, G-M., Baird, A., and Böhlen, P.,** Isolation of brain fibroblast growth factor by heparin-sepharose affinity chromatography: identity with pituitary fibroblast growth factor, *Proc. Natl. Acad. Sci. U.S.A.,* 81, 6963, 1984.
4. **Gospodarowicz, D., Massoglia, S., Cheng, J., Lui, G-M., and Böhlen, P.,** Isolation of pituitary fibroblast growth factor by fast protein liquid chromatography (FPLC): partial chemical and biological characterization, *J. Cell. Physiol.,* 122, 323, 1985.
5. **Böhlen, P., Esch, F., Baird, A., Jones, K. L., and Gospodarowicz, D.,** Human brain fibroblast growth factor: isolation and partial chemical characterization, *FEBS Lett.,* 185, 177, 1985.
6. **Gospodarowicz, D., Baird, A., Cheng, J., Lui, G. M., Esch, F., and Böhlen, P.,** Isolation of fibroblast growth factor from bovine adrenal gland: physicochemical and biological characterization, *Endocrinology,* 118, 82, 1986.
7. **Böhlen, P., Esch, F., Baird, A., and Gospodarowicz, D.,** Acidic fibroblast growth factor (FGF) from bovine brain: amino-terminal sequence and comparison with basic FGF, *EMBO J.,* 4, 1951, 1985.
8. **Thomas, K. A., Rios-Candelore, M., Gimenez-Gallego, G., DiSalvo, J., Bennett, C., Rodkey, J., and Fitzpatrick, S.,** Pure brain-derived acidic fibroblast growth factor is a potent angiogenic vascular endothelial cell mitogen with sequence homology to interleukin 1, *Proc. Natl. Acad. Sci. U.S.A.,* 82, 6409, 1985.
9. **Schreiber, A. B., Kenney, J., Kowalski, J., Thomas, K. A., Gimenez-Gallego, G., Rios-Candelore, M., DiSalvo, J., Barritault, D., Courty, J., Courtois, Y., Moenner, M., Loret, C., Burgess, W. H., Mehlman, T., Friesel, R., Johnson, W., and Maciag, T.,** A unique family of endothelial cell polypeptide mitogens: the antigenic and receptor cross-reactivity of bovine endothelial cell growth factor, brain-derived acidic fibroblast growth factor, and eye-derived growth factor-II, *J. Cell Biol.,* 101, 1623, 1985.
10. **Esch, F., Ueno, N., Baird, A., Hill, F., Denoroy, L., Ling, N., Gospodarowicz, D., and Guillemin, R.,** Primary structure of bovine brain acidic fibroblast growth factor (FGF), *Biochem. Biophys. Res. Comm.,* 133, 554, 1985.
11. **Stroobant, P., Gullick, W. J., Waterfield, M. D., and Rozengurt, E.,** Highly purified fibroblast-secreted mitogen, is closely related to platelet-derived growth factor, *EMBO J.,* 4, 1945, 1985.
12. **Baird, A., Mormède, P., and Böhlen, P.,** Immunoreactive fibroblast growth factor in cells of peritoneal exudate suggests its identity with macrophage-derived growth factor, *Biochem. Biophys. Res. Comm.,* 126, 358, 1985.
13. **Baird, A., Mormède, P., Ying, S-Y., Wehrenberg, W. B., Ueno, N., Ling, N., and Guillemin, R.,** A nonmitogenic pituitary function of fibroblast growth factor: regulation of thyrotropin and prolactin secretion, *Proc. Natl. Acad. Sci. U.S.A.,* 82, 5545, 1985.
14. **Lathrop, B., Olson, E., and Glaser, L.,** Control by fibroblast growth factor of differentiation in the BC$_3$H1 muscle cell line, *J. Cell Biol.,* 100, 1540, 1985.
15. **Tsuda, T., Kaibuchi, K., Kawahara, Y., Fukuzaki, H., and Takai, Y.,** Induction of protein kinase C activation and Ca^{2+} mobilization by fibroblast growth factor in swiss 3T3 cells, *FEBS Lett.,* 191, 205, 1985.
16. **Eisinger, M., Marko, O., Ogata, S-I., and Old, L. J.,** Growth regulation of human melanocytes: mitogenic factors in extracts of melanoma, astrocytoma, and fibroblast cell lines, *Science,* 229, 984, 1985.
17. **Richmond, A., Lawson, D. H., Nixon, D. W., and Chawla, R. K.,** Characterization of autostimulatory and transforming growth factors from human melanoma cells, *Cancer Res.,* 45, 12, 1985.
18. **Thoenen, H. and Edgar, D.,** Neurotrophic factors, *Science,* 229, 238, 1985.
19. **Guenther, J., Nick, H., and Monard, D.,** A glia-derived neurite-promoting factor with protease inhibitory activity, *EMBO J.,* 4, 1963, 1985.
20. **Hayashi, Y. and Miki, N.,** Purification and characterization of a neurite outgrowth factor from chicken gizzard smooth muscle, *J. Biol. Chem.,* 260, 14269, 1985.
21. **Levi-Montalcini, R. and Angeletti, P. U.,** Nerve growth factor, *Physiol. Rev.,* 48, 534, 1968.
22. **Mobley, W. C., Server, A. C., Ishii, D. N., Riopelle, R. J., and Shooter, E. M.,** Nerve growth factor, *N. Engl. J. Med.,* 297, 1096, 1149, and 1211, 1977.
23. **Harper, G. P. and Thoenen, H.,** Nerve growth factor: biological significance, measurement, and distribution, *J. Neurochem.,* 34, 5, 1980.
24. **Zabel, B. U., Eddy, R. L., Lalley, P. A., Scott, J., Bell, G. I., and Shows, T. B.,** Chromosomal locations of the human and mouse genes for precursors of epidermal growth factor and the beta subunit of nerve growth factor, *Proc. Natl. Acad. Sci. U.S.A.,* 82, 469, 1985.

25. **Greene, L. A. and Shooter, E. M.**, The nerve growth factor: biochemistry, synthesis, and mechanism of action, *Ann. Rev. Neurosci.,* 3, 353, 1980.
26. **Feinstein, S. C., Dana, S. L., McConlogue, L., Shooter, E. M., and Coffino, P.**, Nerve growth factor rapidly induces ornithine decarboxylase mRNA in PC12 rat pheochromocytoma cells, *Proc. Natl. Acad. Sci. U.S.A.,* 5761, 1985.
27. **Curran, T. and Morgan, J. I.**, Superinduction of c-*fos* by nerve growth factor in the presence of peripherally active benzodiazepines, *Science,* 229, 1265, 1985.
28. **Chandler, C. E., Cragoe, E. J., Jr., and Glaser, L.**, Nerve growth factor does not activate Na^+/H^+ exchange in PC12 pheochromocytoma cells, *J. Cell. Physiol.,* 125, 367, 1985.
29. **Halegoua, S. and Patrick, J.**, Nerve growth factor mediates phosphorylation of specific proteins, *Cell,* 22, 571, 1980.
30. **Nakanishi, N. and Guroff, G.**, Nerve growth factor-induced increase in the cell-free phosphorylation of a nuclear protein in PC12 cells, *J. Biol. Chem.,* 260, 7791, 1985.
31. **Greenberg, M. E., Greene, L. A., and Ziff, E. B.**, Nerve growth factor and epidermal growth factor induce rapid transient changes in proto-oncogene transcription in PC12 cells, *J. Biol. Chem.,* 260, 14101, 1985.
32. **Kruijer, W., Schubert, D., and Verma, I. M.**, Induction of the proto-oncogene *fos* by nerve growth factor, *Proc. Natl. Acad. Sci. U.S.A.,* 82, 7330, 1985.
33. **Deschamps, J., Meijlink, F., and Verma, I. M.**, Identification of a transcriptional enhancer element upstream from proto-oncogene *fos, Science,* 230, 1174, 1985.
34. **Bar-Sagi, D. and Feramisco, J. R.**, Microinjection of the *ras* oncogene protein into PC12 cells induces morphological differentiation, *Cell,* 42, 841, 1985.
35. **Alemà, S., Casalbore, P., Agostini, E., and Tatò, F.**, Differentiation of PC12 phaeochromocytoma cells induced by v-*src* oncogene, *Nature (London),* 316, 557, 1985.

Chapter 7

TRANSFORMING GROWTH FACTORS

I. INTRODUCTION

Transforming growth factors (TGFs) are a family of polypeptides involved in the regulation of cell growth and cell differentiation. TGFs reversibly induce the expression of a transformed phenotype in cultured cells; in particular, they have the property of stimulating anchorage-independent growth of certain types of nontransformed cells cultured in soft agar.[1-5] However, TGFs have an important role in various types of normal physiological phenomena, including wound healing.[6] TGFs, as well as EGF and PDGF, have been shown to stimulate bone resorption in neonatal mouse calvaria in organ culture via a prostaglandin-mediated mechanism.[7] This effect depends on the stimulation of arachidonic acid synthesis and is inhibited by indomethacin. It seems likely that growth factors such as EGF, PDGF, TGFs, and/or other unidentified factors produced by certain tumors occurring in humans and other animals contribute to stimulate bone resorption and to produce hypercalcemia, which is observed in different types of cancer.[8,9] In addition to TGFs, fetal rat calvarial cultures release another growth factor, celled bone-derived growth factor (BDGF), which stimulates fibroblast replication but has no transforming activity for NRK cells.[10]

TGFs can be detected by different assays, including a convenient microassay where colony-forming activity in soft agar medium is examined by using nontransformed BALB/3T3 or NRK cells.[11] TGFs do not stimulate soft agar colony formation of other cells, such as C3H/10T1/2 cells and human foreskin diploid fibroblasts.

A. Cellular Sources of TGFs

Different TGFs have been isolated from a variety of tumor cell lines or cells transformed by retroviruses and chemical carcinogens.[12-29] TGFs are also produced by flat revertants from cells transformed by acute transforming retroviruses, like Kirsten MuSV.[30,31] These revertants express TGFs and the $p21^{v-ras}$ oncogene protein product but fail to exhibit all of the properties associated with neoplastic transformation, which suggests that the action of $p21^{v-ras}$ maybe blocked at a point distal to its transforming activity and that expression of the oncogene product and TGFs maybe necessary but not sufficient for maintaining the transformed state.

Different types of TGFs are present in apparently all human benign and malignant neoplasms, including leukemic cells and mammary carcinoma cells.[32-35] Certain types of TGFs are present in serum and tissues of embryos and fetuses, as well as in human term placenta, human colostrum, and some adult tissues, including lung and kidney.[32,36-41]

B. TGFs Types and TGFs Receptors

The TGFs family includes different types of polypeptides that have been classified in two main groups: α- or type-1 TGFs and β-1 or type-2 TGFs. The mitogenic action of TGF-α is mediated by direct interaction with the EGF receptor, and a phenomenon of down-regulation of the number of EGF receptors on the cell surface occurs after this interaction.[42] TGF-α competes with EGF for binding to EGF receptors, while TGF-β possesses a unique cell surface receptor and does not complete with the same receptors. Tyrosine-specific protein kinase activity is stimulated after interaction of TGF-α with the EGF receptor whereas, no such type of activity is elicited after interaction of TGF-β and its own receptor.[43]

II. TRANSFORMING GROWTH FACTOR α

The first discovered TGF was a type-1 TGF, transforming growth factor α (TGF-α), which is present in mouse cells transformed by MuSV.[3,12,13] TGF-α, also called sarcoma growth factor (SGF), is not of viral origin but of cellular origin. It was postulated that TGF-α is produced only by tumor cells (retrovirus-transformed cells and certain human tumor cells) but not by normal cells, and it was suggested that autostimulation by endogenous TGFs may contribute to tumor progression.[1,3] However, TGF-α mRNA is present in normal animal tissues, including the brain, liver and kidney,[44] which strongly suggests that TGF-α is not exclusively present in transformed cells but is also produced by at least some types of normal cells. Differential expression of TGF-α is observed during prenatal development of mouse and rat.[45,46] TGF-α, as EGF, is active in promoting eyelid opening in newborn mice.[47]

A. The TGF-α Gene

The human and rat gene for TGF-α have been cloned, and the human gene has been expressed in *Escherichia coli* and the structure of the TGF-α precursor peptide has been determined.[44,48] The human TGF-α gene cloned from a cDNA library prepared using RNA from a human renal carcinoma cell line encodes a precursor polypeptide of 160 amino acids. The 50 amino acid mature human TGF-α produced by expression of the appropriate coding sequences in *E. coli* binds to the EGF receptor and induces anchorage independent cell growth in a soft agar assay, two biological characteristics of EGF and natural TGF-α.[48] The TGF-α gene has been mapped to human chromosome 2, close to the breakpoint of the Burkitt's lymphoma t(2;8) variant translocation.[49] The biological significance of this association, however, is not understood. A cDNA clone encoding rat TGF-α hybridizes to a 4.5-k base mRNA that is 30 times larger than necessary to code for a 50-amino acid TGF-α polypeptide.[44] TGF-α mRNA is present not only in transformed cells but also, although at lower levels, in several normal rat tissues and its nucleotide sequence predicts that TGF-α is synthesized as a larger product and that the larger form may exist as a transmembrane protein. Apparently, the translation product of rat TGF-α mRNA comprises 159 amino acids.[44]

B. Structure of TGF-α

TGF-α purified from human melanoma cell lines and from rat and mouse cells transformed by acute retroviruses, like MuSV, show a high degree of structural homology, differing from each other by only a few amino acid substitutions.[18] TGF-α isolated from the conditioned medium of a human melanoma cell line is a single-chain polypeptide of 7400 mol wt.[50] TGF-α prepared from other sources is a single-chain polypeptide of 5700 daltons with three disulfide bridges in positions homologous to those of EGF.

Larger polypeptides structurally and functionally related to TGF-α have been detected by gel filtration of conditioned medium from cultured retrovirus-transformed rat cells.[51] An antiserum raised against a chemically synthesized oligopeptide corresponding to the carboxy-terminal 17 amino acids of rat TGF-α was used to develop a competitive radioimmunoassay for the immunizing peptide. Immunoblotting analysis revealed three TGF-α-related polypeptides with 24,000, 40,000 and 42,000 mol wt, respectively, which could represent the products of a cleaved TGF-α polypeptide precursor molecule. The precursor maybe, at least partially, processed extracellularly.

The complete amino acid sequence of rat TGF-α has been determined and a striking degree of structural and functional homology between this factor and EGF has been recognized.[52] Rat TGF-α is a single peptide chain containing 50 amino acid residues and 3 disulfide linkages. A synthetic peptide corresponding to rat TGF-α has been produced by a stepwise solid-phase approach, following the general principles of the Merrifield method, and the synthetic product has chemical and biological properties which are indistinguishable from those of natural rat TGF-α purified from rat embryo fibroblasts transformed by FeSV.[53] A synthetic fragment of TGF-α comprising a decapeptide of the third disulfide loop (residues 34 to 43) has no mitogenic activity but prevents the mitogenic effects of EGF and TGF-α on fibroblasts.[54] This fragment acts as an antagonist of the induction of cellular proliferation by EGF and contains an important receptor binding sequence of TGF-α.

C. Cellular Mechanisms of Action of TGF-α

After binding to the EGF receptor on the cell surface, TGF-α stimulates tyrosine phosphorylation at specific acceptor sites of the receptor.[55] TGF-α also stimulates tyrosine phosphorylation in a synthetic peptide whose sequence is related to that of the known site of tyrosine phosphorylation in the oncogene protein product $pp60^{v-src}$.[56] Occupation of the EGF receptor by its ligand is required for both the mitogenic and colony-forming activity of TGF-α.[57] As it occurs with EGF, treatment with TGF-α may cause precaucious eyelid opening and accelerated incisor eruption in newborn mice.[47,58] Such facts strongly suggest that TGF-α, as EGF, may have important functions in the development of immature animals. A concerted action of TGF-α and TGF-β is required for inducing the expression of a transformed phenotype in cultured cells.[20]

III. TRANSFORMING GROWTH FACTOR β

Transforming growth factor β (TGF-β) has been detected in normal and transformed cells and has been purified to homogeneity from human placenta, human platelets, and bovine kidney.[27,38,39,59] TGF-β is operationally defined by its capacity of eliciting anchorage-independent growth (formation of colonies in soft agar medium) by either normal rat kidney (NRK) cells or mouse embryo AKR-2B cells. Substances with this type of activity are released into the culture media by normal chicken, mouse, and human embryo fibroblasts.[41] TGF-β present in conditioned cell culture medium independent of cell transformation may also derive from serum.[60]

A. Structure of TGF-β

TGF-β (type-2 TGF) purified from human placenta has been characterized as a 23,000 to 25,000-dalton protein composed of two polypeptide chains held together by interchain disulfide linkages.[38] TGF-β purified from rat embryo fibroblasts transformed by Snyder-Theilen FeSV has similar physicochemical characteristics.[24] The amino acid sequence of the human TGF-β monomer has been determined from a TGF-β cDNA clone.[61] The 112-amino acid monomeric form of the natural TGF-β homodimer is derived proteolitically from a much larger precursor polypeptide which maybe secreted. TGF-β mRNA is synthesized in various types of normal and transformed cells.[61] cDNA sequences of ovarian follicular fluid inhibin (a hormone involved in the regulation of adenohypophysial function) show precursor structure homology with TGF-β.[62]

B. Cellular Mechanisms of Action of TGF-β

TGF-β binding sites are present in a variety of cultured cells of both mesenchymal and epithelial origin, including normal human fibroblasts and keratinocytes.[63] Lower levels of TGF-β receptors are detected in chemically transformed derivatives of mouse embryo fibroblast-like cell lines as well as in certain human tumor cell lines, which could be associated with a down-regulation phenomenon caused by the production of TGF-β by these cells.[63]

TGF-β modulates the expression of both EGF and TGF-α receptors on the cell surface.[43] An early effect of TGF-β in NRK cells is the induction of a rapid increase in the number of plasma membrane receptors for EGF, and some delayed effects of TGF-β maybe indirect consequences of its ability to regulate EGF receptors and thereby amplify EGF-induced cellular responses.[64] However, in another study using the same cells it was found that exposure to biologically active concentrations of TGF-β induced, not an increase, but a rapid decrease in the binding of EGF and TGF-α.[65] It is possible that TGF-β has a dual action on EGF and TGF-α receptors according to the physiological state of the cell, with an inhibitory effect predominating in situations in which cells evolve from a growth-arrested state to a mitogenically active one.

TGF-β stimulates glucose uptake, glycolysis and amino acid uptake, and these effects are probably mediated by activation of the EGF receptor.[66,67] However, TGF-β does not compete with EGF for binding to the EGF receptor on the cell surface, and a membrane receptor for TGF-β has been characterized in NRK fibroblasts.[43] The subunit structure of TGF-β receptor has been examined and there is evidence for the existence of a disulfide-linked glycosylated receptor complex.[68] The high-affinity TGF-β receptor has a molecular weight of 565 kdaltons in the mouse and 615 kdaltons in the human.

After binding to its receptor, TGF-β is apparently internalized and degraded in the lysosomes, and the TGF-β receptor is subjected to down-regulatory phenomena in a manner which is similar to that of other growth factors. TGF-β does not elicit tyrosine-specific protein kinase activity and has no apparent mitogenic effects by itself. However, in AKR-2B mouse embryo cells TGF-β acts as a potent stimulator of DNA synthesis but only with a prolonged (about 24 hr) prereplicative phase when compared with other growth factors (EGF, PDGF, FGF) that induce DNA synthesis 12 to 14 hr after stimulation.[69]

At least some of the biological effects of TGF-β maybe mediated by modulation of the receptors of other growth factors. TGF-β modulates the high-affinity receptors for EGF and TGF-α.[65] Complex interactions between different types of growth factors are involved in the expression of TGF-β biological actions. For example, stimulation of cells by IGFs is required for transformation induced by TGF-β in vitro.[70]

A high rate of aerobic glycolysis, as measured by the rate of production of lactate, is an almost universal characteristic of tumor cells.[71] This phenomenon is observed in Rat-1 cells transformed with a *ras* oncogene and also, although at a lesser degree, in Rat-1 cells exposed to TGF-β.[72] TGF-β also stimulates aerobic glycolysis in Rat-1 cells transfected with a *myc* oncogene, but even the stimulated rat is considerably lower than that in *ras* transfected cells. The uptake of methylaminoisobutyrate, a specific substrate of system A amino acid transport, is also accelerated after exposure of Rat-1 cells to TGF-β. Methionine was found to cause a marked inhibition of glycolysis in Rat-1 cells exposed to TGF-β, either the cells were transfected or not with c-*ras* or c-*myc*. The results suggested a relationship between the protein product of the *ras* oncogene, $p21^{ras}$, and the protein(s) induced by exposure to TGF-β.[72]

C. Bifunctional Action of TGF-β on Cell Proliferation and/or Transformation

TGF-β is an important regulator of cellular growth.[69,73] The action of TGF-β on cells is bifunctional: in certain instances TGF-β stimulates anchorage-independent growth whereas in others it acts as a growth inhibitor. Moreover, TGF-β can act either synergistically or antagonistically with other growth factors, including EGF and PDGF. TGF-β is a potent stimulator of DNA synthesis in AKR-2B mouse embryo cells with a prolonged lag phase relative to the stimulation obtained with EGF, PDGF or FGF.[69] In these cells, TGF-β inhibits the early peak of DNA synthesis produced by EGF and insulin. In NRK cells, treatment with TGF-β results in biphasic effects on EGF binding. Initially, EGF binding to high affinity sites is inhibited by TGF-β as a result of transient decrease in EGF receptor affinity, but thereafter the binding is stimulated as a result of a persistent effect in EGF receptor number.[74]

Fischer rat 3T3 (FR3T3) fibroblasts transfected with a c-*myc* gene can be induced to grow and form colonies in soft agar by treatment either with EGF alone or the combination of PDGF and TGF-β.[73] In this system, TGF-β can function as either an inhibitor or an enhancer of anchorage-independent growth, depending on the particular set of growth factors operating on the cell together with TGF-β. Induction of anchorage-independent growth by different sets of growth factors involves different cellular pathways which can be distinguished by their sensitivity to retinoic acid.[75] Colony formation induced by the combined action of PDGF and TGF-β is 100-fold more sensitive to inhibition by retinoic acid than is colony formation induced by treatment of c-*myc*-transfected cells with EGF. Moreover, retinoic acid is inhibitory for colony growth whenever TGF-β is present, regardless of whether the effects of TGF-β are stimulatory, as occurs in the presence of PDGF, or inhibitory, as found in the presence of EGF.[75]

An inhibitor factor purified from African green monkey kidney cells (BSC-1 cells) is a polypeptide with similar or identical activity to that of TGF-β purified from human platelets, competing with TGF-β for binding to the same cellular receptors.[76] These findings suggest that similar or identical molecules related to growth factors can either stimulate or inhibit cell proliferation depending on the particular experimental or physiological conditions.

IV. OTHER TRANSFORMING GROWTH FACTORS

In addition to TGF-α and TGF-β, other types of TGFs may exist, for example, a factor termed ND-TGF, which was obtained from a mouse neuroblastoma cell line but is not a specific neuronal growth factor.[77] ND-TGF is a strong mitogen for different cell lines, it does not compete with EGF for receptor binding as does TGF-α, and, unlike TGF-β, it has strong mitogenic activity. Whereas EGF alone is unable to induce any degree of anchorage-independent growth in NRK cells, ND-TGF is able to induce progressively growing colonies of these cells in soft agar, even without the addition of EGF.[78]

A novel TGF has been isolated and purified from an ASV-transformed rat cell line (77N1).[22,23] This factor is composed of a 12,000-dalton protein that does not compete with EGF for binding to EGF membrane receptors and is capable of inducing DNA synthesis in growth-arrested BALB/3T3 cells and promoting anchorage-independent growth of nontransformed BALB/3T3 cells in soft agar. TGFs with unique biological activities have also been isolated from the conditioned medium of ASV-transformed chicken and hamster cells,[16] PC13 embryonal carcinoma cells,[79] and salivary gland epithelial cell lines (CSG 211) chemically transformed in vitro.[28]

It is possible that multiple types of TGFs are produced in transformed cells. Reverse phase high-performance liquid chromatography analyses of acid extracts of urine from normal human donors and cancer patients revealed the presence of 5 EGF-related and 2 non-EGF-related TGFs, the latter having soft-agar colony-stimulating activity only in the presence of added EGF.[80] In these studies it was demonstrated that two of the five EGF-related TGFs were consistently elevated only in the urine of cancer patients, which could be the result of changes in normal TGF metabolism.

Vaccinia virus-infected cells release an acid-stable mitogen that competes with EGF for binding to EGF receptors and shares biological properties similar to those of both EGF and TGFs.[81] The novel factor, however, is immunologically unrelated to EGF and known TGFs, and would consist of a much larger protein.

V. SUMMARY

TGFs are mitogenic polypeptides involved in the regulation of normal processes of cell growth and cell differentiation. TGFs reversibly induce the expression of a transformed phenotype in cultured cells. Several species of TGFs have been isolated from a diversity of normal and transformed cells. Binding of TGF-α, but not TGF-β, to its cellular receptor is associated with stimulation of tyrosine-specific protein kinase activity. At least some of the biological activities of TGFs may be mediated by modulation of the receptors for other growth factors, especially the EGF receptor. The possible role of TGFs in the regulation of proto-oncogene expression is little understood. However, ther is evidence that the mitogenic effects of TGF-β in certain cellular systems are mediated by the expression of the c-*sis* proto-oncogene.[82] The production of PDGF, or a PDGF-like molecule, may therefore be associated with the secondarily induced expression of other proto-oncogenes involved in the development of a mitogenic response, including c-*fos* and c-*myc*.

REFERENCES

1. **Sporn, M. B. and Todaro, G. J.**, Autocrine secretion and malignant transformation of cells, *N. Engl. J. Med.*, 303, 878, 1980.
2. **Todaro, G. J., De Larco, J. E., Fryling, C., Johnson, P. A., and Sporn, M. B.**, Transforming growth factors (TGFs): properties and possible mechanisms of action, *J. Supramol. Struct.*, 15, 287, 1981.
3. **Todaro, G. J., De Larco, J. E., and Fryling, C. M.**, Sarcoma growth factor and other transforming peptides produced by human cells: interactions with membrane receptors, *Fed. Proc.*, 41, 2996, 1982.
4. **Roberts, A. B., Frolik, C. A., Anzano, M. A., and Sporn, M. B.**, Transforming growth factors from neoplastic and nonneoplastic cells, *Fed. Proc.*, 42, 2621, 1983.
5. **Lawrence, D. A.**, Transforming growth factors — an overview, *Biol. Cell*, 53, 93, 1985.
6. **Sporn, M. B., Roberts, A. B., Shull, J. H., Smith, J. M., Ward, J. M., and Sodek, J.**, Polypeptide transforming growth factors isolated from bovine sources and used for wound healing *in vivo*, *Science*, 219, 1329, 1983.
7. **Tashjian, A. H., Jr., Voelkel, E. F., Lazzaro, M., Singer, F. R., Roberts, A. B., Derynck, R., Winkler, M. E., and Levine, L.**, Alpha and beta human transforming growth factors stimulate prostaglandin production and bone resorption in cultured mouse calvaria, *Proc. Natl. Acad. Sci. U.S.A.*, 82, 4535, 1985.
8. **Ibbotson, K. J., D'Souza, S. M., Ng, K. W., Osborne, C. K., Niall, M., Martin, T. J., and Mundy, G. R.**, Tumor-derived growth factor increases bone resorption in a tumor associated with humoral hypercalcemia of malignancy, *Science*, 221, 1292, 1983.
9. **Mundy, G. R., Ibbotson, K. J., and D'Souza, S. M.**, Tumor products and the hypercalcemia of malignancy, *J. Clin. Invest.*, 76, 391, 1985.
10. **Centrella, M. and Canalis, E.**, Transforming and nontransforming growth factors are present in medium conditioned by fetal rat calvariae, *Proc. Natl. Acad. Sci. U.S.A.*, 82, 7335, 1985.
11. **Morita, H., Noda, K., Umeda, M., and Ono, T.**, Activities of transforming growth factors on cell lines and their modification by other growth factors, *Gann*, 75, 403, 1984.
12. **Todaro, G. J., De Larco, J. E., and Cohen, S.**, Transformation by murine and feline sarcoma viruses specifically blocks binding of epidermal growth factor to cells, *Nature (London)*, 264, 26, 1976.
13. **De Larco, J. E. and Todaro, G. J.**, Growth factors from murine sarcoma virus-transformed cells, *Proc. Natl. Acad. Sci. U.S.A.*, 75, 4001, 1978.
14. **Roberts, A. B., Lamb, L. C., Newton, D. L., Sporn, M. B., De Larco, J. E., and Todaro, G. J.**, Transforming growth factors: isolation of polypeptides from virally and chemically transformed cells by acid/ethanol extraction, *Proc. Natl. Acad. Sci. U.S.A.*, 77, 3494, 1980.
15. **Moses, H. L., Branum, E. L., Proper, J. A., and Robinson, R. A.**, Transforming growth factor production by chemically transformed cells, *Cancer Res.*, 41, 2842, 1981.
16. **Kryceve-Martinerie, C., Lawrence, D. A., Crochet, J., Julien, P., and Vigier, P.**, Cells transformed by Rous sarcoma virus release transforming growth factors, *J. Cell. Physiol.*, 113, 365, 1982.
17. **Roberts, A. B., Anzano, M. A., Lamb, L. C., Smith, J. M., Frolik, C. A., Marquardt, M., Todaro, G. J., and Sporn, M. B.**, Isolation from the murine sarcoma cells of novel transforming growth factors potentiated by EGF, *Nature (London)*, 295, 417, 1982.
18. **Marquardt, H., Hunkapiller, M. W., Hood, L. E., Twardzik, D. R., De Larco, J. E., Stephenson, J. R., and Todaro, G. J.**, Transforming growth factors produced by retrovirus-transformed rodent fibroblasts and human melanoma cells: amino acid sequence homology with epidermal growth factor, *Proc. Natl. Acad. Sci. U.S.A.*, 80, 4684, 1983.

19. **Chua, C. C., Geiman, D., and Ladda, R. L.**, Transforming growth factors released from Kirsten sarcoma virus transformed cells do not complete for epidermal growth factor membrane receptors, *J. Cell. Physiol.*, 117, 116, 1983.
20. **Anzano, M. A., Roberts, A. B., Smith, J. M., Sporn, M. B., and De Larco, J. E.**, Sarcoma growth factor from conditioned medium of virally transformed cells is composed of both type alpha and type beta transforming growth factors, *Proc. Natl. Acad. Sci. U.S.A.*, 80, 6264, 1983.
21. **Massagué, J.**, Epidermal growth factor-like transforming growth factor. I. Isolation, chemical characterization, and potentiation by other transforming growth factors from feline sarcoma virus-transformed cells, *J. Biol. Chem.*, 258, 13606, 1983.
22. **Hirai, R., Yamaoka, K., and Mitsui, H.**, Isolation and partial purification of a new class of transforming growth factors from an avian sarcoma virus-transformed rat cell line, *Cancer Res.*, 43, 5742, 1983.
23. **Yamaoka, K., Hirai, R., Tsugita, A., and Mitsui, H.**, The purification of an acid- and heat-labile transforming growth factor from an anvian sarcoma virus-transformed rat cell line, *J. Cell. Physiol.*, 119, 307, 1984.
24. **Massagué, J.**, Type beta transforming growth factor from feline sarcoma virus-transformed rat cells: isolation and biological properties, *J. Biol. Chem.*, 259, 9756, 1984.
25. **Pircher, R., Lawrence, D. A., and Julien, P.**, Latent beta-transforming growth factor in nontransformed and Kirsten sarcoma virus-transformed normal rat kidney cells, clone 49F, *Cancer Res.*, 44, 5538, 1984.
26. **Anzano, M. A., Roberts, A. B., De Larco, J. E., Wakefield, L. M., Assoian, R. K., Roche, N. S., Smith, J. M., Lazarus, J. E., and Sporn, M. B.**, Increased secretion of transforming growth factor accompanies viral transformation of cells, *Mol. Cell. Biol.*, 5, 242, 1985.
27. **Kryceve-Martinerie, C., Lawrence, D. A., Crochet, J., Jullien, P., and Vigier, P.**, Further study of beta-TGFs released by virally transformed and non-transformed cells, *Int. J. Cancer*, 35, 553, 1985.
28. **Wigley, C. B., Trejdosiewicz, L. K., Southgate, J., Coventry, R., and Ozanne, B.**, Growth factor production during multistage transformation of epithelium in vitro, I. Partial purification and characterisation of the factor(s) from a fully transformed epithelial cell line, *J. Cell. Physiol.*, 125, 156, 1985.
29. **De Larco, J. E., Pigott, D. A., and Lazarus, J. A.**, Ectopic peptides released by a human melanoma cell line that modulate the transformed phenotype, *Proc. Natl. Acad. Sci. U.S.A.*, 82, 5015, 1985.
30. **Noda, M., Selinger, Z., Scolnick, E. M., and Bassin, R. H.**, Flat revertants isolated from Kirsten sarcoma virus-transformed cells are resistant to the action of specific oncogenes, *Proc. Natl. Acad. Sci. U.S.A.*, 80, 5602, 1983.
31. **Salomon, D. S., Zwiebel, J. A., Noda, M., and Bassin, R. H.**, Flat revertants derived from Kirsten murine sarcoma virus-transformed cells produce transforming growth factors, *J. Cell. Physiol.*, 121, 22, 1984.
32. **Nickell, K. A., Halper, J., and Moses, H. L.**, Transforming growth factors in solid human malignant neoplasms, *Cancer Res.*, 43, 1966, 1983.
33. **Nakamura, H., Komatsu, K., Akedo, H., Hosokawa, M., Shibata, H., and Masaoka, T.**, Human leukemic cells contain transforming growth factor, *Cancer Lett.*, 21, 133, 1983.
34. **Salomon, D. S., Zwiebel, J. A., Bano, M., Losonczy, I., Fehnel, P., and Kidwell, W. R.**, Presence of transforming growth factors in human breast cancer cells, *Cancer Res.*, 44, 4069, 1984.
35. **Hamburger, A. W., White, C. P., and Dunn, F. E.**, Production of transforming growth factors by primary human tumour cells, *Br. J. Cancer.*, 51, 9, 1985.
36. **Roberts, A. B., Anzano, M. A., Lamb, L. C., Smith, J. M., and Sporn, M. B.**, New class of transforming growth factors potentiated by epidermal growth factor: isolation from nonneoplastic tissues, *Proc. Natl. Acad. Sci. U.S.A.*, 78, 5339, 1981.
37. **Stromberg, K., Pigott, D. A., Ranchalis, J. E., and Twardzik, D. R.**, Human term placenta contains transforming growth factors, *Biochem. Biophys. Res. Comm.*, 106, 354, 1982.
38. **Frolik, C. A., Dart, L. L., Meyers, C. A., Smith, D. M., and Sporn, M. B.**, Purification and initial characterization of a type beta transforming growth factor from human placenta, *Proc. Natl. Acad. Sci. U.S.A.*, 80, 3676, 1983.
39. **Roberts, A. B., Anzano, M. A., Meyers, C. A., Wideman, J., Blacher, R., Pan, Y-C. E., Stein, S., Lehrman, S. R., Smith, J. M., Lamb, L. C., and Sporn, M. B.**, Purification and properties of a type beta transforming growth factor from bovine kidney, *Biochemistry*, 22, 5692, 1983.
40. **Noda, K., Umeda, M., and Ono, T.**, Transforming growth factor activity in human colostrum, *Gann*, 75, 109, 1984.
41. **Lawrence, D. A., Pircher, R., Kryceve-Martinerie, C., and Jullien, P.**, Normal embryo fibroblasts release transforming growth factors in a latent form, *J. Cell. Physiol.*, 121, 184, 1984.
42. **Massagué, J.**, Epidermal growth factor-like transforming growth factor. II. Interaction with epidermal growth factor receptor in human placental membranes and A431 cells, *J. Biol. Chem.*, 258, 13614, 1983.
43. **Frolik, C. A., Wakefield, L. M., Smith, D. M., and Sporn, M. B.**, Characterization of a membrane receptor for transforming growth factor-beta in normal rat kidney fibroblasts, *J. Biol. Chem.*, 259, 10995, 1984.

44. Lee, D. C., Rose, T. M., Webb, N. R., and Todaro, G. J., Cloning and sequence analysis of a cDNA for rat transforming growth factor-alpha, *Nature (London)*, 313, 489, 1985.
45. Twardzik, D. R., Differential expression of transforming growth factor-alpha during prenatal development of the mouse, *Cancer Res.*, 45, 5413, 1985.
46. Lee, D., Rochford, R., Todaro, G. J., and Villareal, L. P., Developmental expression of rat transforming growth factor-alpha mRNA, *Mol. Cell. Biol.*, 5, 3644, 1985.
47. Smith, J. M., Sporn, M. B., Roberts, A. B., Derynck, R., Winkler, M. E., and Gregory, H., Human transforming growth factor-alpha causes precocious eyelid opening in newborn mice, *Nature (London)*, 315, 515, 1985.
48. Derynck, R., Roberts, A. B., Winkler, M. E., Chen, E. Y., and Goeddel, D. V., Human transforming growth factor-alpha: precursor structure and expression in E. coli, *Cell*, 38, 287, 1984.
49. Brissenden, J. E., Derynck, R., and Francke, U., Mapping of transforming growth factor alpha gene on human chromosome 2 close to the breakpoint of the Burkitt's lymphoma t(2;8) variant translocation, *Cancer Res.*, 45, 5593, 1985.
50. Marquardt, H. and Todaro, G. J., Human transforming growth factor: production by a melanoma cell line, purification, and characterization, *J. Biol. Chem.*, 257, 5220, 1982.
51. Linsley, P. S., Hargreaves, W. R., Twardzik, D. R., and Todaro, G. J., Detection of larger polypeptides structurally and functionally related to type I transforming growth factor, *Proc. Natl. Acad. Sci. U.S.A.*, 82, 356, 1985.
52. Marquardt, H., Hunkapiller, M. W., Hood, L. E., and Todaro, G. J., Rat transforming growth factor type I: structure and relation to epidermal growth factor, *Science*, 223, 1079, 1984.
53. Tam, J. P., Marquardt, H., Rosberger, D. F., Wong, T. W., and Todaro, G. J., Synthesis of biologically active rat transforming growth factor I, *Nature (London)*, 309, 376, 1984.
54. Nestor, J. J., Newman, S. R., DeLustro, B., Todaro, G. J., and Schreiber, A. B., A synthetic fragment of rat transforming growth factor alpha with receptor binding and antigenic properties, *Biochem. Biophys. Res. Comm.*, 129, 226, 1985.
55. Reynolds, F. H., Jr., Todaro, G. J., Fryling, C., and Stephenson, J. R., Human transforming growth factors induce tyrosine phosphorylation of EGF receptors, *Nature (London)*, 292, 259, 1981.
56. Pike, L. J., Marquardt, H., Todaro, G. J., Gallis, B., Casnellie, J. E., Bornstein, P., and Krebs, E. G., Transforming growth factor and epidermal growth factor stimulate the phosphorylation of a synthetic, tyrosine-containing peptide in a similar manner, *J. Biol. Chem.*, 257, 14628, 1982.
57. Carpenter, G., Stoscheck, C. M., Preston, Y. A., and De Larco, J. E., Antibodies to the epidermal growth factor receptor block the biological activities of sarcoma growth factor, *Proc. Natl. Acad. Sci. U.S.A.*, 80, 5627, 1983.
58. Tam, J. P., Physiological effects of transforming growth factor in the newborn mice, *Science*, 229, 673, 1985.
59. Assoian, R. K., Komoriya, A., Meyers, C. A., Miller, D. M., and Sporn, M. B., Transforming growth factor-beta in human platelets: identification of a major storage site, purification and characterization, *J. Biol. Chem.*, 258, 7155, 1983.
60. Stromberg, K. and Twardzik, D. R., A beta-type transforming growth factor, present in conditioned cell culture medium independent of cell transformation, may derive from serum, *J. Cell. Biochem.*, 27, 443, 1985.
61. Derynck, R., Jarrett, J. A., Chen, E. Y., Eaton, D. H., Bell, J. R., Assoian, R. K., Roberts, A. B., Sporn, M. B., and Goeddel, D. V., Human transforming growth factor-beta complementary DNA sequence and expression in normal and transformed cells, *Nature (London)*, 316, 701, 1985.
62. Mason, A. J., Hayflick, J. S., Ling, N., Esch, F., Ueno, N., Ying, S-H., Guillemin, R., Niall, H., and Seeburg, P. H., Complementary DNA sequences of ovarian follicular fluid inhibin show precursor structure homology with transforming growth factor-beta, *Nature (London)*, 318, 659, 1985.
63. Tucker, R. F., Branum, E. L., Shipley, G. D., Ryan, R. J., and Moses, H. L., Specific binding to cultured cells of ^{125}I-labeled type beta transforming growth factor from human platelets, *Proc. Natl. Acad. Sci. U.S.A.*, 81, 6757, 1984.
64. Assoian, R. K., Frolik, C. A., Roberts, A. B., Miller, D. M., and Sporn, M. B., Transforming growth factor-beta controls receptor levels for epidermal growth factor in NRK fibroblasts, *Cell*, 36, 35, 1984.
65. Massagué, J., Transforming growth factor-beta modulates the high-affinity receptors for epidermal growth factor and transforming growth factor-alpha, *J. Cell Biol.*, 100, 1508, 1985.
66. Inman, W. H. and Colowick, S. P., Stimulation of glucose uptake by transforming growth factor beta: evidence for the requirement of epidermal growth factor-receptor activation, *Proc. Natl. Acad. Sci. U.S.A.*, 82, 1346, 1985.
67. Boerner, P., Resnick, R. J., and Racker, E., Stimulation of glycolysis and amino acid uptake in NRK-49F cells by transforming growth factor beta and epidermal growth factor, *Proc. Natl. Acad. Sci. U.S.A.*, 82, 1350, 1985.

68. **Massagué, J.,** Subunit structure of a high-affinity receptor for type-beta transforming growth factor: evidence for a disulfide-linked glycosylated receptor complex, *J. Biol. Chem.,* 260, 7059, 1985.
69. **Shipley, G. D., Tucker, R. F., and Moses, H. L.,** Type beta transforming growth factor/growth inhibitor stimulates entry of monolayer cultures of AKR-2B cells into S phase after a prolonged prereplicative interval, *Proc. Natl. Acad. Sci. U.S.A.,* 82, 4147, 1985.
70. **Massagué, J., Kelly, B., and Mottola, C.,** Stimulation by insulin-like growth factors is required for cellular transformation by type beta transforming growth factor, *J. Biol. Chem.,* 260, 4551, 1985.
71. **Warburg, O.,** On the origin of cancer cells, *Science,* 123, 309, 1956.
72. **Racker, E., Resnick, R. J., and Feldman, R.,** Glycolysis and methylaminoisobutyrate uptake in rat-1 cells transfected with *ras* or *myc* oncogenes, *Proc. Natl. Acad. Sci. U.S.A.,* 82, 3535, 1985.
73. **Roberts, A. B., Anzano, M. A., Wakefield, L. M., Roche, N. S., Stern, D. F., and Sporn, M. B.,** Type beta transforming growth factor: a bifunctional regulator of cell growth, *Proc. Natl. Acad. Sci. U.S.A.,* 82, 119, 1985.
74. **Assoian, R. K.,** Biphasic effects of type beta transforming growth factor on epidermal growth factor receptors in NRK fibroblasts: functional consequences for epidermal growth factor-stimulated mitosis, *J. Biol. Chem.,* 260, 9613, 1985.
75. **Roberts, A. B., Roche, N. S., and Sporn, M. B.,** Selective inhibition of the anchorage-independent growth of *myc*-transfected fibroblasts by retinoic acid, *Nature (London),* 315, 237, 1985.
76. **Tucker, R. F., Shipley, G. D., Moses, H. L., and Holley, R. W.,** Growth inhibitor from BSC-1 cells closely related to platelet type beta transforming growth factor, *Science,* 226, 705, 1984.
77. **van Zoelen, E. J. J., Twardzik, D. R., van Oostwaard, T. M. J., van der Saag, P. T., de Laat, S. W., and Todaro, G. J.,** Neuroblastoma cells produce transforming growth factors during exponential growth in a defined hormone-free medium, *Proc. Natl. Acad. Sci. U.S.A.,* 81, 4085, 1984.
78. **van Zoelen, E. J. J., van Oostwaard, T. M. J., van der Saag, P. T., and de Laat, S. W.,** Phenotypic transformation of normal rat kidney cells in a growth-factor-defined medium: induction by a neuroblastoma-derived transforming growth factor independently of the EGF receptor, *J. Cell. Physiol.,* 123, 151, 1985.
79. **Heath, J. K. and Isacke, C. M.,** PC13 embryonal carcinoma-derived growth factor, *EMBO J.,* 3, 2957, 1984.
80. **Kimball, E. S., Bohn, W. H., Cockley, K. D., Warren, T. C., and Sherwin, S. A.,** Distinct high-performance liquid chromatography pattern of transforming growth factor activity in urine of cancer patients as compared with that of normal individuals, *Cancer Res.,* 44, 3613, 1984.
81. **Twardzik, D. R., Brown, J. P., Ranchalis, J. E., Todaro, G. J., and Moss, B.,** Vaccinia virus-infected cells release a noval polypeptide functionally related to transforming and epidermal growth factors, *Proc. Natl. Acad. Sci. U.S.A.,* 82, 5300, 1985.
82. **Leof, E. B., Proper, J. A., Goustin, A. S., Shipley, G. D., DiCorleto, P. E., and Moses, H. L.,** Induction of c-*sis* mRNA and activity similar to platelet-derived growth factor by transferring growth factor β: a proposed model for indirect mitogenesis involving autocrine activity, *Proc. Natl. Acad. Sci. U.S.A.,* 83, 2453, 1986.

Chapter 8

HEMATOPOIETIC GROWTH FACTORS

I. INTRODUCTION

Normal cells circulating in the blood (leukocytes and erythrocytes) are usually short-lived and need to be replaced constantly throughout life. The process of blood cell formation (hematopoiesis) is not only enormous in scale but is also complex, since cells of nine different hematopoietic cell lineages, each with multiple maturation stages, are admixed apparently at random in the bone marrow. Furthermore, hematopoiesis must be capable of rapid but controlled fluctuations in order to meet a wide variety of emergency situations ranging from blood loss to infections.[1] All blood cells originate from a small common population of multipotential stem cells with immortal-like characteristics associated with an extensive capacity for self-generation. Derangements of these complex hematopoietic processes may result in life-threatening diseases such as leukemias and lymphomas. These diseases are clonal neoplasms originated by deregulation of the maturation and proliferation of the respective blood cell precursors in the bone marrow and lymphoid organs. However, the hematopoietic system usually functions with remarkable fidelity as a consequence of regulation by an overlapping and precise system of control mechanisms.[1]

Clinical and experimental observations suggest that all hematic cells (erythrocytes, granuloyctes, monocytes, lymphocytes, and megakaryocytes) originate from common stem cells. On this basis, the hematopoietic system maybe divided into the three main compartments.

1. Multipotential hematopoietic stem cells that possess extensive capabilities to give rise to new hematopoietic stem cells (self-renewal) and generate primitive progenitors that are programmed to differentiate (commitment).
2. Progenitors committed to differentiation in a single lineage (erythropoiesis, granulopoiesis, or megakaryopoiesis).
3. Mature cells such as erythrocytes and granulocytes, which have lost the ability to proliferate.[2]

It has been postulated that no humoral control would exist for the commitment of undifferentiated stem cells in one of the several lineages and that this process could be controlled by either probabilistic (stochastic) or deterministic phenomena of unknown nature. In normal steady-state hematopoiesis the great majority of stem cells are at a quiescent (G_O) state, serving as a reserve from which the system may be replenished if it is depleted.[3] Only cells forming part of the second compartment, i.e., progenitor cells committed to differentiation in a single lineage, would respond to humoral regulators called hematopoietic growth factors (HGFs).[4]

Different types of factors related to the growth and differentiation of each of the different hematopoietic cell lineages have been described.[5-8] Some HGFs maybe able to stimulate the growth of multiple lineages of hematopoietic cells.[9,10] A number of HGFs with either helper or suppressor activity are released by human T-cell lines grown in long-term culture under the stimulus of the T-cell growth factor, IL-2.[11] A hierarchical modulation in the expression of HGF receptors maybe of great importance for determining the final physiological effects of the different types of HGFs.[12] In addition to the identified factors, there is evidence of the existence of other not yet identified HGFs.[13] Some of the known and unknown HGFs may not be specific for blood cells but maybe involved in the control of growth and differentiation of other types of cells or tissues.

In contrast to other growth factors, HGFs maybe necessary not only for the proliferation and differentiation of the target cells but also for their survival. In the absence of particular HGFs, cultured hematopoietic cells die at a constant rate but addition of the factor prevents cell death and allows the rapid proliferation of the progenitor cells to continue.[8] The possible role of HGFs in hematologic malignant diseases is not understood but, in general, these diseases do not appear to be produced by an autocrine type of mechanism as a consequence of unregulated HGFs production by the malignant cells themselves. To the contrary, HGFs are likely to be mandatory for the proliferation and emergence of leukemic cell clones.[14] Moreover, specific types of HGFs maybe capable of inducing differentiation in leukemic cells.[15,16] The clinical application of particular types of HGFs for the treatment of different forms of leukemia and other hematologic malignant diseases which are characterized by specific blocks in the processes of cellular differentiation may prove to be useful for the treatment of these diseases. However, at least in some cases an exaggerated endogenous production of a particular type of growth factor maybe involved in the complex processes of malignant transformation of hematopoietic cells. Expression of a HGF cDNA in a factor-dependent cell clone may result in autonomous growth and tumorigenicity.[17]

A. Colony-Stimulating Factors

The biological activity of HGFs is generally determined by their ability to stimulate cultured hematopoietic progenitor cells (bone marrow stem cells) to form colonies of differentiated cells in semisolid medium and for this reason many of these factors are designated as colony-stimulating factors (CSFs). The colonies are constituted by true clones in which immature cells may give rise to cells showing morphological characteristics of terminal differentiation.[18,19] In certain cases antigenical or biochemical markers can be used for determining cell proliferation and/or differentiation induced by HGFs in responsive cells. Another approach would be constituted by the measurement of specific peptide growth factor receptors in cells incubated with putative HGFs.[20,21] The in vitro systems used to study the effects of purified CSFs generally contain substantial amounts of animal sera, which may contain substances capable of modulating CSF activity. The assay of purified CSFs may be improved by using a serum-free methylcellulose medium that is able to support the growth of colonies derived from primitive erythroid and nonerythroid progenitors in the same cultures.[22]

CSFs constitute a family of glycoproteins involved in the control of proliferation and differentiation of hematopoietic cells.[6] CSFs are produced by mitogen-activated T-lymphocytes as well as by a variety of other cell types including fibroblasts, macrophages, and endothelial cells. CSFs maybe classified according to the types of mature blood cells found in the resulting colonies.[4] All of these factors act upon binding to specific, high-affinity receptors expressed on the surface of cells responsive to each factor. Unfortunately, many of these factors remain poorly characterized and there is much confusion on their nomenclature.

In addition to CSFs there is evidence that some other factors act as inhibitors of hematopoietic progenitor cell proliferation. One of these factors, called colony-inhibiting lymphokine (CIL), is produced by a human cell line and exerts a potent inhibitory effect on the growth of bone marrow progenitor cells.[23] It is possible that growth-inhibitory factors like CIL play a role in the pathogenesis of some cases of severe human aplastic anemia.

II. HEMOPOIETINS

There is evidence for the existence of humoral factors capable of stimulating the proliferation and differentiation of cells with capability to form more than one blood cell type.[1] Factors involved in stimulating the proliferation and differentiation of cells capable of forming more than one cell lineage are termed multilineage growth factors, in contradistinction to lineage-specific growth factors which act at later stages controlling cells already committed to form a particular type of cell.[5,24] Synergistic interaction between lineage-specific and multilineage growth factors is important for the regulation of hematopoiesis. Two multilineage hematopoietic growth factors have been described, namely, hemopoietin-1 and hemopoietin-2. These two factors are different in several respects, including their physicochemical and biological properties.[20,21]

A. Hemopoietin-1

Hemopoietin-1 has been purified from a serum-free medium conditioned by the human urinary bladder carcinoma cell line 5637 and is apparently constituted by two molecular species of peptides with similar size and different charges, as determined by chromatofocusing analysis.[25] Hemopoietin-1 would act on very primitive hematopoietic cells, generating cells bearing receptors for CSF-1, which is a monocyte/phagocyte lineage-specific growth factor. It is not known whether or not the two species of hemopoietin-1 have different biological properties.

B. Hemopoietin-2 (IL-3)

Hemopoietin-2 is probably identical to a factor, called multicolony-stimulating factor (multi-CSF) or interleukin-3 (IL-3), which is capable of stimulating growth and differentiation of progenitor cells for most of the hematopoietic cell lineages.[26] IL-3 is probably also identical to a factor with erythroid burst-promoting activity,[24] as well as to a factor called mast-cell growth factor (MCGF).[27] IL-3 is able to stimulate in vitro the full development of many erythroid burst-forming units even in the absence of serum and detectable erythropoietin.[28] IL-3 acts directly on multipotential hematopoietic progenitors in culture, exerting a permissive role in their proliferation and differentiation.[29] However, IL-3 does not trigger hematopoietic progenitors into active cell proliferation (this role would correspond to hemopoietin-1) but is necessary for their continued proliferation.[29] IL-3 does not seem to act on mature B-cells but supports the growth the B-cell precursors, favoring the development of pre-B-cell clones that can be induced to mature into antibody-secreting cells in vitro.[30] Activated T-cells are believed to be the major physiological source of IL-3 but the myelomonocytic leukemia cell line WEHI-3B produces IL-3 constitutively.[31] The alteration occurring in WEHI-3B cells is due to the insertion of an intracisternal A particle (IAP) close to the promoter region of the IL-3 gene.[32] The constitutive expression of IL-3 in WEHI-3B cells may contribute to the maintenance of their transformed phenotype by an autostimulatory mechanism.

1. IL-3 Structure and the IL-3 Gene

IL-3 purified to homogeneity is a glycoprotein of 28,000 mol wt.[33] The IL-3 gene cloned from either mouse inducer T-lymphocytes or a mouse sperm DNA library has been sequenced and expressed in transfected COS monkey cells.[9,27,34,35] The primary IL-3 precursor structure, as deduced from its cDNA, is composed of 166 amino acids and show no significant homology to either human IL-2 or human immune interferon (IFN-γ).[34] The automated chemical synthesis of IL-3 has been reported recently.[36]

Repeated DNA sequences present in the mouse IL-3 gene share extensive homology with similar sequences from the human genome which was shown to have enhancer activity. These repeated sequences may play a role in the expression of the IL-3 gene in concanavalin A- or antigen-stimulated T-lymphocytes.[27]

2. IL-3-Like Proteins

A pluripotent hematopoietic CSF is constitutively produced by the human carcinoma cell line 5637.[37] The purified protein has a 18,000 mol wt and supports the growth of human mixed hematopoietic cell colonies, granulocyte/macrophage colonies, and early erythroid colonies. In addition, it induces differentiation of the human promyelocytic cell line HL-60 and the murine myelomonocytic leukemia cell line WEHI-3B(D+). An unidentified factor with IL-3 activity is produced by a number of adherent cell lines.[13] The precise relationships of such IL-3-like proteins to other pluripotent hematopoietic growth factors has not been established.

3. IL-3 and Oncogene Expression

The Abelson virus (A-MuLV), an acute retrovirus transducing the v-*abl* oncogene, potentiates long-term growth of mature B-lymphocytes.[38] A-MuLV induces transformation of nontumorigenic murine blood cells and the transformed cells become independent on the exogenous supply of IL-3. However, the A-MuLV-derived mast cell transformants do not express or secrete detectable levels of IL-3 nor is their growth inhibited by anti-IL-3 serum, which strongly argues against an autocrine type of stimulation as the mechanism of A-MuLV-induced transformation.[39-41] In contrast to the v-*abl*-induced transformants, murine hematic cells infected with other acute transforming retrovirus, H-MuSV, which transduces the v-H-*ras* oncogene, although converted in immortal cell lines do not loss their requirement for IL-3.[39,42] Moreover, in spite of the expression of high level of the v-*ras* p21 protein, the H-MuSV-infected cells (immune mast cells) may retain a differentiated phenotype.

The effects of IL-3 on its target cells may be mediated, at least partially, by induction of c-*myc* proto-oncogene expression. Recombinant murine retroviruses expressing v-*myc* oncogenes abrogate IL-3 and IL-2 dependence in cultured cells and suppress c-*myc* expression.[43] IL-3 could also act through activation of protein kinase C. Interaction of IL-3 with its cellular receptors may result in a rapid and transient redistribution of protein kinase C from cytosol to plasma membrane.[44]

III. LYMPHOKINES

There is strong evidence that both T- and B-lymphocytes and hematopoietic cells are derived from a common progenitor but it is not known whether there is a common progenitor for T- and B-lymphocytes or whether they are direct descendants of pluripotent lymphohematopoietic stem cells.[42] In any case, certain humoral factors are involved in the differentiation and proliferation of cells committed to the lymphoid cell lineage. Soluble factors elaborated by lymphocytes and transmitting signals for growth and differentiation of various cell types constitute a particular type of cytokines called lymphokines. Human blood monocyte subsets may show a differential ability to release various of these substances after stimulation.[45]

In general, for stem cells, renewal and differentiation are incompatible alternatives and differentiation is usually associated with progressive limitation in the proliferative capacity of the cell lineage. A terminally differentiated cell is usually a "dead cell", in relation to its inability for reproduction. However, this general concept does not appear to hold for lymphopoiesis.[3] Cells phenotypically recognized as differentiated T-lymphocytes can be maintained in long-term culture providing a specific growth factor, the T-cell growth factor (TCGF), is included in the medium. A clone of differentiated B-cells, capable of synthesizing a particular molecular species of immunoglobulin, maybe expanded by cell proliferation in vivo under the appropriate antigenic stimulus.

Lymphocyte mytogenic factors (LMFs) or lymphokines were discovered in the supernatants of human lymphocyte cultures.[46,47] A number of different factors have been classified under the category of lymphokines.[48] According to their origin and functions, as well as to the assay systems used for their identification, these factors have been denominated and classified in frequently confusing manners. Two soluble mediators are required for optimal stimulation of T-cell mitogenesis, namely, interleukin-1 (IL-1) and interleukin-2 (IL-2).[49,50]

Following antigen or lectin stimulation, a subset of T-cells responds to IL-1 exposure with the subsequent production of IL-2.[51] IL-1 is unable to support continuous proliferation of T-cells. IL-2 is the actual mitogenic stimulus causing susceptible T-cells to progress through the cell cycle and is the growth factor responsible for supporting long-term proliferation of cultured T-cells. Certain proto-oncogenes, like c-*ets*,[52] are preferentially expressed in lymphoid cells but their precise relationship to lymphokine actions is not yet understood.

A. IL-1

IL-1 is a growth factor with an important role in the modulation of proliferation, maturation, and functional activation of a broad spectrum of blood cell types. IL-1 stimulates the proliferation of T-lymphocytes and the differentiation of B-lymphocytes into antibody-producing cells.[53] IL-1 may also play a role in the initiation and amplification of immune and inflammatory responses.[54-56] IL-1 is a complete secretagogue for human neutrophils, being involved in stimulating the release of granule constituents from these cells.[57] IL-1 and IFN-γ may induce profound modifications in the morphological and physiological properties of endothelial cells.[58] IL-1 maybe, in association with antigen, a signal capable of activating the generation of cytolytic T-lymphocytes through an activation of the helper T-cell pathway which may result in the generation of helper/differentiation factors, and activation and differentiation of precursor cytolytic T-cells. An antiserum that inhibits IL-1-mediated functions is immunosuppressive of T-cell functions both in vivo and in vitro.[59] IL-1 may play an important role in host defense against tumor cells through activation of monocytes.[60,61] monocytes.[60,61]

IL-1 may also exert important physiological effects in nonlymphoid cells, especially in the pathogenesis of the acute phase response to microbial invasion, tissue injury, immunologic reactions, and inflammatory processes.[54] IL-1 maybe responsible for the pronounced changes in hepatic protein synthesis occurring in the acute phase response to inflammation or tissue injury.[62] IL-1 is involved in the regulatory mechanisms of bone remodelation. The osteoblastic cell line MC3T3-E1, derived from newborn mouse calvaria, spontaneously produces IL-1 or an IL-1-like cytokine which enhances thymocyte proliferation in the presence of PHA.[63] IL-1 has independent effects on DNA and collagen synthesis in cultures of rat calvariae.[64] There is evidence that IL-1 may act as an inducer of IFN-β production in cultured human fibroblasts.[65] IL-1 appears to be identical with endogenous pyrogen.[66]

The cellular effects of IL-1 are mediated through IL-1 binding to a specific high-affinity cellular receptor located at the level of the plasma membrane.[67] Phorbol esters can display IL-1-like activity. A PMA-protein complex can mimic the bioactivity of IL-1.[68]

1. IL-1 Structure and the IL-1 Gene

IL-1 has been difficult to purify but it has been characterized from several sources, including a human acute monocytic leukemia cell line.[69] IL-1 is constitutively produced by the human hepatic adenocarcinoma cell line SK-hep-1, in which addition of the calcium ionophore A23187 and lipopolysaccharide results in a 30-fold enhancement in the release of IL-1 activity.[70] Purified IL-1 is a heat-labile protein with a 14,000 mol wt. Both IL-1 and hepatocyte-stimulating factor (HSF) are produced by some leukemia cell lines but these factors are different from each other.[71]

Recently, cDNAs corresponding to the human and murine IL-1 gene have been produced and the cloned genes were expressed thereafter in *Escherichia coli*.[72,73] The constructed murine IL-1 cDNA codes for a IL-1 polypeptide precursor of 270 amino acids and biologically active IL-1 was produced by expressing the carboxy-terminal 156 amino acids of the IL-1 precursor. Apparently, the 37,000-dalton IL-1 precursor is processed by proteolysis and generates a diversity of IL-1-related substances with between 13,000 and 19,000 mol wt, which would explain the microheterogeneity previously detected in IL-1 by physico-chemical procedures. Messenger RNA isolated by hybridization to a human IL-2 cDNA clone has been translated in a reticulocyte cell-free system, yielding immunoprecipitable IL-1.[73] Microinjection of this hybrid-selected mRNA into *Xenopus laevis* oocytes resulted in the production of biologically active IL-1. Increased levels of IL-1 mRNA can be detected in stimulated macrophages.

The amino acid sequence analysis of human IL-1 yielded evidence for the existence of biochemically distinct forms of IL-1.[74] There is evidence, however, that at least two distinct molecules maybe responsible for some of the manyfold physiological activities attributed to IL-1. Two distinct but distantly related cDNAs encoding proteins sharing human IL-1 activity have been isolated from a macrophage cDNA library and have been termed IL-1 alpha and IL-1 beta.[75] The primary translation products of the genes expressed in *E. coli* are 271 and 269 amino acids long, respectively. Both molecules seem to be synthesized as large precursors that are processed to smaller forms. It is not known if all activities attributed to IL-1 are shared by both molecules.

B. IL-2

An important member of the interleukin family of blood cell growth factors is a soluble substance named T-cell growth factor (TCGF) or IL-2. This substance is present in crude phytohemagglutinin (PHA)-stimulated lymphocyte conditioned media and is capable of maintaining normal lectin-activated T-cells in continuous proliferative culture.[76-80] In the constant presence of IL-2, mature T-cells can be grown indefinitely, and this is valid for both, normal and neoplastic human T-cells. Whereas normal T-cells show an absolute requirement for prior antigenic stimulation for response to IL-2, some malignant cells from leukemias and lymphomas of mature T cell origin respond directly to lectin free, partially purified IL-2. IL-2 is also capable of stimulating the growth of large but not small B-cells, and this stimulatory effect does not appear to be mediated through T-cells.[81] Release of IL-3 activity by T4$^+$ human T-cell clones may depend on IL-2.[82] In addition to its proliferation-stimulating properties, IL-2 has several important influences on the determination of the immune response and has remarkable anti-tumor effects both in vivo and in vitro.[83,84]

1. The Human Antigen-Specific T-Cell Receptor

Induction of an immune response to a foreign antigen requires the activation of T-lymphocytes expressing receptors for the specific antigen. The human antigen-specific T-cell receptor is a 75,000 to 90,000-mol wt polymorphic heterodimer composed of an acidic alpha (46,000 mol wt) and a basic beta (40,000 mol wt) chain which are extensively glycosylated.[85] The organization of the antigen-specific T-cell receptor genes have been determined and the genes have been mapped on human chromosomes 7 (chain β) and 14 (chain γ).[85,86] Like the immunoglobulin light and heavy chains, both the antigen receptor alpha and beta chains are composed of variable (v) and constant (c) regions. Specific rearrangements of the antigen receptor peptide chains can be used as markers of lineage and clonality in human lymphoic neoplasms.[87,88] Fusion of an immunoglobulin variable gene and a T-cell receptor constant gene has been detected in the chromosome 14 inversion associated with human T-cell tumors.[89] The isolation of a cDNA clone encoding a T-cell receptor beta-chain from a beef insulin-specific hybridoma has been reported recently.[90]

The antigen receptor alpha and beta peptidic chains are associated with three nonpolymorphic peptide chains of 20,000 to 28,000 mol wt identified by the T-3 monoclonal antibody.[91] The antigen receptor molecule and the T3 molecule are spatially associated on the lymphocyte surface forming the so-called T3/T-cell receptor complex. The predominant association of this complex occurs between the T3 heavy subunit (28,000 mol wt) and the antigen receptor beta subunit.[92] Two of the three subunits of the T3 invariable proteins are glycoproteins and the complete amino acid sequence of one of the human T3 glycoprotein components, the T3-delta chain, has been deduced from a cDNA clone.[93] Activation of protein kinase C down-regulates and phosphorylates the T3/T-cell receptor complex on the surface on human lymphocytes.[94]

Activation of T-cells is initiated following the interaction of antigen with the T-cell receptor which, upon interaction with macrophage-derived IL-1, induces T-cells to synthesize and secrete IL-2. In order to exert its biological effects, IL-2 must interact with high-affinity specific membrane receptors and these receptors are rapidly expressed on T-cells following activation with antigen or mitogen. Thus, both IL-2 and its receptor are absent in resting T-cells but, following activation, the genes for both protein become expressed.[91] It is thus clear that both the production of IL-2 and the expression of IL-2 receptors are pivotal events in the development of the immune response.

2. Structure and Synthesis of IL-2

Human IL-2 has been purified to homogeneity and has been characterized as a protein with an estimated molecular weight of 15,000 daltons released by a subset of mature T-cells upon lectin-antigen activation.[95] The amino acid sequence and post-translational modification of human IL-2 have been determined.[96] Nucleotide sequence analysis of the human IL-2 gene cloned from the Jurkat cell line revealed that the IL-2-encoded protein contains three cysteines located at amino acid residues 58, 105, and 125 of the mature protein.[97] Site-directed mutagenesis procedures indicated that substitution of serine for cysteine at either position 58 or 105 of the IL-2 protein substantially reduced biological activity. These two cysteine residues are thus necessary for maintenance of the biologically active conformation and may therefore be linked by a disulfide bridge. In contrast, the modified IL-2 protein containing a substitution at position 125 retained full biological activity, suggesting that the cysteine at this position is not involved in a disulfide bond and that a free sulfhydril group at that position is not necessary for receptor binding.[97]

The production of IL-2 does not require DNA synthesis and can occur in terminally differentiated T-cells, but it requires binding of a stimulatory agent and protein synthesis.[76] Control of IL-2 synthesis immediately after stimulation occurs exclusively at the level of transcription. The wave of IL-2 mRNA synthesis corresponding to the initial phase of the response to stimulation is followed by an active shutoff phase which depends on protein synthesis, most likely representing synthesis of a short half-life repressor protein.[98] Transcription of the IL-2 gene can be modulated by IL-2 and phorbol ester.[99]

3. The IL-2 Gene

The structural gene of IL-2 is located on human chromosome 4, region 4q26-28, and on the feline chromosome B1.[100,101] The gene coding for human IL-2 has been cloned from normal and leukemic cells and has been expressed in *E. coli,* and the amino acid sequence of the protein product has been deduced from the cDNA nucleotide sequence.[102-107] The human IL-2 gene has also been cloned and expressed in *Streptomyces lividans* using the *E. coli* consensus promoter.[108] The IL-2 cDNA codes for a protein of 153 amino acids, the first 20 of which appear to constitute a signal peptide and are not present in the secreted protein. There is only a single copy of the human IL-2 gene and its organization in a variety of human malignant lymphoid cell types is apparently identical as that of normal human cells, the gene being not rearranged in the malignant cells.[100,102]

The 5′ flanking region of the human IL-2 gene shows sequence homology with HTLV,[109] a virus which is associated with certain forms of human T-cell leukemia/lymphoma.[110–112] A viral LTR has been detected in the IL-2 gene of a gibbon leukemia cell line (MLA 144) that constitutively produces IL-2.[113] This viral LTR is integrated at the 3′ end of the IL-2 gene but sequences related to gibbon leukemia virus (GLV) were also detected at the 5′ end of the rearranged IL-2 gene. The integration event results in transcription of a composite mRNA made up of the protein coding sequences of the IL-2 gene transcript but incorporating the viral LTR in the 3' nontranslated region of the mRNA.[113] The MLA 144 cell line responds to mitogen stimulation by increasing about 30-fold the synthesis and secretion of IL-2 over the constitutive level.

The cloning, expression, and sequence analysis of murine IL-2 cDNA has also been reported.[114] The murine IL-2 cDNA sequence contains an open reading frame which encodes 169 amino acids and its most peculiar feature is the presence of tandem repeats of a CAG sequence (nucleotides 150 to 185) which produces 12 consecutive glutamine residues in the IL-2 polypeptide. As for human IL-2, there is no potential *N*-glycosylation site in the murine IL-2 sequence. Three cysteine residues, two of which seem to have a critical role in maintaining the human IL-2 molecule in the active conformation, are present in the murine IL-2 sequence. There is only one gene for IL-2 in the mouse genome and the molecular heterogeneity of the native IL-2 protein may be due to the different extent of post-translational modifications of a single polypeptide product.[114]

Structural homology has been detected between the flanking regions of the c-*myc* and c-*fos* proto-oncogenes and the region located immediately 5' to the IL-2 gene, where the transcriptional control on the level of IL-2 expression is probably exerted.[115] Moreover, the region of homology, represented by the consensus sequence TGGANNGNANCCAA, is also shared with sequences of viruses with oncogenic properties, including the long terminal repeats (LTRs) of the human T-cell leukemia viruses HTLV-I and HTLV-II, and the E1a region of human adenovirus 5 (Ad5).[115] The presence of a consensus sequence in these cellular and viral genomes suggests the existence of common mechanisms related to the control of cell proliferation.

4. The Cellular IL-2 Receptor

The cellular IL-2 receptor protein, also called Tac antigen, has been purified by affinity chromatography using an anti-Tac monoclonal antibody and has been chemically characterized from human T-lymphocytes and human lymphoma cell lines.[116,117] The receptor isolated from normal PHA-activated T-lymphocytes has a 60,000 mol wt whereas the molecular weight of the IL-2 receptor isolated from a human T-cell lymphoma cell line was 55,000. Such difference between normal and malignant T-lymphocytes would not be caused by structural differences in the polypeptide chains of the receptors, which are probably identical, but by differences in post-translational processes of the molecule.

The gene encoding the IL-2 receptor protein is located on human chromosome 10, at region 10p14-15.[118] The short arm of human chromosome 10 is usually not involved in translocations in T-cell lymphomas. The cloning, sequence and expression of the human IL-2 receptor gene has been reported recently.[117,119-121] The gene consists of 8 exons spanning more than 25 kb.[121] The IL-2 receptor protein is composed of 251 amino acids and is separated into two domains by a 19-residue transmembrane region.[117,119,121] The amino acid sequence of the IL-2 receptor does not show significant homology with any known proto-oncogene protein. A cDNA encoding the human IL-2 receptor has been stably expressed in eukaryotic cells.[122]

After stimulation of the resting (G_0) peripheral mononuclear cell with mitogens, T-cells enter the G_1 phase of the cell cycle, where they produce the expression IL-2 receptors. Expression of IL-2 receptors at the surface of T-cells, which is critical to the development of a normal immune response, is regulated in an uncommon way because the receptor is not expressed in the absence of activation determined by immunostimulatory ligands such as antigens, T-cell-specific monoclonal antibodies, mitogenic lectins, or phorbol esters, and the receptor levels decline progressively upon removal of the immunostimulatory signal.[123] Once a T-cell population is exposed to immunostimulatory ligands, individual cells express IL-2 receptors at different rates, giving rise to the asynchronous entry of cells into the S phase of the cell cycle. If IL-2 is excluded from an antigen-activated cell population, IL-2 receptor expression still occurs, but DNA synthesis does not take place. In the thymus, IL-2 receptors are expressed preferentially by a small subpopulation of T-cells characterized by the absence of two differentiation antigens, Lyt-2 and L3T4.[124,125] Such cells maybe considered as intrathymic stem cells and it seems appropriate to accept the hypothesis that IL-2 plays a role in thymus cell ontogeny.

IL-2 has an important role in augmenting the expression of its own receptor and the synthesis of IFN-γ by human T-lymphocytes.[123,126] IL-2 stimulation of human peripheral blood mononuclear cells results in increased accumulation of IL-2 receptor mRNA within 4 hr, while an increase in IL-2 receptor transcription is observed within 30 min in isolated nuclei.[127] However, IL-2 alone is insufficient to maintain a continuous state of maximal transcriptional activity of the IL-2 receptor gene. Stimulation of peripheral blood lymphocytes with PHA induces IL-2 mRNA precursor forms within 1 hr after stimulation.[128] Phorbol ester (PMA) also stimulates IL-2 receptor mRNA and protein expression by the same cells. IL-2 is a sufficient signal to induce the expression of its own receptor on PHA-stimulated T-cells, with subsequent cell proliferation, but it is not sufficient to cause endogenous IL-2 synthesis and secretion.[129] Interaction of IL-2 with its receptor on activated T-cells is required for expression of transferrin receptors, and the binding of transferrin to its receptor during late phase G_1 of the cell cycle allows T-cells to make the G_1 to S transition. IL-2 receptors are expressed not only in T-cells but also in B-cells, and interaction of IL-2 with its receptor on B-cell surface induces a proliferative B-cell response.[130] This response is sharply inhibited by anti-Tac antibodies. IL-2 induces T-cell-dependent immunoglobulin (IgM) production in human B-cells.[131]

5. IL-2 Receptor in Leukemic Cells

In contrast to normal cells, neoplastically transformed T-cells may grown independently from the presence of IL-2. Acute T-cell leukemic populations and lines derived from them, as well as most populations of Sézary leukemic T-cells, do not express IL-2 receptors.[91] IL-2-independent cell lines with tumorigenic properties may arise during prolonged cultivation of IL-2-dependent cell lines, probably as a result of exposure to chemicals, viruses, or other contaminants.[132] However, not all neoplastically transformed lymphoid cells become IL-2 independent. Populations of leukemic cells from patients with adult T-cell leukemia associated with HTLV-I infection spontaneously and continuously express high amounts of the Tac antigen without requiring prior activation.[91,133] Moreover, HTLV-I-infected continuous T-cell lines may display aberrantly sized IL-2 receptors which are not modulated (downregulated) by anti-Tac (unlike normal activated cells) and which are spontaneously (IL-2-independently) phosphorylated.[91] The constant presence of high numbers of IL-2 receptors on the adult T-cell leukemic cells and/or the aberrancy of these receptors could play a role in the pathogenesis of neoplastic growth of these malignant cells. Most interestingly, clinical remission of the disease can be obtained in patients with adult T-cell leukemia upon treatment with intravenously administered anti-Tac monoclonal antibody.[91]

Increased production of IL-2 has also been detected in B-cell chronic lymphocytic leukemia (CLL),[134] but there is evidence that IL-2 maybe rapidly removed and absorbed by the neoplastic B-cells in CLL.[135] IL-2 can act on B-cell progenitors of CLL which express IL-2 membrane receptors upon activation with PHA or TPA.[136] Non-T acute lymphoblastic leukemia (ALL) cells can also respond to IL-2 and proliferate in culture.[137]

6. Functions of IL-2 in Normal and Neoplastic Cells

It appears that the main function of IL-2 is to provide the stimulus necessary for cell proliferation. IL-2-directed T-cell mitosis is a quantal response determined by a critical threshold of signals generated by the interaction between IL-2 and the IL-2 receptor.[77] Three factors are involved in T-cell cycle progression: IL-2 concentration, IL-2 receptor density, and the duration of the interaction of IL-2 with its receptor.

Purified IL-2 does not stimulate growth of T-lymphocytes unless they are first activated by lectin-antigen. In contrast, T-cells from patients with T-cell malignancies can be grown in long-term culture directly with IL-2.[138] These cells express IL-2 receptors on the cell surface constitutively in the absence of known activators that induce IL-2 receptor on normal resting T-cells. Peripheral blood leukemic T-cells in adult T-cell leukemia (ATL) associated with HTLV-I infection express IL-2 receptors (Tac antigen) of similar weight to that of normal activated T-cells and can bind IL-2 but, paradoxically, these cells respond very poorly to exogenous IL-2.[133] In contrast, cells from patients with chronic lymphocytic leukemia (CLL) express Tac antigen and show good proliferative response to IL-2. Constitutive expression of the IL-2 receptor occurs in rat lymphoid cell lines producing the human T-cell leukemia-associated virus HTLV-I,[139] and there is direct evidence for IL-2 receptor induction of HTLV-I in Epstein-Barr virus (EBV)-transformed human B-cell lines.[140] Deregulation of the IL-2 receptor gene expression may occur in HTLV-I-induced adult T-cell leukemia.[141] Mitogenic stimuli (PHA and PMA) activate IL-2 receptor expression in normal T-cells, whereas these stimuli paradoxically inhibit IL-1 receptor gene transcription in HTLV-I-infected leukemic cells. HTLV-I can infect normal mature B-cells in vitro, inducing in thse cells the expression of IL-2 receptors.[142] HTLV-I induces the differentiation of B-cells into immunoglobulin-secreting cells, without affecting their proliferation. IL-1 and tumor promoters such as phorbol diesters may induce the expression and phosphorylation of the human IL-2 receptor.[99,143]

It has been suggested that a mechanism for lymphomagenesis may consist in transformation of IL-2-dependent T-lymphoblasts into IL-2-independent T-lymphoma cells.[144,145] This change could be attributed to an endogenous production of IL-2, or an IL-2-like growth factor, by the transformed cells, which would be consistent with the autocrine hypothesis of neoplasia. However, a simple autostimulation model would probably not be realistic for human T-cell leukemia associated with infection by HTLV-I because extremely low levels of IL-2 mRNA transcripts are detected in HTLV-I-infected human cell lines and IL-2 is not expressed in fresh leukemic cells from patients with T-cell leukemia.[140,146]

Altered expression of IL-2 receptors may occur in nonmalignant diseases. Autoimmune MRL-*lpr/lpr* mice manifest clinical features characteristic of systemic lupus erythematosus (SLE), including autoantibody production, arthritis, vasculitis, and glomerulonephritis.[147] Rapidly growing T-cell lines and clones developed from these mice exhibit a defect in IL-2 receptor expression and unusually high levels of expression of the proto-oncogenes c-*myb* and c-*raf*.[148] All long-term lines and clones derived from MRL-*lpr/lpr* T-cells bear large numbers of IL-2 receptors continuously and without stimulation and, paradoxically, they are poorly inhibited by anti-IL-2 receptor antibody, which suggests that the rapidly growing abnormal T-cells maybe stimulated by some kind of autocrine mechanism.

IL-2 maybe involved in the cytotoxic T-cell responses occurring in the vicinity of tumors. Preliminary evidence suggests tumor regression after intra-lesional injection of IL-2 in human bladder cancer,[84] but the mechanisms involved in this phenomenon have not been characterized. Production of IL-2 inhibitors by tumor cells may facilitate tumor growth by inhibiting cytotoxic T-cell responses. An inhibitor has been detected, for example, in the supernatants from melanoma cell cultures.[149]

7. IL-2-Mediated Changes in Phosphoinositide Metabolism and Protein Kinase C Activity

The mechanisms of IL-2-induced lymphoid cell proliferation are not understood but IL-2 can stimulate association of protein kinase C with the plasma membrane.[150] In turn, this association may stimulate phosphorylation on serine and threonine of some cellular proteins,[151] probably including the IL-2 receptor.[152] Tumor promoters like phorbol esters, which directly activate protein kinase C, may induce the expression of IL-2 receptor in IL-2 receptor-negative human T-leukemic cell lines and may also induce the rapid phosphorylation of the receptor on serine and threonine residues.[143] Protein kinase C is also activated by action of 1,2-diacylglycerol, which functions as an important messenger within the plane of the plasma membrane to induce a marked increase in the affinity for calcium ions.[153] Activation of protein kinase C maybe associated with the control of DNA synthesis and with cell proliferation and differentiation.[154] The calcium ionophores A23187 and ionomycin do not activate mouse T-lymphocytes but either one in combination with phorbol ester (TPA) induces in lymphoid cell populations the expression of receptors for IL-2, the secretion of IL-2, and cell proliferation.[155] Diacylglycerol can mimic phorbol ester induction of leukemic cell differentiation in vitro.[156]

8. IL-2 Action on Monovalent Ion Transport

There is evidence that IL-2 induces a rapid increase in intracellular pH through activation of a Na^+/H^+ antiport, but cytoplasmic alkalinization is not required for lymphocyte proliferation.[157]

9. IL-2-Mediated Induction of Proto-Oncogene Expression

IL-2 may have an important role in the expression of cell cycle-related genes in human T-lymphocytes.[158] However, the possible action of IL-2 on proto-oncogene expression is still little known. There is evidence that expression of c-*myc* may be regulated by IL-2 stimulation of T-cells. Stimulation of murine splenocytes with the lectin mitogen concanavalin A (a T-cell-specific mitogen), or with lipopolysaccharide (a B-cell-specific mitogen), results in a rapid and marked increase in the levels of c-*myc* mRNA.[159] Similar results have been obtained for expression of c-*myc* protein in PHA-stimulated human peripheral blood mononuclear cells.[160] It is known that T-cells stimulated with specific T-cell mitogens produce and express receptors for IL-2.[77] Recombinant IL-2 has little effect on c-*myc* expression but, in combination with PHA, it augments levels of c-*myc* transcripts measured at 24 hr but not at 3 hr after stimulation.[161,162] Moreover, anti-Tac antibody, which interferes with IL-2-mediated events by binding to the IL-2 receptor, inhibits the increase of c-*myc* mRNA levels at 24 hr, but not at 3 hr, after PHA stimulation.[161] The induction of c-*myc* expression by IL-2 in lymphoid cells was confirmed in another independent study.[127] Incubation of human peripheral blood mononuclear cells with IL-2 purified from Jurkat cells or obtained by recombinant DNA technology induces an increase of c-*myc* mRNA levels.[127] Since binding of IL-2 to its receptor on T-cells leads to protein kinase C activation,[150,152] and since recombinant IL-2 increases c-*myc* mRNA accumulation in PHA-stimulated T-cells,[161] it is likely that protein kinase C is involved in regulating the levels of c-*myc* gene expression in normal T-cells.

Expression of the gene coding for the p53 cellular protein, which has been considered as a proto-oncogene,[163] is increased by IL-2.[164] Moreover, in IL-2-stimulated cells the expression of p53 protein is inhibited by monoclonal antibody to the IL-2 receptor (anti-Tac).

10. IL-2 and the Acquired Immune Deficiency Syndrome

The acquired immune deficiency syndrome (AIDS) has been observed with increased frequency since 1981, when an unprecedented occurrence of Kaposi's sarcoma, malignant lymphomas, and opportunistic infections, including *Pneumocystis carinii* pneumonia, was observed among homosexual men, intravenous drug abusers, hemophiliacs and other persons undergoing blood transfusions, immunosuppressed patients and other groups of persons.[165-171] In addition to Kaposi's sarcoma a diversity of malignant diseases have been observed in patients with AIDS, including Hodgkin's disease, nonHodgkin lymphomas, Burkitt-like and immunoblastic lymphomas, cloacogenic and squamous cell carcinomas of the rectum, and squamous cell carcinomas of the tongue.[172]

Epidemiological data point to the existence of an infectious etiologic agent of AIDS and a new type of retrovirus, termed HTLV-III or LAV, has been suggested as the agent etiologically associated with AIDS. Patients with AIDS usually have a severe immune deficiency with a reduction in the number of helper T-lymphocytes (OKT4+). The prodromes of AIDS, corresponding to a "pre-AIDS syndrome", frequently include unexplained chronic lymphadenopathy and leukopenia with diminished number of circulating helper T-lymphocytes. In general, patients with AIDS have reduced T-helper:T-suppressor (OKT4/OKT8) ratios in peripheral blood.

The nature of the T-cell defect in AIDS is not understood. T-cells from these patients produce adequate amounts of IL-2 following stimulation with lectins but their T-cell blasts have a diminished response to exogenously supplied IL-2.[173] This defect maybe related to the failure of expression of the IL-2 receptor on the cell surface following lectin stimulation, which is recognized by the anti-Tac (T-cell activation) antibody in AIDS patients. Expression of the IL-2 receptor is significantly diminished on lectin-stimulated T-cells from AIDS patients.[173]

C. Glial Growth Promoting Factor

In addition to growth factor acting on hematic cells, antigen- or lectin-stimulated T-lymphocytes and HTLV-infected human T-lymphocyte cell lines secrete lymphokines that can influence a variety of cell types. One of these lymphokines, purified from the HTLV-II-infected Mo-T-cell line, stimulates the proliferation of rat brain oligodendrocytes and has been called glial growth-promoting factor (GGPF).[174] The biological activity of GGPF is assayed by its ability to stimulate DNA synthesis in oligodendrocytes, as measured by (^{3}H)thymidine uptake. The purified GGPF protein has an apparent 30,000 mol wt under nonreducing conditions but under reducing conditions GGPF appears as a single compound of 18,000 mol wt. Both reduced and unreduced forms have biological activity, which suggests that GGPF exists in both a functional monomeric and dimeric form.[174] GGPF would function in vivo during trauma to the central nervous system associated with inflammation to induce proliferation of oligodendrocytes and subsequent regulation of remyelinization.

D. B-Cell Growth Factors

Gene rearrangements are critically involved in B-cell differentiation. Mature B-cells synthesize immunoglobulins (Igs), which are molecules composed of light-chain and heavy-chain polypeptides, and the genes coding for these two types of chains are not situated in continuous stretches of DNA at the genomic level in the most primitive B-cell precursors but are separated from each other on the chromosomes.[175,176] Specific rearrangements of the IgG genes occur during differentiation of the B-cell lineage, thus allowing an appropriate synthesis of the different molecular species and subspecies of immunoglobulins. It is not known whether stochastic or deterministic phenomena are responsible for B-cell differentiation processes and little is known about the role of humoral factors in B-cell differentiation.

Several soluble factors are involved in the regulation of B-cell activation, proliferation, and differentiation.[177] There exists a cascade of sequential steps from a resting B-cell to an Ig-secreting cell in the growth factor-dependent differentiation of human B-cells.[177] This cascade may be divided into three phases: the activation step (G_0 to G_1), the proliferation step (G_1 to S, G_2/M), and the final maturation step to Ig-secreting cells.

Immunoregulatory factors acting on B-lymphocytes include B-cell growth factor (BCGF), B-cell differentiation factor (BCDF), IL-1 and IL-2, and other factors whose respective activities reside in different molecules.[178] However, the precise chemical differences among several of these factors and the exact role that each plays in B-cell activation and function remains unclear. The following general model has been proposed for B-cell activation and differentiation.[177]

1. The initial activation of B-cells is associated with the expression of receptors for growth factors, including BCGF receptors.
2. Activated B-cells, in turn, can respond by proliferation to BCGF, probably via binding of this factor to the cellular receptors.
3. These cells continue to proliferate without terminal differentiation if BCGF remains present and BCDF is absent.
4. This phenomenon may result in the clonal expansion of B-cells of various specificities.
5. If BCDF is supplied to these proliferating B-cells, they preferentially be driven to differentiate and secrete Ig. As little as 1 pM purified BCGF induces Ig secretion in activated B-cells without inducing any cell growth.[179]

Activated B-cells express IL-2 receptors and at least some of the effects of BCGFs on B-cells would depend on IL-2 action.[130,180] Immune interferon (IFN-γ) may also be one of several BCDF molecules with a direct role in driving the maturation of resting B-cells to active Ig secretion.[181]

Some BCGFs, like B-cell growth factor type I (B-cell stimulating factor 1 or BSF-1), act as differentiation or activation factors for resting B-cells and may not induce cell growth.[182,183] While BSF-1 acts in the initial stages leading to the functional activation of B-cells, another growth factor, the B-cell growth factor II (BCGF-II or BSF-2), also called interleukin 4 (IL-4), is required for the final maturation process of B-cells. By means of a monoclonal antibody to BSF-1 it has been demonstrated that this molecule is distinct from IL-1, IL-2, and IL-3.[184] BSF-1 is identical to B-cell differentiation factor (BCDF) for IgG1.[185] Immortalized B-lymphocyte cell lines derived from Burkitt's lymphomas and EBV-induced transformation produce autostimulatory B-cell growth factors in vitro.[186,187] Recently, a B-cell growth factor produced by a human B-cell lymphoma cell line (Namalva) and a human T-cell line (T-ALL) has been purified to apparent homogeneity and a monoclonal antibody has been produced to the purified protein.[188] This protein (60,000 mol wt) binds specifically to activated but not to resting B-cells. Although only some of the B-cell growth factors have been characterized, the production of some of them in neoplastic cells would be in accordance to the autocrine hypothesis of neoplastic cell transformation.

IV. GRANULOCYTE-MACROPHAGE COLONY-STIMULATING FACTORS

Factors have been identified which specifically stimulate the proliferation of committed progenitor cells of the granulocyte-macrophage lineage.[1,4] A factor, called macrophage colony-stimulating factor (M-CSF) or colony-stimulating factor 1 (CSF-1), stimulates the proliferation of progenitor cells committed to the macrophage lineage, whereas other factor, called granulocyte colony-stimulating factor (G-CSF) stimulates the proliferation of progenitor cells committed to the granulocyte lineage. Another factor, called granulocyte/macrophage colony-stimulating factor (GM-CSF) or colony-stimulating factor 2 (CSF-2), is not a lineage-specific but a multilineage growth factor, capable of stimulating the proliferation of both the macrophage and granulocyte lineages. There is evidence that CSF-1 and CSF-2 may act independently, influencing the proliferative capacity of a single bone marrow progenitor cell population but the molecular events involved in these control phenomena are not yet understood.[189]

A. Colony-Stimulating Factor 1

Colony-stimulating factor 1 is a lineage-specific growth factor which stimulates primarily the survival, proliferation, and differentiation of mononuclear phagocytes and their precursors.[190] CSF-1 has been isolated in microgram amounts and has been purified from murine L-cells and human urine.[191,192] CSF-1 binds specifically to mononuclear phagocytes and their precursors as well as to macrophages and myelomonocytic cell lines in culture.

CSF-1 is a heterogeneous glycoprotein of 65,000 to 70,000 mol wt and the active molecule appears to consist of two subunits of equal molecular weights, linked by disulfide bonds. Mild reduction results in subunit dissociation and the loss of all biological activity. The amino acid composition, the amino-terminal sequence, and the tryptic map of murine L-cell CSF-1 have been determined.[193] CSF-1 contains four or five cysteine residues per subunit, with the unusual configuration of three contiguous cysteine residues (Cys-Cys-Cys) located in the amino-terminal sequence. The CSF-1 molecule contains approximately 30 to 40% carbohydrate by weight and shows limited homology to CSF-2 or IL-3.[193]

1. The CSF-1 Receptor and the c-fms Proto-Oncogene Protein

The murine cellular receptor for CSF-1 is a glycoprotein composed of a single polypeptide chain of approximately 165,000 daltons and has an associated tyrosine-specific protein kinase activity.[194] It has been demonstrated that the McDonough strain of feline sarcoma virus (FeSV) transduces the oncogene v-*fms* whose product is a glycoprotein, gp140^{v-fms}, and that the structure and topology of this product resembles that of known cell surface receptors.[195,196] Like the v-*erb*-B product, the v-*fms* protein has an amino terminal portion orientated to the cell surface, a membrane-spanning sequence, and a carboxy-terminal signal transducing function corresponding to the protein kinase domain. Transformation by the v-*fms* oncogene protein depends on glycosylational processing and cell surface expression.[197] It is thus most interesting that the CSF-1 receptor protein is closely related to the c-*fms* proto-oncogene.[194]

The product of the cat c-*fms* proto-oncogene is a normal cellular protein of approximate 170,000 mol wt which serves as a substrate for associated tyrosine-specific protein kinase activity in vitro.[198] Although the expression of the CSF-1 receptor is restricted to mononuclear phagocytes and their precursors, c-*fms* transcripts have been detected in a diversity of mammalian organs including spleen, brain, and liver.[198] The human c-*fms* gene product has been characterized and its expression in cells of the monocyte-macrophage lineage has been determined.[199] The same product is expressed in different types of human tumors.[200] The possible relationship between the c-*fms* protein and the CSF-1 receptor in cells other than monocyte-macrophages and their precursors is not understood.

The human c-*fms* proto-oncogene sequences are dispersed over a DNA region of 32 kbp. A restriction fragment length polymorphism (RFLP) of the c-*fms* gene was detected in a patient with acute lymphocytic leukemia and congenital hypothyroidism.[201] The abnormal allele of the c-*fms* gene present in the patient contained a deletion 426 bp in size located in close proximity to a putative c-*fms* exon. The patient was homozygous for the abnormal allele which was inherited from each one of his parents.[202] The possible relationship between the genetic finding and the clinical condition of the patient remained undetermined. RFLP of the c-*fms* locus maybe common in the human population with the existence of apparent selective pressure in favor of heterozygotes.[203]

The c-*fms* gene has been assigned to human chromosome 5, region 5q34.[204] This chromosome region is deleted in the 5q-syndrome, an acquired cytogenetic abnormality associated with refractory macrocytic or aplastic anemia or with polycytemia vera.[205-207] The 5q-deletion removes the c-*fms* gene and it has been postulated that hemizygosity at the c-*fms* locus leads to abnormalities in the maturation of hematopoietic cells.[208]

2. Tumor Promoters and c-fms/CSF-1 Receptor Expression

Phorbol ester (TPA) can stimulate mouse bone marrow cells to form myeloid colonies in agar cultures without the addition of exogenous CSFs, suggesting that TPA can mimic the action of CSFs.[209] Phorbol ester cause a rapid but transient decrease in the expression of its own receptor (homologous down-regulation) as well as in the expression of CSF-1 receptors (heterologous down-regulation) present on the membrane of murine peritoneal exudate macrophages (PEM).[210] Contrarily, CSF-1, which causes homologous down-regulation of its own receptor, fails to induce heterologous down-regulation of phorbol ester receptors in PEM cells. Whether TPA induces phosphorylation of the CSF-1 receptor through activation of protein kinase C is not yet clear.

Cells from the human HL-60 promyelocytic leukemia cell line mature morphologically and functionally towards granulocytes after induction with DMSO or hexamethylene bisacetamide (HMBA), and can be induced to differentiate along a monocyte/macrophage pathway when treated with TPA or calcitriol. Although no c-*fms* transcripts are detected in uninduced HL-60 cells, these transcripts are detectable after 24 hr of TPA or calcitriol treatment.[211] In contrast, c-*fms* mRNA was not detected following induction of HL-60 cells with DMSO or HMBA. Thus, c-*fms* expression is induced when the HL-60 cells mature along the monocyte/macrophage, but not the granulocyte, pathway. The c-*fms* transcripts appear 6 hr after TPA addition and the maximum levels of c-*fms* mRNA are present 24 to 48 hr after induction initiation.[211] In contrast, c-*fos* mRNA is detectable within 1 hr of treating HL-60 cells with TPA, and c-*myc* expression is not detected before 24 hr after TPA induction.

3. CSF-1 and Proto-Oncogene Expression

CSF-1 is able to stimulate macrophage proliferation.[212] When primary cultures of bone marrow-drived macrophages are deprived of CSF-1 for 18 hr proliferation ceases and the cells become quiescent. This phenomenon is accompanied by a marked reduction in levels of both c-*fos* and c-*myc* mRNAs, which suggests that expression of both proto-oncogenes depends on the presence of the growth factor, or on cellular proliferation.[213] When CSF-1 -deprived cultured macrophages in the G_0 phase of the cell cycle are stimulated by addition of the growth factor, induction of c-*myc* transcription is observed after 1 hr, reaching a maximum at 2 to 4 hr, and this is followed by a 10-fold induction of c-*fos* mRNA levels, which remain high for the following 18 hr.[213]

Transformation of mouse bone marrow cells by transfection with a human proto-oncogene related to c-*myc* (R-*myc*) is associated with the endogenous production of CSF-1.[214] R-*myc* has homology to the 5' and 3' coding exons of c-*myc*, but its restriction map is different.

B. Colony-Stimulating Factor 2

Colony-stimulating factor 2 (CSF-2), also called granulocyte/macrophage colony-stimulating factor (GM-CSF), is required for the proliferation of neutrophile granulocyte/macrophage progenitor cells in soft agar cultures.[215,216] Removal of CSF-2 from the medium results in accumulation of the precursor cells in the G_1 phase of the cell cycle whereas readdition of CSF-2 to such quiescent cells is followed by progression of the cells from G_1 to S phase with a lag period of 10 hr.[217] The requirement for the presence of CSF-1 is limited to the first 6 hr of the 10-hr duration of the G_1 phase, which suggests that CSF-1 should be considered as a competence factor acting in the early part of G_1 and allowing the cells to respond to other factors (progression factors).[217] The progression factors, acting in the last part of the G_1 phase would signal the cells to continue through the cell cycle and replicate. However, CSF-2 acts not only on hematopoietic precursor cells but also on mature effector cells where it behaves as a factor inhibiting the migration of human peripheral blood neutrophils under agarose and as a potent activator of human neutrophils.[218,219] The leukemic cells from patients with acute or chronic myelogenous leukemia are absolutely dependent on stimulation by CSF-2 for proliferation in vitro.[215]

In the mouse, there is only one gene located on chromosome 11 encoding CSF-2, and cDNAs corresponding to this gene have been constructed.[220] These cDNAs are capable of directing CSF-2 synthesis in *Xenopus* oocytes. The primary structure of murine CSF-2, deduced from the cDNA clones, is composed of a single chain polypeptide of 13,500 mol wt containing 118 amino acids. The higher mol wt forms of CSF-2 detected in mouse tissues would correspond to post-translational modifications of the polypeptide. Recombinant murine CSF-2, like the native molecule, stimulates the growth of granulocyte and macrophage colonies in serum-free cultures of mouse bone marrow cells.[221] Antibodies raised against recombinant murine CSF-2 neutralize the biological activity of both native and recombinant CSF-2.

The human CSF-2 gene has also been cloned, sequenced, and expressed.[222-224] In addition to granulocyte-monocyte colony-stimulating activity, recombinant human CSF-2 has burst-promoting activity, stimulating also the formation of colonies derived from multipotent (mixed) progenitors.[225] Thus, human CSF-2 (GM-CSF) possesses multilineage colony-stimulating activity. The predicted human CSF-2 polypeptide weights 16,293 daltons, contains 144 amino acids, and shows 54% homology with mouse CSF-2. The naturally occurring human CSF-2 protein has a molecular mass of approximately 22,000 daltons and could contain up to 34% carbohydrate by weight.[224] The organization of the mouse and human CSF-2 genes are highly homologous, both being composed of three introns and four exons.[226] Interestingly, the CSF-2 gene includes two promoters giving rise to alternative mRNAs which encode pre-CSF-2 polypeptides with different amino terminal sequences.[227] The human CSF-2 gene is approximately 2.5 kbp in length with at least three intervening sequences. The gene has been localized to human chromosome region 5q21-32, which is involved in interstitial deletions in the 5q- syndrome associated with refractory anemia, low leukocyte count, and elevated platelet counts.[228] Monosomy of human chromosome 5 has been described in acute nonlymphocytic leukemia.

The A-MuLV retrovirus is capable, by action of its v-*abl* oncogene, of transforming CSF-2-dependent, nontumorigenic myeloid cell lines into CSF-2-independent, tumorigenic cell lines, but this transformation is not associated with the production of the specific growth factor by a kind of autocrine type of stimulation.[40] CSF-2 is able to stimulate differentiation in promonocytic leukemia cell lines derived from A-MuLV-infected mice,[229] which shows that malignant cells expressing the v-*abl* oncogene are not irreversibly blocked in their capacity to differentiate.

1. CSF-2 and Proto-Oncogene Expression

CSF-2 induces cloned myelomonocytic leukemia WEHI-3B cells to differentiate to both monocytes (macrophages) and granulocytes in the presence of a low concentration of actinomycin D. This differentiation is accompanied by a decreased expression of the proto-oncogenes c-*myc* and c-*myb* and a markedly increased expression of c-*fos*, whereas the expression of other proto-oncogenes examined remains unaltered (c-*abl*, c-K-*ras*, and c-*fes*) or is not detectable (c-*src*, c-*sis*, c-*mos*, c-*erb*-A, and c-*erb*-B).[230] Apparently, c-*fos* expression is tightly correlated with monocyte differentiation in both normal and leukemic cells and it may also be required for the expression of some macrophage-specific functions.

The proto-oncogene c-*fes* could be involved in regulatory phenomena associated with the action of HGFs, in particular with the action of CSF-2.[231] The KG-1 human myeloblastoid cell line express the c-*fes* protein and can be induced to differentiate in response to CSF-2, while the KG-1a variant of the same line do not express the c-*fes* protein and, concomitantly, has lost the capacity to differentiate and respond to CSF-2 and/or pluripotent CSFs. Two other hematopoietic cell lines, LTBM and IO-3, have an absolute dependence on CSF-2, IL-3, or both for their growth and c-*fes* protein is expressed by both cell lines. These correlations suggest that the c-*fes* protein may be involved in the cellular response of myeloid cells to specific growth factors.[231]

During the first 3 hr of stimulation of bone marrow cells by CSF-2 there is both creation of DNA breaks and closure of pre-existing DNA breaks, as demonstrated with inhibitors of ADP-ribosyl transferase activity.[232] These results suggest that DNA rearrangements may occur not only during B-lymphocyte differentiation but also in other cell lineages like the monocyte-macrophage lineage. Such DNA rearrangements might provide a molecular basis for differentiation of the different cell lineages. The possible relationship between DNA rearrangements and proto-oncogene activation in different cell lineages during the respective differentiation processes remains uncharacterized.

C. Granulocyte Colony-Stimulating Factor

Mouse peritoneal cells may be used to obtain granulocyte colony-stimulating factor (G-CSF).[233] Purified G-CSF is a glycoprotein of approximate 25,000 mol wt with a capacity to stimulate, by direct action, the proliferation of committed precursor cells in the bone marrow to form colonies of differentiated granulocytes.[233,234] In addition, G-CSF may exert a strong stimulus for the functional activity of mature granulocytes.[235] At higher concentrations, G-CSF can also stimulate the formation of some granulocyte and macrophage colonies. G-CSF has an exceptionally high capacity to suppress leukemic stem cell self-regeneration and induces differentiation in murine myeloid leukemia cells.[236]

As is true for other HGFs, G-CSF acts by binding to specific, saturable, and high-affinity receptors expressed on the surface of G-CSF-responsive cells present in the bone marrow.[237] No gross difference exists in G-CSF binding capacity between normal blast cells and the blast cells of myelomonocytic leukemia (cell line WEHI-3B D+). Murine and human G-CSFs show almost complete biological and receptor-binding cross-reactivities to normal and leukemic murine or human cells.[238]

D. Leukemia-Derived Growth Factor

A factor designated leukemia-derived growth factor (LDGF) has been detected in supernatants of the human T-cell leukemia cell line MOLT4f and in AKR murine T-cell lymphoma lines.[239,240] LDGF is a polypeptide different from IL-2 and other known growth factors and is able to stimulate the growth of human leukemia T-cells. The findings suggest that LDGF is an autocrine growth factor and that continued production of the factor maybe responsible for the malignant transformation of lymphoid cells. Since the phenotype of lymphoma T-cells is similar to that of immature thymocytes, LDGF, or a normal counterpart of LDGF, would be produced by certain thymic progenitor cells in the thymus of normal animals.

E. Myelomonocytic Growth Factor

Avian leukemia retroviruses (ALVs) containing the v-*myb* or v-*myc* oncogenes are able to induce acute transformation in chicken hematopoietic cells of the myelomonocytic lineage.[241] Both v-*myb*-transformed myeloblasts and v-*myc*-transformed macrophages are dependent for in vitro proliferation on the presence of a chicken myelomonocytic growth factor (MMGF).[242,243] This factor also stimulates the proliferation of normal, bone marrow-derived myeloid cells. Avian retroviruses containing oncogenes of the v-*src* family, but not those transducing other oncogenes, render the v-*myb*-transformed myeloblasts and v-*myc*-transformed macrophages independent of exogenous MMGF.[243] In v-*myb*-transformed myeloid cells the independence is achieved via the induced secretion and utilization, in an autocrine fashion, of a MMGF-like activity. The induction of MMGF in RSV-superinfected myeloblasts is dependent on the continuous expression of a functional $pp60^{v-src}$ viral oncogene product but it is not clear whether the factor(s) produced by *src*-type virus-superinfected myeloblasts is identical with MMGF or is only an antigenically related molecule.

F. Macrophage-Derived Growth Factors

Monocytes and macrophages are the blood and tissue phases, respectively, of the same cell lineage that has its origin in the bone marrow. Monocytes give origin to macrophages both in vivo and in vitro, and the macrophage can be considered as more highly differentiated and functionally more active than its less mature precursor.[18] Macrophages are cells directly involved in inflammatory processes, host defense mechanisms and injury repair. At least in in vitro systems, activated macrophages are capable of destroying neoplastic cells while they leave non-neoplastic cells unharmed even under conditions of cocultivation.[244]

Several types of growth factors are present in macrophages but only some of them have been characterized. There is evidence that the action of some of the growth factors associated with macrophages is not limited to these cells but may stimulate a broad spectrum of cells. One of these factors is IL-1, and another factor, called macrophage-derived growth factor (MDGF) is a potent mitogen for nonlymphoid mesenchymal cells, including fibroblasts, smooth muscle cells and endothelial cells. This factor may be identical to FGF.[245] However, a significant part of MDGF activity consists of at least two forms of PDGF-like proteins.[246] Activation of macrophages in vivo may result from their interaction with a lymphokine termed macrophage-activating factor (MAF). This substance is constitutively produced by an HTLV-positive human T-cell line (C10/MJ-2).[247] MAF is distinct from IFN-γ and is capable of activating human monocytes to lyse tumor cells.

V. INTERFERONS

The interferons (IFNs) are a group of proteins which where initially identified by their ability to protect cells against viral infections. They are synthesized and secreted by a variety of cell types in response to different substances and factors acting as interferon inducers and exert their effects in vivo as a result of interaction with other cells in distant parts of the body in a way similar to that of hormones. A functional similarity with hormones is further emphasized by the existence in target cells of specific membrane receptors for IFNs and by the multiplicity of biological effects which result from the interaction of IFNs with their cellular receptors.[248]

IFNs are constituted by a family of protein including at least three different classes: α, β, and γ. The α and β IFNs are also called leukocyte and fibroblast IFNs, respectively, and are known as type-I IFNs. Both α and β IFNs are synthesized in response to viral infection. In contrast, IFN-γ is a lymphokine produced in T-lymphocytes following mitogenic or antigenic stimulation and is known for this reason as immune or type-II IFN. Production of IFN-γ is regulated by many factors, including growth factors like PDGF, EGF, and FGF.[249,250] In humans there are many molecular species of IFN-α, but probably only one of IFN-β and IFN-γ.[248]

A. IFN Action on Cell Proliferation and Differentiation

IFNs have well known antiviral and antiproliferative cellular effects. In addition to these effects, IFN-γ promote the differentiation of cytotoxic T-cells and natural killer cells and may possibly be involved in the regulation of T-cell growth. Apparently, there is no absolute tissue specificity of the different IFN classes, and no single stage of the cell cycle is uniquely sensitive to IFN action. The antiproliferative effect of interferon and the mitogenic action of growth factors are independent cell cycle events, as demonstrated in studies with vascular smooth muscle cells and endothelial cells.[251] In general, IFNs lower the probability of cells to exit from quiescent G_0 state under adverse growth conditions. The G_1 phase is also prolonged in actively proliferating cells following IFN treatment. In other types of cells, particularly transformed cells, other phases of the cycle may be prolonged.[248] In certain types of cells IFN action may lead to cessation of DNA replication. In addition to cytostatic effects, IFNs may have a cytocidal effect, especially when acting on certain types of tumor cells. IFNs may also have inhibitory effects on cellular protein synthesis. The effects of IFNs on cell differentiation, however, are variable, with either inhibition, stimulation, or no effect on the differentiation processes. The molecular mechanisms responsible for the manyfold effects of IFNs have not been characterized, but there is evidence that the initial event in IFNs action consists in their binding to specific high-affinity cell surface receptors. The receptors for IFN-α and IFN-β are different from those for IFN-γ.

B. IFN Action on Proto-Oncogene Expression

The effects of IFNs on the processes of cell proliferation and/or differentiation could be mediated, at least partially, by effects on the levels of expression of particular proto-oncogenes or other cellular genes. There is evidence that the concentrations of mRNA coding for several discrete protein products may change in cells exposed to discrete protein products may change in cells exposed to exogenous IFNs.[248] The changes consist in either increased or decreased expression of particular genes, including proto-oncogenes, after different protocols of IFN treatment.

A selective reduction in c-*myc* mRNA levels is observed in the Daudi human lymphoblastoid cell line after treatment with human IFN-β, and this reduction is accompanied by inhibition of cell growth.[252] A close link would exist between reduction of c-*myc* expression and IFN-induced arrest of cells at a specific phase (G_0/G_1) of the cell cycle.[253] Apparently, IFN-β regulates c-*myc* expression in Daudi cells not at the transcriptional level but at the post-transcriptional level by reducing the half-life of c-*myc* mRNA.[254,255] Although the steady-state level of c-*myc* mRNA is reduced by 60% in IFN-β-treated cells within 3 hr, the rate of c-*myc* transcription is virtually unaffected even at 24 hr of IFN-β exposure.[254] It is not known at which post-transcriptional level the c-*myc* mRNA transcripts are regulated. In other human cell lines (HL-60 and U937) IFN-β fails to reduce the c-*myc* mRNA level.[253] The basis of this difference between Daudi cells and other leukemic cell lines is unknown. HL-60 and U937 cells display normal induction of other IFN-regulated activities and show a decline in c-*myc* gene expression when they become arrested in the G_0/G_1 phase of the cycle as part of their terminal differentiation.[253] IFN and calcitriol may reduce the expression of c-*myc* in HL-60 cells in a cooperative manner related to inhibition of cell proliferation.[256]

Mouse IFN inhibits the development of transformed foci in NIH/3T3 cell cultures transfected with v-H-*ras*, v-*mos*, or EJ/T24 c-*ras* DNA.[257] IFN treatment may reduce the amount of c-H-*ras* mRNA and p21 protein expressed in NIH/3T3 cells transformed by an activated (mutant) c-*ras* gene or may inhibit the stabilization or integration of proto-oncogene sequences transfected into the same cells.[258,259] In general, IFN is capable of inhibiting the transformation of mouse cells by exogenous cellular or viral genes.[260] Treatment of RSV-transformed rat cells with rat IFN-α causes a 50% reduction in intracellular pp60^{v-src}-associated protein kinase activity, which is accompanied by a reduction in the growth rate of the transformed cells.[261] IFN is also capable of lowering the steady-state concentrations of c-H-*ras* and c-*src* mRNAs in human bladder carcinoma cells.[262] The mechanisms responsible for such IFN inhibitory effects have not been characterized. A major difficulty in interpreting date related to inhibition of proto-oncogene expression induced by agents inhibiting cell proliferation is to know whether such changes are causes or effects of the modulation of cell proliferation or cell differentiation. The same difficulty exists in relation to differentiation induced by exogenous agents like IFNs. Although mouse IFN may induce differentiation and phenotypic reversion of transformation in NIH/3T3 cells transfected with a normal human c-H-*ras* gene activated with retroviral LTR, mouse IFN does not reduce reversion when the same cells are transformed with a mutant human c-*ras* oncogene.[257]

VI. CYTOTOXINS

Proteins with selective cytotoxic properties for transformed cells are known under the generic name of cytotoxins. However, these proteins may have growth-stimulating properties in certain types of cells or under certain physiological conditions and the generic name "cytokines" should be more conveniently applied to them. A member of this family of proteins is lymphotoxin, which is produced by mitogen-stimulated lymphocytes.[263] Other members of the family are tumor necrosis factor α (TNF-α) and IFN-γ.

A. Lymphotoxin (TNF-β)

Lymphotoxin, also called tumor necrosis factor β (TNF-β), is produced by mitogen-activated lymphocytes.[264] Lymphotoxin has been isolated from a human lymphoblastoid cell line (RPMI-1788) and has been characterized as a glycoprotein of 25,000 mol wt containing 171 amino acid residues.[265] Human lymphotoxin is encoded by a single gene. A cDNA copy of this gene was expressed in *E. coli* and no significant homology was found between lymphotoxin and other proteins secreted by stimulated lymphocytes, such as IL-2, IL-3, and IFN-γ.[265] However, lymphotoxin is highly homologous to tumor necrosis factor α (TNF-α), especially in the last exons which code for more than 80% of the secreted protein, and the term TNF-β has been proposed to designate lymphotoxin.[266] Moreover, the genes of both TNF-α and TNF-β are syntenic, being located on human chromosome 6.[267]

B. Tumor Necrosis Factor α (TNF-α)

Activated macrophages produce and release a factor, called tumor necrosis factor α (TNF-α), which is selectively cytotoxic to certain neoplastic cells, but not normal cells, in vitro and in vivo.[268] TNF-α was originally detected in sera from endotoxin-treated mice, rats, and rabbits that had been previously sensitized with *Mycobacterium bovis* strain of bacillus Calmette-Guérin (BCG).[269] Serum from such animals causes hemorrhagic necrosis and in some cases complete regression of certain types of transplanted tumors such as methylcholanthrene (MCA)-induced tumors in mice.

TNF-α is a cytokine antigenically distinct from TNF-β and IFN-γ.[270,271] The differences between TNF-α and lymphotoxin have been confirmed recently. TNF-α has been purified to homogeneity from serum-free tissue culture supernatants of the HL-60 human promyelocytic leukemia cell line induced by a phorbol ester.[272,273] The protein has a molecular weight of approximately 17,000 and shows about 50% homology with another cytolytic lymphokine, lymphotoxin. The gene encoding TNF-α has been identified in a human genomic library, has been cloned from HL-60 cells and the cDNA was thereafter expressed in bacteria (*E. coli*) and eukaryotic cells (COS cells).[274-277] The product of this expression, isolated in pure form, was shown to produce inhibition of the proliferation of transformed cells in vitro and necrosis of murine tumors in vivo. The gene coding for murine TNF-α has also been cloned and expressed in mammalian cells as well as in *E. coli*.[278,279] The 235 amino acid murine pre-TNF-α polypeptide is 79% homologous to the human pre-TNF-α protein. The murine pre-TNF-α polypeptide consists of a amino acid per sequence followed by a mature TNF-α sequence of 156 amino acids. The mature mouse TNF-α is a glycosylated dimer.

Recombinant TNF-α has cytostatic or cytolytic effects on only some tumor cell lines; other tumor cell lines may not respond to the growth-inhibitory action of TNF-α, and the growth of normal human fibroblasts maybe enhanced when TNF-α is added to the culture medium.[277] TNF-α may act synergistically with IFN-γ in some cellular systems. A high-affinity binding site for TNF-α may determine the susceptibility of target cells to the action of TNF-α.[280,281] TNF-α receptors maybe regulated by IFN-γ.[282] However, the presence of a TNF-α receptor is apparently necessary but insufficient to explain the sensitivity of target cells to the cytotoxic activity of TNF-α.[277] Apparently, TNF-α is highly homologous or identical to cachectin, a hormone secreted by macrophages which produces a specific suppression of the enzyme lipoprotein lipase (LPL) in cultured adipocytes.[283] Cachectin is a potent pyrogen and is responsible for a hypertriglyceridemic state occurring in mammals during certain infections. Cachectin may be involved as a mediator in the wasting that accompanies chronic invasive diseases of both infectious and neoplastic origin.

VII. ERYTHROPOIESIS-STIMULATING FACTORS

Erythropoietin is apparently the most important regulator of erythropoiesis but other chemical agents maybe involved in stimulating erythroid progenitors. Testosterone maybe one of such substances because it cooperates with erythropoietin in the regulation of RNA synthesis in bone marrow cells.[284,285] Purified CSF-2, G-CSF and IL-3 are all able to initiate proliferation of the earlier erythroid precursors, and other HGFs maybe involved in controlling the proliferation of erythroid precursors through the subsequent stages of differentiation.[286] IL-3 (hemopoietin-2) is capable of stimulating the full development of many erythroid burst-forming units in vitro, even in the absence of serum and detectable amounts of erythropoietin.[28] The respective physiological roles of testosterone, IL-3 and erythropoietin and other factors in erythropoiesis occurring in the intact animal are unknown but it is possible that IL-3 is the stimulus for the early erythroid progenitor and that erythropoietin acts only later in the maturation progression.

A. Erythroid-Potentiating Activity

Erythroid-potentiating activity (EPA) is a factor with burst-promoting activity (BPA), i.e., it stimulates the growth of early erythroid progenitors referred to as burst-forming units-erythroid (BFU-E), which give rise to colonies of hemoglobinized cells. EPA is a glycoprotein of approximate 28,000 mol wt which specifically stimulates cells of the erythroid lineage. According to studies performed with an EPA cDNA clone, human EPA appears to be encoded by a single gene approximately 3 kb in length interrupted by at least two intervening sequences.[287] Human EPA specifically stimulates the growth of peripheral blood- and bone marrow-derived erythroid precursors from mouse and man. However, the physiological activities of EPA in vivo are unknown. A major retroviral core protein has been identified as structurally related to EPA.[288]

B. Erythropoietin

Under normal conditions the rate of red cell production is adjusted to the oxygen demand and supply in the peripheral tissues. This adjustment is accomplished by a feedback circuit linking oxygen sensors in the tissues to the erythroid cells in the bone marrow and mediated by the red cell in one direction and the hormone, erythropoietin, in the opposite direction. Consequently, the level of erythropoietin in blood must be inversely related to the number of circulating red cells.[289] The major sites of oxygen sensing and erythropoietin production are the liver during fetal life and the kidney in the adult.

1. Structure and Function and Erythropoietin

Erythropoietin is the primary factor regulating red blood cell formation in mammals and some other animals.[290-292] The concentration of erythropoietin in the blood is of approximately 0.01 nM under normal conditions and is increased under conditions of hypoxia. Erythropoietin exerts its physiological activities by attaching to specific binding sites on erythroid progenitor cells, which induces their differentiation into mature erythrocytes. Kidney mass loss associated with chronic renal failure and other renal diseases may result in anemia due to decreased production of erythropoietin.

Erythropoietin is a glycoprotein of approximate 34,000 mol wt which has been purified from the plasma of animals made artificially anemic and from the urine of severely anemic patients. The gene coding for human erythropoietin has been cloned and its expression in a mammalian system (CHO cells) yielded a secreted product with biological activity.[293,294] The complete nucleotide sequence of erythropoietin cDNA allowed to deduce that the human hormone is composed of 166 amino acid residues and that it possesses at least two disulfide bonds and three sites of N-linked glycosylation. The calculated molecular weight of the polypeptide is 18,399. Erythropoietin mRNA transcripts have been detected in human fetal and adult liver.[293]

Erythropoietin has no significant homology with any known protein. However, the amino-terminal portion of an erythropoietin-like polypeptide isolated from bovine fetal serum shows sequence homology with the low density lipoprotein (LDL) receptor as well as with two proteins of the Epstein-Barr virus.[295]

2. Erythropoietin in Neoplastic Diseases

Erythropoietin-like activity is produced in tumors cells like human renal cell and hepatocellular carcinomas.[296] Erythropoiesis in chronic myelogenous leukemia is generally erythropoietin-dependent.[297] Harvey and Kirsten sarcoma retroviruses, which contain the v-H-*ras* and v-K-*ras* oncogenes, respectively, can alter the growth of erythroid cells in vitro, inducing erythropoietin-dependent proliferation.[298] In this system self-renewal appears to be restricted despite the presence of the acute transforming retrovirus, and the cells eventually undergo extensive hemoglobinization, indicating that they can reach a state of terminal differentiation. Infection of mice with the polycythemia-inducing strain of Friend murine leukemia virus (F-MuLV, a chronic transforming retrovirus, not containing oncogenes) causes a rapid emergence of new erythroid precursor cells which, in the absence of erythropoietin, proliferate in vitro to colonies and even become differentiated but that do not require the addition of erythropoietin to proliferate and synthesize hemoglobin in vitro.[293]

VIII. RELATIONSHIPS BETWEEN HEMATOPOIETIC CELL PROLIFERATION AND DIFFERENTIATION, HGFs AND PROTO-ONCOGENE EXPRESSION

The relationships between the mitogenic factors produced by different blood cells (B-lymphocytes, T-lymphocytes, macrophages) and proto-oncogene functions remain little understood. Distinct tyrosine-specific protein kinase activities have been found in B- and T-lymphocytes and the expression of these kinases maybe related either to cellular differentiation along different lymphocytic pathways or to lymphocyte activation in response to different stimuli.[300] It is interesting to look for the possible relationships between these kinase activities and similar activities present in the protein products of proto-oncogenes of the c-*src* family.

A. Activation of c-*myc* Expression

Activation of resting lymphocytes by mitogens maybe associated with changes in proto-oncogene expression. In resting cultured mouse lymphocytes the levels of c-*myc* messenger RNA are almost undetectable but are induced approximately 20-fold between 1 and 2 hr after the addition of the T-cell specific mitogen concanavalin A to the cell culture.[159] Similar results on c-*myc* induction are obtained with stimulation of mouse spleen cells by the B-cell specific mitogen lipopolysaccharide (LPS). Stimulation of normal human peripheral blood mononuclear cells with the T-cell mitogen PHA results in a definite increase in the level of expression of both c-*myc* and c-*myb* proto-oncogenes.[160,301] Antibody-induced B-cell activation is accompanied by a specific induction of c-*myc* expression in the early G_1 phase of the cell cycle but BCGF is required for the entry of cells into the S phase of the cycle.[302]

In addition to c-*myc*, the expression of genes encoding IL-2, the receptor for IL-2, and IFN-γ are rapidly and independently stimulated in normal human peripheral blood T-lymphocytes after stimulation with either PHA or PMA.[303] The induction of all of these genes occurs in the presence of cycloheximide, thus indicating that protein synthesis is not required for the induction. Moreover, the presence of IL-2 protein is not absolutely required for transcriptional activation of either c-*myc*, IL-2 receptor, or IFN-γ gene. In contrast, transferrin receptor gene transcription is initiated later in the course of T-cell activation and its expression required *de novo* protein synthesis.[303]

A detailed study has demonstrated that the c-*myc* proto-oncogene is regulated in normal human lymphocytes at several points in the cell cycle.[304] Peripheral blood mononuclear cells stimulated with PHA, phorbol ester (PMA), calcium ionophore (ionomycin), or the monoclonal antibody OKT3 (anti-antigen receptor complex) show marked increases in c-*myc* mRNA levels within 3 hr. Recombinant IL-2 has little effect on c-*myc* expression but, in combination with PHA, it augments levels of c-*myc* transcripts measured at 24 hr but not at 3 hr. Adding various inhibitors of lymphocyte proliferation to PHA-stimulated cultures reveals that cyclosporin A (a fungal metabolite that selectively inhibits T-cell proliferation), dexamethasone (a synthetic glucocorticoid), and OKT11A antibody (which binds to the sheep erythrocyte receptor present on all human T-lymphocytes) diminish levels of c-*myc* mRNA measured at 3 and 24 hr, whereas anti-Tac antibody (which binds to IL-2 receptors) inhibits at 24 hr but not at 3 hr.[304] The IL-2 induced increase of c-*myc* transcriptional activity in human peripheral blood mononuclear cells was confirmed in an independent study.[127] Since binding of IL-2 to its receptor on T-cells leads to activation of protein kinase C,[150,152] the results further suggest that protein kinase C maybe involved in regulating the expression of the c-*myc* proto-oncogene in normal human cells.

Another study demonstrated that recombinant murine retroviruses expressing v-*myc* oncogenes abrogate the requirement for IL-2 or IL-3 in cell lines dependent from these factors.[43] The expression of c-*myc* induced by IL-3 is suppressed in the cells infected with the recombinant vector but the mechanism of this phenomenon is not understood. The results suggest that the action of peptide growth factors in the respective target cells maybe mediated partially by their ability to induce expression of the c-*myc* proto-oncogene.

B. Activation of c-*fos* Expression

Expression of the c-*fos* proto-oncogene is a common property of mature hematopoietic cells and in the mononuclear phagocyte lineage it is restricted to differentiated cells.[305] The levels of c-*fos* messenger RNA increase between 30 min and 1 hr after the addition of PHA to human peripheral blood lymphocytes.[306] In general, c-*fos* expression appears to be specifically correlated with signals that stimulate a G_0-to-G_1 cell cycle transition. A rapid increase in c-*fos* expression occurs during human monocyte differentiation induced by phorbol ester (TPA) in hematopoietic cell lines.[307] In such cell lines c-*fos* transcripts increase within 10 min after treatment with the inducer and attain maximum levels by 30 min. Expression of c-*fos* occurs only when the cell lines are induced to enter the macrophage differentiation pathway, not when the cells differentiate to granulocytes. Expression of the c-*fms* proto-oncogene is also regulated during human monocytic differentiation.[211]

C. Proto-Oncogene Expression and Induction of Differentiation in Hematic Neoplastic Cells

Expression of the c-*myc* and c-*myb* proto-oncogenes decreases when differentiation is induced in neoplastic human hemopoietic cell lines by treatment with different compounds like calcitriol, dimethyl sulfoxide, retinoic acid, phorbol diester (TPA) or inhibitors of poly(ADP)-ribose polymerase.[308-315] It should be noticed that c-*myc* and c-*myb* are structurally related and that their protein products are predominanctly located in the nucleus.

Human promyelocytic leukemia cells HL-60 can be induced to differentiate in vitro along either the monocytic or the myeloid pathways, depending on the chemical inducer used.[316] Induction of differentiation of HL-60 cells by either TPA (monocyte pathway) or DMSO (myeloid pathway) is accompanied by a decrease in the rate of c-*myc* transcription.[314] Changes in c-*myc* expression in HL-60 cells induced to differentiate by treatment with dimethyl sulfoxide are apparently caused by the simultaneously occurring inhibition of cell proliferation since c-*myc* expression remains elevated in HL-60 growth-inhibited but undifferentiated cells.[313]

In cultured mouse erythroleukemia (MEL) cells induced to differentiate into mature erythroid cells by DMSO the level of c-*myc* mRNA decreases approximately 15-fold after 2 hr of DMSO treatment, then remains low until 12 hr but increases to pretreatment levels by 18 hr.[317] Whereas the DMSO-induced decline does not absolutely require new protein synthesis, the reappearance of c-*myc* mRNA is dependent on continued protein synthesis, which suggests the existence of both negative and positive regulatory mechanisms of c-*myc* expression. A cell cycle-dependent change in the levels of c-*myc* expression is observed in chemically induced differentiation of MEL cells.[318] Prior to inducer treatment the level of c-*myc* mRNA is relatively constant throughout the cell cycle but when the c-*myc* mRNA level is restored in inducer-treated cells it is more highly restricted to cells in the G_1 phase of the cycle. Thus, treatment with inducers of differentiation may lead to a change in the cell cycle regulation of c-*myc* gene transcriptional activity.

A selective reduction of c-*myc* expression is also observed in Daudi cells (a line of lymphoblastoid cells derived from an African Burkitt's lymphoma) under the influence of human IFN-β and a close link would exist between reduction of c-*myc* expression and interferon-induced arrest of cells at a specific phase (G_0/G_1) of the cell cycle.[252,253] Interferon regulates c-*myc* gene expression in Daudi cells at the post-transcriptional level.[253]

The levels of c-*myc* mRNA are regulated in normal human lymphocytes by modulators of cell proliferation.[304] In a human leukemia cell line (K562) hemin-mediated erythroid induction and cell proliferation are associated with changes in the expression of c-*abl* and c-*myc*.[319] In the murine myeloid leukemia cell line WEHI-3B the expression of c-*myc* and c-*myb* decreases after the cells reach the monocyte stage of differentiation induced by treatment with G-CSF plus a low concentration of actinomycin D.[230] Transcripts of other proto-oncogenes are either unaltered (c-*abl*, c-K-*ras*, and c-*fes*) or are not detectable (c-*src*, c-*sis*, c-*mos*, c-*erb*-A, and c-*erb*-B) during the process of differentiation induced in the same cells. In contrast, there is a marked increase in the expression of the c-*fos* proto-oncogene, which is apparently associated with monocyte differentiation. A similar increase in c-*fos* expression is observed when HL-60 cells are induced to differentiate by treatment with TPA but not when the induction is performed by treatment with calcitriol or retinoic acid, which indicates that c-*fos* expression is not an obligatory step in the differentiation of leukemic cells.[320] In quiescent terminally differentiated macrophages, expression of c-*fos* is inducible by CSF-1.[320]

F-MuLV is a replication-competent, chronic transforming retrovirus capable of inducing acute nonlymphocytic leukemias in mice. F-MuLV-induced leukemias can be divided into two stages based on the growth properties of the leukemia cells. In the early stage (stage I disease) of leukemia, the blast cells are unable to grow outside their normal hematopoietic environment (bone marrow or spleen), whereas in the late stage of leukemia (stage II disease), leukemia cells can grow at any site in the mouse and will form continuous cell lines in vitro. However, hematopoietic cells obtained from mice with stage I disease will grow as immortal cell lines if cultured in the presence of hemopoietin-2 (IL-3) or in a WEHI-3 cell conditioned medium (WEHI-3 cells produce IL-3 constitutively).[321] Superinfection of F-MuLV-induced stage I disease cells with A-MuLV, which carries the v-*abl* oncogene, but not with H-MuSV, which carries the v-H-*ras* oncogene, determines independence from IL-3-containing medium (WEHI-3 conditioned medium) and proliferation in culture occurs in the absence of exogenous growth factors.[39] Concomitant with the loss of growth factor dependence in culture, the A-MuLV-superinfected lines become tumorgenic in syngeneic mice. Neither A-MuLV- nor H-MuSV-infected normal mouse myeloid cell cultures produce growth factor-independent or tumorigenic cell lines. The molecular basis of these results is unclear. In particular, it is not known which factor induced IL-3 independence in the A-MuLV infected cells.

Analysis of these cells fails to support the autocrine model of tumorigenesis since the conditioned medium obtained from these cells does not support the growth of HGF-dependent cell lines. Moreover, no evidence of altered IL-3 transcription or rearranged IL-3 DNA sequences is found in such cells.[39] It would appear that instead of triggering the production of IL-3, action of the v-*abl* protein makes IL-3 superfluous for the control of cell proliferation.

IX. SUMMARY

A diversity of HGFs is produced by different types of hematopoietic cells during the complex processes associated with their differentiation and proliferation. Each HGF is involved in the control of one or more distinct stages of cell differentiation and proliferation in the different hematopoietic cell lineages. The cellular receptor for CSF-1, is structurally identical to the protein product of the c-*fms* proto-oncogene. At least some HGFs are involved in regulating the expression of proto-oncogenes, in particular the expression of c-*myc* and c-*fos*. The role of HGFs in the origin and/or development of hematologic neoplasms is still not understood. There is little evidence that HGFs participate in the development of these neoplasms through a kind of autocrine mechanism as has been described for certain solid tumors. Constitutive production of HGFs is observed in only a few hematologic neoplasms and most of these tumors require an exogenous supply of the respective factors for growing. Certain hematologic malignancies are associated with specific chromosome translocations involving proto-oncogenes, for example, c-*abl* and c-*sis* in chronic myelogenous leukemia and c-*myc* in B cell lymphomas.[322] However, the exact role of proto-oncogene alterations in the origin and/or development of spontaneous tumors affecting the hematopoietic tissues is not yet understood. Further studies are required for a better characterization of the complex relationships between HGFs and proto-oncogene products in normal and malignant cells of the hematopoietic tissues.

REFERENCES

1. **Metcalf, D.**, The granulocyte-macrophage colony-stimulating factors, *Science,* 229, 16, 1985.
2. **Ogawa, M., Porter, P. N., and Nakahata, T.**, Renewal and commitment to differentiation of hemopoietic stem cells (an interpretative review), *Blood,* 61, 823, 1983.
3. **McCulloch, E. A.**, Stem cells in normal and leukemic hemopoiesis, *Blood,* 62, 1, 1983.
4. **Golde, D. W. and Marks, P. A.**, Eds., *Normal and Neoplastic Hematopoiesis,* Alan R. Liss, New York, 1983.
5. **Stanley, E. R. and Jubinsky, P. T.**, Factors affecting the growth and differentiation of haemopoietic cells in culture, *Clin. Haematol.,* 13, 329, 1984.
6. **Nicola, N. A. and Vadas, M.**, Hemopoietic colony-stimulating factors, *Immunol. Today,* 5, 76, 1984.
7. **Dexter, T. M., Heyworth, C., and Whetton, A. D.**, The role of growth factors in haemopoiesis, *BioEssays,* 2, 154, 1985.
8. **Burgess, A. W.**, Hematopoietic growth factors, in *Mediators in Cell Growth and Differentiation,* Ford, R. J. and Maizel, A. L., Eds., Raven Press, New York, 1985, 159.
9. **Clark-Lewis, I., Kent, S. B. H., and Schrader, J. W.**, Purification to apparent homogeneity of a factor stimulating the growth of multiple lineages of hemopoietic cells, *J. Biol. Chem.,* 259, 7488, 1984.
10. **Platzer, E., Welte, K., Gabrilove, J. L., Lu, L., Harris, P., Mertelsmann, R., and Moore, M. A. S.**, Biological activities of a human pluripotent hemopoietic colony stimulating factor on normal and leukemic cells, *J. Exp. Med.,* 162, 1788, 1985.
11. **Lanfrancone, L., Ferrero, D., Gallo, E., Foa, R., and Tarella, C.**, Release of hemopoietic factors by normal human T cell lines with either suppressor or helper activity, *J. Cell. Physiol.,* 122, 7, 1985.
12. **Walker, F., Nicola, N. A., Metcalf, D., and Burgess, A. W.**, Hierarchical down-modulation of hemopoietic growth factor receptors, *Cell,* 43, 269, 1985.

13. **Li, C. L. and Johnson, G. R.**, Stimulation of multipotential, erythroid and other murine haematopoietic progenitor cells by adherent cell lines in the absence of detectable multi-CSF (IL-3), *Nature (London)*, 316, 633, 1985.
14. **Metcalf, D. and Nicola, N. A.**, Role of the colony stimulating factors in the emergence and suppression of myeloid leukemia populations, in *Molecular Biology of Tumor Cells*, Wahren, B., Ed., Raven Press, New York, 1985, 215.
15. **Metcalf, D.**, Clonal analysis of the response of HL60 human myeloid leukemia cells to biological regulators, *Leuk. Res.*, 7, 117, 1983.
16. **Metcalf, D.**, Regulation of self-replication in normal and leukemic stemm cells, in *Normal and Neoplastic Hematopoiesis*, Alan R. Liss, New York, 1983, 141.
17. **Lang, R. A., Metcalf, D., Gough, N. M., Dunn, A. R., and Gonda, T. J.**, Expression of a hemopoietic growth factor cDNA in a factor-dependent cell line results in autonomous growth and tumorigenicity, *Cell*, 43, 531, 1985.
18. **Golde, D. W., Finley, T. N., and Cline, M. J.**, Production of colony-stimulating factor by human macrophages, *Lancet*, ii, 1397, 1972.
19. **Stanley, E. R., Hansen, G., Woodcock, J., and Metcalf, D.**, Colony stimulating factor and the regulation of granulopoiesis and macrophage production, *Fed. Proc.*, 34, 2272, 1975.
20. **Bartelmez, S. H., Saaca, R., and Stanley, E. R.**, Lineage specific receptors used to identify a growth factor for developmentally early hemopoietic cells: assay of hemopoietin-2, *J. Cell. Physiol.*, 122, 362, 1985.
21. **Bartelmez, S. H. and Stanley, E. R.**, Synergism between hemopoietic growth factors (HGFs) detected by their effects on cells bearing receptors for a lineage specific HGF: assay of hemopoietin-1, *J. Cell. Physiol.*, 122, 370, 1985.
22. **Eliason, J. F. and Odartchenko, N.**, Colony formation by primitive hemopoietic progenitor cells in serum-free medium, *Proc. Natl. Acad. Sci. U.S.A.*, 82, 775, 1985.
23. **Trucco, M., Rovera, G., and Ferrero, D.**, A novel human lymphokine that inhibits haematopoietic progenitor cell proliferation, *Nature (London)*, 309, 166, 1984.
24. **Iscove, N. N., Roitsch, C. A., Williams, N., and Guilbert, L. J.**, Molecules stimulating early red cell, granulocyte, macrophage, and megakaryocyte precursor in culture: similarity in size, hydrophobicity, and charge, *J. Cell. Physiol.*, Suppl. 1, 65, 1982.
25. **Jubinsky, P. T. and Stanley, E. R.**, Purification of hemopoietin 1: a multilineage hemopoietic growth factor, *Proc. Natl. Acad. Sci. U.S.A.*, 82, 2764, 1985.
26. **Metcalf, D.**, Multi-CSF-dependent colony formation by cells of a murine hemopoietic cell line: specificity and action of multi-CSF, *Blood*, 65, 357, 1985.
27. **Miyatake, S., Yokota, T., Lee, F., and Arai, K.**, Structure of the chromosomal gene for murine interleukin 3, *Proc. Natl. Acad. Sci. U.S.A.*, 82, 316, 1985.
28. **Goodman, J. W., Hall, E. A., Miller, K. L., and Shinpock, S. G.**, Interleukin 3 promotes erythroid burst formation in "serum-free" cultures without detectable erythropoietin, *Proc. Natl. Acad. Sci. U.S.A.*, 82, 3291, 1985.
29. **Suda, T., Suda, J., Ogawa, M., and Ihle, J. N.**, Permissive role of interleukin 3 (IL-3) in proliferation and differentiation of multipotential hemopoietic progenitors in culture, *J. Cell. Physiol.*, 124, 182, 1985.
30. **Palacios, R., Henson, G., Steinmetz, M., and McKearn, J. P.**, Interleukin-3 supports growth of mouse pre-B-cell clones *in vitro*, *Nature (London)*, 309, 126, 1984.
31. **Lee, J. C., Hapel, A. J., and Ihle, J. N.**, Constitutive production of a unique lymphokine (IL-3) by the WEHI-3 cell line, *J. Immunol.*, 128, 2393, 1982.
32. **Ymer, S., Tucker, W. Q. J., Sanderson, C. J., Hapel, A. J., Campbell, H. D., and Young, I. G.**, Constitutive synthesis of interleukin-3 by leukaemia cell line WEHI-3B is due to retroviral insertion near the gene, *Nature (London)*, 317, 255, 1985.
33. **Ihle, J. W., Keller, J., Henderson, L., Klein, F., and Palazinski, E.**, Procedure for the purification of interleukin-3 to homogeneity, *J. Immunol.*, 129, 2431, 1982.
34. **Fung, M. C., Hapel, A. J., Ymer, S., Cohen, D. R., Johnson, R. M., Campbell, H. D., and Young, I. G.**, Molecular cloning of cDNA for murine interleukin-3, *Nature (London)*, 307, 233, 1984.
35. **Greenberger, J. S., Humphries, R. K., Messner, H., Reid, D. M., and Sakakeeny, M. A.**, Molecularly cloned and expressed murine T-cell gene product is similar to interleukin-3, *Exp. Hematol.*, 13, 249, 1985.
36. **Clark-Lewis, I., Aebersold, R., Ziltener, H., Schrader, J. W., Hood, L. E., and Kent, S. B. H.**, Automated chemical synthesis of a protein factor for hemopoietic cells, interleukin 3, *Science*, 231, 134, 1986.
37. **Welte, K., Platzer, E., Lu, L., Gabrilove, J. L., Levi, E., Mertelsmann, R., and Moore, M. A. S.**, Purification and biochemical characterization of human pluripotent hematopoietic colony-stimulating factor, *Proc. Natl. Acad. Sci. U.S.A.*, 82, 1526, 1985.
38. **Serunian, L. A. and Rosenberg, N.**, Abelson virus potentiates long-term growth of mature B lymphocytes, *Mol. Cell. Biol.*, 6, 183, 1986.

39. **Oliff, A., Agranovsky, O., McKinney, M. D., Murty, V. V. V. S., and Bauchwitz, R.**, Friend murine leukemia virus-immortalized myeloid cells are converted into tumorigenic cell lines by Abelson leukemia virus, *Proc. Natl. Acad. Sci. U.S.A.*, 82, 3306, 1985.
40. **Cook, W. D., Metcalf, D., Nicola, N. A., Burgess, A. W., and Walker, F.**, Malignant transformation of a growth factor-dependent myeloid cell line by Abelson virus without evidence of an autocrine mechanism, *Cell*, 41, 677, 1985.
41. **Pierce, J. H., Di Fiore, P. P., Aaronson, S. A., Potter, M., Pumphrey, J., Scott, A., and Ihle, J. N.**, Neoplastic transformation of mast cells by Abelson-Mu-LV: abrogation of IL-3 dependence by a nonautocrine mechanism, *Cell*, 41, 685, 1985.
42. **Rein, A., Keller, J., Schultz, A. M., Holmes, K. L., Medicus, R., and Ihle, J. N.**, Infection of immune mast cells by Harvey sarcoma virus: immortalization without loss of requirement for interleukin-3, *Mol. Cell. Biol.*, 5, 2257, 1985.
43. **Rapp, U. R., Cleveland, J. L., Brightman, K., Scott, A., and Ihle, J. N.**, Abrogation of IL-3 and IL-2 dependence by recombinant murine retroviruses expressing v-*myc* oncogenes, *Nature (London)*, 317, 434, 1985.
44. **Farrar, W. L., Thomas, T. P., and Anderson, W. B.**, Altered cytosol/membrane enzyme redistribution on interleukin-3 activation of protein kinase C, *Nature (London)*, 315, 235, 1985.
45. **Akiyama, Y., Stevenson, G. W., Schlick, E., Matsushima, K., Miller, P. J., and Stevenson, H. C.**, Differential ability of human blood monocyte subsets to release various cytokines, *J. Leukocyte Biol.*, 37, 519, 1985.
46. **Kasakura, S. and Lowenstein, L.**, A factor stimulating DNA synthesis derived from the medium of leukocyte cultures, *Nature (London)*, 208, 794, 1965.
47. **Gordon, J. and Maclean, L. D.**, A lymphocyte-stimulating factor produced in vitro, *Nature (London)*, 208, 795, 1965.
48. **Gearing, A. J. H., Johnstone, A. P., and Thorpe, R.**, Production and assay of the interleukins, *J. Immunol. Methods*, 83, 1, 1985.
49. **Larsson, E. L., Iscove, N. N., and Coutinho, A.**, Two distinct factors are required for induction of T-cell growth, *Nature (London)*, 283, 664, 1980.
50. **Smith, K. A., Lachman, L. B., Oppenheim, J. J., and Favata, M. F.**, The functional relationship of the interleukins, *J. Exp. Med.*, 151, 1551, 1980.
51. **Maizel, A. L. and Mehta, S. R.**, Effect of interleukin 1 on human thymocytes and purified human T cells, *J. Exp. Med.*, 153, 470, 1980.
52. **Chen, J. H.**, The proto-oncogene c-*ets* is preferentially expressed in lymphoic cells, *Mol. Cell. Biol.*, 5, 2993, 1985.
53. **Pike, B. L. and Nossal, G. J. V.**, Interleukin 1 can act as a B-cell growth and differentiation factor, *Proc. Natl. Acad. Sci. U.S.A.*, 82, 8153, 1985.
54. **Dinarello, C. A.**, Interleukin-1 and the pathogenesis of the acute-phase response, *N. Engl. J. Med.*, 311, 1413, 1984.
55. **Dinarello, C. A.**, Interleukin 1, *Rev. Infect. Dis.*, 6, 51, 1984.
56. **Durum, S. K., Schmidt, J. A., and Oppenheim, J. J.**, Interleukin 1: an immunological perspective, *Ann. Rev. Immunol.*, 3, 263, 1985.
57. **Smith, R. J., Speziale, S. C., and Bowman, B. J.**, Properties of interleukin-1 as a complete secretagogue for human neutrophils, *Biochem. Biophys. Res. Commun.*, 130, 1233, 1985.
58. **Montesano, R., Orci, L., and Vassalli, P.**, Human endothelial cell cultures: phenotypic modulation by leukocyte interleukins, *J. Cell. Physiol.*, 122, 424, 1985.
59. **McMannis, J. D. and Plate, J. M. D.**, Xenogeneic antiserum to soluble products from activated lymphoid cells inhibits interleukin 1-mediated functions in the helper pathway of cytolytic-effector-cell differentiation, *Proc. Natl. Acad. Sci. U.S.A.*, 82, 1513, 1985.
60. **Onozaki, K., Matsushima, K., Kleinerman, E. S., Saito, T., and Oppenheim, J. J.**, Role of interleukin 1 in promoting human monocyte-mediated tumor cytotoxicity, *J. Immunol.*, 135, 314, 1985.
61. **Lovett, D., Kozan, B., Hadam, M., Resch, K., and Gemsa, D.**, Macrophage cytotoxicity: interleukin 1 is a mediator of tumor cytostasis, *J. Immunol.*, 136, 340, 1986.
62. **Ramadori, G., Sipe, J. D., Dinarello, C. A., Mizel, S. B., and Colten, H. R.**, Pretranslational modulation of acute phase hepatic protein synthesis by murine recombinant interleukin 1 (IL-1) and purified human IL-1, *J. Exp. Med.*, 162, 930, 1985.
63. **Hanazawa, S., Ohmori, Y., Amano, S., Miyoshi, T., Kumegawa, M., and Kitano, S.**, Spontaneous production of interleukin-1-like cytokine from a mouse osteoblastic cell line (MC3T3-E1), *Biochem. Biophys. Res. Commun.*, 131, 774, 1985.
64. **Canalis, E.**, Interleukin-1 has independent effects on deoxyribonucleic acid and collagen synthesis in cultures of rat calvariae, *Endocrinology*, 118, 74, 1986.

65. **Van Damme, J., De Ley, M., Opdenakker, G., Billiau, A., De Somer, P., and Van Beeumen, J.**, Homogeneous interferon-inducing 22K factor is related to endogenous pyrogen and interleukin-1, *Nature (London)*, 314, 266, 1985.
66. **Murphy, P. A., Simon, P. L., and Willoughby, W. F.**, Endogenous pyrogens made by the rabbit peritoneal exudate cells are identical with lymphocyte activating factor made by rabbit alveolar macrophages, *J. Immunol.*, 124, 2498, 1980.
67. **Dower, S. K., Kronheim, S. R., March, C. J., Conlon, P. J., Hopp, T. P., Gillis, S., and Urdal, D. L.**, Detection and characterization of high affinity plasma membrane receptors for human interleukin 1, *J. Exp. Med.*, 162, 501, 1985.
68. **Williams, J. M., Dinarello, C. A., Rosenwasser, L. J., Kelley, V., Reddish, M., and Strom, T. B.**, Phorbol myristate acetate-protein complex mimics bioactivity of human IL-1, *Lymphokine Res.*, 4, 275, 1985.
69. **Krakauer, T.**, Biochemical characterization of interleukin 1 from a human monocytic cell line, *J. Leukocyte Biol.*, 37, 511, 1985.
70. **Doyle, M. V., Brindley, L., Kawasaki, E., and Larrick, J.**, High level human interleukin 1 production by a hepatoma cell line, *Biochem. Biophys. Res. Comm.*, 130, 768, 1985.
71. **Woloski, B. M. R. N. J. and Fuller, G. M.**, Identification and partial characterization of hepatocyte-stimulating factor from leukemia cell lines: comparison with interleukin 1, *Proc. Natl. Acad. Sci. U.S.A.*, 82, 1443, 1985.
72. **Lomedico, P. T., Gubler, U., Hellmann, C. P., Dukovich, M., Giri, J. G., Pan, Y-C. E., Collier, K., Semionov, R., Chua, A. O., and Mizel, S. B.**, Cloning and expression of murine interleukin-1 cDNA in *Escherichia coli*, *Nature (London)*, 312, 458, 1984.
73. **Auron, P. E., Webb, A. C., Rosenwasser, L. J., Mucci, S. F., Rich, A., Wolff, S. M., and Dinarello, C. A.**, Nucleotide sequence of human monocyte interleukin 1 precursor cDNA, *Proc. Natl. Acad. Sci. U.S.A.*, 81, 7907, 1984.
74. **Cameron, P., Limjuco, G., Rodkey, J., Bennett, C., and Schmidt, J. A.**, Amino acid sequence analysis of human interleukin 1 (IL-1): evidence for biochemically distinct forms of IL-1, *J. Exp. Med.*, 162, 790, 1985.
75. **March, C. J., Mosley, B., Larsen, A., Cerretti, D. P., Braedt, G., Price, V., Gillis, S., Henney, C. S., Kronheim, S. R., Grabstein, K., Conlon, P. J., Hopp, T. P., and Cosman, D.**, Cloning, sequence and expression of two distinct human interleukin-1 complementary DNAs, *Nature (London)*, 315, 641, 1985.
76. **Ruscetti, F. W. and Gallo, R. C.**, Human T-lymphocyte growth factor: regulation of growth and function of T lymphocytes, *Blood*, 57, 379, 1981.
77. **Cantrell, D. A. and Smith, K. A.**, The interleukin-2 T-cell system: a new cell growth model, *Science*, 224, 1312, 1984.
78. **Sarin, P. S. and Gallo, R. C.**, Human T-cell growth factor (TCGF), *CRC Crit. Rev. Immunol.*, 4, 279, 1984.
79. **Smith, K. A.**, Interleukin 2, *Ann. Rev. Immunol.*, 2, 319, 1984.
80. **Robb, R. J.**, Interleukin-2: the molecule and its function, *Immunol. Today*, 5, 203, 1984.
81. **Mond, J. J., Thompson, C., Finkelman, F. D., Farrar, J., Schaefer, M., and Robb, R. J.**, Affinity-purified interleukin 2 induces proliferation of large but not small B cells, *Proc. Natl. Acad. Sci. U.S.A.*, 82, 1518, 1985.
82. **Ythier, A. A., Abbud-Filho, M., Williams, J. M., Lertscher, R., Schuster, M. W., Nowill, A., Hansen, J. A., Maltezos, D., and Strom, T. B.**, Interleukin 2-dependent release of interleukin 3 activity by T4$^+$ human T-cell clones, *Proc. Natl. Acad. Sci. U.S.A.*, 82, 7020, 1985.
83. **Hersey, P., Bindon, C., Edwards, A., Murray, E., Phillips, G., and McCarthy, W. H.**, Induction of cytotoxic activity in human lymphocytes against autologous and allogeneic melanoma cells *in vitro* by culture with interleukin 2, *Int. J. Cancer*, 28, 695, 1981.
84. **Pizza, G., Severini, G., Menniti, D., De Vinci, C., and Corrado, F.**, Tumour regression after intralesional injection of interleukin 2 (IL-2) in bladder cancer. Preliminary report, *Int. J. Cancer*, 34, 359, 1984.
85. **Yagüe, J. and Palmer, E.**, Antigen specific T cell receptor: isolation, structure and genomic organization, *Immunologia*, 4, 89, 1985.
86. **Croce, C. M., Isobe, M., Palumbo, A., Puck, J., Ming, J., Tweardy, D., Erikson, J., Davis, M., and Rovera, G.**, Gene for alpha-chain of human T-cell receptor: location on chromosome 14 region involved in T-cell neoplasms, *Science*, 227, 1044, 1985.
87. **Murre, C., Waldmann, R. A., Morton, C. C., Bongiovanni, K. F., Waldmann, T. A., Shows, T. B., and Seidman, J. G.**, Human gamma-chain genes are rearranged in leukaemic T cells and map to the short arm of chromosome 7, *Nature (London)*, 316, 549, 1985.
88. **Waldmann, T. A., Davis, M. M., Bongiovanni, K. F., and Korsmeyer, S. J.**, Rearrangements of genes for the antigen receptor on T cells as markers of lineage and clonality in human lymphoid neoplasms, *N. Engl. J. Med.*, 313, 776, 1985.

89. **Baer, R., Chen, K.-C., Smith, S. D., and Rabbitts, T. H.,** Fusion of an immunoglobulin variable gene and a T cell receptor constant gene in the chromosome 14 inversion associated with T cell tumors, *Cell,* 43, 705, 1985.
90. **Morinaga, T., Fotedar, A., Singh, B., Wegmann, T. G., and Tamaoki, T.,** Isolation of cDNA clones encoding a T-cell receptor beta-chain from a beef insulin-specific hybridoma, *Proc. Natl. Acad. Sci. U.S.A.,* 82, 8163, 1985.
91. **Waldmann, T. A., Longo, D. L., Leonard, W. J., Depper, J. M., Thompson, C. B., Krönke, M., Goldman, C. K., Sharrow, S., Bongiovanni, K., and Greene, W. C.,** Interleukin 2 receptor (Tac antigen) expression in HTLV-I-associated adult T-cell leukemia, *Cancer Res. (Suppl.),* 45, 4559s, 1985.
92. **Brenner, M. B., Trowbridge, I. S., and Strominger, J. L.,** Cross-linking of human T cell receptor proteins: association between the T cell idiotype beta subunit and the T3 glycoprotein heavy subunit, *Cell,* 40, 183, 1985.
93. **van den Elsen, P., Shepley, B-A., Borst, J., Coligan, J. E., Markham, A. F., Orkin, S., and Terhorst, C.,** Isolation of cDNA clones encoding the 20K T3 glycoprotein of human T-cell receptor complex, *Nature (London),* 312, 413, 1984.
94. **Cantrell, D. A., Davies, A. A., and Crumpton, M. J.,** Activation of protein kinase C down-regulate and phosphorylate the T3/T-cell antigen receptor complex of human T lymphocytes, *Proc. Natl. Acad. Sci. U.S.A.,* 82, 8158, 1985.
95. **Mier, J. W. and Gallo, R. C.,** Purification and characteristics of human T-cell growth factor, *Proc. Natl. Acad. Sci. U.S.A.,* 77, 6134, 1980.
96. **Robb, R. J., Kutny, R. M., Panico, M., Morris, H. R., and Chowdhry, V.,** Amino acid sequence and post-translational modification of human interleukin 2, *Proc. Natl. Acad. Sci. U.S.A.,* 81, 6486, 1984.
97. **Wang, A., Lu, S-D., and Mark, D. F.,** Site-specific mutagenesis of the human interleukin-2 gene: structure-function analysis of the cysteine residues, *Science,* 224, 1431, 1984.
98. **Efrat, S. and Kaempfer, R.,** Control of biologically active interleukin 2 messenger RNA formation in induced human lymphocytes, *Proc. Natl. Acad. Sci. U.S.A.,* 81, 2601, 1984.
99. **Arya, S. K. and Gallo, R. C.,** Transcriptional modulation of human T-cell growth factor gene by phorbol ester and interleukin 1, *Biochemistry,* 23, 6685, 1984.
100. **Shows, T., Eddy, R., Haley, L., Byers, M., Henry, M., Fujita, T., Matsui, H., and Tanaguchi, T.,** Interleukin 2 (*IL2*) is assigned to human chromosome 4, *Somat. Cell Mol. Genet.,* 10, 315, 1984.
101. **Seigel, L. J., Harper, M. E., Wong-Staal, F., Gallo, R. C., Nash, W. G., and O'Brien, S. J.,** Gene for T-cell growth factor: location on human chromosome 4q and feline chromosome B1, *Science,* 223, 175, 1984.
102. **Holbrook, N. J., Smith, K. A., Fornace, A. J., Jr., Comeau, C. M., Wiskocil, R. L., and Crabtree, G. R.,** T-cell growth factor: complete nucleotide sequence and organization of the gene in normal and malignant cells, *Proc. Natl. Acad. Sci. U.S.A.,* 81, 1634, 1984.
103. **Clark, S. C., Arya, S. K., Wong-Staal, F., Matsumoto-Kobayashi, M., Kay, R. M., Kaufman, R. J., Brown, E. L., Shoemaker, C., Copeland, T., Oroszlan, S., Smith, K., Sarngadharan, M. G., Lindner, S. G., and Gallo, R. C.,** Human T-cell growth factor: partial amino acid sequence, cDNA cloning, and organization and expression in normal and leukemic cells, *Proc. Natl. Acad. Sci. U.S.A.,* 81, 2543, 1984.
104. **Fuse, A., Fujita, T., Yasumitsu, H., Kashima, N., Hasegawa, K., and Taniguchi, T.,** Organization and structure of the mouse interleukin-2 gene, *Nucleic Acids Res.,* 12, 9323, 1984.
105. **Kato, K., Yamada, T., Kawahara, K., Onda, H., Asano, T., Sugino, H., and Kakinuma, A.,** Purification and characterization of recombinant human interleukin-2 produced in *Escherichia coli, Biochem. Biophys. Res. Commun.,* 130, 692, 1985.
106. **Ishida, N., Kanomori, H., Noma, T., Nikaido, T., Sabe, H., Suzuki, N., Shimizu, A., and Honjo, T.,** Molecular cloning and structure of the human interleukin 2 gene, *Nucleic Acids Res.,* 13, 7579, 1985.
107. **Lindenmaier, W., Dittmar, K. E. J., Hauser, H., Necker, A., and Sebald, W.,** Isolation of a functional human interleukin 2 gene from a cosmid library by recombination in vivo, *Gene,* 39, 33, 1985.
108. **Muñoz, A., Perez-Aranda, A., and Barbero, J. L.,** Cloning and expression of human interleukin 2 in *Streptomyces lividans* using the *Escherichia coli* consensus promoter, *Biochem. Biophys. Res. Commun.,* 133, 511, 1985.
109. **Holbrook, N. J., Lieber, M., and Crabtree, G. R.,** DNA sequence of the 5' flanking region of the human interleukin 2 gene: homologies with adult T-cell leukemia virus, *Nucleic Acids Res.,* 12, 5005, 1984.
110. **Sugamura, K. and Hinuma, Y.,** Human retrovirus in adult T-cell leukemia/lymphoma, *Immunol. Today,* 6, 83, 1985.
111. **Gallo, R. C.,** The human T-cell leukemia/lymphotropic retroviruses (HTLV) family: past, present, and future, *Cancer Res.,* 45 (Suppl), 4524s, 1985.
112. **Wong-Staal, F. and Gallo, R. C.,** Human T-lymphotropic retroviruses, *Nature (London),* 317, 395, 1985.

113. **Chen, S. J., Holbrook, N. J., Mitchell, K. F., Vallone, C. A., Greengard, J. S., Crabtree, G. R., and Lin, Y.**, A viral long terminal repeat in the interleukin 2 gene of a cell line that constitutively produces interleukin 2, *Proc. Natl. Acad. Sci. U.S.A.*, 82, 7284, 1985.
114. **Kashima, N., Nishi-Takaoka, C., Fujita, T., Taki, S., Yamada, G., Hamuro, J., and Taniguchi, T.**, Unique structure of murine interleukin-2 as deduced from cloned cDNAs, *Nature (London)*, 313, 402, 1985.
115. **Renan, M. J.**, Sequence homologies in the control regions of c-*myc*, c-*fos*, HTLV and the interleukin 2 receptor, *Cancer Lett.*, 28, 69, 1985.
116. **Urdal, D. L., March, C. J., Gillis, S., Larsen, A., and Dower, S. K.**, Purification and chemical characterization of the receptor for interleukin 2 from activated human T lymphocytes and from a human T-cell lymphoma cell line, *Proc. Natl. Acad. Sci. U.S.A.*, 81, 6481, 1984.
117. **Nikaido, T., Shimizu, A., Ishida, N., Sabe, H., Teshigawara, K., Maeda, M., Uchiyama, T., Yodoi, J., and Honjo, T.**, Molecular cloning of cDNA encoding human interleukin-2 receptor, *Nature (London)*, 311, 631, 1984.
118. **Leonard, W. J., Donlon, T. A., Lebo, R. V., and Greene, W. C.**, Localization of the gene encoding the human interleukin-2 receptor on chromosome 10, *Science*, 228, 1547, 1985.
119. **Leonard, W. J., Depper, J. M., Crabtree, G. R., Rudikoff, S., Pumphrey, J., Robb, R. J., Krönke, M., Svetlik, P. B., Peffer, N. J., Waldmann, T. A., and Greene, W. C.**, Molecular cloning and expression of cDNAs for the human interleukin-2 receptor, *Nature (London)*, 311, 626, 1984.
120. **Cosman, D., Cerretti, D. P., Larsen, A., Park, L., March, C., Dower, S., Gillis, S., and Urdal, D.**, Cloning, sequence and expression of human interleukin-2 receptor, *Nature (London)*, 312, 768, 1984.
121. **Leonard, W. J., Depper, J. M., Kanehisa, M., Krönke, M., Peffer, N. J., Svetlik, P. B., Sullivan, M., and Greene, W. C.**, Structure of the human interleukin-2 receptor gene, *Science*, 230, 633, 1985.
122. **Greene, W. C., Robb, R. J., Svetlik, P. B., Rusk, C. M., Depper, J. M., and Leonard, W. J.**, Stable expression of cDNA encoding the human interleukin 2 receptor in eukaryotic cells, *J. Exp. Med.*, 162, 363, 1985.
123. **Smith, K. A. and Cantrell, D. A.**, Interleukin 2 regulates its own receptors, *Proc. Natl. Acad. Sci. U.S.A.*, 82, 864, 1985.
124. **Ceredig, R., Lowenthal, J. W., Nabholz, M., and MacDonald, H. R.**, Expression of interleukin-2 receptors as a differentiation marker on intrathymic stem cells, *Nature (London)*, 314, 98, 1985.
125. **Raulet, D. H.**, Expression and function of interleukin-2 receptors on immature thymocytes, *Nature (London)*, 314, 101, 1985.
126. **Reem, G. H. and Yeh, N-H.**, Interleukin 2 regulates expression of its receptor and synthesis of gamma interferon by human lymphocytes, *Science*, 225, 429, 1984.
127. **Depper, J. M., Leonard, W. J., Drogula, C., Krönke, M., Waldmann, T. A., and Greene, W. C.**, Interleukin 2 (IL-2) augments transcription of the IL-2 receptor gene, *Proc. Natl. Acad. Sci. U.S.A.*, 82, 4230, 1985.
128. **Leonard, W. J., Krönke, M., Peffer, N. J., Depper, J. M., and Greene, W. C.**, Interleukin 2 receptor expression in normal human T lymphocytes, *Proc. Natl. Acad. Sci. U.S.A.*, 82, 6281, 1985.
129. **Katzen, D., Chu, E., Terhost, C., Leung, D. Y., Gesner, M., Miller, R. A., and Geha, R. S.**, Mechanisms of human T cell response to mitogens: IL 2 receptor expression and proliferation but not IL 2 synthesis in PHA-stimulated T cells, *J. Immunol.*, 135, 1840, 1985.
130. **Mingari, M. C., Gerosa, F., Carra, G., Accolla, R. S., Moretta, A., Zubler, R. H., Waldmann, T. A., and Moretta, L.**, Human interleukin-2 promotes proliferation of activated B cells via surface receptors similar to those of activated T cells, *Nature (London)*, 312, 641, 1984.
131. **Sauerwein, R. W., Van der Meer, W. G. J., Dräger, A., and Aarden, L. A.**, Interleukin 2 induces T cell dependent IgM production in human B cells, *Eur. J. Immunol.*, 15, 611, 1985.
132. **Giglia, J. S., Ovak, G. M., Yoshida, M. A., Twist, C. J., Jeffery, A. R., and Pauly, J. L.**, Isolation of mouse T-cell lymphoma lines from different long-term interleukin 2-dependent cultures, *Cancer Res.*, 45, 5027, 1985.
133. **Uchiyama, T., Hori, T., Tsudo, M., Wano, Y., Umadome, H., Tamori, S., Yodoi, J., Maeda, M., Sawami, H., and Uchino, H.**, Interleukin-2 receptor (Tac antigen) expressed on adult T cell leukemia cells, *J. Clin. Invest.*, 76, 446, 1985.
134. **Rossi, J.-F., Klein, B., Commes, T., and Jourdan, M.**, Interleukin 2 production in B cell chronic lymphocytic leukemia, *Blood*, 66, 840, 1985.
135. **Foa, R., Giovarelli, M., Jemma, C., Fierro, M. T., Lusso, P., Ferrando, M. L., Lauria, F., and Forni, G.**, Interleukin 2 (IL 2) and interferon-gamma production by T lymphocytes from patients with B-chronic lymphocytic leukemia: evidence that normally released IL 2 is absorbed by the neoplastic B cell population, *Blood*, 66, 614, 1985.

136. **Touw, I. and Löwenberg, B.,** Interleukin 2 stimulates chronic lymphocytic leukemia colony formation in vitro, *Blood,* 66, 237, 1985.
137. **Touw, I., Delwel, R., Bolhuis, R., van Zanen, G., and Löwenberg, B.,** Common and pre-B acute lymphoblastic lsukemia cells express interleukin 2 receptors, and interleukin 2 stimulates in vitro colony formation, *Blood,* 66, 556, 1985.
138. **Poiesz, B. J., Ruscetti, F. W., Mier, J. W., Woods, A. M., and Gallo, R. C.,** T-cell lines from human T-lymphocytic neoplasias by direct response to T-cell growth factor, *Proc. Natl. Acad. Sci. U.S.A.,* 77, 6815, 1980.
139. **Yodoi, J., Okada, M., Tagawa, Y., Teshigawara, K., Fukui, K., Ishida, N., Ikuta, K-I., Maeda, M., Honjo, T., Osawa, H., Diamantstein, T., Tateno, M., and Yoshiki, T.,** Rat lymphoid cell lines producing human T-cell leukemia virus, II. Constitutive expression of rat interleukin 2 receptor, *J. Exp. Med.,* 161, 924, 1985.
140. **Sugamura, K., Fujii, M., Kobayashi, N., Sakitani, M., Hatanaka, M., and Hinuma, Y.,** Retrovirus-induced expression of interleukin 2 receptors on cells of human B-cell lineage, *Proc. Natl. Acad. Sci. U.S.A.,* 81, 7441, 1984.
141. **Krönke, M., Leonard, W. J., Depper, J. M., and Greene, W. C.,** Deregulation of interleukin-2 receptor gene expression in HTLV-I-induced adult T-cell leukemia, *Science,* 228, 1275, 1985.
142. **Tomita, S., Ambrus, J. L., Jr., Volkman, D. J., Longo, D. L., Mitsuya, H., Reitz, M. S., Jr., and Fauci, A. S.,** Human T cell leukemia/lymphoma virus I infection and subsequent cloning of normal human B cells: direct responsiveness of cloned cells to recombinant interleukin 2 by differentiation in the absence of enhanced proliferation, *J. Exp. Med.,* 162, 393, 1985.
143. **Shackelford, D. A. and Trowbridge, I. S.,** Induction of expression and phosphorylation of the human interleukin 2 receptor by a phorbol diester, *J. Biol. Chem.,* 259, 11706, 1984.
144. **Haas, M., Altman, A., Rothenberg, E., Bogart, M. H., and Jones, O. W.,** Mechanism of T-cell lymphomagenesis: transformation of growth-factor-dependent T-lymphoblastoma cells to growth-factor-independent T-lymphoma cells, *Proc. Natl. Acad. Sci. U.S.A.,* 81, 1742, 1984.
145. **Duprez, V., Lenoir, G., and Dautry-Varsat, A.,** Autocrine growth stimulation of a human T-cell lymphoma line by interleukin 2, *Proc. Natl. Acad. Sci. U.S.A.,* 82, 6932, 1985.
146. **Broder, S., Bunn, P. A., Jr., Jaffe, E. S., Blattner, W., Gallo, R. C., Wong-Staal, F., Waldmann, T. A., and DeVita, V. T.,** T-cell lymphoproliferative syndrome associated with human T-cell leukemia/lymphoma virus, *Ann. Int. Med.,* 100, 543, 1984.
147. **Theofilopoulos, A. N. and Dixon, F. J.,** Etiopathogenesis of murine SLE, *Immunol. Rev.,* 55, 179, 1981.
148. **Rosenberg, Y. J., Malek, T. R., Schaeffer, D. E., Santoro, T. J., Mark, G. E., Steinberg, A. D., and Mountz, J. D.,** Unusual expression of IL 2 receptors and both the c-*myb* and c-*raf* oncogenes in T cell lines and clones derived from autoimmune MRL-*lpr/lpr* mice, *J. Immunol.,* 134, 3120, 1985.
149. **Hersey, P., Bindon, C., Czerniecki, M., Spurling, A., Wass, J., and McCarthy, W. H.,** Inhibition of interleukin 2 production by factors released from tumor cells, *J. Immunol.,* 131, 2837, 1983.
150. **Farrar, W. L. and Anderson, W. B.,** Interleukin-2 stimulates association of protein kinase C with plasma membrane, *Nature (London),* 315, 233, 1985.
151. **Mire, A. R., Wickremasinghe, R. G., Michalevicz, R., and Hoffbrand, A. V.,** Interleukin-2 induces rapid phosphorylation of an 85 kilodalton protein in permeabilized lymphocytes, *Biochim. Biophys. Acta,* 847, 159, 1985.
152. **Farrar, W. L. and Taguchi, M.,** Interleukin 2 stimulation of protein kinase C membrane association: evidence for IL-2 receptor phosphorylation, *Lymphokine Res.,* 4, 87, 1985.
153. **Berridge, M. J.,** Inositol trisphosphate and diacylglycerol as second messengers, *Biochem. J.,* 220, 345, 1984.
154. **Donnelly, T. E., Jr., Sittler, R., and Scholar, E. M.,** Relationship between membrane-bound protein kinase C activity and calcium-dependent proliferation of BALB/c 3T3 cells, *Biochem. Biophys. Res. Commun.,* 126, 741, 1985.
155. **Trunch, A., Albert, F., Golstein, P., and Schmitt-Verhulst, A.-M.,** Early steps of lymphocyte activation bypassed by synergy between calcium ionophores and phorbol ester, *Nature (London),* 313, 318, 1985.
156. **Ebeling, J. G., Vandenbark, G. R., Kuhn, L. J., Ganong, B. R., Bell, R. M., and Niedel, J. E.,** Diacylglycerols mimic phorbol diester induction of leukemic cell differentiation, *Proc. Natl. Acad. Sci. U.S.A.,* 82, 815, 1985.
157. **Mills, G. B., Cragoe, E. J., Jr., Gelfand, E. W., and Grinstein, S.,** Interleukin 2 induces a rapid increase in intracellular pH through activation of a Na^+/H^+ antiport. Cytoplasmic alkalinization is not required for lymphocyte proliferation, *Proc. Natl. Acad. Sci. U.S.A.,* 260, 12500, 1985.
158. **Kaczmarek, L., Calabretta, B., and Baserga, R.,** Effect of interleukin-2 on the expression of cell cycle genes in human T lymphocytes, *Biochem. Biophys. Res. Commun.,* 133, 410, 1985.
159. **Kelly, K., Cochran, B. H., Stiles, C. D., and Leder, P.,** Cell-specific regulation of the c-*myc* gene by lymphocyte mitogens and platelet-derived growth factor, *Cell,* 35, 603, 1983.

160. **Persson, H., Hennighausen, L., Taub, R., DeGrado, W., and Leder, P.,** Antibodies to human c-*myc* oncogene product: evidence of an evolutionary conserved protein induced during cell proliferation, *Science,* 225, 687, 1984.
161. **Reed, J. C., Nowell, P. C., and Hoover, R. G.,** Regulation of c-*myc* mRNA levels in normal human lymphocytes by modulators of cell proliferation, *Proc. Natl. Acad. Sci. U.S.A.,* 82, 4221, 1985.
162. **Reed, J. C., Sabath, D. E., Hoover, R. G., and Prystowsky, M. B.,** Recombinant interleukin 2 regulates levels of c-*myc* mRNA in a cloned murine T lymphocyte, *Mol. Cell. Biol.,* 5, 3361, 1985.
163. **Rotter, V. and Wolf, D.,** Biological and molecular analysis of p53 cellular-encoded tumor antigen, *Adv. Cancer Res.,* 113, 1985.
164. **Mercer, W. E. and Baserga, R.,** Expression of the p53 protein during the cell cycle of human peripheral blood lymphocytes, *Exp. Cell Res.,* 160, 31, 1985.
165. **Gottlieb, M. S., Groopman, J. E., Weinstein, W. M., Fahey, J. L., and Detels, R.,** The acquired immunodeficiency syndrome, *Ann. Int. Med.,* 99, 208, 1983.
166. **Fauci, A. S., Macher, A. M., Longo, D. L., Lane, H. C., Rook, A. H., Masur, H., and Gelmann, E. P.,** Acquired immunodeficiency syndrome: epidemiologic, clinical, immunologic, and therapeutic considerations, *Ann. Int. Med.,* 100, 92, 1984.
167. **Wofsy, C. B. and Mills, J.,** The acquired immune deficiency syndrome: an international health problem of increasing importance, *Klin. Wochenschr.,* 62, 512, 1984.
168. **Fauci, A. S. and Lane, H. C.,** The acquired immunodeficiency syndrome (AIDS): an update, *Int. Arch. Allergy Appl. Immunol.,* 77, 81, 1985.
169. **Fauci, A. S., Masur, H., Gelmann, E. P., Markham, P. D., Hahn, B. H., and Lane, H. C.,** The acquired immunodeficiency syndrome: an update, *Ann. Int. Med.,* 102, 800, 1985.
170. **Curran, J. W., Morgan, W. M., Hardy, A. M., Jaffe, H. W., Darrow, W. W., and Dowdle, W. R.,** The epidemiology of AIDS: current status and future prospects, *Science,* 229, 1352, 1985.
171. **Gallo, R. C. and Wong-Staal, F.,** A human T-lymphotropic retrovirus (HTLV-III) as the cause of the acquired immunodeficiency syndrome, *Ann. Int. Med.,* 103, 679, 1985.
172. **Dorfman, R. F.,** Kaposi's sarcoma revisited, *Human Pathol.,* 15, 1013, 1984.
173. **Reuben, J. M., Hersh, E. M., Murray, J. L., Munn, C. G., Mehta, S. R., and Mansell, P. W. A.,** IL 2 production and response in vitro by the leukocytes of patients with acquired immune deficiency syndrome, *Lymphokine Res.,* 4, 103, 1985.
174. **Benveniste, E. N., Merrill, J. E., Kaufman, S. E., Golde, D. W., and Gasson, J. C.,** Purification and characterization of a human T-lymphocyte-derived glial growth-promoting factor, *Proc. Natl. Acad. Sci. U.S.A.,* 82, 3930, 1985.
175. **Korsmeyer, S. J. and Waldmann, T. A.,** Immunoglobulin genes: rearrangement and translocation in human lymphoid malignancy, *J. Clin. Immunol.,* 4, 1, 1984.
176. **Waldmann, T. A., Korsmeyer, S. J., Bakhshi, A., Arnold, A., and Kirsch, I. R.,** Molecular genetic analysis of human lymphoid neoplasms: immunoglobulin genes and the c-*myc* oncogene, *Ann. Int. Med.,* 102, 497, 1985.
177. **Muraguchi, A., Kehrl, J. H., Butler, J. L., and Fauci, A. S.,** Regulation of human B-cell activation, proliferation, and differentiation by soluble factors, *J. Clin. Immunol.,* 4, 337, 1984.
178. **Butler, J. L., Falkoff, R. J. M., and Fauci, A. S.,** Development of a human T-cell hybridoma secreting separate B-cell growth and differentiation factors, *Proc. Natl. Acad. Sci. U.S.A.,* 81, 2475, 1984.
179. **Hirano, T., Taga, T., Nakano, N., Yasukawa, K., Kashiwamura, S., Shimizu, K., Nakajima, K., Pyun, K. H., and Kishimoto, T.,** Purification to homogeneity and characterization of human differentiation factor (BCDF or BSFp-2), *Proc. Natl. Acad. Sci. U.S.A.,* 82, 5490, 1985.
180. **Muraguchi, A., Kehrl, J. H., Longo, D. L., Volkman, J., Smith, K. A., and Fauci, A. S.,** Interleukin 2 receptors on human B cells: implications for the role of interleukin 2 in human B cell function, *J. Exp. Med.,* 161, 181, 1985.
181. **Sidman, C. L., Marshall, J. D., Shultz, L. D., Gray, P. W., and Johnson, H. M.,** Gamma-interferon is one of several direct B cell-maturing lymphokines, *Nature (London),* 309, 801, 1984.
182. **Oliver, K., Noelle, R. J., Uhr, J. W., Krammer, P. H., and Vitetta, E. S.,** B-cell growth factor (B-cell growth factor I or B-cell-stimulating factor, provisional 1) is a differentiation factor for resting B cells and may not induce cell growth, *Proc. Natl. Acad. Sci. U.S.A.,* 82, 2465, 1985.
183. **Rabin, E. M., Ohara, J., and Paul, W. E.,** B-cell stimulatory factor 1 activates resting B cells, *Proc. Natl. Acad. Sci. U.S.A.,* 82, 2935, 1985.
184. **Ohara, J. and Paul, W. E.,** Production of a monoclonal antibody to and molecular characterization of B-cell stimulatory factor-1, *Nature (London),* 315, 333, 1985.
185. **Vitetta, E. S., Ohara, J., Myers, C. D., Layton, J. F., Krammer, P. H., and Paul, W. E.,** Serological, biochemical, and functional identity of B cell-stimulatory factor 1 and B cell differentiation factor for IgG1, *J. Exp. Med.,* 162, 1726, 1985.

186. **Blazar, B. A., Sutton, L. M., and Strome, M.,** Self-stimulating growth factor production by B-cell lines derived from Burkitt's lymphomas and other lines transformed *in vitro* by Epstein-Barr virus, *Cancer Res.*, 43, 4562, 1983.
187. **Gordon, J., Ley, S. C., Melamed, M. D., English, L. S., and Hughes-Jones, N. C.,** Immortalized B lymphocytes produce B-cell growth factor, *Nature (London)*, 310, 145, 1984.
188. **Ambrus, J. L., Jr., Jurgensen, C. H., Brown, E. J., and Fauci, A. S.,** Purification to homogeneity of a high molecular weight human B cell growth factor: demonstration of specific binding to activated B cells; and development of a monoclonal antibody to the factor, *J. Exp. Med.*, 162, 1319, 1985.
189. **Lazar, G. S., Quon, D. H., and Lusis, A. J.,** A gene-controlling response of bone marrow progenitor cells to granulocyte-macrophage colony stimulating factors, *J. Cell. Physiol.*, 124, 293, 1985.
190. **Stanley, E. R., Guilbert, L. J., Tushinski, R. J., and Bartelmez, S. H.,** CSF-1 — a mononuclear phagocyte lineage-specific hemopoietic growth factor, *J. Cell. Biochem.*, 21, 151, 1983.
191. **Waheed, A. and Shadduck, R. K.,** Purification of colony-stimulating factor by affinity chromatography, *Blood*, 60, 238, 1982.
192. **Das, S. K. and Stanley, E. R.,** Structure-function studies of a colony-stimulating factor (CSF-1), *J. Biol. Chem.*, 257, 13679, 1982.
193. **Ben-Avram, C. M., Shively, J. E., Shadduck, R. K., Waheed, A., Rajavashisth, T., and Lusis, A. J.,** Amino-terminal amino acid sequence of murine colony-stimulating factor 1, *Proc. Natl. Acad. Sci. U.S.A.*, 82, 4486, 1985.
194. **Sherr, C. J., Rettenmier, C. W., Sacca, R., Roussel, M. F., Look, A. T., and Stanley, E. R.,** The c-*fms* proto-oncogene product is related to the receptor for the mononuclear phagocyte growth factor, CSF-1, *Cell*, 41, 665, 1985.
195. **Manger, R., Najita, L., Nichols, E. J., Hakomori, S., and Rohrschneider, L.,** Cell surface expression of the McDonough strain of feline sarcoma virus *fms* gene product (gp140^{fms}), *Cell*, 39, 327, 1984.
196. **Rettenmier, C. W., Roussel, M. F., Quinn, C. O., Kitchingman, G. R., Look, A. T., and Sherr, C. J.,** Transmembrane orientation of glycoproteins encoded by the v-*fms* oncogene, *Cell*, 40, 971, 1985.
197. **Nichols, E. J., Manger, R., Hakomori, S., Herscovics, A., and Rohrschneider, L. R.,** Transformation by the v-*fms* oncogene product: role of glycosylational processing and cell surface expression, *Mol. Cell. Biol.*, 5, 3467, 1985.
198. **Rettenmier, C. W., Chen, J. H., Roussel, M. F., and Sherr, C. J.,** The product of the c-*fms* proto-oncogene: a glycoprotein with associated tyrosine kinase activity, *Science*, 228, 320, 1985.
199. **Woolford, J., Rothwell, V., and Rohrschneider, L.,** Characterization of the human c-*fms* gene product and its expression in cells of the monocyte-macrophage lineage, *Mol. Cell. Biol.*, 5, 3458, 1985.
200. **Slamon, D. J., de Kernion, J. B., Verma, I. M., and Cline, M. J.,** Expression of cellular oncogenes in human malignancies, *Science*, 224, 256, 1984.
201. **Verbeek, J. S., Roebroek, A. J. M., van den Ouwenland, A. M. W., Bloemers, H. P. J., and Van de Ven, W. J. M.,** Human c-*fms* proto-oncogene: comparative analysis with an abnormal allele, *Mol. Cell. Biol.*, 5, 422, 1985.
202. **Verbeek, J. S., van Heerikhuizen, H., de Pauw, B. E., Haanen, C., Bloemers, H. P. J., and Van de Ven, W. J. M.,** A hereditary abnormal c-*fms* proto-oncogene in a patient with acute lymphocytic leukaemia and congenital hypothyroidism, *Br. J. Haematol.*, 61, 135, 1985.
203. **Xu, D. Q., Guilhot, S., and Galibert, F.,** Restriction fragment length polymorphism of the human c-*fms* gene, *Proc. Natl. Acad. Sci. U.S.A.*, 82, 2862, 1985.
204. **Groffen, J., Heisterkamp, N., Spurr, N., Dana, S., Wasmuth, J. J., and Stephenson, J. R.,** Chromosomal localization of the human c-*fms* oncogene, *Nucleic Acids Res.*, 11, 6331, 1983.
205. **Kerkhofs, H., Hagemeijer, A., Leeksma, C. H. W., Abels, J., den Ottolander, G. J., Somers, R., Gerrits, W. B. J., Langenhuÿen, M. M. A. C., von dem Borne, A. E. G. K., Van Hemel, J. O., and Geraedts, J. P. M.,** the 5q-chromosome abnormality in haematological disorders: a collaborative study of 34 cases from the Netherlands, *Br. J. Haematol.*, 52, 365, 1982.
206. **Tinegate, H., Gaunt, L., and Hamilton, P. J.,** The 5q-syndrome: an underdiagnosed form of macrocytic anaemia, *Br. J. Haematol.*, 54, 103, 1983.
207. **Wisniewski, L. P. and Hirschhorn, K.,** Acquired partial deletions of the long arm of chromosome 5 in hematologic disorders, *Am. J. Hematol.*, 15, 295, 1983.
208. **Nienhuis, A. W., Bunn, H. F., Turner, P. H., Gopal, T. V., Nash, W. G., O'Brien, S. J., and Sherr, C. J.,** Expression of the human c-*fms* proto-oncogene in hematopoietic cells and its deletion in the 5q-syndrome, *Cell*, 42, 421, 1985.
209. **Stuart, R. K. and Hamilton, J. A.,** Tumor-promoting esters stimulate hematopoietic colony formation *in vitro*, *Science*, 208, 402, 1980.
210. **Chen, B. D.-M. and Wilkins, K. L.,** Role of phorbol ester receptors in the 12-O-tetradecanoyl-phorbol-13-acetate (TPA)-induced down-regulation of colony-stimulating factor (CSF-1) binding to murine peritoneal exudate macrophates, *J. Cell. Physiol.*, 124, 305, 1985.

211. **Sariban, E., Mitchell, T., and Kufe, D.,** Expression of the c-*fms* proto-oncogene during human monocytic differentiation, *Nature (London)*, 316, 64, 1985.
212. **Tushinski, R. J., Oliver, I. T., Guilbert, L. J., Tynan, P. W., Warner, J. R., and Stanley, E. R.,** Survival of mononuclear phagocytes depends on a lineage-specific growth factor that the differentiated cells selectively destroy, *Cell*, 28, 71, 1982.
213. **Muller, R., Curran, T., Müller, D., and Guilbert, L.,** Induction of c-*fos* during myelomonocytic differentiation and macrophage proliferation, *Nature (London)*, 314, 546, 1985.
214. **Sklar, M. D., Tereba, A., Chen, B. D.-M., and Walker, W. S.,** Transformation of mouse bone marrow cells by transfection with a human oncogene related to c-myc is associated with the endogenous production of macrophage colony stimulating factor 1, *J. Cell. Physiol.*, 125, 403, 1985.
215. **Burgess, A. W. and Metcalf, D.,** The nature and action of granulocyte-macrophage colony stimulating factors, *Blood*, 56, 947, 1980.
216. **Metcalf, D.,** The granulocyte-macrophage colony-stimulating factors, *Cell*, 43, 5, 1985.
217. **Pluznik, D. H., Cunningham, R. E., and Noguchi, P. D.,** Colony-stimulating factor (CSF) controls proliferation of CSF-dependent cells by acting during the G_1 phase of the cell cycle, *Proc. Natl. Acad. Sci. U.S.A.*, 81, 7451, 1984.
218. **Gasson, J. C., Weisbart, R. H., Kaufman, S. E., Clark, S. C., Hewick, R. M., Wong, G. G., and Golde, D. W.,** Purified human granulocyte-macrophage colony-stimulating factor: direct action on neutrophils, *Science*, 226, 1339, 1984.
219. **Weisbart, R. H., Golde, D. W., Clark, S. C., Wong, G. G., and Gasson, J. C.,** Human granulocyte-macrophage colony-stimulating factor is a neutrophil activator, *Nature (London)*, 314, 361, 1985.
220. **Gough, N. M., Gough, J., Metcalf, D., Kelso, A., Grail, D., Nicola, N. A., Burgess, A. W., and Dunn, A. R.,** Molecular cloning of cDNA encoding a murine haematopoietic growth regulator, granulocyte-macrophage colony-stimulating factor, *Nature (London)*, 309, 763, 1984.
221. **DeLarmarter, J. F., Mermod, J.-J., Liang, C.-M., Eliason, J. F., and Thatcher, D. R.,** Recombinant murine GM-CSF from *E. coli* has biological activity and is neutralized by a specific antiserum, *EMBO J.*, 4, 2575, 1985.
222. **Wong, G. G., Witek, J. S., Temple, P. A., Wilkens, K. M., Leary, A. C., Luxenberg, D. P., Jones, S. S., Brown, E. L., Kay, R. M., Orr, E. C., Shoemaker, C., Golde, D. W., Kaufman, R. J., Hewick, R. M., Wang, E. A., and Clark, S. C.,** Human GM-CSF: molecular cloning of the complementary DNA and purification of the natural and recombinant proteins, *Science*, 228, 810, 1985.
223. **Lee, F., Yokota, T., Otsuka, T., Gemmell, L., Larson, N., Luh, J., Arai, K., and Rennick, D.,** Isolation of cDNA for a human granulocyte-macrophage colony-stimulating factor by functional expression in mammalian cells, *Proc. Natl. Acad. Sci. U.S.A.*, 82, 4360, 1985.
224. **Cantrell, M. A., Anderson, D., Cerretti, D. P., Price, V., McKereghan, K., Tushinski, R. J., Mochizuki, D. Y., Larsen, A., Grabstein, K., Gillis, S., and Cosman, D.,** Cloning, sequence, and expression of a human granulocyte/macrophage colony-stimulating factor, *Proc. Natl. Acad. Sci. U.S.A.*, 82, 6250, 1985.
225. **Sieff, C. A., Emerson, S. G., Donahue, R. E., Nathan, D. G., Wang, E. A., Wong, G. G., and Clark, S. C.,** Human recombinant granulocyte-macrophage colony-stimulating factor: a multilineage hematopoietin, *Science*, 230, 1171, 1985.
226. **Miyatake, S., Otsuka, T., Yokota, T., Lee, F., and Arai, K.,** Structure of the chromosomal gene for granulocyte-macrophage colony stimulating factor: comparison of the mouse and human genes, *EMBO J.*, 4, 2561, 1985.
227. **Stanley, E., Metcalf, D., Sobieszczuk, P., Gough, N. M., and Dunn, A. R.,** The structure and expression of the murine gene encoding granulocyte-macrophage colony stimulating factor: evidence for utilisation of alternative promoters, *EMBO J.*, 4, 2569, 1985.
228. **Huebner, K., Isobe, M., Croce, C. M., Golde, D. W., Kaufman, S. E., and Gasson, J. C.,** The human gene encoding GM-CSF is at 5q21-q32, the chromosome region deleted in the 5q-anomaly, *Science*, 230, 1282, 1985.
229. **Hines, D. L.,** Differentiation of Abelson murine leukemia virus-infected promonocytic leukemia cells, *Int. J. Cancer*, 36, 233, 1985.
230. **Gonda, T. J. and Metcalf, D.,** Expression of *myb*, *myc* and *fos* proto-oncogenes during the differentiation of a murine myeloid leukaemia, *Nature (London)*, 310, 249, 1984.
231. **Feldman, R. A., Gabrilove, J. L., Tam, J. P., Moore, M. A. S., and Hanafusa, H.,** Specific expression of the human cellular *fps/fes*-encoded protein NCP92 in normal and leukemic myeloid cells, *Proc. Natl. Acad. Sci. U.S.A.*, 82, 2379, 1985.
232. **Francis, G. E., Ho, A. D., Gray, D. A., Berney, J. J., Wing, M. A., Yaxley, J. J., Ma, D. D. F., and Hoffbrand, A. V.,** DNA strand breakage and ADP-ribosyl transferase mediated DNA ligation during stimulation of human bone marrow cells by granulocyte-macrophage colony stimulating activity, *Leuk. Res.*, 8, 407, 1984.

233. **Metcalf, D. and Nicola, N. A.**, Synthesis by mouse peritoneal cells to G-CSF, the differentiation inducer for myeloid leukemia cells: stimulation by endotoxin, M-CSF and multi-CSF, *Leuk. Res.*, 9, 35, 1985.
234. **Metcalf, D. and Nicola, N. A.**, Proliferative effects of purified granulocyte colony-stimulating factor (G-CSF) on normal mouse hemopoietic cells, *J. Cell. Physiol.*, 116, 198, 1983.
235. **Lopez, A. F., Nicola, N. A., Burgess, A. W., Metcalf, D., Battye, F. L., Sewell, W. A., and Vadas, M.**, Activation of granulocyte cytotoxic function by purified mouse colony-stimulating factors, *J. Immunol.*, 131, 2938, 1984.
236. **Nicola, N. A., Metcalf, D., Matsumoto, M., and Johnson, G. R.**, Purification of a factor inducing differentiation in murine myelomonocytic leukemia cells: identification as granulocyte colony-stimulating factor, *J. Biol. Chem.*, 258, 9017, 1983.
237. **Nicola, N. A. and Metcalf, D.**, Binding of ^{125}I-labeled granulocyte colony-stimulating factor to normal murine hemopoietic cells, *J. Cell. Physiol.*, 124, 313, 1985.
238. **Nicola, N. A., Begley, C. G., and Metcalf, D.**, Identification of the human analogue of a regulator that induces differentiation in murine leukaemic cells, *Nature (London)*, 314, 625, 1985.
239. **Uittenbogaart, C. H. and Fahey, J. L.**, Leukemia-derived growth factor (non-interleukin 2) produced by a human malignant T lymphoid cell line, *Proc. Natl. Acad. Sci. U.S.A.*, 79, 7004, 1982.
240. **Hays, E. F., Goodrum, D., Bessho, M., Kitada, S., and Uittenbogaart, C. H.**, Leukemia-derived growth factor (non-interleukin-2) produced by murine lymphoma T-cell lines, *Proc. Natl. Acad. Sci. U.S.A.*, 81, 7807, 1984.
241. **Graf, T. and Beug, H.**, Avian leukemia viruses: interaction with their target cells in vivo and in vitro, *Biochim. Biophys. Acta*, 516, 269, 1978.
242. **Beug, H., Hayman, M. J., and Graf, T.**, Myeloblasts transformed by avian acute leukemia virus E26 are hormone-dependent for growth and for the expression of a putative *myb*-containing protein, p135 E26, *EMBO J.*, 1, 1069, 1982.
243. **Adkins, B., Leutz, A., and Graf, T.**, Autocrine growth induced by *src*-related oncogenes in transformed chicken myeloid cells, *Cell*, 39, 439, 1984.
244. **Fidler, I. J. and Raz, A.**, The induction of tumoricidal capabilities in mouse and rat macrophages by lymphokines, in *Lymphokines*, Vol. 3, Pick, E., Ed., Academic Press, New York, 1981, 345.
245. **Baird, A., Mormede, P., and Böhlen, P.**, Immunoreactive fibroblast growth factor in cells of peritoneal exudate suggests its identity with macrophage-derived growth factor, *Biochem. Biophys. Res. Comm.*, 126, 358, 1985.
246. **Shimokado, K., Raines, E. W., Madtes, D. K., Barrett, T. R., Benditt, E. P., and Ross, R.**, A significant part of macrophage-derived growth factor consists of at least two forms of PDGF, *Cell*, 43, 277, 1985.
247. **Kleinerman, E. S., Zicht, R., Sarin, P. S., Gallo, R. C., and Fidler, I. J.**, Constitutive production and release of a lymphokine with macrophage-activating factor activity distinct from gamma-interferon by a human T-cell leukemia virus-positive cell line, *Cancer Res.*, 44, 4470, 1984.
248. **Clemens, M. J. and McNurlan, M. A.**, Regulation of cell proliferation and differentiation by interferon, *Biochem. J.*, 226, 345, 1985.
249. **Trinchieri, G. and Perussia, B.**, Immune interferon: a pleiotropic lymphokine with multiple effects, *Immunol. Today*, 6, 131, 1985.
250. **Johnson, H. M. and Torres, B. A.**, Peptide growth factors PDGF, EGF, and FGF regulate interferon-gamma production, *J. Immunol.*, 134, 2824, 1985.
251. **Heyns, A. du P., Eldor, A., Vlodavsky, I., Kaiser, N., Fridman, R., and Panet, A.**, The antiproliferative effect of interferon and the mitogenic action of growth factors are independent cell cycle events: studies with vascular smooth muscle cells and endothelial cells, *Exp. Cell Res.*, 161, 297, 1985.
252. **Jonak, G. J. and Knight, E., Jr.**, Selective reduction of c-*myc* mRNA in Daudi cells by human beta interferon, *Proc. Natl. Acad. Sci. U.S.A.*, 81, 1747, 1984.
253. **Einat, M., Resnitzky, D., and Kimchi, A.**, Close link between reduction of c-*myc* expression by interferon and G_0/G_1 arrest, *Nature (London)*, 313, 597, 1985.
254. **Knight, E., Jr., Anton, E. D., Fahey, D., Friedland, B. K., and Jonak, G. J.**, Interferon regulates c-*myc* gene expression in Daudi cells at the post-transcriptional level, *Proc. Natl. Acad. Sci. U.S.A.*, 82, 1151, 1985.
255. **Dani, C., Mechti, N., Piechaczyk, M., Lebleu, B., Jeanteur, P., and Blanchard, J. M.**, Increased rate of degradation of c-myc mRNA in interferon-treated Daudi cells, *Proc. Natl. Acad. Sci. U.S.A.*, 42, 4896, 1985.
256. **Matsui, T., Takahashi, R., Mihara, K., Nakagawa, T., Koizumi, T., Nakao, Y., Sugiyama, T., Fujita, T.**, Cooperative regulation of c-*myc* expression in differentiation of human promyelocytic leukemia induced by recombinant gamma-interferon and 1,25-dihydroxyvitamin D_3, *Cancer Res.*, 45, 4366, 1985.
257. **Samid, D., Chang, E. H., and Friedman, R. M.**, Development of transformed phenotype induced by a human *ras* oncogene is inhibited by interferon, *Biochem. Biophys. Res. Comm.*, 126, 509, 1985.

258. **Samid, D., Chang, E. H., and Friedman, R. M.**, Biochemical correlates of phenotypic reversion in interferon-treated mouse cells transformed by a human oncogene, *Biochem. Biophys. Res. Comm.*, 119, 21, 1984.
259. **Perucho, M. and Esteban, M.**, Inhibitory effect of interferon on the genetic and oncogenic transformation by viral and cellular genes, *J. Virol.*, 54, 229, 1985.
260. **Dubois, M.-F., Vignal, M., Le Cunff, M., and Chany, C.**, Interferon inhibits transformation of mouse cells by exogenous cellular or viral oncogenes, *Nature (London)*, 303, 433, 1983.
261. **Lin, S. L., Garber, E. A., Wang, E., Caliguiri, L. A., Schellekens, H., Goldberg, A. R., and Tamm, I.**, Reduced synthesis of pp60^{src} and expression of the transformation-related phenotype in interferon-treated Rous sarcoma virus-transformed rat cells, *Mol. Cell. Biol.*, 3, 1656, 1983.
262. **Soslau, G., Bogucki, A. R., Gillespie, D., and Hubbell, H. R.**, Phosphoproteins altered by antiproliferative doses of human interferon-beta in a human bladder carcinoma cell line, *Biochem. Biophys. Res. Comm.*, 119, 941, 1984.
263. **Fair, D. S., Jeffes, E. W. B., III, and Granger, G. A.**, Release of LT molecules with restricted physical heterogeneity by a continuous human lymphoid cell line *in vitro*, *Cell. Immunol.*, 16, 185, 1979.
264. **Evans, C. H.**, Lymphotoxin — an immunologic hormone with anticarcinogenic and antitumor activity, *Cancer Immunol. Immunother.*, 12, 181, 1982.
265. **Gray, P. W., Aggarwal, B. B., Benton, C. V., Bringman, T. S., Henzel, W. J., Jarrett, J. A., Leung, D. W., Moffat, B., Ng, P., Svedersky, L. P., Palladino, M. A., and Nedwin, G. E.**, Cloning and expression of cDNA for human lymphotoxin, a lymphokine with tumour necrosis activity, *Nature (London)*, 312, 721, 1984.
266. **Nedwin, G. E., Naylor, S. L., Sakaguchi, A. Y., Smith, D., Jarrett-Nedwin, J., Pennica, D., Goeddel, D. V., and Gray, P. W.**, Human lymphotoxin and tumour necrosis factor genes: structure, homology and chromosomal localization, *Nucleic Acids Res.*, 13, 6361, 1985.
267. **Nedwin, G. E., Jarrett-Nedwin, J., Smith, D. H., Naylor, S. L., Sakaguchi, A. Y., Goeddel, D. V., and Gray, P. W.**, Structure and chromosomal localization of the human lymphotoxin gene, *J. Cell. Biochem.*, 29, 171, 1985.
268. **Old, L. J.**, Tumor necrosis factor, *Science*, 230, 630, 1985.
269. **Carswell, E. A., Old, L. J., Kassel, R. L., Green, S., Fiore, N., and Williamson, B.**, An endotoxin-induced serum factor that causes necrosis of tumors, *Proc. Natl. Acad. Sci. U.S.A.*, 72, 3666, 1975.
270. **Stone-Wolff, D. S., Yip, Y. K., Chroboczek Kelker, H., Le, J., Henriksen-DeStefano, D., Rubin, B. Y., Rinderknecht, E., Aggarwal, B. B., and Vilcek, J.**, Interrelationships of human interferon-gamma with lymphotoxin and monocyte cytotoxin, *J. Exp. Med.*, 159, 828, 1984.
271. **Chroboczek Kelker, H., Oppenheim, J. D., Stone-Wolff, D., Henriksen-deStefano, D., Aggarwal, B. B., Stevenson, H. C., and Vilcek, J.**, Characterization of human tumor necrosis factor produced by peripheral blood monocytes and its separation from lymphotoxin, *Int. J. Cancer*, 36, 69, 1985.
272. **Pennica, D., Nedwin, G. E., Hayflick, J. S., Seeburg, P. H., Derynck, R., Palladino, M. A., Kohr, W. J., Aggarwal, B. B., and Goeddel, D. V.**, Human tumour necrosis factor: precursor structure, expression and homology to lymphotoxin, *Nature (London)*, 312, 724, 1984.
273. **Aggarwal, B. B., Kohr, W. J., Hass, P. E., Moffat, B., Spencer, S. A., Henzel, W. J., Bringman, T. S., Nedwin, G. E., Goeddel, D. V., and Harkins, R. N.**, Human tumor necrosis factor: production, purification, and characterization, *J. Biol. Chem.*, 260, 2345, 1985.
274. **Shirai, T., Yamaguchi, H., Ito, H., Todd, C. W., and Wallace, R. B.**, Cloning and expression in *Escherichia coli* of the gene for human tumour necrosis factor, *Nature (London)*, 313, 803, 1985.
275. **Wang, A. M., Creasey, A. A., Ladner, M. B., Lin, L. S., Strickler, J., Van Arsdell, J. N., Yamamoto, R., and Mark, D. F.**, Molecular cloning of the complementary DNA for human tumor necrosis factor, *Science*, 228, 149, 1985.
276. **Marmenout, A., Fransen, L., Tavernier, J., Van der Heyden, J., Tizard, R., Kawashima, E., Shaw, A., Johnson, M.-J., Semon, D., Müller, R., Ruysschaert, M.-R., Van Vliet, A., and Fiers, W.**, Molecular cloning and expression of human tumor necrosis factor and comparison with mouse tumor necrosis factor, *Eur. J. Biochem.*, 152, 515, 1985.
277. **Sugarman, B. J., Aggarwal, B. B., Hass, P. E., Figari, I. S., Palladino, M. A., Jr., and Shepard, H. M.**, Recombinant human tumor necrosis factor-alpha: effects on proliferation of normal and transformed cells in vitro, *Science*, 230, 943, 1985.
278. **Fransen, L., Müller, R., Marmenout, A., Tavernier, J., Van der Heyden, J., Kawashima, E., Chollet, A., Tizard, R., Van Heuverswyn, H., Van Vliet, A., Ruysschaert, M.-R., and Fiers, W.**, Molecular cloning of mouse tumour necrosis factor cDNA and its eukaryotic expression, *Nucleic Acids Res.*, 13, 4417, 1985.
279. **Pennica, D., Hayflick, J. S., Bringman, T. S., Palladino, M. A., and Goeddel, D. V.**, Cloning and expression in *Escherichia coli* of the cDNA for murine tumor necrosis factor, *Proc. Natl. Acad. Sci. U.S.A.*, 82, 6060, 1985.

280. **Kull, F. C., Jr., Jacobs, S., and Cuatrecasas, P.,** Cellular receptor for ^{125}I-labeled tumor necrosis factor: specific binding, affinity labeling, and relationship to sensitivity, *Proc. Natl. Acad. Sci. U.S.A.*, 82, 5756, 1985.
281. **Baglioni, C., McCandless, S., Tavernier, J., and Fiers, W.,** Binding of human tumor necrosis growth factor to high affinity receptors in HeLa and lymphoblastoid cells sensitive to growth inhibition, *J. Biol. Chem.*, 260, 13395, 1985.
282. **Aggarwal, B. B., Eessalu, T. E., and Hass, P. E.,** Characterization of receptors for human tumour necrosis factor and their regulation by gamma-interferon, *Nature (London)*, 318, 665, 1985.
283. **Beutler, B., Greenwald, D., Hulmes, J. D., Chang, M., Pan, Y.-C. E., Mathison, J., Ulevitch, R., and Cerami, A.,** Identity of tumour necrosis factor and the macrophage-secreted factor cachectin, *Nature (London)*, 316, 552, 1985.
284. **Perretta, M., Waissbluth, L., Ludwig, U., and Garrido, F.,** Hormonal control of RNA polymerases in rat bone marrow nuclei. The action of erythropoietin and testosterone, *Arch. Biol. Med. Exp.*, 13, 247, 1980.
285. **Perretta, M., Waissbluth, L., Ludwig, U., Garrido, F., Garrido, A., and Ronco, A. M.,** Different RNA species stimulated by testosterone and erythropoietin in isolated bone marrow nuclei obtained from polycythemic and anemic rats, *J. Steroid Biochem.*, 14, 537, 1981.
286. **Metcalf, D. and Nicola, N. A.,** The regulatory factors controlling murine erythropoiesis in vitro, in *Aplastic Anemia: Stem Cell Biology and Advances in Treatment,* Alan R. Liss, New York, 1984, 93.
287. **Gasson, J. C., Golde, D. W., Kaufman, S. E., Westbrook, S. E., Westbrook, C. A., Hewick, R. M., Kaufman, R. J., Wong, G. G., Temple, P. A., Leary, A. C., Brown, E. L., Orr, E. C., and Clark, S. C.,** Molecular characterization and expression of the gene encoding human erythroid-potentiating activity, *Nature (London)*, 315, 768, 1985.
288. **Patarca, R. and Haseltine, W. A.,** A major retroviral core protein related to EPA and TIMP, *Nature (London)*, 318, 390, 1985.
289. **Erslev, A. J., Caro, J., Birgegard, G., Silver, R., and Miller, O.,** The biogenesis of erythropoietin, *Exp. Hematol.*, 8 (Suppl. 8), 1, 1980.
290. **Goldwasser, E.,** Erythropoietin and its mode of action, *Blood Cells*, 10, 147, 1984.
291. **Goldwasser, E.,** Erythropoietin, *Blut*, 33, 135, 1976.
292. **Finch, C. A.,** Erythropoiesis, erythropoietin, and iron, *Blood*, 60, 1241, 1982.
293. **Jacobs, K., Shoemaker, C., Rudersdorf, R., Neill, S. D., Kaufman, R. J., Mufson, A., Seehra, J., Jones, S. S., Hewick, R., Fritsch, E. F., Kawakita, M., Shimizu, T., and Miyake, T.,** Isolation and characterization of genomic and cDNA clones of human erythropoietin, *Nature (London)*, 313, 806, 1985.
294. **Lin, F-K., Suggs, S., Lin, C-H., Browne, J. K., Smalling, R., Egrie, J. C., Chen, K. K., Fox, G. M., Martin, F., Stabinsky, Z., Badrawi, S. M., Lai, P-H., and Goldwasser, E.,** Cloning and expression of the human erythropoietin gene, *Proc. Natl. Acad. Sci. U.S.A.*, 82, 7580, 1985.
295. **Congote, L. F.,** The N-terminal portion of an erythropoietin-like peptide from fetal bovine serum has sequence homology with the LDL receptor and two proteins of the Epstein-Barr virus, *Biochem. Biophys. Res. Comm.*, 133, 404, 1985.
296. **Okabe, T., Urabe, A., Kato, T., Chiba, S., and Takaku, F.,** Production of erythropoietin-like activity by human renal and hepatic carcinomas in cell culture, *Cancer*, 55, 1918, 1985.
297. **Greenberg, B. R., Hirasuna, J. D., and Woo, L.,** In vitro response to erythropoietin in erythroblastic transformation of chronic myelogenous leukemia, *Exp. Hematol.*, 8, 52, 1980.
298. **Hankins, W. D. and Scolnick, E. M.,** Harvey and Kirsten sarcoma viruses promote the growth and differentiation of erythroid precursor cells *in vitro*, *Cell*, 26, 91, 1981.
299. **Horoszewicz, J. S., Leong, S. S., and Carter, W. A.,** Friend leukemia: rapid development of erythropoietin-independent hematopoietic precursors, *J. Natl. Cancer Inst.*, 54, 265, 1975.
300. **Harrison, M. L., Low, P. S., and Geahlen, R. L.,** T and B lymphocytes express distinct tyrosine protein kinases, *J. Biol. Chem.*, 259, 9348, 1984.
301. **Ferrari, S., Torelli, U., Selleri, L., Donelli, A., Venturelli, D., Narni, F., Moretti, L., and Torelli, G.,** Study of the levels of expression of two oncogenes, c-*myc* and c-*myb*, in acute and chronic leukemias of both lymphoid and myeloid lineage, *Leuk. Res.*, 9, 833, 1985.
302. **Smeland, E., Godal, T., Ruud, E., Beiske, K., Funderud, S., Clark, E. A., Pfeifer-Ohlsson, S., and Ohlsson, R.,** The specific induction of *myc* protooncogene expression in normal human B cells is not a sufficient event for acquisition of competence to proliferate, *Proc. Natl. Acad. Sci. U.S.A.*, 82, 6255, 1985.
303. **Krönke, M., Leonard, W. J., Depper, J. M., and Greene, W. C.,** Sequential expression of genes involved in human T lymphocyte growth and differentiation, *J. Exp. Med.*, 161, 1593, 1985.
304. **Reed, J. C., Nowell, P. C., and Hoover, R. G.,** Regulation of c-*myc* mRNA levels in normal human lymphocytes by modulators of cell proliferation, *Proc. Natl. Acad. Sci. U.S.A.*, 82, 4221, 1985.
305. **Müller, R., Müller, D., and Guilbert, L.,** Differential expression of c-*fos* in hematopoietic cells: correlation with differentiation of monomyelocytic cells *in vitro*, *EMBO J.*, 3, 1887, 1984.

306. **Kelly, K. and Siebenlist, U.**, The role of c-*myc* in the proliferation of normal and neoplastic cells, *J. Clin. Immunol.*, 5, 65, 1985.
307. **Mitchell, R. L., Zokas, L., Schreiber, R. D., and Verma, I. M.**, Rapid induction of the expression of proto-oncogene *fos* during human monocytic differentiation, *Cell*, 40, 209, 1985.
308. **Westin, E. H., Wong-Staal, F., Gelmann, E. P., Dalla Favera, R., Papas, T. S., Lautenberger, J. A., Eva, A., Reddy, E. P., Tronick, S. R., Aaronson, S. A., and Gallo, R. C.**, Expression of cellular homologues of retroviral *onc* genes in human hematopoietic cells, *Proc. Natl. Acad. Sci. U.S.A.*, 79, 2490, 1982.
309. **Reitsma, P. H., Rothberg, P. G., Astrinb, S. M., Trial, J., Bar-Shavit, Z., Hall, A., Teitelbaum, S. L., and Kahn, A. J.**, Regulation of *myc* gene expression in HL-60 leukaemia cells by a vitamin D metabolite, *Nature (London)*, 306, 492, 1983.
310. **Craig, R. W. and Bloch, A.**, Early decline in c-*myb* oncogene expression in the differentiation of human myeloblastic leukemia (ML-1) cells induced with 12-*O*-tetradecanoylphorbol 13-acetate, *Cancer Res.*, 44, 442, 1984.
311. **Grosso, L. E. and Pitot, H. C.**, The expression of the *myc* proto-oncogene in a dimethylsulfoxide resistant HL-60 cell line, *Cancer Lett.*, 22, 55, 1984.
312. **Grosso, L. E. and Pitot, H. C.**, Modulation of c-*myc* expression in the HL-60 cell line, *Biochem. Biophys. Res. Comm.*, 119, 473, 1984.
313. **Filmus, J. and Buick, R. N.**, Relationship of c-*myc* expression to differentiation and proliferation of HL-60 cells, *Cancer Res.*, 45, 822, 1985.
314. **Grosso, L. E. and Pitot, H. C.**, Transcriptional regulation of c-*myc* during chemically induced differentiation of HL-60 cultures, *Cancer Res.*, 45, 847, 1985.
315. **Watanabe, T., Sariban, E., Mitchell, T., and Kufe, D.**, Human c-*myc* and N-*ras* expression during induction of HL-60 cellular differentiation, *Biochem. Biophys. Res. Comm.*, 126, 999, 1985.
316. **Koeffler, H. P.**, Induction of differentiation of human acute myelogenous leukemia cells: therapeutic implications, *Blood*, 62, 709, 1983.
317. **Lachman, H. M. and Skoultchi, A. I.**, Expression of c-*myc* changes during differentiation of mouse erythroleukaemia cells, *Nature (London)*, 310, 592, 1984.
318. **Lachman, H. M., Hatton, K. S., Skoultchi, A. I., and Schildkraut, C. L.**, c-*myc* mRNA levels in the cell cycle change in mouse erythroleukemia cells following inducer treatment, *Proc. Natl. Acad. Sci. U.S.A.*, 82, 5323, 1985.
319. **Gambari, R., del Senno, L., Piva, R., Barbieri, R., Amelotti, F., Bernardi, F., Marchetti, G., Citarella, F., Tripodi, M., and Fantoni, A.**, Human leukemia K562 cells: relationship between hemin-mediated erythroid induction, cell proliferation, and expression of c-*abl* and c-*myc* oncogenes, *Biochem. Biophys. Res. Comm.*, 125, 90, 1984.
320. **Müller, R., Curran, T., Müller, D., and Guilbert, L.**, Induction of c-*fos* during myelomonocytic differentiation and macrophage proliferation, *Nature (London)*, 314, 546, 1985.
321. **Oliff, A., Oliff, I., Schmidt, B., and Famulari, N.**, Isolation of immortal cell lines from the first stage of murine leukemia virus-induced leukemia, *Proc. Natl. Acad. Sci. U.S.A.*, 81, 5464, 1984.
322. **Pimentel, E.**, *Oncogenes*, CRC Press, Boca Raton, Fla., 1986.

Chapter 9

PLATELET-DERIVED GROWTH FACTOR

I. INTRODUCTION

Platelets contain and release several products (PDGF, platelet factor-4, beta-thromboglobulin, platelet basic proteins, platelet-derived vascular permeability factor) capable of exerting direct stimulatory actions on normal and tumor cells and may influence tumor cell growth and metastasization.[1] PDGF is a major growth factor present in clotted blood serum but not in plasma.[2-10] It is synthesized in the megakaryocyte, transported in blood by the alpha-granules of platelets and is released during blood clotting. Small amounts of PDGF (average of 17.5 ng/mℓ) are present in human serum,[11] where it is bound to alpha$_2$-macroglobulin (alpha$_2$M) in form of a PDGF-alpha$_2$M complex.[12,13] The possible physiological role of this complex has not been established but it may contribute to regulate the action of PDGF upon binding of PDGF to its cellular receptor and to prevent the entrance of high amounts of free PDGF into the systemic circulation from sites of injury. The levels of PDGF in plasma are undetectable. PDGF is cleared very rapidly from the plasma compartment after release from the platelet. The half-life of radio-iodinated purified human PDGF injected intravenously into baboons is less than 2 min.[11]

A polypeptide of approximately 35,000 daltons, named colostrum basic growth factor (CBGF), has been identified in the colostrum of goats, cows, and sheep.[14] Although the functions of CBGF are unknown, its chemical and biological properties are very similar to those of PDGF. A PDGF-like substance, termed PDGF-c, secreted by cultured rat aortic smooth muscle cells (ASMC), is developmentally regulated, being secreted by ASMC isolated from 13- to 18-day-old rats (pups) but not from 3-month-old animals (adults).[15] A similar or identical PDGF-like product is synthesized by adult rat arterial smooth muscle cells in primary culture when modulated from contractile to synthetic phenotype, which gives the cells the ability to synthesize DNA and divide upon stimulation with serum or growth factors.[16] The possibility that an altered regulation of PDGF-c expression contributes to the pathogenesis of arteriosclerosis should be considered.

II. STRUCTURE OF PDGF

Human PDGF is a glycoprotein of approximately 30,000 mol wt which is composed of two peptide chains, termed PDGF A chain or PDGF-1 and PDGF B chain or PDGF-2, respectively, linked by disulfide bonds.[7,8,17-19] The two chains of PDGF are of about the same size and sequence homology of approximately 60% exists between them, which suggests that their gene sequences have a common ancestral origin.[20,21] The dimer structure is important for the biological activity of the molecule since reduction irreversibly inactivates PDGF.

PDGF may consist of several forms of different molecular weight. Two active fractions, termed PDGF-I and PDGF-II, have been obtained by gel filtration,[22] but additional forms of PDGF may exist.[4,23,24] PDGF-I is a 31,000-dalton protein containing approximately 7% carbohydrate, and PDGF-II is also a glycoprotein but has a molecular weight of about 28,000 daltons and contains approximately 4% carbohydrate.[8,22] Both forms of PDGF have essentially equal mitogenic activity, amino acid composition, and immunological reactivity. Dissociation of the chemotactic and mitogenic activities of PDGF may be produced by human neutrophil elastase.[25] Porcine and human PDGF have similar physicochemical characteristics.[26,27]

III. FUNCTIONS OF PDGF

PDGF is indispensable for the growth in vitro of cells derived from connective tissue and it is the most potent mitogenic agent for cells of mesenchymal origin, including fibroblasts, vascular smooth muscle cells, and glial cells. The in vivo effects of PDGF are only partially understood but it stimulates different metabolic and functional cellular activities, including commitment to DNA synthesis and induction of proliferation in different types of cells under certain physiological conditions. PDGF and PDGF-like molecules are subjected to developmental regulation.[15] PDGF alone cannot initiate growth of quiescent cells but transient treatment of cells with PDGF followed by plasma results in DNA synthesis and cell growth.[28] The plasma requirement of cultured mouse BALB/c 3T3 cells can be replaced by EGF and either insulin of IGF-I.[29] In the BALB/c 3T3 cell system PDGF can be considered as a "competence factor" since it stimulates the initial events in the replicative response by priming cells to respond to other peptide growth factors, called "progression factors".[6,30] The amount of PDGF per cell, rather than PDGF concentration, would regulate the mitogenic response.[31]

PDGF also stimulates chemotaxis of human neutrophils and monocytes,[32] and may have an important role in vital processes such as tissue regeneration, wound healing and inflammatory response at sites of injury or thrombosis. Expression of PDGF receptors at the cell surface is of fundamental importance for the initiation of the cellular actions of PDGF. In the bone marrow, stromal cells constitute a structural and functional support for hematopoiesis, and PDGF receptors, as well as EGF receptors, are expressed in these cells.[33] Interestingly, PDGF acts as a chemoattractant for the ciliated protozoan *Tetrahymena*, where it induces a rapid increase in RNA and DNA synthesis.[34] The results obtained in the latter experiments suggest that *Tetrahymena* may contain receptors for PDGF on the cell surface and that PDGF may represent part of an evolutionary ancient regulatory mechanism.

IV. THE PDGF RECEPTOR AND THE TRANSDUCTION OF THE PDGF SIGNAL

The actions of PDGF at the cellular level are initiated by interaction of the native PDGF molecule with specific receptors which are phosphotyrosine-containing glycoproteins of 160,000 to 180,000 mol wt located at the cell surface.[35,36] Interaction of PDGF with its cellular receptor is associated with activation of tyrosine-specific protein kinase in the PDGF-receptor complex.[37-40]

A. Structure of the PDGF Receptor

The PDGF receptor (purified to homogeneity by using an antiphosphotyrosine antibody in conjunction with lectin affinity chromatography) is a 180,000-dalton plasma membrane glycoprotein containing phosphotyrosine.[41] Different numbers of PDGF receptors are present in a wide variety of cells, including vascular smooth muscle cells and many types of connective tissue cells (human skin fibroblasts, mouse 3T3 fibroblasts), as well as in human tumor cells but no PDGF receptors have been detected in blood cells or epithelial cells.[9,42,43] Binding of PDGF to its cellular receptors on either normal or transformed cells can be blocked by protamine sulfate by competitive inhibition.[44] PDGF and protamine sulfate share several properties in common, both being extremely basic proteins with a strong predominance of basic amino acids.

B. Phosphorylation of the PDGF Receptor

It has been proposed that the PDGF receptor has a ligand-binding domain on the outside of the cell membrane and an effector domain on its inside.[9] The cytoplasmic part of the receptor complex has an associated tyrosine-specific kinase activity.[9,38] The PDGF-stimulated kinase activity is not dependent on Ca^{2+} or cyclic nucleotides but requires Mn^{2+}.[38] A 185-kdalton cellular protein, which probably represents the PDGF receptor itself, is rapidly autophosphorylated on tyrosine, and possibly also on serine, after PDGF stimulation.[45] Autophosphorylation of the PDGF receptor on tyrosine residues after PDGF binding has been ascertained by using monoclonal antibody to phosphotyrosine.[46]

C. Internalization and Processing of the PDGF Receptor

After binding to its receptor, PDGF is internalized and degraded in the lysosomes but the fate of the PDGF receptor after internalization is not understood.[47] PDGF receptors are subjected to down-regulation after incubation of cells with PDGF, which may serve to impede excessive cell proliferation by an eventually prolonged exposure to the growth factor. In addition, there are plasma PDGF-binding proteins capable of inhibiting binding of PDGF to its cell surface receptors.[13]

D. Transductional Mechanisms of Action of PDGF

It is generally accepted that tyrosine phosphorylation is critically involved in the transductional mechanisms of action of PDGF.[38,48] However, other cellular changes elicited by the formation of a PDGF-receptor complex could also have a role in this transduction, especially changes in phosphoinositide metabolism, Na^+/H^+ exchange, and Ca^{2+} mobilization. PDGF stimulates a rapid breakdown of polyphosphoinositides in fetal human fibroblasts and Swiss mouse 3T3 cells, with a severalfold increase in intracellular levels of inositol 1,4,5-trisphosphate.[49,50] This increase may induce mobilization of Ca^{2+} from intracellular stores. In fact, there is evidence that calcium maybe involved in mediating the cellular actions of PDGF. Release of Ca^{2+} from intracellular sequestration sites maybe a mechanism by which PDGF stimulates cell growth.[51] Addition of PDGF to quiescent cultures of mouse fibroblasts induces a rapid increase in the cytoplasmic free Ca^{2+} concentrations, and the calcium ionophores A23187 and ionomycin can mimic the actions of PDGF on these cells, including an increased expression of the c-*myc* proto-oncogene.[52] In addition to these effects, PDGF-induced generation of 1,2-diacylglycerol may activate a Na^+/H^+ antiport, which may result in alkalinization of the cytoplasm.[49] Phorbol ester (PMA) blocks the activation of Na^+/H^+ exchange by PDGF or serum.[53] Apparently, activation of Na^+/H^+ exchange by PDGF does not require the intermediate activation of protein kinase C. Further studies are required for a better characterization of the respective roles of tyrosine phosphorylation, changes in phosphoinositide metabolism, Ca^{2+} mobilization, protein kinase C activation, Na^+/H^+ exchange, and other cellular changes in the transductional mechanisms of action of PDGF.

Different determinants of the PDGF molecule may be related to different cellular responses after interaction with the specific cell-surface receptors. For example, reduction of the native 32,000-dalton PDGF to its constituent polypeptide chains (14,000- and 17,000-dalton components) causes a loss of the ability to stimulate cell proliferation but all the activity as a chemotactic agent for human neutrophiles and monocytes is wholly retained.[54] Neutrophil elastase abolishes the chemotactic activity of PDGF for fibroblasts but has no effect on its chemotactic activity for monocytes, or on its mitogenic activity for 3T3 cells or its capacity to bind to 3T3 cells.[25] The results obtained in these experiments suggest that the biological effects of PDGF can be modulated selectively by factors that might be released at the same sites as PDGF.

E. Effects of PDGF on EGF Action

A fundamental effect of PDGF in some cellular systems is to increase the sensitivity of cells to EGF.[55] PDGF mimics phorbol ester action on EGF receptor phosphorylation at threonine-654 in WI-38 human fetal lung fibroblasts.[56] Since threonine-654 is the major site of phosphorylation of the EGF receptor catalyzed by protein kinase C, these results are consistent with the hypothesis that PDGF does stimulate the activity of protein kinase C and that some of the effects of PDGF may be mediated by a rise in the cellular levels of 1,2-diacylglycerol and free Ca^{2+} resulting from the hydrolysis of phosphatidylinositol 4,5-bisphosphate. The effect of PDGF on the phosphorylation state of the EGF receptor is similar to that observed when cells are treated with phorbol ester or exogenous diacylglycerols. However, PDGF, phorbol esters and 1,2-diacylglycerol increase the phosphorylation of the EGF receptor at sites in addition to threonine-654, which suggests that these agents may affect the activity of cellular protein kinases other than kinase C but also capable of phosphorylating the EGF receptor. The physiological role of the phosphorylation of the EGF receptor at different sites is not understood. The effect of PDGF to cause phosphorylation of the EGF receptor is an example of the complex interactions existing between growth factors and proto-oncogene protein products.[56] The B chain of PDGF is derived from the c-*sis* proto-oncogene and the EGF receptor probably represents the product of the c-*erb*-B proto-oncogene.

Addition of PDGF or phorbol ester to cultured human fetal lung fibroblasts causes an inhibition of the EGF-dependent phosphorylation of the EGF receptor on tyrosine residues.[56] This result is in apparent paradox with data indicating that tyrosine phosphorylation of the EGF receptor is important for its activation and that PDGF and phorbol ester enhance the mitogenic effects of EGF.

In mouse BALB/c 3T3 cells and human foreskin fibroblasts PDGF action correlates with a substantial loss of EGF receptor expression on the cell surface.[6,57,58] This effect is unidirectional since EGF does not decrease PDGF receptors. It has been speculated that the synergistic action of PDGF and EGF may provide an important method of physiologic growth control in vivo through a PDGF-induced alteration of the sensitivity of cells to EGF.[6] However, there is no good evidence for the general validity of this assumption and at least some cellular effects of PDGF may be independent from EGF.[59]

V. POST-TRANSDUCTIONAL MECHANISMS OF ACTION OF PDGF

After binding to its cellular receptor PDGF initiates a cascade of events that may lead cells competent to DNA synthesis and proliferation under the stimulus of several hormones and growth factors. Biochemical, functional, and morphological changes maybe observed at both extranuclear and nuclear sites in susceptible cells after stimulation with PDGF.[4,9]

A. Extranuclear Effects of PDGF

PDGF induces pleiotropic effects in responsive cells such as cultured mouse fibroblasts, stimulating many different types of biochemical phenomena like ion transport, amino acid transport, glycolysis, and protein and prostaglandin synthesis. The cellular actions of PDGF are probably mediated by the tyrosine-specific protein kinase activity of the growth factor-receptor complex.[37-40] A number of cellular proteins are phosphorylated on tyrosine residues in response to PDGF.[60,61] Proteins phosphorylated after cell stimulation with PDGF have 300,000, 200,000, 115,000, 72,000, 54,000, 45,000, and 35,000 mol wt[45] but their respective functional properties are unknown. As it happens with other peptide growth factors, phosphorylation of the ribosomal protein S6 on serine residues is stimulated by PDGF.[62] Some of the cellular proteins phosphorylated by the action of PDGF may be involved in the intracellular transmission of the PDGF-induced mitogenic signal and may also be substrates for the protein products of viral or cellular oncogenes with a similar type of kinase activity.[40]

PDGF induces structural changes in the cytoskeleton, including rearrangement of preformed microfilaments. The rearrangement of microfilaments induced by PDGF in

BALB/c 3T3 cells is accompanied by phosphorylation of the myosin light chain, which could depend on the PDGF-associated kinase activity.[63] Rearrangements of cytoskeletal elements, such as microfilaments and microtubules, may be important in transducing the mitogenic response to PDGF and other growth factors. Rapid and reversible time- and dose-dependent alterations in the distribution of vinculin and actin are observed in 3T3 cells stimulated by PDGF.[64] Within 2.5 min after PDGF exposure, vinculin disappears from adhesion plaques while actin, in the form of stress fibers, becomes disrupted a few minutes thereafter. Adhesion plaques (focal contacts) are regions where actin microfilaments terminate in cells such as cultured fibroblasts and where the plasma membrane comes close to the underlaying substrates.[64] Actin microfilaments terminate in vinculin contained in adhesion plaques or focal contacts.[65] Human skin collagenase expression is stimulated by PDGF in vitro.[66]

B. Nuclear Effects of PDGF

Some actions of PDGF seem to be exerted at the transcriptional level, and it has been estimated that between 0.1 and 1.0% of 3T3 cell genes are regulated by PDGF.[67] Unfortunately, the functions of the PDGF-inducible genes are poorly understood. A particular species of RNA, termed 28H6 RNA, is increased when resting cultured mouse cells are either infected with SV40 or stimulated with PDGF.[68]

A protein, named pI (29,000 mol wt), is also induced in response to PDGF.[29] The synthesis of pI is dependent on RNA synthesis and reaches a maximum at 2 to 4 hr after stimulation with PDGF. pI is localized in the nucleus and could be responsible for some of the growth regulatory actions of PDGF. Spontaneously transformed BALB/c 3T3 cells (ST3T3) show constitutive synthesis of pI and do not have requirement for PDGF,[69] which also suggests that pI may mediate at least some actions of PDGF.

1. Effects of PDGF on c-fos Expression

There is evidence that mitogens, including PDGF, may be involved in cell-specific regulation of proto-oncogene expression. Some gene sequences regulated by PDGF have been cloned in molecular vectors,[67] and it has been demonstrated that the expression of both the c-*fos* oncogene and a c-*fos*-related gene in quiescent BALB/c 3T3 mouse fibroblasts is stimulated in a rapid and transient manner by PDGF.[70,71] Expression of the c-*fos* proto-oncogene is induced in cultured mouse fibroblasts by treatment with PDGF and is super-induced by a combined treatment of PDGF and cycloheximide.[70]

2. Effects of PDGF on c-myc Expression

Expression of c-*myc* may also be regulated by PDGF.[72,73] Quiescent BALB/c and NIH/3T3 cells show a rapid, approximately 40-fold increase in c-*myc* messenger RNA levels subsequent to treatment with PDGF but not with either EGF or insulin.[72,74] PDGF has been shown to stimulate phosphoinositide turnover to produce inositol trisphosphate, which then mobilizes Ca^{2+} from intracellular stores to the cytoplasm,[49] and this mechanism may be responsible for the increase of c-*myc* expression induced by PDGF.[52] The calcium ionophores, A23187 and ionomycin, can mimic the actions of PDGF and increase the c-*myc* mRNA levels. These results suggest that Ca^{2+} may serve as a messenger for PDGF-induced expression of the c-*myc* proto-oncogene.[52] Enhanced expression of c-*myc* is also induced very soon following the activation of T-lymphocytes with LPS.[72] In general, c-*myc* gene expression is stimulated by agents that activate protein kinase C.[75] Induction of c-*fos* expression by growth factors, including PDGF, precedes the activation of c-*myc*.[76]

Apparently, the c-*myc* protein acts as an intracellular competence factor, i.e., as a factor rendering cells competent for undergoing the G_0-S phase transition. Microinjection of the purified c-*myc* protein into mouse cells (Swiss 3T3 fibroblasts) stimulates DNA synthesis when the cells are exposed to platelet-poor plasma.[77] The presence in the medium of an antibody to PDGF abolishes DNA synthesis induced by PDGF, but the microinjected c-*myc* protein stimulates DNA synthesis even when its own antibody is present in the medium. Expression of c-*myc*, however, may not account for the mitogenic effect of PDGF, which would depend on other types of molecular mechanisms.[75]

3. Effects of PDGF on c-src Expression

The product of the c-*src* gene is modified during the cellular response to PDGF.[78] Addition of PDGF (or phorbol ester) to quiescent, serum-deprived chicken embryo cells stimulates the entering to cell cycle, which is accompanied by a two- to fourfold increase in phosphorylation of pp60^{c-src} on serine.[79] The amino-terminal end of the viral oncogene protein counterpart, pp60^{v-src}, is considered as essential for transformation, not necessarily in relation to effects on protein kinase activity. A possible candidate for the kinase responsible for the phosphorylation of pp60^{c-src} on serine is protein kinase C.

4. Effects of PDGF on the Expression of Non Proto-Oncogene Genes

In addition to proto-oncogenes like c-*fos* and c-*myc*, PDGF probably contributes to the regulation of other cellular genes involved in the control of cell growth. However, PDGF alone may not optimally stimulate DNA synthesis, and other growth factors present in plasma are required for the fulfillment of the complex biochemical events involved in cell cycle control. PDGF and other peptide growth factors, including EGF, may contribute to the regulation of IFN-γ production.[80] In turn, IFN-γ may inhibit the expression of genes regulated by PDGF.[81]

5. PDGF as a Competence Factor

A disection of the prereplicative phase of BALB/c 3T3 cells has led to the identification of two separate, peptide growth factor-controlled cell cycle events, called competence and progression.[6,28,30] A brief exposure to PDGF renders BALB/c 3T3 cells competent to DNA replication but such cells do not actually progress through the prereplicative phase unless exposed to plasma or a second set of peptide growth factors (progression factors). In the BALB/c 3T3 system PDGF acts as a competence factor and EGF or IGF-I may act as progression factors. It is not known, however, if this scheme can be applied to other cell systems, in particular to complex in vivo systems.

In cultured human foreskin fibroblasts, PDGF and EGF are interchangeable in the prereplicative phase: a subliminal pulse of one factor, followed by a subliminal pulse of the other factor leads to stimulation of DNA synthesis.[82] The latter findings suggest that, at least in the prereplicative phase of cultured human skin fibroblasts, PDGF and EGF may act through the induction of similar intracellular signals. Since both the PDGF receptor and the EGF receptor are protein kinases, one possibility is that both growth factors phosphorylate a common substrate(s) that is involved in the transmission of the mitogenic signals. Alternatively, PDGF and EGF may initiate different early events which converge to a common pathway in the prereplicative phase.[82] Activation of particular proto-oncogenes may be involved in the final pathway leading to the mitogenic action of growth factors.

The biochemical mechanisms responsible for the PDGF-induced competence state are unknown. Cyclic AMP may mediate some cellular actions of PDGF, especially those related to cellular proliferation. PDGF induces a striking accumulation of cyclic AMP in confluent and quiescent cultures of mouse 3T3 cells, which is probably mediated by increased synthesis of E-type prostaglandins.[83] Other possible mechanisms to be considered for PDGF action would consist in induction of the expression of specific proteins, like the pI protein,[29] or a proto-oncogene protein product. The c-*myc* protein, whose expression is induced by PDGF,[72,74] is a good candidate for mediating the competence state induced by PDGF. This suggestion is reinforced by the fact that microinjected c-*myc* protein can act as a competence factor.[77]

However, cells may remain competent in spite of expressing c-*myc* and c-*fos* genes at very low levels (i.e., at levels similar to those of quiescent cells), which suggests that these proto-oncogene products are not directly involved in the maintenance of the competence state.[84]

VI. PDGF AND THE *sis* ONCOGENE PROTEIN PRODUCT

Unlike normal fibroblasts, mouse and human fibroblasts transformed by the simian sarcoma virus (SSV) do not require exogenous PDGF for growth.[85] A region of virtual identity has been found in the amino acid sequences of purified human PDGF and p28$^{v\text{-}sis}$, the transforming protein product of SSV.[86-89] This finding suggests that SSV acquired primate cellular sequences which encode part of PDGF. Antisera to a synthetic peptide representing amino acid residues 139 to 155 of the predicted v-*sis* oncogene product recognize human PDGF.[90] SSV-transformed cells secrete a PDGF-like product into conditioned media and the secreted protein has been identified as p28$^{v\text{-}sis}$ or its cellular processed product.[8,91-93]

When normal cells are co-cultured with SSV-transformed cells, the PDGF receptors of the normal cells are down-regulated by factors released from the transformed cells, which suggests that SSV-transformed cells release a material that is functionally similar to PDGF.[94] A component that blocks binding of PDGF to its cellular receptors is suramin.[95] SSV-transformed cells have no detectable PDGF receptors on the surface but when they are exposed to suramin the receptors reappear on the cell surface and within 8 hr are present at the same levels as in control cells.[94] These results strongly suggest that the absence of PDGF receptors in SSV-transformed cells is due to down-regulation of the receptor by an autocrine mechanism involving the endogenous production of PDGF or a PDGF-like compound, which is the p28$^{v\text{-}sis}$ oncogene protein product. Antibodies against PDGF inhibit both cell proliferation and SSV-induced morphological transformation in human diploid fibroblasts,[96] which suggests that v-*sis*-induced transformation is due solely to the autocrine action of a PDGF agonist.

A. Comparative Structures of PDGF and p28$^{v\text{-}sis}$

Reduced forms of PDGF contain two polypeptides, PDGF A chain (PDGF-1) and PDGF B chain (PDGF-2), and p28$^{v\text{-}sis}$ has been shown to correspond in size and amino acid sequence to a PDGF B chain monomer.[89] Moreover, p28$^{v\text{-}sis}$ undergoes dimer formation and subsequent processing to a form analogous in structure to that of biologically active PDGF. While the entire PDGF-2 sequence is encompassed within the v-*sis* gene product, there are three regions of the predicted v-*sis* protein that are unrelated to PDGF.[97] These include the amino terminal portion of the viral oncogene product, which is encoded by sequences derived from the *env* gene of the helper virus, and two v-*sis* cell-derived regions immediately flanking the PDGF-2 homologous sequence.

The SSV-encoded oncogene protein, p28$^{v\text{-}sis}$, consists of 226 amino acid residues. The *N*-terminal 109 amino acid residues of PDGF-2 are virtually identical to residues 67 to 175 of p28$^{v\text{-}sis}$.[21,86,87] The region corresponding to PDGF starts at the serine residue in position 67, which follows a double basic (Lys-Arg) sequence at positions 65 to 66 and appears to be the processing point yielding a 18,056-dalton polypeptide of 160 residues, essentially the same size estimated for the PDGF-2 chain on the basis of sodium dodecyl sulfate (SDS)-polyacrylamide gel electrophoresis.[19] In further studies it was concluded that the biologically active SSV oncogene protein product p28$^{v\text{-}sis}$ is a homodimer consisting of two PDGF-2 chains linked together by disulfide bonds, and that its immunologic properties are identical to those of PDGF.[7] Predictions have been made on the conformation and antigenic determinants of the v-*sis* oncogene protein product homologous with human PDGF.[98] According to such predictions, the state of the art would be sufficient to aid artificial vaccine design by anti-p28^{sis} antibody-inducing synthetic peptides.

The v-*sis* oncogene has been cloned in bacteria where its p28$^{v\text{-}sis}$ product can be expressed with high efficiency under the control of strong phage transcriptional and translational signals.[99] Colonies of genetically manipulated *E. coli* expressing the v-*sis* oncogene product can be detected with monoclonal antibodies made against specific synthetic peptides.[100] p28$^{v\text{-}sis}$ protein produced by genetically manipulated bacteria inhibits PDGF binding to its receptor, presumably by directly competing for the binding sites.[101] The latter finding suggests that the predominant biological activity of the p28$^{v\text{-}sis}$ protein is the result of its interaction with the PDGF receptor, and it has been postulated that p28$^{v\text{-}sis}$ stimulates autocrine cell growth through PDGF cell-surface receptors.[102] SSV-transformed NIH/3T3 mouse fibroblasts respond mitogenically to PDGF and have a limited number of PDGF receptors on the cell surface, which appear to recognize p28$^{v\text{-}sis}$ as a ligand because partially purified p28$^{v\text{-}sis}$ competes with labeled PDGF for receptor binding.[103] It has been demonstrated that v-*sis* translational products are capable of specifically binding PDGF receptors, stimulating tyrosine phosphorylation of the receptors and inducing DNA synthesis in quiescent mouse fibroblasts.[104]

B. The c-*sis* Proto-Oncogene

The complete nucleotide sequence of a cDNA clone corresponding to the normal human c-*sis* gene and its 3' flanking cellular sequences has been reported.[105] In previous studies nucleotide sequence analysis demonstrated that the human c-*sis* proto-oncogene contains five exons and is a structural gene for PDGF.[106] DNA sequences of the isolated and cloned human c-*sis* gene are homologous to sequences present in five regions of the SSV v-*sis* oncogene.[21] The study of nucleotide sequences of the six regions within the normal human cellular proto-oncogene locus c-*sis* that corresponds to the entire transforming region of v-*sis* indicates that the predicted protein products of v-*sis* and c-*sis* are 93% homologous and suggest that c-*sis* would encode the B-chain of PDGF.[107] Analysis of the B-chain of PDGF demonstrated identity of amino acid sequences predicted for the human c-*sis* gene product through 109 residues, indicating that the human c-*sis* gene encodes a polypeptide precursor of the B-chain of PDGF.[8]

Comparative analysis of the human and feline c-*sis* proto-oncogenes indicates the existence of 5' human c-*sis* coding sequences that are not homologous to v-*sis*.[108] In both loci, similar unique DNA sequences were found upstream of the v-*sis* homologous region, and these sequences hybridized to a 4.2 kbp c-*sis* transcript in human lung tumor cells. *sis*-related mRNA transcripts of 4.2 kbp are present in a relatively high proportion of sarcoma cell lines and, less frequently, in glioma cell lines.[109] However, c-*sis* transcripts may be present in some normal tissues. The c-*sis* proto-oncogene is actively transcribed in the first trimester human placenta, especially in the highly proliferative and invasive cytotrophoblastic shell, where it parallels the distribution of c-*myc* transcripts.[110] The translation product of the c-*sis* gene transcripts in human placenta is a PDGF-like protein, as shown by the release of a PDGF receptor-competing activity into media conditioned by fresh explants of first trimester placenta. Moreover, cultured cytotrophoblasts display abundant high-affinity PDGF receptors and respond to exogenous PDGF by an activation of c-*myc* gene expression and induction of DNA synthesis.[110] The results suggest that the developing human placenta represents a case of autocrine growth regulation in a normal tissue. The c-*sis* gene is also normally expressed in human and bovine endothelial cells.[111] Cultured human endothelial cells express 4.3-kbp c-*sis*/PDGF-related RNA transcripts whose levels are increased in conditions favoring proliferation and decreased in conditions favoring differentiation.[112] The function(s) of c-*sis*/PDGF gene product(s) in endothelial cells is unknown but it could have some role in the normal development of the vessel wall and in physiological processes related to vascularization, as well as in the pathogenesis of atherosclerosis.

C. v-*sis*- and c-*sis*-Induced Transformation

The crucial question as to whether the transforming ability of the v-*sis* gene protein product, as compared with the apparent lack of such an ability in the PDGF-2 (PDGF B-chain) molecule, is due to qualitative or quantitative differences between the two molecules cannot be answered as yet. The of the v-*sis* protein differs in several structural aspects from PDGF-2, and such differences could be important for its oncogenic potential. The protein product of v-*sis* differs from PDGF in the presence of a short amino-terminal sequence derived from the viral *env* gene, in a number of amino acid substitutions, and in the lack of the PDGF A chain. Essential regions for c-*sis*-induced transformation are those encoding the amino-terminal hydrophobic sequence, which is presumably a signal sequence for insertion into or across the plasma membrane, and the region of predicted homology with PDGF.[105] Removal of the signal sequence from the v-*sis* protein results in loss of biological activity,[113] which could be hypothetically attributed to impossibility for localization of the oncogene protein at the cell membrane and interaction with the PDGF receptor. In contrast to PDGF, which is a classical secretory protein, the vast majority of the v-*sis* protein product cannot be secreted by the cell and remains cell associated.[114]

Expression of the normal human c-*sis*/PDGF-2 coding sequence can induce cellular transformation in specific experimental conditions.[115,116] A recombinant vector containing the PDGF chain B sequence (in the referred report PDGF chain B is termed PDGF chain A) is capable of inducing transformation of NIH/3T3 cells in contrast to the inability of PDGF itself to transform cells when applied externally, which may be due to the intracellular overproduction of PDGF chain B as a consequence of transcriptional activation by the SV40 promoter contained in the vector.[115] It is possible that the expression of PDGF-2 sequences in abnormal sites or abnormal tissues may contribute to the induction of malignant transformation. Further work is needed to elucidate the factors responsible for the oncogenic activity of the v-*sis* protein product and the apparent lack of this activity in the normal PDGF-2 protein product under natural conditions.

VII. PDGF AND PDGF-LIKE FACTORS IN RELATION TO NEOPLASIA

Platelets probably have an important role in the growth of primary and metastatic tumors,[1,117] and several types of antiplatelet substances have been tested in experimental model systems as possible agents for the prevention of metastasis.[118] Growth factors related to PDGF are present in tumor tissues and may be released by some human tumor cell lines, and the suggestion has been advanced that "an autocrine activation of the PDGF receptor may be operational in the growth of human tumors of mesenchymal or glial origin".[119]

A. Production of PDGF-Like Proteins by Normal Cells

A diversity of PDGF-like molecules is produced in a broad spectrum of cells from different types of tumors but similar or identical substances are produced by cells that are not oncogenically transformed, such as endothelial cells, normal diploid rat smooth muscle cells and human lung fibroblasts.[24,120] Cultured human endothelial cells express 4.3-kb c-*sis*/PDGF-related RNA transcripts, whose levels are increased in conditions favoring cellular proliferation and decreased in conditions favoring cellular differentiation.[112] Consequently, the possible role of c-*sis*/PDGF-related polypeptides in tumor growth remains little understood.

B. Production of PDGF or PDGF-Like Proteins by Transformed Cells

Cultured neoplastic cells may require less exogenous PDGF to grow to high saturation densities.[121] The molecular events related to this phenomenon have not been characterized but an obvious possibility is the endogenous production of PDGF or PDGF-like substances by the transformed cells. A protein similar or identical to PDGF is produced by the breast cancer cell line T47D.[122] Proteins resembling PDGF but with molecular weights ranging from 16 to 140 kdaltons are synthesized and secreted by human glioblastoma (A172) and fibrosarcoma (HT-1080) cells in culture.[123] These cell lines synthesize a 4.4-kbp mRNA that contains sequences from all the six identified exons of the human c-*sis* gene. Mouse neuroblastoma cells (Neuro-2A cells) express c-*sis* transcripts with concomitant production and secretion of a PDGF-like growth factor.[124]

A PDGF-like substance is also produced by a human osteosarcoma cell line (U-2 OS) that does not respond to exogenous PDGF.[125,126] However, U-2 OS cells possess PDGF receptors that can be unmasked by treatment of the cells with the polyanionic compound, suramin.[43] Apparently, U-2 OS cells satisfy their requirement for PDGF by endogenous production of the factor or similar factors,[127] which is consistent with the autocrine hypothesis of neoplastic transformation.[128] Moreover, mRNA from the U-2 OS cells hybridize with cDNA probes corresponding to the v-*sis* oncogene sequence.[129] These results suggest that the appearance and/or maintenance of a transformed phenotype in U-2 OS osteosarcoma cells involves a transcriptional activation of the c-*sis* proto-oncogene which results in the production of biologically active PDGF.

However, production of PDGF is not universal in human osteogenic sarcomas because cells from another human osteosarcoma cell line (MG-63) do not show endogenous secretion of PDGF-like growth factors.[129] MG-63 cells produce mitogenic activity which is not PDGF-like, have membrane receptors for PDGF, and respond to exogenous PDGF increasing the amino acid transport, DNA synthesis and cellular proliferation. Moreover, a 160,000 to 180,000-dalton protein present in the membrane of MG-63 cells, and probably corresponding to the EGF receptor, is phosphorylated on tyrosine residues after addition of exogenous PDGF.[129] In absence of PDGF a prominent alkali-stable phosphoprotein of 116,000 daltons is detected in MG-63 cells but the origin of the latter protein and its possible relationship to the activity of non-c-*sis*-related oncogene protein products remain uncharacterized.

PDGF-like growth factors are produced by mouse embryonal carcinoma cells.[130] These cells do not bind exogenously added PDGF, which could be attributed to either a lack of expression of PDGF receptors on the cell surface or to the occupancy and down-regulation of the receptors by the PDGF-like growth factors produced by the same cells in an autocrine fashion. In contrast, endoderm-like cells derived from embryonal carcinoma cells do not release PDGF-like growth factors into the medium but their process of differentiation is accompanied by a marked increase in ability to bind exogenously added PDGF.

Production of a PDGF-like substance, termed PDGF-c, maybe a common result of transformation induced by retroviruses (K-MuSV, M-MuSV, SSV) and DNA viruses (SV40 and adenovirus). PDGF-c is also present in cells spontaneously transformed in culture as well as in some human carcinoma cell lines (bladder carcinoma T24 and hepatoma HepG2) but is absent in methylcholanthrene-transformed mouse cells.[131] A marked decrease in the number of PDGF receptors is observed in the same cells.

NIH/3T3 cells transformed with the DNA oncogenic virus SV40 produce a factor which shares many biological properties with PDGF but which may be different from PDGF itself.[132] In contrast to PDGF, which is heat-stable basic protein, the SV40-NIH/3T3-derived factor is a heat-labile acidic protein and no c-*sis* transcripts are detectable in SV40-infected NIH/3T3 cells. The results suggest the existence of a family of polypeptides capable of acting via the PDGF receptor but which are the products of cellular genes different from the c-*sis*/PDGF gene. An FGF secreted by SV40-transformed baby hamster kidney cell line (Sv28) has been purified to almost homogeneity. The coincidence of the physical, biological, and immunological characteristics of this factor with PDGF strongly suggests that they are closely related in structure and maybe produced by activation of the c-*sis* proto-oncogene. The precise relationship between the PDGF-like substances produced by normal or transformed cells and p28$^{c\text{-}sis}$ remains to be established.

VIII. NON-PDGF-LIKE FACTORS PRODUCED BY PLATELETS

In addition to PDGF and PDGF-like substances, other peptide growth factors are present in platelets, including EGF and EGF-like peptides, TGF-β, a hepatocyte growth factor, an endothelial growth factor, and other growth factors which remain poorly identified.[9,133-137] Although PDGF alone does not induce transformation in non-neoplastic, contact-inhibited NRK fibroblasts, it is capable of inducing such transformation when it acts, in the presence of plasma, in concert with the other two growth factors derived from platelets (TGF-β and EGF-like peptides).[136] Platelets also contain factor(s) capable of producing specific inhibition of cellular proliferation and induction of cellular differentiation in certain human carcinoma cells under the conditions of in vitro culture.[138] Such factor(s) should be different of PDGF and their possible relation to oncogene products has not been determined.

IX. SUMMARY

Platelet-derived growth factor (PDGF) has been recognized as a major mitogenic factor present in serum. It is a dimer molecule composed of two polypeptide chains termed A (or 1) and B (or 2). PDGF is released into the blood by the alpha granules of platelets. PDGF and/or PDGF-like polypeptides are also produced by several types of normal and neoplastic cells. A close structural homology exists between PDGF chain B (PDGF-2) and the protein product of the v-*sis* oncogene. The factors responsible for the high oncogenic activity of the v-*sis* oncogene protein and the apparent lack of this activity in PDGF chain B may be qualitative and/or quantitative in nature but are still unknown. PDGF may be involved in regulating the expression of proto-oncogenes, in particular, the expression of c-*fos* and c-*myc*, which apparently involves a Ca^{2+}-mediated mechanism. Expression of c-*fos* and c-*myc*, however, may not account for the mitogenic effect of PDGF, which would depend on other mechanisms. Protein kinase activity with specificity for tyrosine residues, similar to the activity present in the protein products of the *src* gene family, is present in the cellular PDGF receptor protein.

REFERENCES

1. **Mehta, P.,** Potential role of platelets in the pathogenesis of tumor metastasis, *Blood,* 63, 55, 1984.
2. **Ross, R. and Vogel, A.,** The platelet-derived growth factor, *Cell,* 14, 203, 1978.
3. **Scher, C. D., Shepard, R. C., Antoniades, H. N., and Stiles, C. D.,** Platelet-derived growth factor and the regulation of the mammalian fibroblast cell cycle, *Biochim. Biophys. Acta,* 560, 217, 1979.
4. **Antoniades, H. N. and Williams, L. T.,** Human platelet-derived growth factor: structure and function, *Fed. Proc.,* 42, 2630, 1983.
5. **Stiles, C. D.,** The molecular biology of platelet-derived growth factor, *Cell,* 33, 653, 1983.
6. **O'Keefe, E. J. and Pledger, W. J.,** A model of cell cycle control: sequential events regulated by growth factors, *Mol. Cell. Endocrinol.,* 31, 167, 1983.
7. **Antoniades, H. N.,** Platelet-derived growth factor and malignant transformation, *Biochem. Pharmacol.,* 33, 2823, 1984.
8. **Deuel, T. F. and Huang, J. S.,** Platelet-derived growth factor: structure, function, and roles in normal and transformed cells, *J. Clin. Invest.,* 74, 669, 1984.
9. **Heldin, C.-H., Wasteson, A., and Westermark, B.,** Platelet-derived growth factor, *Mol. Cell. Endocrinol.,* 39, 169, 1985.
10. **Stiles, C. D.,** The biological role of oncogenes: insights from platelet-derived growth factor, *Cancer Res.,* 45, 5215, 1985.
11. **Bowen-Pope, D. F., Malpass, T. W., Foster, D. M., and Ross, R.,** Platelet-derived growth factor in vivo: levels, activity, and rate of clearance, *Blood,* 64, 458, 1984.

12. **Huang, J. S., Huang, S. S., and Deuel, T. F.,** Specific covalent binding of platelet-derived growth factor to human plasma alpha$_2$-macroglobulin, *Proc. Natl. Acad. Sci. U.S.A.,* 81, 342, 1984.
13. **Raines, E. W., Bowen-Pope, D. F., and Ross, R.,** Plasma binding proteins for platelet-derived growth factor that inhibit its binding to cell-surface receptors, *Proc. Natl. Acad. Sci. U.S.A.,* 81, 3424, 1984.
14. **Brown, K. D. and Blakeley, D. M.,** Partial purification and characterization of a growth factor present in goat's colostrum, *Biochem. J.,* 219, 609, 1984.
15. **Seifert, R. A., Schwartz, S. M., and Bowen-Pope, D. F.,** Developmentally regulated production of platelet-derived growth factor-like molecules, *Nature (London),* 311, 669, 1984.
16. **Nilsson, J., Sjölund, M., Palmberg, L., Thyberg, J., and Heldin, C.-H.,** Arterial smooth muscle cells in primary culture produce a platelet-derived growth factor-like protein, *Proc. Natl. Acad. Sci. U.S.A.,* 82, 4418, 1985.
17. **Heldin, C. H., Westermark, B., and Wasteson, A.,** Platelet-derived growth factor: purification and partial characterization, *Proc. Natl. Acad. Sci. U.S.A.,* 76, 3722, 1979.
18. **Johnsson, A., Heldin, C.-H., Westermark, B., and Wasteson, A.,** Platelet-derived growth factor. Identification of constituent polypeptide chain, *Biochem. Biophys. Res. Comm.,* 104, 66, 1982.
19. **Antoniades, H. N. and Hunkapiller, M. W.,** Human platelet-derived growth factor (PDGF): amino-terminal amino acid sequence, *Science,* 220, 963, 1983.
20. **Heldin, C.-H., and Westermark, B.,** Growth factors: mechanism of action and relation to oncogenes, *Cell,* 37, 9, 1984.
21. **Johnsson, A., Heldin, C.-H., Wasteson, A., Westermark, B., Deuel, T. F., Huang, J. S., Seeburg, P. H., Gray, A., Ullrich, A., Scrace, G., Stroobant, P., and Waterfield, M. D.,** The c-*sis* gene encodes a precursor of the B chain of platelet-derived growth factor, *EMBO J.,* 3, 921, 1984.
22. **Deuel, T. F., Huang, J. S., Profitt, R. T., Baenziger, J. U., Chang, D., and Kennedy, B. B.,** Human platelet-derived growth factor: purification and resolution into two active fractions, *J. Biol. Chem.,* 256, 8896, 1981.
23. **Raines, E. and Ross, R.,** Platelet-derived growth factor. I. High yield purification and evidence for multiple forms, *J. Biol. Chem.,* 257, 5154, 1982.
24. **Niman, H. L., Houghten, R. A., and Bowen-Pope, D. F.,** Detection of high molecular weight forms of platelet derived growth factor by sequence-specific antisera, *Science,* 226, 701, 1984.
25. **Senior, R. M., Huang, J. S., Griffin, G. L., and Deuel, T. F.,** Dissociation of the chemotactic and mitogenic activities of platelet-derived growth factor by human neutrophil elastase, *J. Cell Biol.,* 100, 351, 1985.
26. **Poggi, A., Rucinski, B., James, P., Holt, J. C., and Niewiarowski, S.,** Partial purification and characterization of porcine platelet-derived growth factor (PDGF), *Exp. Cell Res.,* 150, 436, 1984.
27. **Stroobant, P. and Waterfield, M. D.,** Purification and properties of porcine platelet-derived growth factor, *EMBO J.,* 3, 2963, 1984.
28. **Pledger, W. J., Stiles, C. D., Antoniades, H. N., and Scher, C. D.,** Induction of DNA synthesis in BALB/c-3T3 cells by serum components: re-evaluation of the commitment process, *Proc. Natl. Acad. Sci. U.S.A.,* 74, 4481, 1977.
29. **Olashaw, N. E. and Pledger, W. J.,** Association of platelet-derived growth factor-induced protein with nuclear material, *Nature (London),* 306, 272, 1983.
30. **Pledger, W. J., Stiles, C. D., Antoniades, H. N., and Scher, C. D.,** An ordered sequence of events is required before BALB/c-3T3 cells become committed to DNA synthesis, *Proc. Natl. Acad. Sci. U.S.A.,* 75, 2839, 1978.
31. **Scher, C. D., Whipple, A. P., Singh, J. P., and Pledger, W. J.,** Modulation of the platelet-derived growth factor induced replicative response, *J. Cell. Physiol.,* 123, 10, 1985.
32. **Deuel, T. F., Senior, R. M., Huang, J. S., and Griffin, G. L.,** Chemotaxis of monocytes and neutrophils to platelet-derived growth factor, *J. Clin. Invest.,* 69, 1046, 1982.
33. **Rosenfeld, M., Keating, A., Bowen-Pope, D. F., Singer, J. W., and Ross, R.,** Responsiveness of the *in vitro* hematopoietic microenvironment to platelet-derived growth factor, *Leuk. Res.,* 9, 427, 1985.
34. **Andersen, H. A., Flodgaard, H., Klenow, H., and Leick, V.,** Platelet-derived growth factor stimulates chemotaxis and nucleic acid synthesis in the protozoan *Tetrahymena, Biochim. Biophys. Acta,* 782, 437, 1984.
35. **Heldin, C. H., Westermark, B., and Wasteson, A.,** Specific receptors for platelet-derived growth factor on cells derived from connective tissue and glia, *Proc. Natl. Acad. Sci. U.S.A.,* 78, 3664, 1981.
36. **Bowen-Pope, D. F. and Ross, R.,** Platelet-derived growth factor. II. Specific binding to cultured cells, *J. Biol. Chem.,* 257, 5161, 1982.
37. **Ek, B., Westermark, B., and Heldin, C.-H.,** Stimulation of tyrosine-specific phosphorylation of platelet-derived growth factor, *Nature (London),* 295, 419, 1982.
38. **Ek, B. and Heldin, C.-H.,** Characterization of a tyrosine-specific kinase activity in human fibroblast membranes stimulated by platelet-derived growth factor, *J. Biol. Chem.,* 257, 10486, 1982.

39. **Nishimura, J., Huang, J. S., and Deuel, T. F.**, Platelet-derived growth factor stimulates tyrosine-specific protein kinase activity in Swiss mouse 3T3 cell membranes, *Proc. Natl. Acad. Sci. U.S.A.*, 79, 4303, 1982.
40. **Cooper, J. A., Bowen-Pope, D. F., Raines, E., Ross, R., and Hunter, T.**, Similar effects of platelet-derived growth factor and epidermal growth factor on the phosphorylation of tyrosine in cellular proteins, *Cell*, 31, 263, 1982.
41. **Daniel, T. O., Tremble, P. M., Frackelton, A. R., Jr., and Williams, L. T.**, Purification of the platelet-derived growth factor receptor by using an anti-phosphotyrosine antibody, *Proc. Natl. Acad. Sci. U.S.A.*, 82, 2684, 1985.
42. **Graves, D. T., Antoniades, H. N., Williams, S. R., and Owen, A. J.**, Evidence for functional platelet-derived growth factor receptors on MG-63 human osteosarcoma cells, *Cancer Res.*, 44, 2966, 1984.
43. **Graves, D. T., Owen, A. J., and Antoniades, H. N.**, Demonstration of receptors for a PDGF-like mitogen on human osteosarcoma cells, *Biochem. Biophys. Res. Comm.*, 129, 56, 1985.
44. **Huang, J. S., Nishimura, J., Huang, S. S., and Deuel, T. F.**, Protamine inhibits platelet derived growth factor receptor activity but not epidermal growth factor activity, *J. Cell Biochem.*, 26, 205, 1984.
45. **Ek, B. and Heldin, C.-H.**, Use of an antiserum against phosphotyrosine for the identification of phosphorylated components in human fibroblasts stimulated by platelet-derived growth factor, *J. Biol. Chem.*, 259, 11145, 1984.
46. **Frackelton, A. R., Jr., Tremble, P. M., and Williams, L. T.**, Evidence for the platelet-derived growth factor-stimulated tyrosine phosphorylation of the platelet-derived growth factor receptor *in vivo*: immunopurification using a monoclonal antibody to phosphotyrosine, *J. Biol. Chem.*, 259, 7909, 1984.
47. **Rosenfeld, M. E., Bowen-Pope, D. F., and Ross, R.**, Platelet-derived growth factor: morphologic and biochemical studies of binding, internalization, and degradation, *J. Cell. Physiol.*, 121, 263, 1984.
48. **Heldin, C.-H., Ek, B., and Rönnstrand, L.**, Characterization of the receptor for platelet-derived growth factor on human fibroblasts: demonstration of an intimate relationship with 185,000 dalton substrate for the PDGF receptor-kinase, *J. Biol. Chem.*, 258, 10054, 1983.
49. **Berridge, M. J., Heslop, J. P., Irvine, R. F., and Brown, K. D.**, Inositol trisphosphate formation and calcium mobilization in Swiss 3T3 cells in response to platelet-derived growth factor, *Biochem. J.*, 222, 195, 1984.
50. **Chu, S-H. W., Hoban, C. J., Owen, A. J., and Geyer, R. P.**, Platelet-derived growth factor stimulates polyphosphoinositide breakdown in fetal human fibroblasts, *J. Cell. Physiol.*, 124, 391, 1985.
51. **Frantz, C. N.**, Effects of platelet-derived growth factor on Ca^{2+} in 3T3 cells, *Exp. Cell Res.*, 158, 287, 1985.
52. **Tsuda, T., Kaibuchi, K., West, B., and Takai, Y.**, Involvement of Ca^{2+} in platelet-derived growth factor-induced expression of c-*myc* oncogene in Swiss 3T3 fibroblasts, *FEBS Lett.*, 187, 43, 1985.
53. **Whiteley, B., Deuel, T., and Glaser, L.**, Modulation of the activity of the platelet-derived growth factor receptor by phorbol myristate acetate, *Biochem. Biophys. Res. Comm.*, 129, 854, 1985.
54. **Williams, L. T., Antoniades, H. N., and Goetzl, E. J.**, Platelet-derived growth factor stimulates mouse 3T3 cell mitogenesis and leukocyte chemotaxis through different structural determinants, *J. Clin. Invest.*, 72, 1759, 1983.
55. **Wharton, W., Leof, E., Olashaw, N., O'Keefe, E. J., and Pledger, W. J.**, Mitogenic response to epidermal growth factor (EGF) modulated by platelet-derived growth factor in cultured fibroblasts, *Exp. Cell Res.*, 147, 443, 1983.
56. **Davis, R. J. and Czech, M. P.**, Platelet-derived growth factor mimics phorbol diester action on epidermal growth factor receptor phosphorylation at threonine-654, *Proc. Natl. Acad. Sci. U.S.A.*, 82, 4080, 1985.
57. **Wrann, M., Fox, C. F., and Ross, R.**, Modulation of epidermal growth factor receptors on 3T3 cells by platelet-derived growth factor, *Science*, 210, 1363, 1980.
58. **Heldin, C.-H., Wasteson, A., and Westermark, B.**, Interaction of platelet-derived growth factor with its fibroblast receptor: demonstration of ligand degradation and receptor modulation, *J. Biol. Chem.*, 257, 4216, 1982.
59. **Pledger, W. J., Hart, C. A., Locatell, K. L., and Scher, C. D.**, Platelet-derived growth factor-modulated proteins: constitutive synthesis by a transformed cell line, *Proc. Natl. Acad. Sci. U.S.A.*, 78, 4358, 1981.
60. **Nakamura, K. D., Martinez, R., and Weber, M. J.**, Tyrosine phosphorylation of specific proteins after mitogen stimulation of chicken embryo fibroblasts, *Mol. Cell. Biol.*, 3, 380, 1983.
61. **Harrington, M. A., Estes, J. E., Leof, E., and Pledger, W. J.**, PDGF stimulates transient phosphorylation of 180,000 dalton protein, *J. Cell Biochem.*, 27, 67, 1985.
62. **Nishimura, J. and Deuel, T. F.**, Stimulation of protein phosphorylation in Swiss mouse 3T3 cells by human platelet-derived growth factor, *Biochem. Biophys. Res. Comm.*, 103, 355, 1981.
63. **Bockus, B. J. and Stiles, C. D.**, Regulation of cytoskeletal architecture by platelet-derived growth factor, insulin and epidermal growth factor, *Exp. Cell Res.*, 153, 186, 1984.
64. **Herman, B. and Pledger, W. J.**, Platelet-derived growth factor-induced alterations in vinculin and actin distribution in BALB/c-3T3 cells, *J. Cell Biol.*, 100, 1031, 1985.

65. **Geiger, B.**, Membrane-cytoskeletal interaction, *Biochim. Biophys. Acta,* 737, 305, 1983.
66. **Bauer, E. A., Cooper, T. W., Huang, J. S., Altman, J. A., and Deuel, T. F.**, Stimulation of *in vitro* human skin collagenase expression by platelet-derived growth factor, *Proc. Natl. Acad. Sci. U.S.A.,* 82, 4132, 1985.
67. **Cochran, B. H., Reffel, A. C., and Stiles, C. D.**, Molecular cloning of gene sequences regulated by platelet-derived growth factor, *Cell,* 33, 939, 1983.
68. **Linzer, D. I. H. and Nathans, D.**, Growth-related changes in specific mRNAs of cultured mouse cells, *Proc. Natl. Acad. Sci. U.S.A.,* 80, 4271, 1983.
69. **LoBue, J. and LoBue, P. A.**, Control of cell growth, *Transplant. Proc.,* 16, 341, 1984.
70. **Cochran, B. H., Zullo, J., Verma, I. M., and Stiles, C. D.**, Expression of the c-*fos* gene and of an *fos*-related gene is stimulated by platelet-derived growth factor, *Science,* 226, 1080, 1984.
71. **Kruijer, W., Cooper, J. A., Hunter, T., and Verma, I. M.**, Platelet-derived growth factor induces rapid but transient expression of the c-*fos* gene and protein, *Nature (London),* 312, 711, 1984.
72. **Kelly, K., Cochran, B. H., Stiles, C. D., and Leder, P.**, Cell-specific regulation of the c-*myc* gene by lymphocyte mitogens and platelet-derived growth factor, *Cell,* 35, 603, 1983.
73. **Armelin, H. A., Armelin, M. C. S., Kelly, K., Stewart, T., Leder, P., Cochran, B. H., and Stiles, C. D.**, Functional role for c-*myc* in mitogenic response to platelet-derived growth factor, *Nature (London),* 310, 655, 1984.
74. **Kelly, K. and Siebenlist, U.**, The role of c-myc in the proliferation of normal and neoplastic cells, *J. Immunol.,* 5, 65, 1985.
75. **Coughlin, S. R., Lee, W. M. F., Williams, P. W., Giels, G. M., and Williams, L. T.**, c-*myc* gene expression is stimulated by agents that activate protein kinase C and does not account for the mitogenic action of PDGF, *Cell,* 43, 243, 1985.
76. **Müller, R., Bravo, R., Burckhardt, J., and Curran, T.**, Induction of c-*fos* gene and protein by growth factors precedes activation of c-*myc*, *Nature (London),* 312, 716, 1984.
77. **Kaczmarek, L., Hyland, J. K., Watt, R., Rosenberg, M., and Baserga, R.**, Microinjected c-*myc* as a competence factor, *Science,* 228, 1313, 1985.
78. **Ralston, R. and Bishop, J. M.**, The product of the proto-oncogene c-*src* is modified during the cellular response to platelet-derived growth factor, *Proc. Natl. Acad. Sci. U.S.A.,* 82, 7845, 1985.
79. **Tamura, T., Friis, R. R., and Bauer, H.**, pp60$^{c\text{-}src}$ is a substrate for phosphorylation when cells are stimulated to enter cycle, *FEBS Lett.,* 177, 151, 1984.
80. **Johnson, H. M. and Torres, B. A.**, Peptide growth factors PDGF, EGF, and FGF regulate interferon-gamma production, *J. Immunol.,* 134, 2824, 1985.
81. **Einat, M., Resnitzky, D., and Kimchi, A.**, Inhibitory effects of interferon on the expression of genes regulated by platelet-derived growth factors, *Proc. Natl. Acad. Sci. U.S.A.,* 82, 7608, 1985.
82. **Westermark, B. and Heldin, C.-H.**, Similar action of platelet-derived growth factor and epidermal growth factor in the prereplicative phase of human fibroblasts suggests a common intracellular pathway, *J. Cell. Physiol.,* 124, 43, 1985.
83. **Rozengurt, E., Stroobant, P., Waterfield, M. D., Deuel, T. F., and Keehan, M.**, Platelet-derived growth factor elicits cyclic AMP accumulation in Swiss 3T3 cells: role of prostaglandin production, *Cell,* 34, 265, 1983.
84. **Bravo, R., Burckhardt, J., and Müller, R.**, Persistence of the competent state in mouse fibroblasts is independent of c-*fos* and c-*myc* expression, *Exp. Cell Res.,* 160, 540, 1985.
85. **Scher, C. D., Pledger, W. J., Martin, P., Antoniades, H. N., and Stiles, C. D.**, Transforming viruses directly reduce the cellular growth requirement for a platelet-derived growth factor, *J. Cell. Physiol.,* 97, 371, 1978.
86. **Waterfield, M. D., Scrace, G. T., Whittle, N., Stroobant, P., Johnsson, A., Wasteson, A., Westermark, B., Heldin, C.-H., Huang, J. S., and Deuel, T. F.**, Platelet-derived growth factor is structurally related to the putative transforming protein p28^{sis} of simian sarcoma virus, *Nature (London),* 304, 35, 1983.
87. **Doolittle, R. F., Hunkapiller, M. W., Hood, L. E., Devare, S. G., Robbins, K. C., Aaronson, S. A., and Antoniades, H. N.**, Simian sarcoma *onc* gene, v-*sis*, is derived from the gene (or genes) encoding a platelet-derived growth factor, *Science,* 221, 275, 1983.
88. **Deuel, T. F., Huang, J. S., Huang, S. S., Stroobant, P., and Waterfield**, Expression of a platelet-derived growth factor-like protein in simian sarcoma virus transformed cells, *Science,* 221, 1348, 1983.
89. **Robbins, K. C., Antoniades, H. N., Devare, S. G., Hunkapiller, M. W., and Aaronson, S. A.**, Structural and immunological similarities between simian sarcoma virus gene product(s) and human platelet-derived growth factor, *Nature (London),* 305, 605, 1983.
90. **Niman, H. L.**, Antisera to a synthetic peptide of the *sis* viral oncogene product recognize human platelet-derived growth factor, *Nature (London),* 307, 180, 1984.
91. **Thiel, H.-J. and Hafenrichter, R.**, Simian sarcoma virus transformation-specific glycopeptide: immunological relationship to human platelet-derived growth factor, *Virology,* 136, 414, 1984.

92. **Owen, A. J., Pantazis, P., and Antoniades, H. N.**, Simian sarcoma virus-transformed cells secrete a mitogen identical to platelet-derived growth factor, *Science*, 225, 54, 1984.
93. **Johnsson, A., Betsholtz, C., von der Helm, K., Heldin, C.-H., and Westermark, B.**, Platelet-derived growth factor agonist activity of a secreted form of the v-*sis* oncogene product, *Proc. Natl. Acad. Sci. U.S.A.*, 82, 1721, 1985.
94. **Garrett, J. S., Coughlin, S. R., Niman, H. L., Tremble, P. M., Giels, G. M., and Williams, L. T.**, Blockade of autocrine stimulation in simian sarcoma virus-transformed cells reverses down-regulation of platelet-derived growth factor receptors, *Proc. Natl. Acad. Sci. U.S.A.*, 81, 7466, 1984.
95. **Hosang, M.**, Suramin binds to platelet-derived growth factor and inhibits its biological activity, *J. Cell Biochem.*, 29, 265, 1985.
96. **Johnsson, A., Betsholtz, C., Heldin, C.-H., and Westermark, B.**, Antibodies against platelet-derived growth factor inhibit acute transformation by simian sarcoma virus, *Nature (London)*, 317, 438, 1985.
97. **King, C. R., Giese, N. A., Robbins, K. C., and Aaronson, S. A.**, In vitro mutagenesis of the v-*sis* transforming gene defines functional domains of its growth factor-related product, *Proc. Natl. Acad. Sci. U.S.A.*, 82, 5295, 1985.
98. **Robson, B., Platt, E., Finn, P. W., Millard, P., Gibrat, J.-F., and Garnier, J.**, Predictions of the conformation and antigenic determinants of the v-sis viral oncogene product homologous with human platelet-derived growth factor, *Int. J. Pept. Protein Res.*, 25, 1, 1985.
99. **Devare, S. G., Shatzman, A., Robbins, K. C., Rosenberg, M., and Aaronson, S. A.**, Expression of the PDGF-related transforming protein of simian sarcoma virus in E. coli, *Cell*, 36, 43, 1984.
100. **Kennett, R. H., Leunk, R., Meyer, B., and Silenzio, V.**, Detection of *E. coli* colonies expressing the v-sis oncogene product with monoclonal antibodies made against synthetic peptides, *J. Immunol. Methods*, 85, 169, 1985.
101. **Wang, J. Y. J. and Williams, L. T.**, A v-*sis* oncogene protein produced in bacteria competes for platelet-derived growth factor binding to its receptor, *J. Biol. Chem.*, 259, 10645, 1984.
102. **Huang, J. S., Huang, S. S., and Deuel, T. F.**, Transforming protein of simian sarcoma virus stimulates autocrine growth of SSV-transformed cells through PDGF cell-surface receptors, *Cell*, 39, 79, 1984.
103. **Deuel, T. F. and Huang, J. S.**, Roles of growth factor activities in oncogenesis, *Blood*, 64, 951, 1984.
104. **Leal, F., Williams, L. T., Robbins, K. C., and Aaronson, S. A.**, Evidence that the v-*sis* gene product transforms by interaction with the receptor for platelet-derived growth factor, *Science*, 230, 327, 1985.
105. **Ratner, L., Josephs, S. F., Jarrett, R., Reitz, M. S., Jr., and Wong-Staal, F.**, Nucleotide sequence of transforming human c-*sis* cDNA clones with homology to platelet-derived growth factor, *Nucleic Acids Res.*, 13, 5007, 1985.
106. **Chiu, I-M., Reddy, E., Givol, D., Robbins, K. C., Tronick, S. R., and Aaronson, S. A.**, Nucleotide sequence analysis identifies the human c-*sis* proto-oncogene as a structural gene for platelet-derived growth factor, *Cell*, 37, 123, 1984.
107. **Josephs, S. F., Guo, C., Ratner, L., and Wong-Staal, F.**, Human proto-oncogene nucleotide sequences corresponding to the transforming region of simian sarcoma virus, *Science*, 223, 487, 1984.
108. **van den Ouweland, A. M. W., Breuer, M. L., Steenbergh, P. H., Schalken, J. A., Bloemers, H. P. J., and Van de Ven, W. J. M.**, Comparative analysis of the human and feline c-*sis* proto-oncogenes. Identification of 5' human c-*sis* coding sequences that are not homologous to the transforming gene of simian sarcoma virus, *Biochim. Biophys. Acta*, 825, 140, 1985.
109. **Eva, A., Robbins, K. C., Andersen, P. R., Srinivasan, A., Tronick, S. R., Reddy, E. P., Ellmore, N. W., Galen, A. T., Lautenberger, J. A., Papas, T. S., Westin, E. H., Wong-Staal, F., Gallo, R. C., and Aaronson, S. A.**, Cellular genes analogous to retroviral *onc* genes are transcribed in human tumour cells, *Nature (London)*, 295, 116, 1982.
110. **Goustin, A. S., Betsholtz, C., Pfeifer-Ohlsson, S., Persson, H., Rydnert, J., Bywater, M., Holmgren, G., Heldin, C.-H., Westermark, B., and Ohlsson, R.**, Coexpression of the *sis* and *myc* proto-oncogenes in developing human placenta suggests autocrine control of trophoblast growth, *Cell*, 41, 301, 1985.
111. **Barrett, T. B., Gajdusek, C. M., Schwartz, S. M., McDougall, J. K., and Benditt, E. P.**, Expression of the *sis* gene by endothelial cells in culture and *in vivo*, *Proc. Natl. Acad. Sci. U.S.A.*, 81, 6772, 1984.
112. **Jaye, M., McConathy, E., Drohan, W., Tong, B., Deuel, T., and Maciag, T.**, Modulation of the *sis* gene transcript during endothelial cell differentiation in vitro, *Science*, 228, 882, 1985.
113. **Hannink, M. and Donoghue, D. J.**, Requirement for a signal sequence in biological expression of the v-*sis* oncogene, *Science*, 226, 1197, 1984.
114. **Robbins, K. C., Leal, F., Pierce, J. H., and Aaronson, S. A.**, The v-*sis*/PDGF-2 transforming gene product localizes to cell membranes but is not a secretory protein, *EMBO J.*, 4, 1783, 1985.
115. **Josephs, S. F., Ratner, L., Clarke, M. F., Westin, E. H., Reitz, M. S., and Wong-Staal, F.**, Transforming potential of human c-*sis* nucleotide sequences encoding platelet-derived growth factor, *Science*, 225, 636, 1984.

116. **Gazit, A., Igarashi, H., Chiu, I-M., Srinivasan, A., Yaniv, A., Tronick, S. R., Robbins, K. C., and Aaronson, S. A.**, Expression of the normal human *sis*/PDGF-2 coding sequence induces cellular transformation, *Cell,* 39, 89, 1984.
117. **Gasic, G. J.**, Role of plasma, platelets, and endothelial cells in tumor metastasis, *Cancer Metast. Rev.,* 3, 99, 1984.
118. **Tsubura, E., Yamashita, T., and Sone, S.**, Inhibition of the arrest of hematogeneously disseminated tumor cells, *Cancer Metast. Rev.,* 2, 223, 1983.
119. **Nistér, M., Heldin, C.-H., Wasteson, A., and Westermark, B.**, A glioma-derived analog to platelet-derived growth factor: demonstration of receptor competing activity and immunological crossreactivity, *Proc. Natl. Acad. Sci. U.S.A.,* 81, 926, 1984.
120. **Di Corleto, P. E. and Bowen-Pope, D. F.**, Cultured endothelial cells produce a platelet-derived growth factor-like protein, *Proc. Natl. Acad. Sci. U.S.A.,* 80, 1919, 1983.
121. **Powers, S., Fisher, P. B., and Pollack, R.**, Analysis of the reduced growth factor dependency of simian virus 40-transformed 3T3 cells, *Mol. Cell. Biol.,* 4, 1572, 1984.
122. **Rozengurt, E., Sinnett-Smith, J., and Taylor-Papadimitriou, J.**, Production of PDGF-like growth factor by breast cancer cell lines, *Int. J. Cancer,* 36, 247, 1985.
123. **Pantazis, P., Pelicci, P. G., Dalla-Favera, R., and Antoniades, H. N.**, Synthesis and secretion of proteins resembling platelet-derived growth factor by human glioblastoma and fibrosarcoma cells in culture, *Proc. Natl. Acad. Sci. U.S.A.,* 82, 2404, 1985.
124. **van Zoelen, E. J. J., van de Ven, W. J. M., Franssen, H. J., van Oostwaard, T. M. J., van der Saag, P. T., Heldin, C.-H., and de Laat, S. W.**, Neuroblastoma cells express c-*sis* and produce a transforming growth factor antigenically related to the platelet-derived growth factor, *Mol. Cell. Biol.,* 5, 2289, 1985.
125. **Heldin, C. H., Westermark, B., and Wasteson, A.**, Chemical and biological properties of a growth factor from human cultured osteosarcoma cells, resembling PDGF, *J. Cell. Physiol.,* 105, 235, 1980.
126. **Graves, D. T., Owen, A. J., and Antoniades, H. N.**, Evidence that a human osteosarcoma cell line secretes a mitogen similar to platelet-derived growth factors present in platelet-poor plasma, *Cancer Res.,* 43, 83, 1983.
127. **Betsholtz, C., Westermark, B., Ek, B., and Heldin, C.-H.**, Coexpression of a PDGF-like growth factor and PDGF receptors in a human osteosarcoma cell line: implications for autocrine receptor activation, *Cell,* 39, 447, 1984.
128. **Sporn, M. B. and Todaro, G. J.**, Autocrine secretion and malignant transformation of cells, *N. Engl. J. Med.,* 303, 878, 1980.
129. **Graves, D. T., Owen, A. J., Barth, R. K., Tempst, P., Winoto, A., Fors, L., Hood, L. E., and Antoniades, H. N.**, Detection of c-*sis* transcripts and synthesis of PDGF-like proteins by human osteosarcoma cells, *Science,* 226, 972, 1984.
130. **Rizzino, A. and Bowen-Pope, D. F.**, Production of PDGF-like growth factors by embryonal carcinoma cells and binding of PDGF to their endoderm-like differentiated cells, *Dev. Biol.,* 110, 15, 1985.
131. **Bowen-Pope, D. F., Vogel, A., and Ross, R.**, Production of platelet-derived growth factor-like molecules and reduced expression of platelet-derived growth factor receptors accompany transformation by a wide spectrum of agents, *Cancer Res.,* 81, 2396, 1984.
132. **Bleiberg, I., Harvey, A. K., Smale, G., and Grotendorst, G. R.**, Identification of a PDGF-like mitoattractant produced by NIH/3T3 cells after transformation with SV40, *J. Cell. Physiol.,* 123, 161, 1985.
133. **Heldin, C. H., Wasteson, A., and Westermark, B.**, Partial purification and characterization of platelet factors stimulating the multiplication of normal human glial cells, *Exp. Cell Res.,* 109, 429, 1977.
134. **Cowan, D. H. and Graham, J.**, Stimulation of human tumor colony formation by platelet lysate, *J. Lab. Clin. Med.,* 102, 973, 1983.
135. **Paul, D. and Piasecki, A.**, Rat platelets contain growth factor(s) distinct from PDGF which stimulate DNA synthesis in primary adult rat hepatocyte cultures, *Exp. Cell Res.,* 154, 95, 1984.
136. **Assoian, R. K., Grotendorst, G. R., Miller, D. M., and Sporn, M. B.**, Cellular transformation by coordinated action of three peptide growth factors from human platelets, *Nature (London),* 309, 804, 1984.
137. **Bauer, G., Birnbaum, U., Höfler, P., and Heldin, C.-H.**, EBV-inducing factor from platelets exhibits growth-promoting activity for NIH 3T3 cells, *EMBO J.,* 4, 1957, 1985.
138. **Lechner, J. F., McClendon, I. A., LaVeck, M. A., Shamsuddin, A. M., and Harris, C. C.**, Differential control by platelet factors of squamous differentiation in normal and malignant human bronchial epithelial cells, *Cancer Res.,* 43, 5915, 1983.

Chapter 10

TRANSFERRINS

I. INTRODUCTION

Transferrins are a family of proteins with iron-binding properties, carrying ferric iron from the intestine, reticuloendothelial system, and liver parenchymal cells to all proliferating cells in the body.[1] Transferrins have molecular weights of approximately 80,000, and the amino acid sequences of different transferrins of human and nonhuman origin indicate a relatively high degree of structural homology.

Transferrins are present in human serum and milk, and in many embryonic and adult tissues.[2] Liver is the major source of serum transferrin but a similar or identical molecule may be synthesized in other organs, including the chick oviduct and the testis.[3] Ovotransferrin, which is a transferrin-like protein present in the chicken oviduct, is the iron-binding protein found in avian egg white and is near identical to transferrin in its polypeptide structure but both molecules differ in their carbohydrate moieties. Ovotransferrin is a substrate for protein kinase C-dependent phosphorylation in vitro.[4] The physiological significance of this phosphorylation remains to be established. Transferrins are necessary components of almost all serum-free tissue culture media, acting as requisite factors.

Transferrin-like proteins may be present in tumors, and a protein identified as the human melanoma surface antigen p97 is a member of the transferrin family.[5] Antigen p97 was detected in 90% of different human melanoma cell lines, but was also present in more than half of cell lines established from a wide diversity of other human tumors.[6] Antigen p97 was not detected in B-lymphoblastoid cell lines or in cultivated fibroblasts from human donors.

II. SYNTHESIS AND STRUCTURE OF TRANSFERRINS

The transferrin family of proteins is the result of intragenic duplication followed by a series of independent gene duplications. Human and rat transferrins have been studied in more detail.

A. Human Transferrin

The major transferrin gene is located on human chromosome 3q15-q25, and 3q21-qter is a region which also contains the gene for the transferrin cellular receptor.[7] The same chromosomal region would contain the gene for human melanoma-associated antigen p97.[8] The complete c-DNA sequence of a human transferrin gene has been determined.[9] This gene is composed of 2324 bp and a single reading frame with a leader sequence encoded by 57 nucleotides that encode homologous amino-terminal and carboxyl-terminal domains of the protein. During evolution three areas of the homologous two domains have been strongly conserved, possibly reflecting functional constraints associated with iron binding.[9] The human transferrin gene contains at least 12 exons, ranging from 33 to 181 bp, separated by introns of 0.7 to 4.9 kbp.[10] The gene can be divided into two unequal parts corresponding to the known domains of the protein. The organization of the human transferrin gene family is more easily explained on the basis of gene duplication during evolution.[10]

Amino acid sequence analysis shows that the human transferrin polypeptide is composed of two homologous regions (residues 1-336 and 337-679), which is reflected in the presence of two discrete structural domains in the transferrin molecule.[11] Transferrin and albumin are both synthesized in the liver but have marked differences in their secretion kinetics. Transferrin is a glycoprotein while albumin is not and mammalian transferrin has a more complex tertiary structure and disulfide bond arrangement than does albumin. An adequate tertiary structure may be required for transferrin secretion but the mechanisms responsible for the relatively delayed transferrin secretion are not understood.[12] A possibility is that some form of rate-limiting receptor-associated transfer mechanism may be involved in transferrin transport and release.

B. Rat Transferrin

The rat transferrin gene is apparently transcribed into a single mRNA species of 2400 bp which is present at a high level in the liver and, at a lower concentration, in various other fetal and adult rat tissues.[2] In most extrahepatic rat tissues (lung, heart, spleen, kidney, muscle) transferrin mRNA content increases progressively during fetal development to reach a maximum between day 3 and 1 before birth, then drops quickly after birth and remains stable at a very low level during adult life. In the rat brain, however, the concentration of transferrin mRNA is very low during fetal life and then increases after birth to reach a maximum in the adult, where it remains constant at 1:10 of the value found in adult liver.[2] Cyclic AMP may induce a transient inhibition of the transcriptional activity of the transferrin gene.[13] Rat transferrin has been synthesized in *Escherichia coli* by means of a recombinant phage constructed from a c-DNA library derived from rat liver mRNA.[14]

C. Transferrins and the B-*lym*-1 Proto-Oncogene

The chicken and human proto-oncogene B-*lym*-1 encodes proteins that are homologous to the amino terminus of transferrins.[15-17] The predicted amino acid sequence of the chicken B-*lym*-1 proto-oncogene product is 36% homologous to sequences of proteins of the transferrin family.[15] The human B-*lym*-1 product is composed of 58 amino acids, with six identities of 39 aligned from the amino-terminal region of the protein and ten residues of the human B-*lym*-1 proteins conserved in at least one of the sequences of the transferrin family.[16] These findings suggest the existence of a common ancestry for the B-*lym*-1 proto-oncogene and genes of the transferrin family. They also suggest that the B-*lym*-1 products may function via a pathway related to transferrin.

III. THE TRANSFERRIN RECEPTOR

The transferrin-transferrin receptor system is apparently involved in the control of cell growth and differentiation in many types of tissues. Transferrin receptors are widely distributed in various types of cells, including both normal cells and tumor cells. These receptors are involved in iron uptake by a process of iron-mediated endocytosis and the subsequent recycling of the receptor to the cell surface. During internalization, iron is released from transferrin and apotransferrin is dissociated from the receptor upon its return to the cell surface.[18]

A. The Transferrin Receptor Gene

The transferrin receptor gene has been cloned, by a gene transfer approach, from human cell lines expressing the transferrin receptor.[19,20] The cloned genomic DNA of the receptor is close to 31 kbp and contains 19 distinct coding sequences. With the exception of the exon at the 3' end, these coding sequences are each less than 200 bp long. A large nontranslated, but transcribed, region at the 3' end of the gene may be involved in the regulated expression of the transferrin receptor. A protein-blotting procedure and a specific DNA probe have been used to identify nuclear proteins that recognize the promoter region of the transferrin receptor gene.[21]

B. Structure of the Transferrin Receptor

The transferrin receptor on the plasma membrane has been identified as a 180,000-dalton phosphorylated glycoprotein in its nonreduced homodimeric form.[22,23] The receptor is apparently composed of two identical subunits of approximately 90,000 daltons linked as a dimer by a disulfide bridge.[24] As deduced from the nucleotide sequence of the respective mRNA, the human transferrin receptor is composed of 760 amino acid residues and contains a stretch of 26 predominantly nonpolar amino acids with a sequence of nine hydrophobic residues in the center (residues 66 to 88), which would correspond to the transmembrane region of the receptor molecule.[20,24] Since the latter sequence resides at the amino-terminal end of the protein and no hydrophobic stretch is detected at the carboxy-terminal portion, it can be deduced that the receptor must be unusually oriented, with its amino-terminus on the cytoplasmic side and with a large carboxy-terminal extracellular domain of 672 amino acids. No strong sequence homology was detected between the transferrin receptor and any known protein.[20,24]

Apparently, transferrin binding to the receptor does not alter the extent of basal receptor phosphorylation. However, hemin rapidly inhibits the incorporation of transferrin iron into reticulocytes, and phosphorylation of transferrin receptors is increased by hemin treatment of the cells in the presence of transferrin.[25] These results are consistent with the hypothesis that a cycle of phosphorylation and dephosphorylation may be involved in the mechanism of transferrin receptor internalization.[26] In contrast to the receptors of other peptide hormones and growth factors, the transferrin receptor is apparently not phosphorylated on tyrosine but on serine and threonine residues. Protein kinase C maybe involved in phosphorylation of the transferrin receptor.[27,28]

C. Expression of the Transferrin Receptor on the Cell Surface

Expression of the transferrin receptor at the cell surface has been considered as a specific marker for rapidly growing cells. In general, this expression is closely linked to the proliferative state of the cell and differentiated or nondividing cells have a reduced level of surface transferrin receptors. Several factors would contribute to regulating the number of transferrin receptors on the cell surface. Hemin, iron, and protoporphyrin IX may represent the main molecules involved in the regulation of transferrin receptors.[29] In particular, the expression of transferrin receptors on the surface depends on the amount of iron accumulated into the cells. When cells accumulate large amounts of iron, they reduce the number of transferrin receptors in order to prevent further accumulation of iron; in contrast, when the intracellular iron concentration is low and the cells need more iron, they induce an increased expression of transferrin receptors on the cell surface to permit rapid accumulation of iron.[29] Transferrin receptor regulation is coupled to intracellular ferritin in proliferating and differentiating HL-60 human leukemia cells.[30] The addition of EGF to human fibroblasts results in induction of a rapid but transient increase in cell surface transferrin receptors, which is apparently due to a translocation of intracellular transferrin receptors to the cell surface without change in the total cellular transferrin receptor contain.[31] INF-α inhibits the expression of transferrin receptors in human lymphoblastoid cells and mitogen-induced lymphocytes.[32] Inhibition of transferrin receptor expression by IFN-α may be at least one of the mechanisms for IFN-induced inhibition of cell proliferation.

Transferrin receptor-diferric transferrin complexes are internalized by endocytosis, which requires the action of a specific trigger.[33] Whether the receptor undergoes a conformational change is not yet known. After release of iron from transferrin, both apotransferrin and the transferrin receptor recirculate back to the cell surface, not being degraded, as it usually occurs with other ligand-receptor complexes, in lysosomes.[34,35] A rapid endocytosis of the transferrin receptor may also occur in the absence of bound transferrin and it has been suggested that binding of the ligand may not be required for an endocytic/exocytic cycle of the transferrin receptor.[36] It is possible that transferrin receptors, and perhaps also other types of receptors, are recognized as being ligand-occupied, not at the cell surface, but at some other site in the recycling pathway within the cell. Down regulation of the receptors on the cell surface would occur not by speeding up the entry of receptors from the cell surface but by slowing down, or abolishing, the return of internalized receptors to the cell surface.[36] Down-regulation of the surface transferrin receptor is induced by phorbol ester, which is associated with rapid hyperphosphorylation of the receptor by the action of protein kinase C.[26-28]

D. Transferrin Receptors in Normal and Neoplastic Cells

Transferrin receptors are especially abundant in hemoglobin synthesizing cells and in the placental trophoblast cells. They are generally detectable on dividing cells, including tumor cells and established cell lines but are usually undetectable in fully differentiated, nondividing cells and tissues.[37]

Transferrin receptor induction is required for human B-lymphocyte activation but not for immunoglobulin secretion.[38] After stimulation of resting (G_0) peripheral blood mononuclear cells with mitogens, T-cells enter the G_1 phase of the cell cycle where they produce and express receptors for IL-2.[39] Interaction of IL-2 with IL-2 receptors on activated T-cells is then required for expression of transferrin receptors.[40] The subsequent binding of transferrin to its receptor during late phase G_1 of the cell cycle allows T-cells to make the G_1 to S transition. Expression of the c-*myc* proto-oncogene is regulated at several points of the cycle in normal lymphocytes and may contribute to the control of lymphocyte proliferation.[41] Incubation of human lymphocytes with purified IL-2 results in increased expression of both c-*myc* and transferrin receptor mRNAs.[42]

Monoclonal anti-transferrin receptor antibodies of the IgM type, but not of the IgG type, are able to induce a complete inhibition of growth in most cell lines.[43,44] The results of these experiments suggest that the profound effects of the IgM anti-transferrin receptor antibodies on cell growth are due to extensive cross-linking of cell surface transferrin receptors which may interfere in some way with transferrin receptor function of iron delivery into the cell.

Monoclonal antibodies recognizing T-cells (Leu 1) and transferrin receptors (OKT9) cross-react with malignant B-cells of distinct differentiation stages.[45] A marked reduction of the surface transferrin receptors is observed in human neoplastic cells of the line A431 when they enter mitosis and this situation persists until telophase when receptors reappear to a level that exceeds the original interphase value.[46] Transferrin receptors are present in human hematopoietic cell lines and high levels of expression of these receptors have been detected in human leukemic cells.[29,47]

Transferrin binding and iron uptake of hematopoietic cell lines is inhibited by phorbol esters.[48] These compounds induce differentiation of the human promyelocytic leukemic cell line HL-60 and concomitantly decrease the number of transferrin receptors on the cell surface.[49] Phorbol esters induce hyperphosphorylation and internalization of transferrin receptors in HL-60 cells and human erythroleukemia cells.[26-28,50] Transferrin receptor phosphorylation occurs on serine and threonine residues and maybe a direct effect of protein kinase C activation. Appearance and internalization of transferrin receptors may be observed at the margins of spreading human tumor cells.[51]

Association between the transferrin receptor and the p21 protein product of c-*ras* oncogenes was detected in extracts of a human bladder carcinoma cell line.[52] This association, however, was recognized later as an artefact of the immunoprecipitation technique.[53]

IV. SUMMARY

Transferrins are a family of glycoproteins with iron binding properties which are present in normal and tumor tissues and are necessary components of almost all serum-free tissue culture media. The chicken and human B-*lym*-1 proto-oncogene encodes proteins that are homologous to the amino terminus of transferrin, suggesting the existence of a common ancestry for the B-*lym*-1 proto-oncogene and genes of the transferrin family. Transferrin and its cellular receptor are apparently involved in the control of cell growth and differentiation in many types of cells both in vivo and in vitro. Expression and phosphorylation of transferrin receptors on the surface of normal or transformed cells are regulated by many factors, including different hormones and growth factors. Phorbol esters induce differentiation in the human promyelocytic cell line HL-60 and concomitantly decrease the number of transferrin receptors expressed on the cell surface, increasing their phosphorylation and internalization. Phosphorylation of the transferrin receptor occurs not on tyrosine but on serine and threonine residues, which maybe attributed to protein kinase C activity.

REFERENCES

1. **Huebers, H. A. and Finch, C. A.**, Transferrin: physiologic behavior and clinical implications, *Blood*, 64, 763, 1984.
2. **Levin, M. J., Tuil, D., Uzan, G., Dreyfus, J.-C., and Kahn, A.**, Expression of the transferrin gene during development of non-hepatic tissues, *Biochem. Biophys. Res. Comm.*, 122, 212, 1984.
3. **Skinner, M. K., Cosand, W. L., and Griswold, M. D.**, Purification and characterization of testicular transferrin secreted by rat Sertoli cells, *Biochem. J.*, 218, 313, 1984.
4. **Horn, F., Gschwendt, M., and Marks, F.**, Partial purification and characterization of the calcium-dependent and phospholipid-dependent protein kinase C from chick oviduct, *Eur. J. Biochem.*, 148, 533, 1985.
5. **Brown, J. P., Hewick, R. M., Hellström, K. E., Doolittle, R. F., and Dreyer, W. J.**, Human melanoma-associated antigen p97 is structurally and functionally related to transferrin, *Nature (London)*, 296, 171, 1982.
6. **Woodbury, R. G., Brown, J. P., Yeh, M-Y., Hellström, I., and Hellström, K. E.**, Identification of a cell surface protein, p97, in human melanomas and certain other neoplasms, *Proc. Natl. Acad. Sci. U.S.A.*, 77, 2183, 1980.
7. **Huerre, C., Uzan, G., Grzeschik, K. H., Weil, D., Levin, M., Hors-Cayla, M.-C., Boué, J., Kahn, A., and Junien, C.**, The structural gene for transferrin (TF) maps to 3q21-3qter, *Ann. Genet.*, 27, 5, 1984.
8. **Plowman, G. D., Brown, J. P., Enns, C. A., Schröder, J., Nikinmaa, B., Sussman, H. H., Hellström, K. E., and Hellström, I.**, Assignment of the gene for human melanoma-associated antigen p97 to chromosome 3, *Nature (London)*, 303, 70, 1983.
9. **Yang, F., Lum, J. B., McGill, J. R., Moore, C. M., Naylor, S. L., van Bragt, P. H., Baldwin, W. D., and Bowman, B. H.**, Human transferrin: cDNA characterization and chromosomal localization, *Proc. Natl. Acad. Sci. U.S.A.*, 81, 2752, 1984.
10. **Park, I., Schaeffer, E., Sidoli, A., Baralle, F. E., Cohen, G. N., and Zakin, M. M.**, Organization of the human transferrin gene: direct evidence that it originated by gene duplication, *Proc. Natl. Acad. Sci. U.S.A.*, 82, 3149, 1985.
11. **MacGillivray, R. T. A., Mendes, E., Shewale, J. G., Sinha, S. K., Lineback-Zins, J., and Brew, K.**, The primary structure of human serum transferrin, *J. Biol. Chem.*, 258, 3543, 1983.
12. **Morgan, E. H. and Peters, T., Jr.**, The biosynthesis of rat transferrin: evidence for rapid glycosylation, disulfide bond formation, and tertiary folding, *J. Biol. Chem.*, 260, 14793, 1985.
13. **Tuil, D., Vaulont, S., Levin, M. J., Munnich, A., Moguilewsky, M., Bouton, M. M., Brissot, P., Dreyfus, J.-C., and Kahn, A.**, Transient transcriptional inhibition of the transferrin gene by cyclic AMP, *FEBS Lett.*, 189, 310, 1985.
14. **Aldred, A. R., Howlett, G. J., and Schreiber, G.**, Synthesis of rat transferrin in Escherichia coli containing a recombinant bacteriophage, *Biochem. Biophys. Res. Comm.*, 122, 960, 1984.

15. **Goubin, G., Goldman, D. S., Luce, J., Neiman, P. E., and Cooper, G. M.,** Molecular cloning and nucleotide sequence of a transforming gene detected by transfection of chicken B-cell lymphoma DNA, *Nature (London),* 302, 114, 1983.
16. **Diamond, A., Cooper, G. M., Ritz, J., and Lane, M.-A.,** Identification and molecular cloning of the human *Blym* transforming gene activated in Burkitt's lymphomas, *Nature (London),* 305, 112, 1983.
17. **Devine, J. M., Diamond, A., Lane, M.-A., and Cooper, G. M.,** Characterization of the Blym-1 transforming genes of chicken and human B-cell lymphomas, *J. Cell. Physiol.,* Suppl. 3, 193, 1984.
18. **Testa, U.,** Transferrin receptors: structure and function, *Curr. Top. Hematol.,* 5, 127, 1985.
19. **Kühn, L. C., McClelland, A., and Ruddle, F. H.,** Gene transfer, expression, and molecular cloning of the human transferrin receptor gene, *Cell,* 37, 95, 1984.
20. **McClelland, A., Kühn, L. C., and Ruddle, F. H.,** The human transferrin receptor gene: genomic organization, and the complete primary structure of the receptor deduced from a cDNA sequence, *Cell,* 39, 267, 1984.
21. **Miskimins, W. K., Roberts, M. P., McClelland, A., and Ruddle, F. H.,** Use of a protein-blotting procedure and a specific DNA probe to identify nuclear proteins that recognize the promoter region of the transferrin receptor gene, *Proc. Natl. Acad. Sci. U.S.A.,* 82, 6741, 1985.
22. **Sutherland, R., Delia, D., Schneider, C., Newman, R., Kemshead, J., and Greaves, M.,** Ubiquitous cell-surface glycoprotein on tumor cells is proliferation-associated receptor for transferrin, *Proc. Natl. Acad. Sci. U.S.A.,* 78, 4515, 1981.
23. **Schneider, C., Sutherland, R., Newman, R., and Greaves, M.,** Structural features of the cell surface receptor for transferrin that is recognized by the monoclonal antibody OKT9, *J. Biol. Chem.,* 257, 8516, 1982.
24. **Schneider, C., Owen, M. J., Banville, D., and Williams, J. G.,** Primary structure of human transferrin receptor deduced from the mRNA sequence, *Nature (London),* 311, 675, 1984.
25. **Cox, T. M., O'Donnell, M. W., Aisen, P., and London, I. M.,** Hemin inhibits internalization of transferrin by reticulocytes and promotes phosphorylation of the membrane transferrin receptor, *Proc. Natl. Acad. Sci. U.S.A.,* 82, 5170, 1985.
26. **May, W. S., Jacobs, S., and Cuatrecasas, P.,** Association of phorbol ester-induced hyperphosphorylation and reversible regulation of transferrin membrane receptors in HL60 cells, *Proc. Natl. Acad. Sci. U.S.A.,* 81, 2016, 1984.
27. **Kohno, H., Taketani, S., and Tokunaga, R.,** Tumor-promoting, phorbol ester-induced phosphorylation of cell-surface transferrin receptors in human erythroleukemia cells, *Cell Struct. Funct.,* 10, 95, 1985.
28. **May, W. S., Sahyoun, N., Jacobs, S., Wolf, M., and Cuatrecasas, P.,** Mechanism of phorbol diester-induced regulation of surface transferrin receptor involves the action of activated protein kinase C and an intact cytoskeleton, *J. Biol. Chem.,* 260, 9419, 1985.
29. **Louache, F., Testa, U., Pelicci, P., Thomopoulos, P., Titeux, M., and Rochant, H.,** Regulation of transferrin receptors in human hematopoietic cell lines, *J. Biol. Chem.,* 259, 11576, 1984.
30. **Rhyner, K., Taetle, R., Bering, H., and To, D.,** Transferrin receptor regulation is coupled to intracellular ferritin in proliferating and differentiating HL60 leukemia cells, *J. Cell. Physiol.,* 125, 608, 1985.
31. **Wiley, H. S. and Kaplan, J.,** Epidermal growth factor rapidly induces a redistribution of transferrin receptor pools in human fibroblasts, *Proc. Natl. Acad. Sci. U.S.A.,* 81, 7456, 1984.
32. **Besancon, F., Bourgeade, M.-F., and Testa, U.,** Inhibition of transferrin receptor expression by interferon-alpha in human lymphoblastoid cells and mitogen-induced lymphocytes, *J. Biol. Chem.,* 260, 13074, 1985.
33. **Larrick, J. W., Enns, C., Raubitschek, A., and Weintraub, H.,** Receptor-mediated endocytosis of human transferrin and its cell surface receptor, *J. Cell. Physiol.,* 124, 283, 1985.
34. **Dautry-Varsat, A., Ciechanover, A., and Lodish, H. F.,** pH and the recycling of transferrin during receptor-mediated endocytosis, *Proc. Natl. Acad. Sci. U.S.A.,* 90, 2258, 1983.
35. **Hopkins, C. R. and Trowbridge, I. S.,** Internalization and processing of transferrin and the transferrin receptor in human carcinoma A431 cells, *J. Cell Biol.,* 97, 508, 1983.
36. **Watts, C.,** Rapid endocytosis of the transferrin receptor in the absence of bound transferrin, *J. Cell Biol.,* 100, 633, 1985.
37. **Trowbridge, I. S. and Omary, M. B.,** Human cell surface glycoprotein related to cell proliferation is the receptor for transferrin, *Proc. Natl. Acad. Sci. U.S.A.,* 78, 3039, 1981.
38. **Neckers, L. M., Yenokida, G., Trepel, J. B., Lipford, E., and James, S.,** Transferrin receptor induction is required for human B-lymphocyte activation but not for immunoglobulin secretion, *J. Cell Biochem.,* 27, 377, 1985.
39. **Cantrell, D. A. and Smith, K. A.,** The interleukin-2 T-cell system: a new cell growth model, *Science,* 224, 1326, 1984.
40. **Neckers, L. M. and Cossman, J.,** Transferrin receptor induction in mitogen-stimulated human T lymphocytes is required for DNA synthesis and cell division and is regulated by interleukin 2, *Proc. Natl. Acad. Sci. U.S.A.,* 80, 3494, 1983.

41. **Reed, J. C., Nowell, P. C., and Hoover, R. G.**, Regulation of c-*myc* mRNA levels in normal human lymphocytes by modulators of cell proliferation, *Proc. Natl. Acad. Sci. U.S.A.*, 82, 4221, 1985.
42. **Depper, J. M., Leonard, W. J., Drogula, C., Krönke, M., Waldmann, T. A., and Greene, W. C.**, Interleukin 2 (IL-2) augments transcription of the IL-2 receptor gene, *Proc. Natl. Acad. Sci. U.S.A.*, 82, 4230, 1985.
43. **Trowbridge, I. S. and Lopez, F.**, Monoclonal antibody to transferrin receptor blocks transferrin binding and inhibits human tumor cell growth in vitro, *Proc. Natl. Acad. Sci. U.S.A.*, 79, 1175, 1982.
44. **Lesley, J. F. and Schulte, R. J.**, Inhibition of cell growth by monoclonal anti-transferrin receptor antibodies, *Mol. Cell. Biol.*, 5, 1814, 1985.
45. **Ludwig, W. D., Kolecki, P., Sieber, G., and Herrmann, F.**, Monoclonal antibodies recognizing T-cells (Leu 1) and transferrin receptors (OKT9) crossreact with malignant B-cells of distinct differentiation stages, *Tumour Biol.*, 5, 321, 1984.
46. **Warren, G., Davoust, J., and Cockroft, A.**, Recycling of transferrin receptors in A431 cells is inhibited during mitosis, *EMBO J.*, 3, 2217, 1984.
47. **Larrick, J. W. and Logue, G.**, Transferrin receptors on leukaemia cells, *Lancet*, ii, 862, 1980.
48. **Pelicci, P. G., Testa, U., Thomopoulos, P., Tabilio, A., Vanchenker, W., Titeux, M., Gourdin, M. F., and Rochant, H.**, Inhibition of transferrin binding and iron uptake of hematopoietic cell lines by phorbol esters, *Leuk. Res.*, 8, 597, 1984.
49. **Rovera, G., Ferreo, D., Pagliardi, G. L., Vartikar, J., Pessano, S., Bottero, L., Abraham, S., and Lebman, D.**, Induction of differentiation of human myeloid leukemias by phorbol diesters: phenotypic changes and mode of action, *Ann. N.Y. Acad. Sci.*, 397, 211, 1982.
50. **Klausner, R. D., Harford, J., and van Renswoude, J.**, Rapid internalization of the transferrin receptor in K562 cells is triggered by ligand binding or treatment with a phorbol ester, *Proc. Natl. Acad. Sci. U.S.A.*, 81, 3005, 1984.
51. **Hopkins, C. R.**, The appearance and internalization of transferrin receptors at the margins of spreading human tumor cells, *Cell*, 40, 199, 1985.
52. **Finkel, T. and Cooper, G. M.**, Detection of a molecular complex between *ras* proteins and transferrin receptor, *Cell*, 36, 1115, 1984.
53. **Harford, J.**, An artefact explains the apparent association of the transferrin receptor with a *ras* gene product, *Nature (London)*, 311, 673, 1984.

Chapter 11

THYROID HORMONES

I. INTRODUCTION

Thyroid hormones play an essential role in the growth, development, and metabolism of vertebrate animals.[1] It is generally accepted that there are two main molecular species of thyroid hormones, namely, 3,3′,5-triiodo-L-thyromine (T3) and L-thyroxine (T4). Although T4 is the main hormone secreted by the thyroid gland, there is much evidence indicating that T3 should be considered as the main active thyroid hormone at the cellular level. Both T4 and T3 circulate in the blood but the levels of T4 are two orders of magnitude higher than those of T3. T3 present in blood is originated partially from direct secretion from the thyroid gland but peripheral intracellular deiodination processes contribute to maintain circulating T3 levels. Thyroglobulin is the protein precursor of thyroid hormones in the thyroid gland. The thyroglobulin gene is located on human chromosome 8 and maps within the region 8q23-24.3, the same region where the c-*myc* proto-oncogene is located.[2] The possible biological significance of this proximity is unknown. In spite of numerous studies performed in the past few decades, the cellular mechanisms of action of thyroid hormones are little understood.[3-5]

II. THYROID HORMONE RECEPTORS

There is much controversy on the subcellular location of thyroid hormone receptors and the chemical structure of the receptors has not been characterized. Thyroid hormone receptors have been identified in plasma membrane,[6] cytosol,[7] mitochondria,[8] and nucleus.[5,9,10]

Two classes of thyroid hormone receptors have been identified in the plasma membrane of GH3 rat pituitary tumor cells as well as in human A431 cells and Swiss 3T3-4 mouse fibroblasts.[6] Apparently, T3 and T4 bind to the same plasma membrane binding sites which have a 2- to 3-fold higher affinity for T3 than T4. A 55,000-dalton (55 k) protein present in GH3 cells plasma membrane is specifically labeled by N-bromoacetyl-3, (^{125}I)3′,5-triiodo-L-thyronine and would correspond to the thyroid hormone receptor on the plasma membrane.[6] Nuclear thyroid hormone receptors are mainly associated with the chromatin and may increase during the DNA synthesis phase (S phase) of the cell growth cycle.[11-13] Thyroid hormone action would be initiated by the binding of T3 to nonhistone and/or histone nuclear receptors.[5,14] The ability of thyroid hormone to interact directly with transcriptionally active chromatin has been demonstrated by using African green monkey kidney cells (CV-1 cells) infected with the SV40 virus.[15] In such cells 7.5% of the total T3-binding specific activity is associated with the nucleosol fraction containing SV40 minichromosomes. In addition, nuclear envelopes and nuclear matrices have a class of binding sites with relatively high affinity for T3.[16] The respective physiological roles of the different cellular thyroid hormone binding sites has not been characterized.

A. Regulation of EGF and Insulin Receptors by Thyroid Hormone

It has been suggested that the effects of thyroid hormone on the growth and development of some tissues are mediated by EGF.[17] There is evidence that thyroid hormone is involved in regulation of the levels of EGF receptors in vivo. EGF binding to liver membrane preparations is markedly reduced in hypothyroid rats and the defect is corrected by administration of a simple dose of T3 to the hypothyroid animals prior to sacrifice.[18] Administration of pharmacological doses of thyroid hormone in rats results in a marked decrease of EGF binding to rat hepatocytes and isolated rat liver membranes, and this reduction is correlated with a decrease of EGF-stimulated phosphorylation of membrane proteins.[19] These somewhat contradictory findings suggest that at least some in vivo effects of thyroid hormone may be mediated through changes in EGF receptor levels.

Thyroid hormone may also be involved in the regulation of insulin receptor autophosphorylation.[20] Hypothyroidism induced by thyroidectomy increases autophosphorylation on tyrosine of the insulin receptor beta subunit without changing the number or affinity of the receptor. Injection of T3 to thyroidectomized rats restores plasma membranes autophosphorylation of the insulin receptor beta subunit to the values observed in control euthyroid rats. The mechanisms involved in these changes are unknown.

III. REGULATION OF GENE EXPRESSION BY THYROID HORMONE

The results of many studies indicate that thyroid hormone is important for the regulation of transcriptional processes but the characterization of genes specifically regulated by thyroid hormone has remained elusive. Only in few instances have thyroid hormone-responsive genes been recognized.[21-23] A specific nuclear mRNA precursor, termed spot 14 mRNA, is rapidly induced by thyroid hormone.[24] Spot 14 mRNA increases within 20 min after T3 administration, representing the earliest known response of an mRNA to thyroid hormone. Moreover, the nuclear precursor of spot 14 mRNA has been found to increase within 10 min after T3 administration.[24] A genomic clone containing the entire rat gene for spot 14 mRNA has been isolated and its sequence has been determined.[22] The gene is present in a single copy per haploid genome but encodes two mRNA species differing by 170 nucleotides in length. The predicted polypeptide product of spot 14 do not share significant homology to any known protein sequences and its physiological role is still unknown.

IV. THYROID HORMONE AND HUMAN CANCER

The involvement of thyroid hormone in processes related to carcinogenesis is suggested by clinical and experimental studies. Nuclear thyroid hormone receptors are present in variable levels in human tumors.[25-27] In human breast cancer the levels of T3 receptors are not correlated with age or endocrine status of the patient or with extension or histological grading of the tumor, and there is also no correlation with estradiol and progesterone receptor concentration.[28] Patients with advanced cancer may be hypothyroid, as judged by the presence of reduced plasma thyroid hormone levels.[29,30]

The effect of hypothyroidism in the development of human tumors is difficult to evaluate because the disease has usually a protracted course and total absence of thyroid hormone is almost never found in clinical situations. There has been much controversy on the possible influence of hypo- or hyperthyroidism on different types of human tumors, especially breast cancer, but no clear-cut relationships have emerged from extensive clinical studies.[31-33] Whereas epidemiological studies have suggested that patients with certain thyroid disorders may constitute a high-risk population for breast cancer,[34] the results obtained in other studies failed to give evidence in support of this assumption.[35,36] Thyroid dysfunction may not be associated with the development of breast cancer in women.[37] However, in a clinical study of 283 women with early breast cancer the mean level of serum free thyroxin, measured as an index of thyroid function, was significantly lower in the patients as compared to controls.[38] Women dying from breast cancer may have thyroid atrophy.[39]

A. Thyroid Hormone in Experimental Tumors

The presence of thyroid hormone is important for tumor development under certain experimental conditions, and tumor implantation may have influence on thyroid function.

Walker carcinoma 256 is an undifferentiated and highly aggressive tumor of rat mammary origin and implantation of this tumor in rats results in a rapid and profound decrease in serum thyroid hormone levels.[40,41] This phenomenon is not mediated by the hypothalamus-hypophysis axis as thyrotropin (TSH) concentrations in blood and thyroid uptake of radioiodine remain unaltered after implantation of the tumor.[42,43] Although the mechanisms involved in the observed changes are not clear, both thyroid and peripheral factors should be considered. A decrease in thyroid hormone levels in the blood is not a general characteristic of transplanted tumors since, to the contrary, a marked increase in the blood T3 levels has been found in rats after transplantation of Morris hepatomas 7777 and 5123tc.[44] Infection of cultured rat thyroid cells with K-MuSV results in an irreversible suppression of iodide uptake and thyroglobulin synthesis by the thyroid cells but these blocking effects do not seem to depend directly from the action of the protein product of the v-K-*ras* oncogene.[45]

1. Effects of Thyroid Function on the Growth of Experimental Tumors

Hyperthyroidism may enhance and hypothyroidism may retard tumor growth and spread in syngeneic mouse tumor systems,[46] but in other studies a reduction in tumor incidence after a single injection of 3-methylcholanthrene in rats is observed when a hyperthyroid-like condition is induced in the animals by feeding thyroid powder.[47] On the other hand, thyroidectomy may determine a reduction in the incidence of tumors induced by chemical carcinogens,[48] and hypothyroidism induced by either radioiodine or methymazole can determine partial or complete suppression of the induction of mammary tumors in rats after administration of carcinogens like 7,12-dimethylbenz(a)anthracene (DMBA).[49,50] Hypothyroidism inhibits the local and metastatic growth of implanted tumors in rats.[51] although this effect is not observed in certain transplanted tumors,[52] including Walker carcinoma.[42] Hypothyroidism induced in rats by either propylthiouracil or radioiodine may increase the incidence of mammary tumors induced by nitrosomethylurea (NMU),[53] whereas in other studies with NMU-induced mammary tumors neither hypothyroidism nor moderate hyperthyroidism significantly increase tumor incidence or tumor growth.[52] The results of experimental studies on this subject are frequently conflictive, probably because total thyroidectomy or total thyroid suppression are technically difficult to perform, and small amounts of residual thyroid hormone production may persist after either operation or treatment with antithyroid drugs.

B. Thyroid Hormone and In Vitro Neoplastic Transformation

More clear results about the importance of thyroid hormone in oncogenic processes have been obtained by means of defined systems in vitro. A total removal of thyroid hormone from the medium can be obtained by adsorption to particular types of resin. Under these conditions a drastic reduction is observed in the yield of transformed cells induced by X-rays in Syrian hamster embryo cell strains and C3H 10T1/2 mouse embryo fibroblasts.[54,55] The results suggest that the presence of thyroid hormone is indispensable for the expression of a transformed phenotype and that in these systems, as well as in other similar systems in vitro, the process of transformation does not appear to be a direct result of X-ray-induced mutational events; rather, X-rays would act by inducing the expression of some cell function(s) required for transformation.[56-58]

Thyroid hormone also modulates the process of cellular transformation and the anchorage-independent growth induced by type-5 adenovirus in cloned populations of rat embryo fibroblasts.[59,60] Thyroid hormone may also play a critical role in the induction of neoplastic transformation by chemical carcinogens in tissue culture.[61] An absolute dependence of transformation on thyroid hormone has been found with both an indirect carcinogen, benzo(a)pyrene, which requires metabolic activation, and a direct carcinogen, *N*-methyl-*N'*-nitro-*N*-nitrosoguanidine (NNNG), which does not require activation. Moreover, in DNA transfection experiments it has been observed that a hypothyroid condition in the cultured recipient cells markedly eliminates the appearance of transformed foci after transfection.[62]

The yield of normal rat kidney (NRK) cells transformed by the Kirsten murine sarcoma virus (K-MuSV), which contains the v-H-*ras* oncogene, is also reduced, although only by about one half, when thyroid hormone (T3) is absent from the culture medium.[63] Under these conditions of thyroid hormone depletion the growth rate or saturation density of NRK cells is not affected. K-MuSV-induced transformation is dose-dependent in relation to thyroid hormone in this system, the maximum yield of transformation being observed at doses 10^{-10} M T3. The maximum inhibition of transformation occurs when the cultures are rendered thyroid hormone deficient 24 or 48 hr before virus infection, which suggests that thyroid hormone may have a critical permissive role for the initiation of neoplastic transformation. These provocative results may contribute to open new vistas on the general complex processes related to the oncogenic transformation of cells.

C. Mechanisms Involved in the Modulation of Neoplastic Transformation by Thyroid Hormone

Thyroid hormone, glucocorticoids, and other hormones may be differentially involved in regulation of the cell cycle, including the entrance into the S phase.[12] There seems to be a brief critical period of the cell cycle where exposure to thyroid hormone commits the cell to progress normally through the G_1 period of the cycle.[64] The effects of thyroid hormone on cell transformation in vitro and tumor formation in vivo are probably mediated by RNA and protein synthesis.

V. SUMMARY

Thyroid hormones are involved in the regulation of the metabolic processes of normal and transformed cells through mechanisms operating at both the genetic and epigenetic levels. Under specific experimental conditions, thyroid hormone may have a critical role in the expression of a transformed phenotype. The molecular phenomena involved in the modulation of malignant transformation by thyroid hormone remain uncharacterized but some epigenetic, regulatory events seem to participate in a critical, undispensable step of the multistage processes leading to transformation. It seems likely that, at least under certain experimental conditions, thyroid hormone induces the synthesis of some protein(s) that is necessary for the initiation of transformation induced by radiation, chemical agents or oncogenic viruses. It is conceivable that such protein(s) may be a proto-oncogene product(s). However, almost nothing is known about the possible role of thyroid hormone in the regulation of proto-oncogene expression. Recently, the protein product of the c-*erb*-A proto-oncogene has been identified with a high-affinity cellular receptor for thyroid hormone.[65,66]

REFERENCES

1. **Kaplan, H. M. and Gass, G. H.**, The thyroid gland, in *Handbook of Endocrinology*, Gass, G. H. and Kaplan, H. M., Eds., CRC Press, Boca Raton, Fla., 1982, 241.
2. **Rabin, M., Barker, P. E., Ruddle, F. H., Brocas, H., Targovnik, H., and Vassart, G.**, Proximity of thyroglobulin and c-*myc* genes on human chromosome 8, *Somat. Cell Mol. Genet.*, 11, 397, 1985.
3. **Sterling, K.**, Thyroid hormone action at the cell level, *N. Engl. J. Med.*, 300, 117, 1979.
4. **Menezes-Ferreira, M. M. and Torresani, J.**, Méchanismes d'action des hormones thyroïdiennes au niveau cellulaire, *Ann. Endocrinol. (Paris)*, 44, 205, 1983.
5. **Oppenheimer, J. H.**, Thyroid hormone action at the nuclear level, *Ann. Int. Med.*, 102, 374, 1985.
6. **Cheng, S.**, Structural similarities between the plasma membrane binding sites for L-thyroxine and 3,3',5-triiodo-L-thyroxine in cultured cells, *J. Receptor Res.*, 5, 1, 1985.

7. **Davis, P. J., Handwerger, B. S., and Glaser, F.,** Physical properties of a dog liver and kidney cytosol that binds thyroid hormone, *J. Biol. Chem.*, 249, 6208, 1974.
8. **Sterling, K. and Milch, P. O.,** Thyroid hormone binding of a component of mitochondrial membrane, *Proc. Natl. Acad. Sci. U.S.A.*, 72, 3225, 1975.
9. **Abdukarimov, A.,** Regulation of genetic activity by thyroid hormones, *Int. Rev. Cytol.*, Suppl. 15, 17, 1983.
10. **Bernal, J., Liewendahl, K., and Lamberg, B.-A.,** Thyroid hormone receptors in fetal and hormone resistant tissues, *Scand. J. Clin. Lab. Invest.*, 45, 577, 1985.
11. **DeFesi, C. R., Fels, E. C., and Surks, M. I.,** Nuclear 3,5,3'-triiodothyronine receptor concentration increases during deoxyribonucleic acid synthesis in partially synchronized GC cell cultures, *Endocrinology*, 111, 1156, 1982.
12. **Surks, M. I. and Kumara-Siri, M. H.,** Increase in nuclear thyroid and glucocorticoid receptors and growth hormone production during deoxyribonucleic acid synthesis phase of the cell growth cycle, *Endocrinology*, 114, 873, 1984.
13. **Wilson, B. D., Wium, C. A., and Gent, W. L.,** The binding of thyroid hormone receptors to DNA, *Biochem. Biophys. Res. Comm.*, 124, 29, 1984.
14. **Apriletti, J. W., David-Inouye, Y., Eberhardt, N. L., and Baxter, J. D.,** Interactions of the nuclear thyroid hormone receptor with core histones, *J. Biol. Chem.*, 259, 10941, 1984.
15. **Savouret, J.-F., Eberhardt, N. L., Cathala, G., and Baxter, J. D.,** Interaction of triiodothyronine-receptor complexes with simian virus 40 minichromosomes in monkey kidney CV-1 cells, *Endocrinology*, 116, 1259, 1985.
16. **Lefebvre, Y. A. and Venkatraman, J. T.,** Characterization of a thyroid-hormone-binding site on nuclear envelopes and nuclear matrices of the male-rat liver, *Biochem. J.*, 219, 1001, 1984.
17. **Fisher, D. A., Hoath, S., and Lakshmanan, J.,** The thyroid hormone effects on growth and development may be mediated by growth factors, *Endocrinol. Exp.*, 16, 259, 1982.
18. **Mukku, V. R.,** Regulation of epidermal growth factor receptor levels by thyroid hormone, *J. Biol. Chem.*, 259, 6543, 1984.
19. **Hayden, L. J. and Severson, D. L.,** Correlation of membrane phosphorylation and epidermal growth factor binding to hepatic membranes isolated from triiodothyronine-treated rats, *Biochim. Biophys. Acta*, 750, 226, 1983.
20. **Correze, C., Pierre, M., Thibout, H., and Toru-Delbauffe, D.,** Autophosphorylation of the insulin receptor in rat adipocytes is modulated by thyroid hormone status, *Biochem. Biophys. Res. Commun.*, 126, 1061, 1985.
21. **Jump, D. B., Narayan, P., Towle, H., and Oppenheimer, J. H.,** Rapid effects of triiodothyronine on hepatic gene expression: hybridization analysis of tissue-specific triiodothyronine regulation of $mRNA_{s14}$, *J. Biol. Chem.*, 259, 2789, 1984.
22. **Liaw, C. W. and Towle, H. C.,** Characterization of a thyroid hormone-responsive gene from rat, *J. Biol. Chem.*, 259, 7253, 1984.
23. **Magnuson, M. A., Dozin, B., and Nikodem, V. M.,** Regulation of specific rat liver messenger ribonucleic acids by triiodothyronine, *J. Biol. Chem.*, 260, 5906, 1985.
24. **Narayan, P., Liaw, C. W., and Towle, H. C.,** Rapid induction of a specific nuclear mRNA precursor by thyroid hormone, *Proc. Natl. Acad. Sci. U.S.A.*, 81, 4687, 1984.
25. **Burke, R. E. and McGuire, W. L.,** Nuclear thyroid hormone receptors in a human breast cancer cell line, *Cancer Res.*, 38, 3769, 1978.
26. **Gupta, M. K., Chiang, T., and Deodhar, S. D.,** Specific triiodothyronine binding by tumor cells and spleen cells in a thyroid hormone dependent mouse tumor system, *Eur. J. Cancer Clin. Oncol.*, 17, 819, 1981.
27. **Sellitti, D. F., Tseng, Y-C. L., and Latham, K. R.,** Nuclear thyroid hormone receptors in C3H/HeN mouse mammary glands and spontaneous tumors, *Cancer Res.*, 43, 1030, 1983.
28. **Cerbon, M.-A., Pichon, M.-F., and Milgrom, E.,** Thyroid hormone receptors in human breast cancer, *Cancer Res.*, 41, 4167, 1981.
29. **Rose, D. P. and Davis, T. E.,** Plasma triiodothyronine concentrations in breast cancer, *Cancer*, 43, 1434, 1979.
30. **Kaptein, E. M., Grieb, D. A., Spencer, C. A., Wheeler, W. S., and Nicoloff, J. T.,** Thyroxine metabolism in the low thyroxine state of critical nonthyroidal illnesses, *J. Clin. Endocrinol. Metab.*, 53, 764, 1981.
31. **Vorherr, H.,** Thyroid disease in relation to breast cancer, *Klin. Wochenschr.*, 56, 1139, 1978.
32. **Hedley, A. J., Jones, S. J., Spigelhalter, D. J., Clements, P., Bewsher, P. D., Simpson, J. G., and Weir, R. D.,** Breast cancer in thyroid disease: fact or fallacy?, *Lancet*, i, 131, 1981.
33. **Hoffman, D. A., McConahey, W. M., Brinton, L. A., and Fraumeni, J. F., Jr.,** Breast cancer in hypothyroid women using thyroid supplements, *J. Am. Med. Assoc.*, 251, 616, 1984.

34. **Itoh, K. and Maruchi, N.**, Breast cancer in patients with Hashimoto's thyroiditis, *Lancet*, ii, 1119, 1975.
35. **Mittra, I., Perrin, J., and Kumaoka, S.**, Thyroid and other auto antibodies in British and Japanese women: an epidemiological study of breast cancer, *Br. Med. J.*, i, 257, 1976.
36. **Maruchi, N., Annegers, J. F., and Kurland, L. T.**, Hashimoto's thyroiditis and breast cancer, *Mayo Clin. Proc.*, 51, 263, 1976.
37. **MacFarlane, I. A., Robinson, E. L., Bush, H., Durning, P., Howat, J. M. T., Beardwell, G. G., and Shalet, S. M.**, Thyroid function in patients with benign and malignant breast cancer, *Br. J. Cancer*, 41, 478, 1980.
38. **Thomas, B. S., Bulbrook, R. D., Russell, M. J., Hayward, J. L., and Millis, R.**, Thyroid function in early breast cancer, *Eur. J. Cancer Clin. Oncol.*, 19, 1213, 1983.
39. **Sommers, S. C.**, Endocrine abnormalities in women with breast cancer, *Lab. Invest.*, 40, 160, 1975.
40. **Pimentel, E., Dávila, F., Sucre, C., and Monteverde, J. A.**, Niveles de tiroxina en la sangre de ratas con carcinoma de Walker, in *Libro de Resúmenes*, IX Congr. Panamer. Endocrinol., Imp. del Estado, Quito, Ecuador, 1978, 52.
41. **Pimentel, E., Dávila, F., Sucre, C., and Monteverde, J. A.**, Niveles de tiroxina en el suero de ratas con carcinoma de Walker, *Rev. Soc. Colomb. Endocrinol.*, 12, 46, 1979.
42. **Pimentel, E. and Contreras, N. E. I. R.**, Effect of Walker carcinoma implantation on thyroid function in the rat, *J. Exp. Clin. Cancer Res.*, 2, 173, 1983.
43. **Pimentel, E.**, Hormones as tumor markers, *Cancer Detect. Prevent.*, 6, 87, 1983.
44. **Short, J., Klein, K., Kibert, L., and Ove, P.**, Involvement of the iodothyronines in liver and hepatoma cell proliferation in the rat, *Cancer Res.*, 40, 2417, 1980.
45. **Colletta, G., Pinto, A., Di Fiore, P. P., Fusco, A., Ferrentino, M., Avvedimento, V. E., Tsuchida, N., and Vecchio, G.**, Dissociation between transformed and differentiated phenotype in rat thyroid epithelial cells after transformation with a temperature-sensitive mutant of the Kirsten sarcoma virus, *Mol. Cell. Biol.*, 3, 2099, 1983.
46. **Kumar, M. S., Chiang, T., and Deodhar, S. D.**, Enhancing effect of thyroxine on tumor growth and metastases in syngeneic mouse tumor systems, *Cancer Res.*, 39, 3515, 1979.
47. **Baker, D. G. and Yaffe, A. H.**, The influence of thyroid stimulation on the incidence of 3-methylcholanthrene-induced tumors, *Cancer Res.*, 35, 528, 1975.
48. **Burnett, A. K.**, Thyroid hormone in 7,12-dimethylbenz(a)anthracene-induced leukemia in rats, *Cancer Res.*, 39, 4252, 1979.
49. **Kellen, J. A.**, Effect of hypothyroidism on induction of mammary tumors in rats by 7,12-dimethylbenz(a)anthracene, *J. Natl. Cancer Inst.*, 48, 1901, 1972.
50. **Jabara, A. G. and Maritz, J. S.**, Effects of hypothyroidism and progesterone on mammary tumours induced by 7,12-dimethylbenz(a)anthracene in Sprague-Dawley rats, *Br. J. Cancer*, 28, 161, 1973.
51. **Mishkin, S. Y., Pollack, R., Yalovsky, M. A., Morris, H. P., and Mishkin, S.**, Inhibition of local and metastatic hepatoma growth and prolongation of survival after induction of hypothyroidism, *Cancer Res.*, 41, 3040, 1981.
52. **Cave, W. T., Jr., Dunn, J. T., and MacLeod, R. M.**, Effects of altered thyroid states on mammary tumor growth and pituitary gland function in rats, *J. Natl. Cancer Inst.*, 59, 993, 1977.
53. **Milmore, J. E., Chandrasekaran, V., and Weisburger, J. H.**, Effects of hypothyroidism on development of nitrosomethylurea-induced tumors of the mammary gland, thyroid gland, and other tissues, *Proc. Soc. Exp. Biol. Med.*, 169, 487, 1982.
54. **Guernsey, D. L., Ong, A., and Borek, C.**, Thyroid hormone modulation of X ray-induced *in vitro* neoplastic transformation, *Nature (London)*, 288, 591, 1980.
55. **Guernsey, D. L., Borek, C., and Edelman, I. S.**, Crucial role of thyroid hormone in x-ray-induced neoplastic transformation in cell culture, *Proc. Natl. Acad. Sci. U.S.A.*, 78, 5708, 1981.
56. **Klein, J. C.**, Evidence against a direct carcinogenic effect of x-rays in vitro, *J. Natl. Cancer Inst.*, 52, 1111, 1974.
57. **Kennedy, A. R., Fox, M., Murphy, G., and Little, J. B.**, Relationship between x-ray exposure and malignant transformation in C3H 10T1/2 cells, *Proc. Natl. Acad. Sci. U.S.A.*, 77, 7262, 1980.
58. **Kennedy, A. R., Cairns, J., and Little, J. B.**, Timing of the steps in transformation of C3H 10T1/2 cells by X-irradiation, *Nature (London)*, 307, 85, 1984.
59. **Fisher, P. B., Guernsey, D. L., Weinstein, I. B., and Edelman, I. S.**, Modulation of adenovirus transformation by thyroid hormone, *Cancer Res.*, 80, 196, 1983.
60. **Babiss, L. E., Guernsey, D. L., and Fisher, P. B.**, Regulation of anchorage-independent growth by thyroid hormone in type 5 adenovirus-transformed rat embryo cells, *Cancer Res.*, 45, 6017, 1985.
61. **Borek, C., Guernsey, D. L., Ong, A., and Edelman, I. S.**, Critical role played by thyroid hormone in induction of neoplastic transformation by chemical carcinogens in tissue culture, *Proc. Natl. Acad. Sci. U.S.A.*, 80, 5479, 1983.

62. **Guernsey, D. L. and Leuthauser, S. W. C.**, Thyroid hormone effects the neoplastic transformation of mammalian cells induced by DNA-mediated gene transfer of DNA from x-ray transformed cells, *Fed. Proc.*, 43, 595, 1984.
63. **Borek, C., Ong, A., and Rhim, J. S.**, Thyroid hormone modulation of transformation induced by Kirsten murine sarcoma virus, *Cancer Res.*, 45, 1702, 1985.
64. **DeFesi, C. R., Fels, E. C., and Surks, M. I.**, Triiodothyronine stimulates growth of cultured GC cells by action early in the G1 period, *Endocrinology*, 114, 293, 1984.
65. **Sap, J., Muñoz, A., Damn, K., Goldberg, Y., Ghysdael, J., Leutz, A., Beug, H., and Vennström, B.**, The c-*erb*-A protein is a high-affinity receptor for thyroid hormone, *Nature (London)*, 324, 635, 1986.
66. **Weinberber, C., Thompson, C. C., Ong, E. S., Lebo, R., Gruol, D. J., and Evans, R. M.**, The c-*erb*-A gene encodes a thyroid hormone receptor, *Nature (London)*, 324, 641, 1986.

Chapter 12

STEROID HORMONES

I. INTRODUCTION

Steroid hormones are androgens, estrogens, gestagens, glucocorticoids, and mineralocorticoids of adrenal or gonadal origin. According to the classical model proposed for their cellular mechanism of action, steroid hormones would penetrate the cell membrane and would bind to specific cytosolic receptors. After binding, the receptor would be activated in some manner and the steroid hormone-receptor complex would be translocated into the nucleus, where it would interact with some component of the chromatin, which may result in the stimulation of specific transcriptional and translational processes leading to the synthesis of specific mRNAs and proteins.[1-5]

DNA sequences involved in the regulation of steroid hormone responses may be recognized and characterized by the use of gene transfer experiments after cloning the steroid hormone-responsive genes and their flanking regions in molecular vectors, which are then introduced into suitable cells.[6] Steroid hormone-receptor complexes are apparently capable to bind directly to a DNA region which modulates some promoter activity and thereby stimulates rate of transcription. In addition, steroid hormones may be involved, at least in some cellular systems, in the regulation of RNA processing, possibly through the production of RNA processing proteins that are required for maturation of primary transcripts.[7]

II. STEROID HORMONE RECEPTORS

The molecular mechanisms associated with steroid hormone action at the cellular level are only partially understood. The classical model proposed for the mechanism of action of steroid hormones has been challenged recently. According to the results of some studies, mammalian estrogen receptors are restricted to the cell nucleus when appropriate assay methods are used in order to avoid redistribution artifacts inherent in homogenization procedures.[8,9] In the spiny dogfish *(Squalus acanthias)* an estrogen-binding molecule possessing the essential physicochemical properties of a classical estrogen receptor and having both occupied and unoccupied sites is restricted to the cell nucleus.[10]

At least in some systems (17 beta-estradiol regulation of prolactin gene expression in the rat pituitary) the action of steroid hormone on transcription maybe independent of protein synthesis and may not require the continued presence of activated steroid hormone receptors within the nucleus.[11] The receptors of steroid hormones may be capable of interacting not only with DNA but also with RNA,[12] but the physiological significance of this interaction remains unknown. Cloning of a human estrogen receptor cDNA has been reported recently.[13] The primary structure and expression of a functional human glucocorticoid receptor cDNA has also been communicated.[14] These findings will allow a better characterization of the structure and functional properties of steroid hormone molecules. A domain structure of the human glucocorticoid receptor is related to the v-*erb*-A oncogene protein product.[15] The latter results suggest that steroid receptor genes and the c-*erb*-A proto-oncogene are derived from a common primordial gene.

Steroid hormones may be produced even by unicellular organisms. The yeast *Saccharomyces cerevisiae,* for example, possesses a high-affinity receptor for estrogen and is able to synthesize 17 beta-estradiol.[16] The biological significance of these facts is not understood.

A. Phosphorylation of Steroid Hormone Receptors

Results from several studies suggest that the physiological properties of steroid hormone receptors are modulated by phosphorylation/dephosphorylation processes, which opens the interesting possibility that steroid hormones may be homologous to certain oncogene protein products.[17] This possibility is reinforced by the identification of a 90,000-dalton protein (pp90), a nonhormone-binding phosphoprotein which is associated with both steroid receptors and the pp60[v-src] oncogene product.[18] Phosphorylation of steroid hormone receptors can occur in vivo.[19-21] Calmodulin stimulates phosphorylation of the uterine 17 beta-estradiol receptor exclusively on tyrosine and this modification is necessary for hormone binding of the receptor.[22] The 94,000-mol wt steroid binding component of rat hepatic glucocorticoid receptor undergoes calcium-stimulated, calmodulin-independent phosphorylation in vitro by ATP.[23]

Protein kinase activity has been detected in purified protein components (90 and 110 k) of the chicken oviduct progesterone receptor.[24] One of these components, the 110 k component, is present in the nucleus and could exert its action by phosphorylating chromatin proteins, thereby contributing to the regulation of transcriptional processes. Endogenous protein kinase activity is also present in the purified glucocorticoid receptor of rat liver cytosol.[19,23] It is not yet known whether such kinase activities are related to the action of proto-oncogene protein products but the progesterone receptor from hen oviduct is a substrate for EGF receptor-associated protein kinase activity,[25] and this receptor is homologous to the *erb*-B oncogene protein product. The androgen receptor from rat ventral prostate is also a phosphoprotein and its phosphorylation is mediated by the nuclear N2 type of cyclic AMP-independent protein kinase.[26] This finding suggests that phosphorylation/dephosphorylation of the androgen receptor may occur primarily within the nucleus.[26]

The precise role of steroid hormone receptor phosphorylation in the modulation of receptor function is not understood. Phosphorylation of steroid hormone receptors could be important in determining their functional properties, especially in relation to the regulation of transcriptional processes. Exogenous histones can be phosphorylated by the activated hepatic glucocorticoid receptor in the presence of ATP and in the absence of calcium, which suggests that glucocorticoid receptor-mediated phosphorylation of nuclear proteins could be a component of gene activation related to cellular response to glucocorticoid.[23]

B. Steroid Hormone Receptors in Cancer Cells

Alteration in the expression of receptors for different steroid hormones can occur in cancer cells.[27-31] The lack of expression of steroid hormone receptors in cancer cells of organs and tissues that are normally regulated by the action of steroid hormones, like the breast and the prostate, is usually attributed to alterations in the mechanisms of cell differentiation associated with the expression of a malignant phenotype. In general, steroid hormone-negative tumors are resistant to endocrine manipulations like therapeutic administration of hormones or surgical ablation of endocrine organs. Cancer cells positive for the presence of specific steroid hormone receptors, however, are not always responsive to the respective steroid hormone. The operation of postreceptor events and/or the action of other hormones or nonhormonal factors must be considered in such cases.

III. CELLULAR MECHANISMS OF ACTION OF STEROID HORMONES AND NEOPLASIA

Steroid hormones are recognized as important modulators of tumorigenic processes occurring in their respective target organs and tissues, acting as either enhancers or inhibitors of carcinogenesis. Treatment of female rats with testosterone enhances the carcinogenic effects of dimethylnitrosamine in kidney, lung, and liver.[32] Glucocorticoid hormones have been considered as agents capable of preventing or suppressing processes related to carcinogenesis,[33] but in certain experimental systems, such as radiation-induced transformation of C3H 10T1/2 cells, they have increasing effects in the yield of transformation.[34] Similar results have been obtained with estradiol in the same system.[35]

A. Androgens

Androgens and other sex steroid hormones may have an important role in the growth of tumors from the respective target organs. The cellular mechanisms responsible for the tumor promoting action of sex steroid hormones are not well understood. There is evidence that the mechanism of action of sex steroid hormones in the growth of normal or neoplastic responsive tissues may be indirect. The prostate has been classically considered as an androgen-dependent organ, but the results of experiments with isolated normal prostate epithelial cells clearly show direct mitogenic effects of insulin, EGF, glucocorticoid, and prolactin, whereas no positive effect of androgen on mitogenic response could be demonstrated.[36] These results suggest that the mitogenic activity of androgens on the prostate gland in vivo may be indirect and mediated by one or more of the above-mentioned factors or other similar factors. In any case, the growth of hormone-dependent prostate carcinoma may be controlled frequently either by suppressing the endogenous source of androgen (orchidectomy or hypophysectomy) or by treatment with estrogens or antiandrogens.

B. Estrogens

Estrogens are frequently involved in regulating the growth of tumors occurring in their target organs, especially the uterus and the mammary gland. In the mouse, mammary tumorigenesis depends on the conjoint action of three factors, namely, the genetic predisposition, the hormonal milieu, and the presence of MMTV. These three factors, however, may be interrelated. In different strains of mice a close correlation has been found between estradiol 16 alpha-hydroxylation (which depends on genetic factors and is associated with the metabolic activation of estradiol) and the presence of exogenous MMTV, which contributes in some manner to increase the 16 alpha-hydroxylation of estradiol.[37]

At present it is not known whether the oncogenic effects of estrogens are exerted through their hormonal properties or they behave as chemical carcinogens, although both types of effects are not mutually exclusive.[38] As may be true for androgens, the mechanism of action of estrogens on cell proliferation may also be indirect. Human MCF7 breast tumor cells are estrogen-dependent tumors in "nude" mice but their proliferation is independent on the presence of estrogens in serumless culture media. Addition of charcoal-dextran stripped female human serum to the media results in inhibition of growth of MCF7 cells but this inhibition can be reversed by addition of natural or synthetic estrogens, whereas other steroid hormones are ineffective to induce the reversion.[39] The inhibitory effect of human serum is specific for estrogen-sensitive cells because estrophilin-positive autonomous KLE human endometrial tumor cells are not inhibited. The results suggest that human serum contain an inhibitor of the proliferation of estrogen-sensitive cells, and that estrogens promote cell proliferation by neutralizing this serum-borne inhibitor.[39] The inhibitor has not yet been characterized.

The effects of estrogen on cell proliferation could be mediated by modulation of the receptors for growth factors. Estrogen can regulate acutely the levels of EGF receptor in the uterus, which occurs between 6 and 12 hr after estrogen administration and precedes increases in the uterine DNA synthesis.[40] It is thus possible that the observed increase in EGF receptor levels plays a role in the regulation of uterine DNA synthesis by estrogens. The polyamines (putrescine, spermidine, and spermine) are apparently essential, although not sufficient, in estrogen-stimulated proliferation of estrogen-dependent human breast cancer cells.[41] The polyamines alone are not sufficient to mimic the effect of estradiol in the growth of estradiol-responsive cells.

The genes involved in estrogen-induced cell proliferation have not been characterized. In MCF-7 mammary tumor cells, the estrogen-stimulated increase in thymidine incorporation into DNA following release from growth arrest is dependent on new RNA synthesis. However, the stimulation of thymidine and uridine incorporation caused by estradiol in these cells is uncoupled from hormone-induced increase in the expression of several estrogen-regulated genes.[42]

1. Estrogens and Proto-Oncogene Expression

Mammary tumors induced in female Sprague-Dawley rats by a single feeding of 7,12-dimethylbenz(*a*)anthracene (DMBA) express 10-fold elevated levels of the p21 protein product of the c-H-*ras* oncogene, in comparison with the levels present in the virgin mammary gland.[43] These tumors are hormone-dependent, showing shrinking after ovariectomy and, interestingly, this regression preceded by a reduction of the elevated p21^{ras} protein, which suggests that the expression of this gene is under the control of ovarian hormones and that the protein product of the c-H-*ras* gene may have a role in the origin and/or maintenance of mammary tumors induced in rats by certain carcinogens.

The levels of p21$^{c\text{-}ras}$ expression in carcinogen-induced hormone-dependent rat mammary tumors are much higher than in hormone-independent tumors induced by other carcinogenic substances.[44] A correlation between hormone dependency and c-H-*ras* expression is also present in human mammary carcinomas. The estrogen- and progesterone-positive human tumors contain p21$^{c\text{-}ras}$ levels that are much higher than those present in hormone receptor-negative tumors.[44] Interestingly, transfection of DNA containing a v-H-*ras* oncogene into MCF7 human breast cancer cultured cells bypasses dependence on estrogen for tumorigenicity in susceptible hosts.[45] In contrast to the parental cell line, MCF-7 cells transfested with the v-H-*ras* oncogene no longer responded to exogenous estrogen in culture and their growth was minimally inhibited by exogenously administered antiestrogens. When tested in the nude mouse, these cells were fully tumorigenic in the absence of estrogen supplementation.[45] These results suggest that an altered expression of oncogene products of the *ras* family could be involved in mammary tumorigenic processes occurring in humans and other animal species. However, the relevance of c-*ras* gene activation to human breast cancer development is not understood.

C. Progesterone

The maturation of *Xenopus* oocytes induced by progesterone is associated with increased phosphorylation of the ribosomal protein S6.[46] The oncogene product pp60$^{v\text{-}src}$ causes an increased phosphorylation in both S6 and total cellular proteins of *Xenopus* oocytes, which suggests a possible interaction with a normal cellular pathway utilized by progesterone. In support to this possibility, microinjection of pp60$^{v\text{-}src}$ into the oocytes markedly accerates the time course of germinal vesicle breakdown in response to progesterone when injected oocytes are incubated for 2 hr before progesterone treatment.[47]

D. Glucocorticoids

The cellular mechanisms of action of glucocorticosteroid hormones (glucocorticoids) and synthetic compounds with a similar structure and function, such as dexamethasone, are initiated by a specific interaction with the glucocorticoid receptor.[4] The human gene for the glucocorticoid receptor is located on chromosome 5.[48] Different endogenous and exogenous factors are involved in the regulation and modulation of glucocorticoid receptor expression.[49] Administration of thyroid hormone to adrenalectomized adult rats induces an increase in glucocorticoid receptor level in the liver.[50] The receptor is a good substrate for phosphorylation both in vitro and in vivo.[20,21,23] The purified glucocorticoid receptor contains protein kinase activity which is capable of inducing autophosphorylation of the receptor,[19,23] but the possible effect of this phenomenon in the location and/or activity of the receptor molecule is not understood.

Glucocorticoids are importantly involved in regulating the expression of animal genes. A well characterized gene whose expression is greatly enhanced by glucocorticoids is the structural gene for tyrosine aminotransferase (TAT) in the liver.[51] Expression of the TAT gene in primary cultured rat hepatocytes is inhibited by different types of carcinogens.[52] In other systems glucocorticoids may not increase but may rather inhibit gene transcription. Dexamethasone blocks the transcriptional activation of globin genes in murine erythroleukemia cells induced to terminal differentiation.[53] Even when acting in a given cellular system, dexamethasone may have opposing effects depending on the stage of differentiation of the target cells. For example, dexamethasone has opposing effects on the clonal growth of granulocyte and macrophage progenitor cells and on the phagocytic capability of mononuclear phagocytes at different stages of differentiation.[54] Post-transcriptional mechanisms are involved in the regulation of gene expression by glucocorticoids in the liver.[55] Glucocorticoids are capable of inducing lysis of certain types of cells, especially lymphoid cells, but the complex mechanisms responsible for this phenomena are little understood.[56] Glucocorticoids are also involved in regulation of DNA synthesis in their target cells, probably through the induction of protein synthesis.[57]

1. Glucocorticoid Receptors in Neoplastic Cells

Glucocorticoid receptors are present in a diversity of tumor cells. For several types of leukemias and lymphomas, simple quantitation of glucocorticoid receptors appear to have predictive value in relation to prognosis and response to therapy.[31] Absolute lack of glucocorticoid receptors is rare in leukemic cells. In a series of 18 untreated patients with chronic lymphocytic leukemia (CLL), all had detectable levels of total and nuclear glucocorticoid receptors in their neoplastic lymphocytes but these levels did not correlate with in vitro sensitivity to glucocorticoid.[58] Changes in glucocorticoid receptors of normal or neoplastic cells maybe related to the respective proliferative conditions.[59] Studies with mouse L1210 and human HL-60 cell lines revealed that glucocorticoid receptors accumulate during the G_1 phase of the cell cycle. In human lymphoblastic leukemia, the per cell receptor number is highest in cells in S and G_2 phase and lowest in small, noncycling cells. In normal human white blood cells, the glucocorticoid receptor content is maximal in large lymphocytes and monocytes while no receptors are present in small lymphocytes. According to the results of this study, the glucocorticoid receptor in leukemic blast cells appears to be highly dependent on cell cycle distribution and on the proliferative state of the tumor cells.[59]

A cDNA clone for the rat liver glucocorticoid receptor has been constructed.[60] The receptor appears to be encoded by a single copy gene which specifies RNA transcripts of approximately 6 kb which are altered quantitatively and qualitatively in several mouse lymphoma cell lines with specific defects in receptor function. Wild-type and mutant glucocorticoid receptors present in rat hepatoma and mouse lymphoma have been characterized.[61] A mouse thymoma cell line (S49), which contains greatly reduced amounts of glucocorticoid hormone binding activity, produce a mutant receptor which is of wild type size and is immunologically reactive but is unable to bind hormone. The S49 mutant lymphoma cell line synthesizes a glucocorticoid receptor with 40,000 mol wt, while the wild type normal receptor has a 94,000 mol wt.[61] The mutant 40,000-mol wt receptor produced by these cells would represent a truncated version of the wild type receptor protein, most likely resulting from a nonsense mutation or from a truncated mRNA.

A single common electrophoretic abnormality of glucocorticoid receptors has been detected in human leukemia cells from 10 of 25 patients with different types of the disease.[62] The abnormality was not associated with a particular type of leukemia and was observed in both newly diagnosed patients and patients who had received chemotherapy. A genetic abnormality in the neoplastic cells may be considered among other possible explanations for the presence of an abnormal glucocorticoid receptor in the leukemic cells. The possible relationship between this abnormality and resistance to glucocorticoid treatment in leukemic patients remains to be established.

2. Glucocorticoids and Regulation of Chronic Transforming Retrovirus Expression

Some viral genes may be regulated by steroid hormones including the glucocorticoids. The regulatory action of glucocorticoids on the expression of mouse mammary tumor virus (MMTV) has been clearly shown in several studies and specific viral sequences related to this action have been defined.[63-69] MMTV maybe acquired by either horizontal or milk-borne transmission infection in mice but MMTV proviruses are present in the genome of all inbred strains of mice and are also found in many feral mice. MMTV is one of the principal causative agents in the development of mouse mammary tumors but the incidence and the progression of the disease is multifactorial and is influenced by parameters such as genetic constitution, parity and hormonal status in the infected animals. In BR6 mice, which suffer a mammary tumor incidence of over 90% as a result of milk-borne MMTV, tumors develop to a detectable size only after an average of four pregnancies, and approximately 70% of them are initially hormone-dependent in that they regress completely or partially between pregnancies. In subsequent pregnancies, the tumors become re-established at the same site but eventually progress to autonomous growth, independent of the hormonal status of the animal.[70]

Glucocorticoid regulation of MMTV sequences is confined to 202 nucleotides preceding the LTR-specific RNA initiation site. The human growth hormone (hGH) gene contains within the first intron a specific binding site for activated glucocorticoid receptor and the site shares sequence homology with glucocorticoid receptor binding sites from MMTV and the human metallothionein II gene.[71] However, no simple pattern emerges from comparisons of the positions of the glucocorticoid receptor binding sites in the various genes. Since the glucocorticoid receptor binding sites are in apparently quite disparate locations and far removed from the start site of transcription, they could act more as enhancers than as promoters in the regulation of transcription. Glucocorticoids are capable of regulating the expression of MMTV sequences in lactating mammary glands as well as in the ovarian and testicular tissue of transgenic mice.[72] The results of these experiments demonstrate hormone regulation of an injected gene.

Glucocorticoids regulate the expression of MMTV at the level of initiation of transcription, increasing the number of active RNA polymerase II molecules on MMTV DNA and flanking mouse sequences.[73,74] Specific glucocorticoid receptor-DNA interactions may alter the configuration of DNA or chromatin in the vicinity of the binding sites, thereby creating an active transcriptional enhancer.[75,76] The molecular events responsible for these changes have not been characterized but may consist in alterations of patterns of DNA methylation or modifications of histone and/or nonhistone chromatin proteins.

A high percentage of MMTV-induced mouse mammary tumors contain the acquired MMTV provirus in either of two defined integration regions, *int*-1 and *int*-2, and provirus insertion is accompanied by specific mRNA transcripts from these regions. It is not known if these cellular sequences can be considered as proto-oncogenes, or contain proto-oncogenes, but the protein product of *int*-1 has no apparent homology to any known oncogene product.[77] In any case, pregnancy-dependent mammary tumors provirus integration within *int*-2 occurs already at the earliest appearance of the tumor and may therefore an important event in the multistage developmental processes of this type of neoplasia.[70]

Glucocorticoid administration may also enhance the expression of other chronic retroviruses, increasing, for example, the production of type-A retroviral particles in Ehrlich ascites tumors of mice.[78] Glucocorticoid-responsive Moloney murine leukemia virus (M-MuLV) may be created by insertion of regulatory sequences from MMTV into the LTR.[79]

3. Glucocorticoids and Oncogene Expression

The effects of glucocorticoids on acute transforming retroviruses and oncogene expression have been little studied. Glucocorticoids inhibit DNA synthesis in primary cultures of normal chick embryo fibroblasts but, even in much higher concentrations, they have little inhibitory effect on DNA synthesis in RSV-transformed cells.[80] Glucocorticoids produce a clear enhancement of the transformation of normal rat and human cultured cells by the Kirsten murine sarcoma virus, whereas no effect on transformation is observed when other steroid hormones (estradiol, testosterone, progesterone) are tested in the same systems.[81] The LTRs of MMTV contain regulatory sequences that act as enhancers and are responsible for glucocorticoid hormone-mediated induction of MMTV transcription.[82] MMTV LTRs can be used to promote a steroid hormone-inducible expression of viral oncogenes such as v-*mos*,[83] which corresponds to a proto-oncogene, c-*mos,* that is usually not expressed in normal cells. Addition of the transcriptional enhancers present in LTRs to recombinant chimeras in which the v-H-*ras* oncogene is expressed under the control of the MMTV promoter increases the ability of v-H-*ras* to transform NIH/3T3 cells 50- to 100-fold, and a significant stimulation occurs only when glucocorticoids are present in the culture medium.[84]

Expression of the c-*sis* proto-oncogene, whose product corresponds to a PDGF-like molecule, is inhibited by glucocorticoids in a smooth muscle cell line, the cells entering a G_0/G_1-like state following 18 hr of treatment with glucocorticoids.[85] The inhibition of cell growth may be overcome by PDGF since cells re-enter the cell cycle when exogenous PDGF is applied, even though c-*sis* mRNA transcripts remain at a low level.[85,86] These results are consistent with the hypothesis that glucocorticoids may inhibit cell growth by attenuating production of PDGF, or a PDGF-like molecule, through a reduction in the transcriptional activity of the c-*sis* proto-oncogene. The results are also consistent with the hypothesis that endogenous, autocrine regulation of cell growth may be an important factor in the development of tumors and that proto-oncogene activity may contribute to the partially autonomous growth of neoplastic cells. However, it should be noted that the inhibition of transcriptional activity induced by glucocorticoids is not specific to the proto-oncogenes like c-*sis* because a similar inhibition is induced by these hormones in the expression of other developmentally regulated genes. For example, dexamethasone administration inhibits the elevated levels of alpha-fetoprotein (AFP) mRNA occurring in regenerating rat liver and in certain hepatocellular carcinogenic processes, and this inhibition is organ-specific because it is not observed in the kidney under similar experimental conditions.[87] The results discussed above reinforce the proposal that proto-oncogenes are genes that participate, in concert with other genes, like those coding for oncofetal antigens, in both ontogenic and carcinogenic processes. This participation, however, does not necessarily implicate a causative role of these genes in tumorigenesis.

E. Calcitriol

The active form of vitamin D_3 (cholecalciferol) is calcitriol (1 alpha, 25-dihydroxyvitamin D_3). According to a classical model, the cellular mechanism of action of calcitriol (cholecalciferol) is similar to that of steroid hormones, being mediated by translocation of the steroid hormone-receptor complex from the cytoplasm into the nucleus, which results in specific enhancement of transcriptional processes, followed by protein synthesis.[88] Much information has been accumulated in the last few years about the structure of the calcitriol receptor in mammalian cells. These receptors are represented by polypeptides from 52,000 to 56,000 mol wt.[89] The physicochemical characteristics of calcitriol receptors are apparently identical in normal cells and malignant cells.[90] It is not clear, however, if the hormone-free calcitriol receptor is primarily located in the cytoplasm or the nucleus.

1. Induction of Cell Differentiation by Calcitriol

Several lines of evidence indicate that calcitriol is capable of inducing differentiation in certain types of normal or transformed cells.[91-95] The induction of differentiation in a human myeloid leukemia cell line (HL-60), as well as in other neoplastic human hematopoietic cell lines, by means of calcitriol occurs via specific calcitriol receptor-mediated events.[96-98] Calcitriol-induced differentiation is accompanied by morphological maturation phenomena and by functional and biochemical changes. An increase in the number of plasma membrane insulin receptors in the induced cells has been observed.[99] However, the solely presence of specific steroid receptors is not sufficient to ensure an induced differentiation response since some cloned sublines of the same cells are resistant to the inducing effect despite the fact that they contain specific receptors.[96] Moreover, the differentiation induced by calcitriol in hematopoietic cell lines of different origins occurs preferentially along the monocytic-macrophage pathway.[100] The effect of induction of differentiation by calcitriol in hematopoietic cell lines is mediated by RNA and protein synthesis, being suppressed by specific inhibitors of these processes, but it is not suppressed by inhibitors of DNA synthesis.[96]

2. Calcitriol and Proto-Oncogene Expression

A cooperative effect between calcitriol and dexamethasone has been observed in differentiation induced in some cellular systems in vitro, such as murine myeloid leukemia cells, which suggests that the mechanisms of action of the two steroid compounds in the induction of cell differentiation in malignant cells are, at least partially, different.[101] The possible participation of proto-oncogenes in these phenomena is suggested by the fact that the induction of differentiation of HL-60 cells into monocyte-like cells by calcitriol is preceded by a marked decrement in the expression of the proto-oncogene c-*myc*, as determined by a marked decrease of c-*myc* mRNA levels.[102,103]

3. Summary

The evidence accumulated in the last few years indicates that the mechanisms of action of steroid hormones at both the receptor and postreceptor levels may be closely related to the cellular mechanisms of action of oncogene protein products. Phosphorylation/dephosphorylation processes occurring in steroid hormone receptors could be regulated, at least partially, by proto-oncogene products possessing protein kinase activity. Moreover, the expression of proto-oncogenes at either the transcriptional or post-transcriptional levels could be subjected to the regulatory action of steroid hormones. Steroid hormone receptor genes and the c-*erb*-A proto oncogene may be derived from a common primordial ancestor molecule.

REFERENCES

1. **Pimentel, E.,** Cellular mechanisms of hormone action. I. Transductional events, *Acta Cient. Venez.*, 29, 73, 1978.
2. **Pimentel, E.,** Cellular mechanisms of hormone action. II. Posttransductional events, *Acta Cient. Venez.*, 29, 147, 1978.
3. **Ringold, G. H.,** Steroid hormone regulation of gene expression, *Ann. Rev. Pharmacol. Toxicol.*, 25, 529, 1985.
4. **Rousseau, G. G.,** Control of gene expression by glucocorticoid hormones, *Biochem. J.*, 224, 1, 1984.
5. **Walters, M. R.,** Steroid hormone receptors and the nucleus, *Endocrine Rev.*, 6, 512, 1985.
6. **Parker, M. G. and Page, M. J.,** Use of gene transfer to study expression of steroid-responsive genes, *Mol. Cell. Endocrinol.*, 34, 159, 1984.
7. **Vannice, J. L., Taylor, J. M., and Ringold, G. M.,** Glucocorticoid-mediated induction of alpha$_1$-acid glycoprotein: evidence for hormone-regulated RNA processing, *Proc. Natl. Acad. Sci. U.S.A.*, 81, 4241, 1984.

8. **King, W. J. and Greene, G. L.**, Monoclonal antibodies localize oestrogen receptor in nuclei of target cells, *Nature (London)*, 307, 745, 1984.
9. **Welshons, W. V., Lieberman, M. E., and Gorski, J.**, Nuclear localization of unoccupied oestrogen receptors, *Nature (London)*, 307, 747, 1984.
10. **Callard, G. V. and Mak, P.**, Exclusive nuclear location of estrogen receptors in *Squalus* testis, *Proc. Natl. Acad. Sci. U.S.A.*, 82, 1336, 1985.
11. **Shull, J. D. and Gorski, J.**, Estrogen stimulates prolactin gene transcription by a mechanism independent of pituitary protein synthesis, *Endocrinology*, 114, 1550, 1984.
12. **Rossini, G. P.**, RNA-containing nuclear binding sites for glucocorticoid-receptor complexes, *Biochem. Biophys. Res. Comm.*, 123, 78, 1984.
13. **Walter, P., Green, S., Greene, G., Krust, A., Bornert, J.-M., Jeltsch, J.-M., Staub, A., Jensen, E., Scrace, G., Waterfield, M., and Chambon, P.**, Cloning of human estrogen receptor cDNA, *Proc. Natl. Acad. Sci. U.S.A.*, 82, 7889, 1985.
14. **Hollenberg, S. M., Weinberger, C., Ong, E. S., Cerelli, G., Oro, A., Lebo, R., Thompson, E. B., Rosenfeld, M. G., and Evans, R. M.**, Primary structure and expression of a functional human glucocorticoid receptor cDNA, *Nature (London)*, 318, 635, 1985.
15. **Weinberger, C., Hollenberg, S. M., Rosenfeld, M. G., and Evans, R. M.**, Domain structure of human glucocorticoid receptor and its relationship to the v-*erb*-A oncogene product, *Nature (London)*, 318, 670, 1985.
16. **Feldman, D., Tökées, L. G., Stathis, P. A., Miller, S. C., Kurz, W., and Harvey, D.**, Identification of 17 beta-estradiol as the estrogenic substance in *Saccharomyces cerevisiae*, *Proc. Natl. Acad. Sci. U.S.A.*, 81, 4722, 1984.
17. **Sluyser, M. and Mester, J.**, Oncogenes homologous to steroid receptors?, *Nature (London)*, 315, 546, 1985.
18. **Schuh, S., Yonemoto, W., Brugge, J., Bauer, V. J., Riehl, R. M., Sullivan, W. P., and Toft, D. O.**, A 90,000-dalton binding protein common to both steroid receptors and the Rous sarcoma virus transforming protein, pp60$^{v\text{-}src}$, *J. Biol. Chem.*, 260, 14292, 1985.
19. **Kurl, R. N. and Jacobs, S. T.**, Phosphorylation of purified glucocorticoid receptor from rat liver by an endogenous protein kinase, *Biochem. Biophys. Res. Comm.*, 119, 700, 1984.
20. **Grandics, P., Miller, A., Schmidt, T. J., and Litwack, G.**, Phosphorylation *in vivo* of rat hepatic glucocorticoid receptor, *Biochem. Biophys. Res. Comm.*, 120, 59, 1984.
21. **Singh, V. B. and Moudgil, V. K.**, Phosphorylation of rat liver glucocorticoid receptor, *J. Biol. Chem.*, 260, 3684, 1985.
22. **Migliaccio, A., Rotondi, A., and Auricchio, F.**, Calmodulin-stimulated phosphorylation of 17 beta-estradiol receptor on tyrosine, *Proc. Natl. Acad. Sci. U.S.A.*, 81, 5921, 1984.
23. **Miller-Diener, A., Schmidt, T. J., and Litwack, G.**, Protein kinase activity associated with the purified rat hepatic glucocorticoid receptor, *Proc. Natl. Acad. Sci. U.S.A.*, 82, 4003, 1985.
24. **Garcia, T., Tuohimaa, P., Mester, J., Buchou, T., Renoir, J.-M., and Baulieu, E.-E.**, Protein kinase activity of purified components of the chicken oviduct progesterone receptor, *Biochem. Biophys. Res. Comm.*, 113, 960, 1983.
25. **Ghosh-Dastidar, P., Coty, W. A., Griest, R. E., Woo, D. D. L., and Fox, C. F.**, Progesterone receptor subunits are high-affinity substrates for phosphorylation by epidermal growth factor receptor, *Proc. Natl. Acad. Sci. U.S.A.*, 81, 1654, 1984.
26. **Goueli, S. A., Holtzman, J. L., and Ahmed, K.**, Phosphorylation of the androgen receptor by a nuclear cAMP-independent protein kinase, *Biochem. Biophys. Res. Comm.*, 123, 778, 1984.
27. **Lippman, M.**, Clinical implications of glucocorticoid receptors in human leukemia, *Am. J. Physiol.*, 243, E103, 1982.
28. **Homo-Delarche, F.**, Glucocorticoid receptors and steroid sensitivity in normal and neoplastic human lymphoid tissues: a review, *Cancer Res.*, 44, 431, 1984.
29. **Wittliff, J. L.**, Steroid-hormone receptors in breast cancer, *Cancer*, 53, 630, 1984.
30. **Hubay, C. A., Arafah, B., Gordon, N. H., Guyton, S. P., and Crowe, J. P.**, Hormone receptors: an update and application, *Surg. Clin. N. Am.*, 64, 1155, 1984.
31. **Thompson, E. B., Smith, J. R., Bourgeois, S., and Harmon, J. M.**, Glucocorticoid receptors in human leukemias and related diseases, *Klin. Wochenschr.*, 63, 689, 1985.
32. **Noronha, R. F. X. and Goodall, C. M.**, Enhancement by testosterone of dimethylnitrosamine carcinogenesis in lung, liver and kidney of inbred NZR/Gd female rats, *Carcinogenesis*, 4, 613, 1983.
33. **Greiner, J. W. and Evans, C.**, Temporal dynamics of cortisol and dexamethasone prevention of benzo(*a*)pyrene-induced morphological transformation of Syrian hamster cells, *Cancer Res.*, 42, 4014, 1982.
34. **Kennedy, A. R. and Weichselbaum, R. R.**, Effects of dexamethasone and cortisone with X-ray irradiation on transformation of C3H 10T1/2 cells, *Nature (London)*, 294, 97, 1981.

35. **Kennedy, A. R. and Weichselbaum, R. R.**, Effects of 17 beta-estradiol on radiation transformation in vitro: inhibition of effects by protease inhibitors, *Carcinogenesis,* 2, 67, 1981.
36. **McKeehan, W. L., Adams, P. S., and Rosser, M. P.**, Direct mitogenic effects of insulin, epidermal growth factor, glucocorticoid, cholera toxin, unknown pituitary factors and possibly prolactin, but not androgen, on normal prostate epithelial cells in serum-free, primary cell culture, *Cancer Res.,* 44, 1998, 1984.
37. **Bradlow, H. L., Hershcopf, R. J., Martucci, C. P., and Fishman, J.**, Estradiol 16 alpha-hydroxylation in the mouse correlates with mammary tumor incidence and presence of murine mammary tumor virus: a possible model for the hormonal etiology of breast cancer in humans, *Proc. Natl. Acad. Sci. U.S.A.,* 82, 6295, 1985.
38. **Li, J. J. and Li, S. A.**, Estrogen-induced tumorigenesis in hamsters: roles for hormonal and carcinogenic activities, *Arch. Toxicol.,* 55, 110, 1984.
39. **Soto, A. M. and Sonnenschein, C.**, Mechanism of estrogen action on cellular proliferation: evidence for indirect and negative control on cloned breast tumor cells, *Biochem. Biophys. Res. Comm.,* 122, 1097, 1984.
40. **Mukku, V. R. and Stancel, G. M.**, Regulation of epidermal growth factor receptor by estrogen, *J. Biol. Chem.,* 260, 9820, 1985.
41. **Lima, G. and Shiu, R. P. C.**, Role of polyamines in estradiol-induced growth of human breast cancer cells, *Cancer Res.,* 45, 2466, 1985.
42. **Aitken, S. C., Lippman, M. E., Kasid, A., and Schoenberg, D. R.**, Relationship between the expression of estrogen-regulated genes and estrogen-stimulated proliferation of MCF-7 mammary tumor cells, *Cancer Res.,* 45, 2608, 1985.
43. **Huang, F. L. and Cho-Chung, Y. S.**, Hormone-regulated expression of cellular ras^H oncogene in mammary carcinomas in rats, *Biochem. Biophys. Res. Comm.,* 123, 141, 1984.
44. **DeBortoli, M. E., Abou-Issa, H., Haley, B. E., and Cho-Chung, Y. S.**, Amplified expression of p21 ras protein in hormone-dependent mammary carcinomas of humans and rodents, *Biochem. Biophys. Res. Comm.,* 127, 699, 1985.
45. **Kasid, A., Lippman, M. E., Papageorge, A. G., Lowy, D. R., and Gelmann, E. P.**, Transfection of v-ras^H DNA into MCF-7 human breast cancer cells bypasses dependence on estrogen for tumorigenicity, *Science,* 228, 725, 1985.
46. **Nielsen, P. J., Thomas, G., and Maller, J. L.**, Increased phosphorylation of ribosomal protein S6 during meiotic maturation of *Xenopus* oocytes, *Proc. Natl. Acad. Sci. U.S.A.,* 79, 2937, 1982.
47. **Spivack, J. G., Erikson, R. L., and Maller, J. L.**, Microinjection of pp60^{v-src} into *Xenopus* oocytes increases phosphorylation of ribosomal protein S6 and accelerates the rate of progesterone-induced meiotic maturation, *Mol. Cell. Biol.,* 4, 1631, 1984.
48. **Gehring, U., Segnitz, B., Foellmer, B., and Francke, U.**, Assignment of the human gene for the glucocorticoid receptor to chromosome 5, *Proc. Natl. Acad. Sci. U.S.A.,* 82, 3751, 1985.
49. **Svec, F.**, Glucocorticoid receptor regulation, *Life Sci.,* 36, 2359, 1985.
50. **Naito, K., Isohashi, F., Tsukanaka, K., Horiuchi, M., Okamoto, K., Matsunaga, T., and Sakamoto, Y.**, Effects of D- and L-thyroxine on the glucocorticoid binding capacity of adult rat liver, *Biochem. Biophys. Res. Comm.,* 129, 447, 1985.
51. **Nickol, J. M., Lee, K.-L., and Kenney, F. T.**, Changes in hepatic levels of tyrosine aminotransferase messenger RNA during induction by hydrocortisone, *J. Biol. Chem.,* 253, 4009, 1978.
52. **Gayda, D. P. and Pariza, M. W.**, Effects of carcinogens on hormonal regulation of gene expression in primary cultures of adult rat hepatocytes, *Carcinogenesis,* 4, 1127, 1983.
53. **Kaneda, T., Murate, T., Sheffery, M., Brown, K., Rifkind, R. A., and Marks, P. A.**, Gene expression during terminal differentiation: dexamethasone suppression of inducer-mediated alpha$_1$- and beta^{maj}-globin, *Proc. Natl. Acad. Sci. U.S.A.,* 82, 5020, 1985.
54. **Shezen, E., Shirman, M., and Goldman, R.**, Opposing effects of dexamethasone on the clonal growth of granulocyte and macrophage progenitor cells and on the phagocytic capability of mononuclear phagocytes at different stages of differentiation, *J. Cell. Physiol.,* 124, 545, 1985.
55. **Fulton, R., Birne, G. D., and Knowler, J. T.**, Post-transcriptional regulation of rat liver gene expression by glucocorticoids, *Nucleic Acids Res.,* 13, 6467, 1985.
56. **Weinroth, S. E., MacLeod, C. L., Minning, L., and Hays, E. F.**, Genetic complexity of glucocorticoid-induced lysis of murine T-lymphoma cells, *Cancer Res.,* 45, 4808, 1985.
57. **Levenson, R., Iwata, K., Klagsbrun, M., and Young, D. A.**, Growth factor- and dexamethasone-induced proteins in Swiss 3T3 cells: relationship to DNA synthesis, *J. Biol. Chem.,* 260, 8056, 1985.
58. **Levine, E. G., Peterson, B. A., Smith, K. A., Hurd, D. D., and Bloomfield, C. D.**, Glucocorticoid receptors in chronic lymphocytic leukemia, *Leuk. Res.,* 9, 993, 1985.
59. **Smets, L. A., van der Klooster, P., and Otte, A.**, Glucocorticoid receptors of normal and leukemic cells: role of proliferation conditions, *Leuk. Res.,* 9, 199, 1985.

60. **Miesfeld, R., Okret, S., Wikström, A.-C., Wrange, O., Gustafsson, J.-A., and Yamamoto, K. R.**, Characterization of a steroid hormone receptor gene and mRNA in wild-type and mutant cells, *Nature (London)*, 312, 779, 1984.
61. **Northrop, J. P., Gametchu, B., Harrison, R. W., and Ringold, G. M.**, Characterization of wild type and mutant glucocorticoid receptors from rat hepatoma and mouse lymphoma cells, *J. Biol. Chem.*, 260, 6398, 1985.
62. **Distelhorst, C. W., Benutto, B. M., and Griffith, R. C.**, A single common electrophoretic abnormality of glucocorticoid receptors in human leukemia cells, *Blood*, 66, 679, 1985.
63. **Ringold, G. M.**, Regulation of mouse mammary tumor virus gene expression by glucocorticoid hormones, *Curr. Top. Microbiol. Immunol.*, 106, 79, 1983.
64. **Ucker, D. S., Firestone, G. L., and Yamamoto, K. R.**, Glucocorticoids and chromosomal position modulate murine mammary tumor virus transcription by affecting efficiency of promoter utilization, *Mol. Cell. Biol.*, 3, 551, 1983.
65. **Hynes, N., van Ooyen, A. J. J., Kennedy, N., Herrlich, N., Herrlich, P., Ponta, H., and Groner, B.**, Subfragments of the large terminal repeat cause glucocorticoid-responsive expression of mouse mammary tumor virus and of an adjacent gene, *Proc. Natl. Acad. Sci. U.S.A.*, 80, 3637, 1983.
66. **Groner, B., Kennedy, N., Skroch, P., Hynes, N. E., and Ponta, H.**, DNA sequences involved in the regulation of gene expression by glucocorticoid hormones, *Biochim. Biophys. Acta*, 781, 1, 1984.
67. **Scheidereit, C. and Beato, M.**, Contacts between hormone receptor and DNA double helix within a glucocorticoid regulatory element of mouse mammary tumor virus, *Proc. Natl. Acad. Sci. U.S.A.*, 81, 3029, 1984.
68. **Lee, F., Hall, C. V., Ringold, G. M., Dobson, D. E., Luh, J., and Jacob, P. E.**, Functional analysis of the steroid hormone control region of mouse mammary tumor virus, *Nucleic Acids Res.*, 12, 4191, 1984.
69. **Ponta, H., Kennedy, N., Skroch, P., Hynes, N., and Groner, B.**, Hormonal response region in the mouse mammary tumor virus long terminal repeat can be dissociated from the proviral promoter and has enhancer properties, *Proc. Natl. Acad. Sci. U.S.A.*, 82, 1020, 1985.
70. **Peters, G., Lee, A. E., and Dickson, C.**, Activation of cellular gene by mouse mammary tumour virus may occur early in mammary tumour development, *Nature (London)*, 309, 273, 1984.
71. **Moore, D. D., Marks, A. R., Buckley, D. I., Kapler, G., Payvar, F., and Goodman, H. M.**, The first intron of the human growth hormone gene contains a binding site for glucocorticoid receptor, *Proc. Natl. Acad. Sci. U.S.A.*, 82, 699, 1985.
72. **Ross, S. R. and Solter, D.**, Glucocorticoid regulation of mouse mammary tumor virus sequences in transgenic mice, *Proc. Natl. Acad. Sci. U.S.A.*, 82, 5880, 1985.
73. **Groner, B., Hynes, N. E., Rahmsdorf, U., and Ponta, H.**, Transcription initiation of transfected mouse mammary tumor virus LTR DNA is regulated by glucocorticoid hormone, *Nucleic Acids Res.*, 11, 4713, 1983.
74. **Firzlaff, J. M. and Diggelmann, H.**, Dexamethasone increases the number of RNA polymerase II molecules transcribing integrated mouse mammary tumor virus DNA and flanking mouse sequences, *Mol. Cell. Biol.*, 4, 1057, 1984.
75. **Zaret, K. S. and Yamamoto, K. R.**, Reversible and persistent changes in chromatin structure accompany activation of a glucocorticoid-dependent enhancer element, *Cell*, 38, 29, 1984.
76. **Peterson, D. O.**, Alterations in chromatin structure associated with glucocorticoid-induced expression of endogenous mouse mammary tumor virus genes, *Mol. Cell. Genet.*, 5, 1104, 1985.
77. **van Ooyen, A. and Nusse, R.**, Structure and nucleotide sequence of the putative mammary oncogene *int-1*; proviral insertions leave the protein-encoding domain intact, *Cell*, 39, 233, 1984.
78. **Kodama, T., Kodama, M., and Nishi, Y.**, Enhancing effect of hydrocortisone acetate administration on the content of A-type particles in Ehrlich ascites tumor, *J. Natl. Cancer Inst.*, 73, 227, 1984.
79. **Overhauser, J. and Fan, H.**, Generation of glucocorticoid-responsive Moloney murine leukemia virus by insertion of regulatory sequences from murine mammary tumor virus into the long terminal repeat, *J. Virol.*, 54, 133, 1985.
80. **Fodge, D. W. and Rubin, H.**, Differential effects of glucocorticoids on DNA synthesis in normal and virus-transformed chick embryo cells, *Nature (London)*, 257, 804, 1975.
81. **Rhim, J. S.**, Glucocorticoids enhance viral transformation of mammalian cells, *Proc. Soc. Exp. Biol. Med.*, 174, 217, 1983.
82. **Huang, A. L., Ostrowski, M. C., Berard, D., and Hager, G. L.**, Glucocorticoid regulation of the Ha-MuSV p21 gene conferred by sequences from mouse mammary tumor virus, *Cell*, 27, 245, 1981.
83. **Papkoff, J. and Ringold, G. M.**, Use of the mouse mammary tumor virus long terminal repeat to promote steroid-inducible expression of v-*mos*, *J. Virol.*, 52, 420, 1984.
84. **Ostrowski, M. C., Huang, A. L., Kessel, M., Wolford, R. G., and Hager, G. L.**, Modulation of enhancer activity by the hormone responsive regulatory element from mouse mammary tumor virus, *EMBO J.*, 3, 1891, 1984.

85. **Norris, J. S., Cornett, L. E., Hardin, J. W., Kohler, P. O., MacLeod, S. L., Srivastava, A., Syms, A. J., and Smith, R. G.**, Autocrine regulation of growth. II. Glucocorticoids inhibit transcription of c-*sis* oncogene-specific RNA transcripts, *Biochem. Biophys. Res. Comm.*, 122, 124, 1984.
86. **Syms, A. J., Norris, J. S., and Smith, R. G.**, Autocrine regulation of growth. I. Glucocorticoid inhibition is overcome by exogenous platelet-derived growth factor, *Biochem. Biophys. Res. Comm.*, 122, 68, 1984.
87. **Cote, G. J. and Chiu, J-F.**, Tissue specific control of alpha-fetoprotein gene expression, *Biochem. Biophys. Res. Comm.*, 120, 677, 1984.
88. **Bell, N. H.**, Vitamin D-endocrine system, *J. Clin. Invest.*, 76, 1, 1985.
89. **Pike, J. W.**, Intracellular receptors mediate the biologic actions of 1,25-dihydroxyvitamin D_3, *Nutr. Rev.*, 43, 161, 1985.
90. **Sher, E., Martin, T. J., and Eisman, J. A.**, Hormone-dependent transformation and nuclear localisation of 1,25-dihydroxyvitamin D_3 receptors from human breast cancer cell lines and chick duodenum, *Horm. Metab. Res.*, 17, 147, 1985.
91. **Miyaura, C., Abe, E., Kuribashi, T., Tanaka, H., Konno, K., Nishii, Y., and Suda, T.**, 1 alpha,25-dihydroxy vitamin D_3 induces differentiation of human myeloid leukemia cells, *Biochem. Biophys. Res. Comm.*, 102, 937, 1981.
92. **Abe, E., Miyaura, C., Sakagami, H., Takeda, M., Konno, K., Yamazaki, T., Yoshiki, S., and Suda, T.**, Differentiation of mouse myeloid leukemia cells by 1 alpha,25-dihydroxyvitamin D_3, *Proc. Natl. Acad. Sci. U.S.A.*, 78, 4990, 1981.
93. **Bar-Shavit, Z., Teitelbaum, S. L., Reitsma, P., Hall, A., Pegg, L. E., Trial, J., and Kahn, A. J.**, Induction of monocytic differentiation and bone resorption by 1,25-dihydroxyvitamin D_3, *Proc. Natl. Acad. Sci. U.S.A.*, 80, 5907, 1983.
94. **McCarthy, D. M., San Miguel, J. F., Freake, H. C., Green, P. M., Zola, H., Catovsky, D., and Goldman, J. M.**, 1,25-dihydroxyvitamin D_3 inhibits proliferation of human promyelocytic leukaemia (HL-60) cells and induces monocyte-macrophage differentiation in HL-60 and normal human bone marrow, *Leuk. Res.*, 7, 51, 1983.
95. **Koeffler, H. P., Amatruda, T., Ikekawa, N., Kobayashi, Y., and DeLuca, H. F.**, Induction of macrophage differentiation of human normal and leukemic myeloid stem cells by 1,25-dihydroxyvitamin D_3 and its fluorinated analogues, *Cancer Res.*, 44, 5624, 1984.
96. **Olsson, I., Gullberg, U., Ivhed, I., and Nilsson, K.**, Induction of differentiation of the human histiocytic lymphoma cell line U-937 by 1 alpha,25-dihydroxycholecalciferol, *Cancer Res.*, 43, 5862, 1983.
97. **Dodd, R. C., Cohen, M. S., Newman, S. L., and Gray, T. K.**, Vitamin D metabolites change the phenotype of monoblastic U937 cells, *Proc. Natl. Acad. Sci. U.S.A.*, 80, 7538, 1983.
98. **Mangelsdorf, D. J., Koeffler, H. P., Donaldson, C. A., Pike, J. W., and Haussler, M. R.**, 1,25-dihydroxyvitamin D_3-induced differentiation in a human promyelocytic leukemia cell line (HL-60): receptor-mediated maturation to macrophage-like cells, *J. Cell Biol.*, 98, 391, 1984.
99. **Yamanouchi, T., Tsushima, T., Murakami, H., Sato, Y., Shizume, K., Oshimi, K., and Mizoguchi, H.**, Differentiation of human promyelocytic leukemia cells is accompanied by an increase in insulin receptors, *Biochem. Biophys. Res. Comm.*, 108, 414, 1982.
100. **Tanaka, H., Abe, E., Miyaura, C., Shiina, Y., and Suda, T.**, 1 alpha,25-dihydroxyvitamin D_3 induces differentiation of human promyelocytic leukemia cells (HL-60) into monocyte-macrophages, but not into granulocytes, *Biochem. Biophys. Res. Comm.*, 117, 86, 1983.
101. **Miyaura, C., Abe, E., Honma, Y., Hozumi, M., Nishii, Y., and Suda, T.**, Cooperative effect of 1 alpha,25-dihydroxyvitamin D_3 and dexamethasone in inducing differentiation of mouse myeloid leukemia cells, *Arch. Biochem. Biophys.*, 227, 379, 1983.
102. **Reitsma, P. H., Rothberg, P. G., Astrin, S. M., Trial, J., Bar-Shavit, Z., Hall, A., Teitelbaum, S. L., and Kahn, A. J.**, Regulation of *myc* gene expression in HL-60 leukaemia cells by a vitamin D metabolite, *Nature (London)*, 306, 492, 1983.
103. **Matsui, T., Takahashi, R., Mihara, K., Nakagawa, T., Koizumi, T., Nakao, Y., Sugiyama, T., and Fujita, T.**, Cooperative regulation of c-*myc* expression in differentiation of human promyelocytic leukemia induced by recombinant gamma-interferon and 1,25-dihydroxyvitamin D_3, *Cancer Res.*, 45, 4336, 1985.

INDEX

A

A172 glioblastoma cells, 204
A431 epidermoid carcinoma cells, 7, 49
  epidermal growth factor in, 113—122, 124—127
  thyroid hormones in, 219
  transferrin in, 214
A673 rhabdomyosarcoma cells, 7
AAF, see 2-Acetylamino-fluorene
Abelson murine leukemia virus (A-MuLV), 8, 12—13, 55, 57
  hematopoietic growth factors and, 160, 172
abl gene, 8, 55, 57, 120
2-Acetylamino-fluorene (AAF), 56
Acquired immune deficiency syndrome (AIDS), 167—168
ACTH, see Adrenocorticotropic hormone
Actin, 45, 126—127, 199
Acute phase response, 161
Adenovirus, 164, 204, 221
Adenylate cyclase, 5, 31—38, 42, 50, 60
  calmodulin-dependent, 45
  C component, 31—32
  G component, 31—32
  ras gene protein products and, 37—43
  R component, 31
  regulation of, 32
Adhesion plaque, 199
Adipocytes, 90, 176
Adipose tissue, 49, 91
ADP-ribosyl transferase, 172
Adrenal gland, 141
Adrenocortical carcinoma, 106
Adrenocorticotropic hormone (ACTH), 2, 28—29
Adriamycin, 48
AEV, see Avian erythroblastosis virus
Agrobacterium tumefaciens, 14
AIDS, see Acquired immune deficiency syndrome
A-kinase, see Protein kinase, cAMP-dependent
AKR-2B mouse embryo cells, 7, 126—127, 149—150
AKR T-cell lymphoma cells, 173
Alpha-fetoprotein, 233
Alpha-globin, 119
Alström syndrome, 88
alu family, 85
ALV, see Avian leukemia virus
Amino acid uptake, 92, 150, 198
Aminoacyl-tRNA, 37
Amnion, 116
Amnion cells, 6
A-MuLV, see Abelson murine leukemia virus
AMV, see Avian myeloblastosis virus
Anchorage-independent growth, 7, 42, 94, 147—151, 221
Androgen, 227, 229
Androgen receptor, 228
Anemia, aplastic, 158

Aneuploidy, 41
ANF, see Atrial natriuretic factor
Angiogenesis, 49
Angiotensin, 44, 47—48, 89
Animal tumor, 12—14
Aniridia-Wilms' tumor association, 85
Annelida oligocheta, 86
Annelids, 86
Antiandrogen, 229
Anticalmodulin drug, 44, 53
Antiestrogen, 230
Antigen p97, melanoma surface, 21
Antigen-specific T-cell receptor, 162—163
Antimitogenic peptide factor, 7
Antiplatelet substance, 203
Anti-Tac antibody, 54
Antithyroid drug, 221
Aplastic anemia, 158
Aplysia, 39
Apotransferrin, 213
Arachidonic acid, 46—47, 147
Archaebacteria, 9
ARV, see Avian reticuloendotheliosis virus
Asparagine, 6, 52, 94
Astrocytes, 58
Astrocytoma cells, 116
ASV, see Avian sarcoma virus
Atherosclerosis, 202
Atrial natriuretic factor (ANF), 112
Autocrine growth, 5
Autocrine hypothesis of oncogenesis, 12—13
Autonomous growth, 12
Autophosphorylation, 39, 45—46, 197
Auxin, 14
Avian erythroblastosis virus (AEV), 8, 119
Avian leukemia virus (ALV), 123, 173
Avian leukemia virus E26 (E26-ALV), 8, 13
Avian myeloblastosis virus (AMV), 8
Avian reticuloendotheliosis virus (ARV), 8
Avian sarcoma virus (ASV), 53, 151

B

B16 melanoma cells, 33
Bacillus Calmette-Guerin, 176
BALB/3T3 cells, 53, 94—95, 151
BALB/c 3T3 cells, 1, 5, 33, 47, 49, 104, 196—200
BALB/MK cells, 42, 128
BALB murine sarcoma virus, 42
BCDF, see B-cell differentiation factor
B-cell differentiation factor (BCDF), 2, 12, 168—169
B-cell growth factor (BCGF, B-cell stimulating factor), 2, 168—169, 178
B-cells, 58, 159—162, 165, 178
B-cell stimulating factor (BSF-1), see B-cell growth factor

BC3H1 muscle cells, 141
BCGF, see B-cell growth factor
BDGF, see Bone-derived growth factor
BEN lung cancer cells, 60
Benzo(a)pyrene, 221
Benzodiazepine, 143
Beta-adrenergic receptor, 36
Beta-adrenergic response, 49, 51
Beta-cells, 51, 86
Beta-cell tumor, 86
Beta-globin gene, 85
Bladder cancer, 85, 121, 166
Bladder carcinoma cells, 159, 175, 214
Blast cells, 180, 231
Blast transformation, 43
Blood cells, 157
Blood clotting, 195
Blood coagulation factors, 111
B-*lym*-1 gene, 212
B-lymphocytes, 15, 160—161, 214
Bombesin, 13
Bombesin-like peptides, 13
Bombesin receptor, 13
Bombyx mori, 86
Bone, 121
  remodeling of, 161
  resorption of, 113, 147
Bone cancer, 121
Bone-derived growth factor (BDGF), 2, 147
Bone marrow, 157, 173, 177, 196
Bone marrow cells, 105, 170—171
Brain, 46, 141—142, 212
Brain tumor, 114, 122
Breast cancer, 105—106, 121—122, 220, 228—230
8-Bromo-cyclic AMP, 34
BSC-1 kidney cells, 151
BSF-1, see B-cell growth factor
BSF-2, see B-cell stimulating factor 2
Burst-promoting activity, 172, 177

C

C10/MJ-2 T-cells, 174
C127 fibroblasts, 33
C3H10T/2 embryo cells, 128, 221, 229
c-*abl* gene, 10—11, 14, 16, 34, 56, 59, 180
Cachectin, 176
Cachexia, 92
Calcineurin, 46, 60, 118
Calcitonin, 3
Calcitonin gene, 54
Calcitriol, 17, 95, 107, 233—234
  cell differentiation and, 234
  hematopoietic growth factors and, 171, 179
  proto-oncogene expression and, 234
Calcium, 16, 29—30, 52, 55, 63
  cell proliferation and, 47—48
  DNA synthesis and, 47—48
  in epidermal growth factor action, 118, 126, 128
  extracellular, 53

in fibroblast growth factor action, 142
in insulin action, 90—91
in interleukin-2 action, 167
in platelet-derived growth factor action, 197, 199
Calcium-calmodulin system, 32, 43—46, 52—55
Caldesmon, 45—46
Calmodulin, see also Calcium-calmodulin system, 63, 89—92, 123, 228
Calmodulin antagonist, 123
Calmodulin-binding protein, 31
Calvaria, neonatal mouse, 147, 161
cAMP, see Cyclic AMP
Cancer, see also specific cancers
  proto-oncogenes and, 9—10
  steroid hormone receptors and, 228
  thyroid hormones and, 220—222
CAP, see Catabolite gene activator protein
Carcinogen, 147
  chemical, 221, 229
Carcinogenesis, 10
  epidermal growth factor in, 128—129
5637 Carcinoma cells, 159—160
Cartilage-derived growth factor (CDGF), 2, 6
Casein, 127
Catabolite gene activator protein (CAP), 31
Catecholamines, 32
CBGF, see Colostrum basic growth factor
c-B-*lym*-1 gene, 10—11
CDC28 gene, 34, 115
CDGF, see Cartilage-derived growth factor
Cell cycle, see also specific phases of cell cycle, 1—4, 15, 27, 161
Cell differentiation, 27, 38, 54, 56
  calcitriol and, 234
  insulin and, 93—95
  insulin-like growth factor receptors and, 107
  insulin receptors and, 107
  interferon and, 174—175
  phorbol ester-induced, 50—52
  protein phosphorylation and, 62—63
  ras proteins and, 41—42
  terminal, 41, 45, 57, 158, 175
  transferrin and, 212
Cell differentiation-inducers, 49—52
Cell division, 44, 92
Cell growth, 200, 212
Cell membrane, 28, 36, 51, 124
Cell morphology, 133
Cell proliferation, 1, 4, 15, 27, 38, 53, 61
  calcium and, 47—48
  cAMP and, 32—33
  cGMP and, 35
  epidermal growth factor in, 126, 128
  growth factor regulation of, 4—8
  growth inhibiting factors and, 7
  hormonal regulation of, 4—8
  insulin and, 92—95
  interferon and, 174—175
  protein kinase C and, 47—48
  protein phosphorylation and, 62—63
  ras proteins and, 41—42

transforming growth factor β in, 150—151
Cell surface, 39
Cell volume, 92
Central nervous system, 58, 113
c-*erb* gene, 119, 227
c-*erb*-A gene, 10—11, 234
c-*erb*-B gene, 10—11, 123—124, 198
c-*erb*-B-2 gene, 120
Cerebrospinal fluid, 104
Cervical cancer, 121
c-*ets* gene, 161
c-*ets*-1 gene, 10—11
c-*ets*-2 gene, 10—11
c-*fes* gene, 10—11, 59, 172
c-*fgr* gene, 10—11, 59
c-*fms* gene, 170—171
c-*fos* gene, 10—11, 15—17, 45
    epidermal growth factor and, 127—128
    hematopoietic growth factors and, 164, 171—172, 179—180
    nerve growth factor and, 143
    platelet-derived growth factor and, 199, 201
CGF, see Chondrocyte growth factor
cGMP, see cyclic GMP, 32
CHE cells, see Chinese hamster embryo cells
Chemotaxis, 196—197
Chick embryo, 113
Chick embryo cells, 92
Chinese hamster embryo (CHE) cells, 123
Chinese hamster ovary (CHO) cells, 34, 87, 177
CHO cells, see Chinese hamster ovary cells
Cholecalciferol, 233
Cholera toxin, 32—33, 37, 126
Choline kinase, 49
Chondrocyte growth factor (CGF), 2
Chondrosarcoma cells, 105
Chorion, 116
c-H-*ras* gene, 10—11, 14, 16, 37
    EJ/T24 mutant, 18, 33, 40—41, 126
    epidermal growth factor and, 114
    hematopoietic growth factors and, 175
    steroid hormones and, 230
c-H-*ras*-1 gene, 42, 85
c-H-*ras*-2 gene, 10—11
Chromatin, 6, 62, 219, 227—228
Chromosome abnormality, 5
CIL, see Colony-inhibiting lymphokine
c-*int*-1 gene, 10—11
c-K-*ras* gene, 10—11, 14, 16, 18, 37
c-K-*ras*-1 gene, 10—11
Cloudman S91 melanoma cells, 89—90, 92, 106
c-*met* gene, 10—11, 56
c-*mos* gene, 10—11, 61, 233
c-*myb* gene, 10—11, 14, 166, 172, 178—180
c-*myc* gene, 7, 10—18, 35, 41, 50—55, 127, 151
    hematopoietic growth factors and, 160, 164, 167, 171—172, 175, 178—180
    platelet-derived growth factor and, 197—201
    steroid hormones and, 234
    thyroid hormones and, 219
    transferrin and, 214

c-*neu* gene, 10—11, 120
c-N-*myc* gene, 10—11
c-N-*ras* gene, 10—11, 37
Coated pit, 28
Cocarcinogen, 94
Collagenase, 199
Collagen synthesis, 161
Colon carcinoma cells, 7
Colonic adenocarcinoma, 106
Colony formation, 18, 147, 151
Colony-inhibiting lymphokine (CIL), 158
Colony-stimulating factor (CSF), 158
Colony-stimulating factor 1 (CSF-1, macrophage-colony stimulating factor), 2, 17, 169—171, 180
Colony-stimulating factor 1 (CSF-1) receptor, 11, 159, 170—171
Colony-stimulating factor 2 (CSF-2, granulocyte-macrophage colony-stimulating factor), 2, 119, 169—174, 177
Colony-stimulating factor 2 (CSF-2) gene, 3, 11, 171—172
Colostrum, 147
Colostrum basic growth factor (CBGF), 2, 195
Commitment, 4, 157
Competence, 15, 17
Competence factor, 5, 15, 171, 196, 199—201
Concanavalin A, 15, 28, 167, 178
Connective tissue, 125, 196
Contact inhibition, 33
COS monkey cells, 159, 176
Countertranscript, 51—52
c-*raf* gene, 10—11, 16—17, 166
c-*ras* gene, 33, 38—41, 121, 143, 214
    EJ/T24 mutant of, 175
Creatine phosphokinase, 141
c-*ros* gene, 59
Crown gall, 14—15
CSF, see Colony-stimulating factor
c-*sis* gene, 10—11, 16, 53, 141
    platelet-derived growth factor and, 198, 202—204
    steroid hormones and, 233
c-*src* gene, 19, 50—53, 56, 59, 143
    epidermal growth factor and, 115
    hematopoietic growth factors and, 175
    platelet-derived growth factor and, 200
c-*srs* gene family, 178
c-*src*-1 gene, 10—11
c-*src*-2 gene, 10—11
CV-1 cells, 219
Cyclic AMP (cAMP), 5—6, 15, 29—35, 200, 212, 228
    cell proliferation and, 32—33
    control of proto-oncogene expression by, 35
    transformation and, 32—33
Cyclic AMP-CAP complex, 31
Cyclic GMP (cGMP), 29, 32, 35
Cyclin, 6, 127
Cycloheximide, 16—17
Cyclosporin A, 54, 179
c-*yes* gene, 10—11, 59

Cytokine, 176
Cytokinin, 14
Cytoplasm, 219
Cytoskeleton, 118, 142, 198—199
Cytotoxic T-cells, 161, 166, 174
Cytotoxin, 175—176

D

Daudi cells, 54, 175, 180
Debromoaplysiatoxin, 50
Decidua, 116
DENA, see Diethylnitrosamine
Developmental regulation, 38—39
Dexamethasone, 6, 54, 179, 230—234
Diabetes mellitus, 86, 94
1,2-Diacylglycerol, 46—54, 91, 93, 142
  epidermal growth factor and, 124, 126
  hematopoietic growth factors and, 167
  platelet-derived growth factor and, 197—198
Diacylglycerol kinase, 126
Diacylglycerol lipase, 46, 50
Dibutyryl cyclic AMP, 35
Dictostelium discoideum, 39
Diethylnitrosamine (DENA), 56
Differentiation, see Cell differentiation
Differentiation antigens, 165
7,12-Dimethylbenz(*a*)anthracene (DMBA), 94, 221, 230
Dimethylnitrosamine, 228
Dimethyl sulfoxide (DMSO), 45, 51, 54, 57, 107
  epidermal growth factor and, 122
  hematopoietic growth factors and, 171, 179—180
Dithiothreitol, 88
DMBA, see 7,12-Dimethylbenz(*a*)anthracene
DMSO, see Dimethyl sulfoxide
DNA
  rearrangements of, 172—173
  repeated sequences in, 85, 159
  supercoiling of, 34, 62
DNA binding proteins, 30
DNA gyrase, 62
DNA nicking activity, 128
DNA polymerase, 93
DNA synthesis, see also Cell proliferation, 15, 30, 35, 42—44, 53, 105
  calcium and, 47—48
  epidermal growth factor and, 126
  growth factor regulation of, 4—8
  hormonal regulation of, 4—8
  insulin and, 92—94
  protein kinase C and, 47—48
  transforming growth factor and, 150
DNA topoisomerase, 34, 62—63, 128
Double minute chromosome, 116
Down regulation of receptors, 89, 171
  epidermal growth factor, 118, 128
  platelet-derived growth factor, 197, 201, 204
  transferrin, 214
  transforming growth factor, 147, 150
Drosophila melanogaster, 35, 38, 50, 56, 86, 115

E

E26-ALV, see Avian leukemia virus E26
EBPA, see Hemopoietin-2
EBV, see Epstein-Barr virus
ECGF, see Endothelial cell growth factor
EDF, see Eosinophil differentiation factor
EDGF-II, see Eye-derived growth factor II
EGF, see Epidermal growth factor
EGF-induced growth inhibition, 117
Ehrlich ascites tumor, 232
Elevator model, 28
Elongation factors, 37—38
Embryo fibroblasts, 33, 41, 120, 149, 221, 233
Embryogenesis, 59
Embryonal carcinoma cells, 104, 204
Embryo neuroretina cells, 19
Embryonic growth factor, 104
Endocrine gland, 1
Endocytosis, 28, 118, 212—214
Endoderm cells, 104
Endoplasmic reticulum, 28
Endothelial cell growth factor (ECGF), 2, 141, 205
Endothelial cells, 49, 125, 141, 158, 161, 174, 202—203
Enhancer, 17, 30, 42, 143, 159, 233
Enolase, 126
env gene, 201, 203
Eosinophil differentiation factor (EDF, interleukin-4), 2
EPA, see Erythroid-potentiating activity
Epidermal cells, 42, 128
Epidermal growth factor (EGF, urogastrone), 2—7, 12—13, 28, 39, 49—52, 61—63, 111—129
  biological actions of, 113—114
  biosynthesis of, 111—113
  in carcinogenic processes, 128—129
  in cell proliferation, 42—43, 128
  cyanide bromide-cleaved, 127
  effects on cell membrane, 124
  insulin and, 92—94
  insulin-like growth factors and, 104
  mechanism of action of, 113—114
    at nuclear level, 126—127
    postreceptor, 124—129
  phosphoinositide metabolism and, 126
  platelet-derived growth factor and, 196—200, 205
  precursor of, 61, 112—113
  production of, 113—114
  protein phosphorylation and, 124—125
  in proto-oncogene expression, 16, 18, 127—128
  steroid hormones and, 229
  structure of, 111
  synthesis of, 111—113
  transforming growth factor and, 148—151
  tumor promoters and, 123—124

vaccinia virus growth factor and, 111
v-*ras* protein product and, 125—126
Epidermal growth factor (EGF) gene, 3, 11, 111—112
Epidermal growth factor (EGF) receptor, 7, 46, 49, 57, 59, 63, 114—124
 changes in, 122
 c-*neu* gene and, 120
 degradation of, 118—119
 in different types of cells, 116—117
 epidermal growth factor and, 111, 114, 126—128
 erb-B gene and, 119—120
 functional heterogeneity of, 116—117
 in human tumors, 121—122
 insulin and, 89
 internalization of, 118—119
 in malignant cells, 122—123
 phosphorylation of, 117—118
 platelet-derived growth factor and, 204
 processing of, 117—118
 steroid hormones and, 229
 structure of, 115
 synthesis of, 115—116
 thyroid hormones and, 219—220
 transforming growth factor and, 147—151
 tumor promoters and, 123—124
 v-*src* gene and, 120—121
Epidermal growth factor (EGF) receptor gene, 11, 114—115, 122
Epidermal growth factor receptor-related polypeptide (ERRP), 117
Epidermal growth factor-related growth factors, 2, 113—114, 205
Epidermoid carcinoma cells, 118
Epithelial cells, 42—43
Epstein-Barr virus (EBV), 166, 169, 178
erb-A gene, 8
erb-B gene, 8, 119—120
ERRP, see Epidermal growth factor receptor-related polypeptide
Erythroblastosis, 119, 123
Erythroblasts, 13
Erythrocytes, 157, 177
Erythroid burst-promoting activity (EBPA), see Hemopoietin-2
Erythroid differentiation, 41
Erythroid-potentiating activity (EPA), 2, 177
Erythroid precursor cells, 178
Erythroleukemia cells, 231
Erythropoiesis-stimulating factor, 177—178
Erythropoietin, 2, 177—178
Erythropoietin gene, 177
Escalator model, 28
Escherichia coli, 28, 37, 86
Estradiol, 28, 56, 227—230, 233
Estradiol receptor, 228
Estrogen, 122, 227—230
Estrogen receptor, 121, 227
Ethylnitrosourea, 120
ets gene, 8
Evolution, of *ras* proteins, 38—39

Eye-derived growth factor II (EDGF-II), 141
Eyelid opening, 113, 148—149

F

F2408 fibroblastoid cells, 12
Fat cells, 91
Fatty acids, 47
FBJ-MOV, see FBJ murine osteosarcoma virus
FBJ murine osteosarcoma virus (FBJ-MOV), 8
FDGF, see Fibroblast growth factor
Feline sarcoma virus (FeSV), 149
Ferritin, 213
fes/fgr gene, 127
fes gene, 8, 57, 120
FeSV, see Feline sarcoma virus
Fetal calf serum, 16, 43, 89
Fetal development, 212
Fetal membranes, 116
α-Fetoprotein, 233
FGF, see Fibroblast growth factor
fgr gene, 8
Fibroblast-derived growth factor, see Fibroblast growth factor
Fibroblast growth factor (FGF), 2, 5, 13—16, 43, 49, 51, 63
 functions of, 141—142
 hematopoietic growth factors and, 174
 mechanism of action of, 142
 platelet-derived growth factor and, 204
 types of, 141
Fibroblast interferon, see Interferon-β
Fibroblasts, 13, 17, 43, 141
 epidermal growth factor in, 112, 116, 121
 hematopoietic growth factors in, 158, 176
 platelet-derived growth factor in, 196—201
 transforming growth factor in, 149
Fischer rat 3T3 fibroblasts, 7, 18, 151
fms gene, 8, 120
F-MuLV, see Friend murine leukemia virus
Focal contact, 199
Foreskin fibroblasts, 43, 56, 198, 200
fos gene, 8
fps gene, 57, 120
Friend erythroleukemia cells, 45, 54, 107
Friend murine leukemia virus (F-MuLV), 34, 178, 180
Fujinami sarcoma virus, 53

G

G_0 phase, 4, 6, 48, 199
 hematopoietic growth factors and, 168, 171, 174—175, 179—180
G_1 phase, 1—6, 35, 38, 43, 48, 53
 insulin and, 94
 hematopoietic growth factors and, 164—165, 168, 171, 175, 179—180
 thyroid hormones and, 222

transferrin and, 214
G$_2$ phase, 1—4, 168, 231
GABA, see Gamma-aminobutyric acid
Gamma-aminobutyic acid (GABA), 19
Gardner-Rasheed feline sarcoma virus (GR-FeSV), 8, 127
Gastric acid secretion, 111—113
Gastric carcinoma, 61, 106, 121
Gastrin-17, 125
Gastrin-releasing peptide (GRP), 113
G-CSF, see Granulocyte colony-stimulating factor
Gene duplication, 211
Gene expression, cell-cycle dependent, 6
Genetic code, 28
Genetic recombination, 62
Gestagen, 227
Gestation, 45
GGPF, see Glial growth-promoting factor
GH, see Growth hormone
GH3 pituitary tumor cells, 219
Gibbon leukemia virus (GLV), 163
Glial cells, 196
Glial growth-promoting factor (GGPF), 2, 168
Glioblastoma, 116, 122
Glioblastoma multiforme, 116, 122
Glioma cells, 142, 202
α-Globin, 119
β-Globin gene, 85
Glucagon, 94
Glucagon receptor, 106
Glucocorticoid(s), 7, 222, 227—233
Glucocorticoid receptor, 227—228, 230—232
Glucocorticoid receptor gene, 3, 11, 230
Glucose uptake, 46, 88, 90, 92, 150
Glutamic acid decarboxylase, 19
GLV, see Gibbon leukemia virus
Glycogen, 39, 90, 93
Glycogen synthase, 91, 93, 113
Glycolysis, 150, 198
GM-CSF, see Colony-stimulating factor 2
Golgi apparatus, 28
Gonadotropin, 2
gp68$^{v\text{-}erb\text{-}B}$, 119
gp74$^{v\text{-}erb\text{-}B}$, 119
gp140$^{v\text{-}fms}$, 170
G proteins, 36—38, 41
Granulocyte colony-stimulating factor (G-CSF), 2, 173, 177, 180
Granulocyte colony-stimulating factor (G-CSF) receptor, 173
Granulocyte-macrophage colony-stimulating factor (GM-CSF), see Colony-stimulating factor 2
Granulocytes, 157, 171—173, 231
GR-FeSV, see Gardner-Rasheed feline sarcoma virus
Growth factor(s)
  assay for, 1
  autostimulating, 13
  common functions with hormones and oncogene proteins, 18—19
  effect on proto-oncogene expression, 15—18
  independence of, proto-oncogenes and, 14
  mitogenic action of, 5
  in oncogenic processes, 11—15
  regulation of cell proliferation by, 4—8
  regulation of DNA synthesis by, 4—8
  in v-onc-transformed cells, 18
Growth factor receptors, 28—29
Growth hormone (GH, somatotropin), 2, 31, 103—104, 125
Growth hormone (GH, somatotropin) gene, 3, 11
Growth hormone (GH, somatotropin) receptor, 122
Growth-inhibiting factors, 6—8
Growth inhibitor protein, 126
GRP, see Gastrin-releasing peptide
GTPase
  intrinsic, 41
  Mg^{2+}-dependent, 32
Guanine nucleotide-binding protein, 35—43, 46
Guanosine nucleotides, in adenylate cyclase system, 31—32
Guanylate cyclase, 32, 35, 46

H

H35 hepatoma cells, 105
Hamster embryo cells, 221
Harvey murine sarcoma virus (H-MuSV), 8, 33, 38, 41
  epidermal growth factor and, 125
  hematopoietic growth factors and, 160, 178
  insulin and, 93
hCG, see Human chorionic gonadotropin
Heart mesenchymal cells, 13
Heat-shock genes, 50
Heat-shock proteins, 18, 50
Heat-shock response, 35
Helper T-cells, 161, 168
Hematologic malignant disease, 10, 158
Hematopoiesis, 157
Hematopoietic cells, 173, 214
Hematopoietic growth factors (HGF), 157—181
  cytotoxins, 175—176
  erythropoiesis-stimulating factors, 177—178
  granulocyte-macrophage colony stimulating factors, 169—174
  hemopoietins, 158—160
  interferons, 174—175
  lymphokines, 160—169
Hematopoietic neoplasm, 106
Hemin, 213
Hemopoietin, 158—160
Hemopoietin-1, 2, 159
Hemopoietin-2 (interleukin-3, mast-cell growth factor, erythroid burst-promoting activity), 2, 119, 172, 177—181
  oncogene expression and, 160
  structure of, 159
Hemopoietin-2 gene, 159
Hepatitis B-virus, 85
Hepatocellular carcinoma, 49, 56, 85, 178

Hepatocyte growth factor, 205
Hepatocytes, 49, 91, 93, 128, 220
Hepatocyte-stimulating factor (HSF), 2, 161
Hepatoma, 121, 231
Hepatoma cells, 87, 89, 93—94, 107
Herpes simplex virus type 1, 42
Hexamethylene bisacetamide, 171
hGH, see Human growth hormone
Histones, 6, 30, 34—35, 51, 90, 125, 143, 228
HL-60 promyelocytic leukemia cells, 17, 50—51, 54, 107
   hematopoietic growth factors in, 160, 171, 175—176, 179—180
   steroid hormones in, 231, 234
   transferrins in, 213—214
H-MuSV, see Harvey murine sarcoma virus
Homogeneously staining region, 116
Hormone(s)
   assay for, 1
   common functions with growth factors and oncogene proteins, 18—19
   effect on proto-oncogene expression, 15—18
   mitogenic action of, 5
   in oncogenic processes, 11—15
   in plant tumors, 14
   regulation of cell proliferation by, 4—8
   regulation of DNA synthesis by, 4—8
Hormone receptors, 28—29
Host defense, 174
H-*ras* gene, 8
HSF, see Hepatocyte-stimulating factor
HT-1080 fibrosarcoma cells, 204
HTLV, see Human T-cell leukemia virus
Human chorionic gonadotropin (hCG), 121, 125
Human growth hormone (hGH), 54
Human growth hormone (hGH) gene, 232
Human idiogram, 9—11
Human T-cell leukemia virus (HTLV), 163, 174
Human T-cell leukemia virus I (HTLV-1), 164—166
Human T-cell leukemia virus II (HTLV-II), 164, 168
Human T-cell leukemia virus III (HTLV-III), 168
Hydrogen ions, 30, 48
Hydrogen peroxide, 90
p-Hydroxybenzoate hydroxylase, 14
Hypercalcemia, 147
Hyperthyroidism, 221
Hypothalamus, 7
Hypothyroidism, 170, 220—221
Hypoxia, 177

I

Idiogram, human, 9—11
IFN, see Interferon
IGF, see Insulin-like growth factor
IL-1, see Interleukin-1
IL-2, see Interleukin-2
IL-3, see Hemopoietin-2
IL-3-like proteins, 2, 159—160
IL-4, see Eosinophil differentiation factor
IM-9 cells, 89, 106—107
Immortality of cells, 5, 42
Immune interferon, see Interferon-α
Immune response, 161—162, 164
Immunoglobulin, 162, 165, 168—169
Incisor eruption, 149
Indole alkaloid, 123
Indomethacin, 113, 147
Inflammatory response, 161, 168, 174, 196
Inhibin, 149
Initiation factors, 37—38
Inositol phospholipids, see Phosphoinositides
Inositol 1,4,5-triphosphate, 16, 46—50, 54, 197, 199
Insulin, 2, 4, 12, 28, 49, 51, 56, 62, 85—95
   circulating, 89
   effect on cell differentiation, 93—95
   effect on cell proliferation, 93—94
   effect on DNA synthesis, 93—94
   epidermal growth factor and, 125—126
   mechanism of action of
      post-transductional, 91—94
      transductional, 90—91
   in neoplastic cells, 94—95
   platelet-derived growth factor and, 196
   postreceptor defects of, in neoplastic cells, 93
   protein phosphorylation and, 92—93
   steroid hormones and, 229
   structure of, 85—86
   synthesis of, 86
   transforming growth factor and, 150
Insulin gene, 3, 11, 85—86, 103
Insulin-like growth factor (IGF), 93, 103—107
Insulin-like growth factor (IGF) gene, 103—104
Insulin-like growth factor (IGF) receptor, 105—107
Insulin-like growth factor I (IGF-I, somatomedin C), 2, 49, 56, 103, 196, 200
Insulin-like growth factor I (IGF-I, somatomedin C) gene, 3, 103
Insulin-like growth factor I (IGF-I, somatomedin C) receptor, 49, 105
Insulin-like growth factor II (IGF-II, somatomedin A, multiplication-stimulating activity), 2, 49, 103
Insulin-like growth factor II (IGF-II) gene, 3, 11, 103
Insulin-like growth factor II (IGF-II) receptor, 105
Insulin mediator, 91
Insulin mimicker, 90
Insulinopathy, 86
Insulin receptor, 49, 56—57, 87, 93, 95
   defective, 88
   epidermal growth factor and, 120—121
   functional role of, 90
   internalization of, 90
   in neoplastic cells, 89—90, 106—107
   oncogene protein products and, 89
   phosphorylation of, 87—90
   in primary tumors, 105—106

regulation of expression of, 89
steroid hormones and, 234
structure of, 87
synthesis of, 87
thyroid hormones and, 219—220
in transformed cells, 105—107
Insulin receptor gene, 3, 11, 87
Insulin receptor kinase, 92
Insulin resistance, 88, 93, 95
int-1 gene, 232
int-2 gene, 232
Interferon (IFN), 3, 5, 54, 174—175
Interferon-α (IFN-α), 174—175
Interferon-α (IFN-α) gene, 3, 11
Interferon-α (IFN-α) receptor gene, 3
Interferon-β (IFN-β), 161, 174—175, 180
Interferon-β (IFN-β) gene, 3, 11
Interferon-γ (IFN-γ), 5, 161, 169, 174, 176, 200
Interferon-γ (IFN-γ) gene, 3, 11
Interferon (IFN) receptor, 174—175
Interleukin, 12
Interleukin-1 (IL-1), 2, 7, 141, 160—162, 168
Interleukin-1 alpha, 162
Interleukin-1 beta, 162
Interleukin-1 (IL-1) gene, 3, 161—162
Interleukin-1 (IL-1) receptor, 161
Interleukin-1 (IL-1) receptor gene, 166
Interleukin-2 (IL-2, T-cell growth factor), 2, 7, 12, 28, 30, 54, 157, 160—168, 179
  acquired immune deficiency syndrome and, 167—168
  in ion transport, 167
  in neoplastic cells, 165—166
  in normal cells, 165—166
  phosphoinositide metabolism and, 166—167
  protein kinase C and, 166—167
  proto-oncogene expression and, 167—168
  structure of, 163
  synthesis of, 163
  transferrin and, 214
Interleukin-2 (IL-2) gene, 3, 11, 163—164
Interleukin-2 (IL-2) inhibitor, 166
Interleukin-2 (IL-2) receptor, 164—169, 179, 214
Interleukin-2 (IL-2) receptor gene, 3, 11, 164, 165
Interleukin-3 (IL-3), see Hemopoietin-2
Interleukin-4 (IL-4), see Eosinophil differentiation factor
Intestine, 211
Intracisternal A particle, 159
IO-3 cells, 172
Iodide uptake, 41—42, 127, 221
Ion channel, 29
Ion fluxes, 30, 48, 63
Ionomycin, 16, 54—55, 167, 179, 197, 199
Ionophore A23187, 16, 43, 55
  epidermal growth factor and, 128
  hematopoietic growth factors and, 161, 167
  platelet-derived growth factor and, 197, 199
Ion transport, 113, 167, 198
Iron-binding protein, 211—215
Islets of Langerhans, 86

J

JEG-3 choriocarcinoma cells, 121
Jurkat cells, 163, 167

K

K562 leukemia cells, 180
KD diploid fibroblasts, 112
Ke 37 lymphoma cells, 58
Keratinocytes, 6, 149
KG-1 myeloblastoid cells, 172
Kidney, cancer of, 94
Kidney cells, 113, 123, 178
Kirsten murine sarcoma virus (K-MuSV), 8, 12—13, 33, 41—42, 204
  epidermal growth factor and, 123, 128
  hematopoietic growth factors and, 178
  steroid hormones and, 233
  thyroid hormones and, 221—222
  transforming growth factor and, 147
KLE endometrial tumor cells, 229
K-MuSV, see Kirsten murine sarcoma virus
K-*ras* gene, 8

L

L1210 cells, 231
Lactate dehydrogenase, 126
Laryngeal cancer, 121
L-cells, 169
LDGF, see Leukemia-derived growth factor
Lectins, 28, 90, 164
Leishmania, 18
LepA protein, 37
Leukemia, 35, 157, 162, 231
  acute lymphoblastic, 106, 165
  acute lymphocytic, 170
  acute myelogenous, 171
  acute nonlymphocytic, 172, 180
  adult T-cell, 165—166
  chronic lymphocytic, 165—166, 231
  chronic lymphoid, 106
  chronic myelogenous, 56, 171, 178
  lymphoblastic, 231
  lymphoid, 58
  pre-B-cell, 87
Leukemia-derived growth factor (LDGF), 2, 173
Leukemic cells, 7, 50, 58—59, 63
  interleukin-2 receptors in, 165
  hematopoietic growth factors in, 158, 161, 172
  transforming growth factors in, 147
Leukocyte interferon, see Interferon-α
Leydig cells, 51
Lipopolysaccharide (LPS), 15, 48, 161, 167, 178, 199
Lipoprotein lipase, 176
Liver, 211

Liver cells, 44
Liver regeneration, 53, 57
LMF, see Lymphocyte mitogenic factor
Long terminal repeats, 17, 164
Low-density lipoprotein receptor, 112, 178
LPS, see Lipopolysaccharide
LSTRA lymphosarcoma cells, 53, 58—59
L3T4 antigen, 165
LTBM cells, 172
Lung, fetal, 113
Lung fibroblasts, 17, 203
Lung tumor cells, 202
Lymphoblastoid cells, 63, 213
Lymphoblasts, 106
Lymphocyte mitogenic factor (LMF), 160
Lymphocytes, 43, 53—54, 106, 157, 176—180, 213, 231
Lymphoid cells, 12, 161, 166, 173
Lymphoid organs, 157
Lymphoid tumor, 106
Lymphokines, 160—169
Lymphoma, 106, 157, 162, 231
    Burkitt's, 148, 169
    histiocytic, 90
    nonHodgkin, 12
    T-cell, 164
Lymphoma cells, 164
Lymphomagenesis, 166
Lymphopoiesis, 160
Lymphotoxin, see Tumor necrosis factor β
Lysosomes, 28, 90, 118, 197, 213
Lyt-2 antigen, 165

M

alpha$_2$-Macroglobulin, 195
Macrophage-activating factor (MAF), 2, 174
Macrophage-colony stimulating factor (MCSF), see Colony-stimulating factor 1
Macrophage-derived growth factor (MDGF), 2, 141, 173—174
Macrophages, 13, 158, 162, 171—176, 231
MAF, see Macrophage-activating factor
Magnesium, 43
Malignant cells
    epidermal growth factor receptor in, 122—123
    tyrosine-specific protein kinase in, 58—60
Malignant transformation, 59, 158
Mammary gland, 113, 229
Mammary gland cancer, 33, 95, 120
Mammary gland carcinoma cells, 51, 147
Mammary gland epithelium, 124, 127
Mammary gland explant, 93, 127
Mammary gland tumor, 128—129, 221, 230, 232
Manganese, 197
Mast-cell growth factor (MCGF), see Hemopoietin-2
Mast cells, 13, 160
MC29-AMCV, see MC29 avian myelocytomatosis virus

MC29 avian myelocytomatosis virus (MC29-AMCV), 8
MC3T3-E1 osteoblastic cells, 161
MCA, see 3-Methylcholanthrene
McDonough feline sarcoma virus (MD-FeSV), 8, 170
MCF-7 breast cancer cells, 60, 89, 94, 106, 229—230
MCGF, see Hemopoietin-2
MCSF, see Colony-stimulating factor 1
MDA-468 breast cancer cells, 114, 116—117, 121—122
MDCK cells, 90
MD-FeSV, see McDonough feline sarcoma virus
MDGF, see Macrophage-derived growth factor
Mediator, 27—31, 43, 91
Medullary thyroid carcinoma, 54
Megakaryocytes, 157, 195
Melanocyte growth factor (MGF), 2, 142
Melanocyte-stimulating hormone (MSH), 2, 33, 94
Melanoma cells, 122—123, 142, 148, 166, 211
MEL cells, 180
Meningioma, 122
Mesenchymal cells, 127, 174
Messenger RNA (mRNA), 16—18
    epidermal growth factor and, 126
    growth factor, 28
    spot, 14, 220
Metallothionein II gene, 232
Metazoa, insulin synthesis in, 86
Methylamine, 118
3-Methylcholanthrene (MCA), 94, 128, 176, 221
N-Methyl-N'-nitro-N-nitrosoguanidine, 221
Methylnitrosourea (MNU), 94
Methymazole, 221
MG-63 osteosarcoma cells, 204
MGF, see Melanocyte growth factor
MH2-ACV, see Mill Hill 2 avian carcinoma virus
mht gene, 8
Microenvironment, 5
Microfilaments, 127, 198—199
Microorganisms, 28
    insulin synthesis in, 86
Microtubules, 93
Milk, 111, 211
Mill Hill 2 avian carcinoma virus (MH2-ACV), 8
Mineralocorticoids, 227
Mitochondria, 219
Mitogenesis, 5, 93
    protein phosphorylation and, 63
Mitogenic agent, 59
Mitogenic signal, 198, 200
Mitosis, 30, 35, 43, 168, 214
MLA 144 leukemia cells, 163
MMGF, see Myelomonocytic growth factor
MMTV, see Mouse mammary tumor virus
M-MuLV, see Moloney murine leukemia virus
M-MuSV, see Moloney murine sarcoma virus
MNU, see Methylnitrosourea
Modulator, 29—30

Moloney murine leukemia virus (M-MuLV), 58, 232
Moloney murine sarcoma virus (M-MuSV), 8, 33, 61, 113, 204
MOLT4f cells, 173
Monensin, 118
Monoacylglycerol, 46
Monocytes
  hematopoietic growth factors in, 157, 161, 173—174, 179—180
  platelet-derived growth factor in, 196—197
  steroid hormones in, 234
Mononuclear phagocytes, 169—170
Monosomy, 172
Morris hepatoma, 106, 221
mos gene, 8, 120
Mouse mammary tumor virus (MMTV), 229, 232—233
mRNA, see Messenger RNA
MSA, see Insulin-like growth factor II
MSCF, see Colony-stimulating factor 1
MSH, see Melanocyte-stimulating hormone
Multilineage colony-stimulating activity, 172
Multilineage growth factor, 158
Multiple myeloma, 106
Multiplication-stimulating activity, see Insulin-like growth factor II
3611 Murine sarcoma virus (3611-MuSV), 8
3611-MuSV, see 3611 Murine sarcoma virus
myb gene, 8
myc gene, 8, 53—55, 150
Mycobacterium bovis, 176
Myeloblasts, 13, 173
Myeloid cells, 42, 172
Myeloid colonies, 170
Myeloid leukemia cells, 13, 173
Myelomonocytic growth factor (MMGF), 2, 13, 173
Myoblasts, 54
Myosin light chain, 199
Myristic acid, 58

N

132N1 astrocytoma cells, 50
77N1 cells, 151
Namalva cells, 169
Natural killer cells, 174
Nb2 lymphoma cells, 54
ND-TGF, see Neuronal-derived transforming growth factor
Neomycin, 48
Neoplastic cells
  erythropoietin in, 178
  glucocorticoid receptor in, 231—232
  hematic, proto-oncogene expression in, 179—181
  insulin in, 93—95
  insulin-like growth factor receptor in, 106—107
  insulin receptor in, 89—90, 106—107
  interleukin-2 in, 165—166
  platelet-derived growth factor in, 203—204
  steroid hormones in, 228—234
  transferrin receptor in, 214
Neoplastic transformation, 8, 41, 63
  protein phosphorylation and, 63—64
  thyroid hormones in, 221—222
Nerve growth factor (NGF), 2, 6, 52, 94
  c-fos expression and, 143
  mechanism of action of, 142—143
  protein phosphorylation and, 143
  structure of, 142
Nerve growth factor (NGF) gene, 3, 11, 142—143
Neurite growth, 94, 142—143
Neurite promoting factor (NPF), 142
Neuroblastoma, 122
Neuroblastoma cells, 54, 56, 94, 142, 151
Neuroendocrine cells, 113
Neuroglioblastoma cells, 120
Neuron, 19, 58
Neuronal-derived transforming growth factor (ND-TGF), 2
Neurotransmitters, 19, 29, 49, 113, 142
Neurotrophic factors, 142
Neutrophil elastase, 195, 197
Neutrophils, 161, 171, 196—197
Nevus, 122
NGF, see Nerve growth factor
NIH/3T3 cells, 16—17, 33, 38, 40—41, 51
  epidermal growth factor in, 120, 127
  platelet-derived growth factor in, 199—204
  steroid hormones in, 233
Nitrosomethylurea, 221
N-*myc* gene, 60
Nonsuppressible insulin-like activity (NSILA), see Insulin-like growth factor
Norepinephrine, 128
Normal human bronchial epithelial cells, 42
Normal rat kidney cells, see NRK cells
NPF, see Neurite promoting factor
NRK cells, 7, 33, 38, 53, 60
  epidermal growth factor in, 113, 125
  insulin in, 93
  thyroid hormones in, 222
  transforming growth factor in, 147—151
NRK fibroblasts, 122, 205
Nuclear membrane, 28, 90
Nuclear proteins, 6, 30, 48, 62, 143
Nucleus, 29, 54, 199—201, 219, 227—228
  epidermal growth factor in, 126—127
  regulation of functions of, 30

O

Obesity, 89
ODGF, see Osteosarcoma-derived growth factor
OKT3 antibody, 54, 179
OKT11A antibody, 54, 179
Oleate, 47
Oligodendrocytes, 168
Oncodevelopmental proteins, 45
Oncofetal antigen, 233

Oncogene(s), see also specific genes, 8—11
  expression of
    glucocorticoids and, 233
    hemopoietin-2 and, 160
  in plant tumors, 14
Oncogene protein products
  common functions with hormones and growth factors, 18—19
  insulin receptor and, 89
Oncogenesis, 11—15
Oncomodulin, 45
Oocytes
  Rana pipiens, 94
  rat, 48
  Xenopus, 162, 230
Oral cancer, 121
Ornithine decarboxylase, 5—6, 52, 94, 143
Osteosarcoma, 121
Osteosarcoma-derived growth factor (ODGF), 2
Ovotransferrin, 211

P

p21, 121
p21$^{c\text{-}ras}$, 39, 41—42, 214, 230
p21^{ras}, 37—38, 230
p21$^{v\text{-}ras}$, 33, 39, 42, 125—126, 160
p28$^{c\text{-}src}$, 204
p28$^{v\text{-}sis}$, 201—202
p37$^{v\text{-}mos}$, 61
p53 cellular protein, 167
p140$^{gag\text{-}fps}$, 53
p185, 120
PAK II, see Protease-activated kinase II
Palate formation, 113
Palmitic acid, 39
PANC-1 pancreatic carcinoma cells, 118
Pancreatic islets, 51
Parathyroid hormone (PTH), 3
Parathyroid hormone (PTH) gene, 3, 11, 85
Parvalbumin, 44
PC12 pheochromocytoma cells, 17, 41, 128, 142—143
PC13 embryonal carcinoma cells, 151
PCNA, see Proliferating cell nuclear antigen
PDGF, see Platelet-derived growth factor
Peptide growth factor, see Growth factor(s)
Peripheral blood lymphocytes, 50, 63, 179
Peripheral blood mononuclear cells, 54, 165, 167, 178—179, 214
Pertussis toxin, 36—37
pH, intracellular, 30, 167
PHA, see Phytohemagglutinin
Pheochromocytoma, 106, 117
Philadelphia chromosome, 56
Phorbol ester receptor, 50—51, 124
Phorbol esters, 48—54, 57, 107
  epidermal growth factor and, 118
  hematopoietic growth factors and, 161, 163—164
  insulin and, 89, 91, 95
  platelet-derived growth factor and, 200
  transferrin and, 214
Phorbol 12-myristate 13-acetate (PMA), 49, 51, 54, 61
  epidermal growth factor and, 123
  hematopoietic growth factors and, 165—166, 178—179
  platelet-derived growth factor and, 197
Phosphatidate, 91
Phosphatidic acid, 49, 52, 91, 126
Phosphatidylcholine, 49
Phosphatidylethanolamine methyltransferase, 91
Phosphatidylinositol, 46, 48—50, 54, 91
Phosphatidylinositol 4,5-bisphosphate, 48—49, 52, 91, 198
Phosphatidylinositol kinase, 49, 53, 126
Phosphatidylinositol 4-phosphate, 48—49, 52, 91
Phosphodiesterase, 31, 90
  calmodulin-sensitive, 32
  cAMP, 35
  cGMP, 35—37
Phosphoinositides, 16, 29, 46—55, 91, 199
  in insulin action, 91
  metabolism of, 48—52
    epidermal growth factor and, 126
    interleukin-2 and, 166—167
Phosphoinositol, 47
Phospholipase A$_2$, 47
Phospholipase C, 46, 48, 93
Phospholipids, methylation of, 91
Phosphotyrosine, 55
Phytohemagglutinin (PHA), 15, 52, 54, 161—162, 166, 178—179
Phytohormones, 14
Pinealocytes, 49
PI protein, 199—200
Pituitary gland, 104, 113, 141
PL, see Placental lactogen
Placenta, 5, 53, 103, 105
  epidermal growth factor in, 114, 116, 119
  insulin in, 87, 90
  platelet-derived growth factor in, 202
  transforming growth factor in, 147, 149
Placental lactogen (PL), 2
Placental membranes, 46
Plant cells, 44
Plant tumor, 14—15
Plasma membrane, 29, 219
Plasmid Ti, 14
Plasminogen activator, 52
Platelet basic protein, 195
Platelet-derived growth factor (PDGF), 2—6, 12—18, 43, 49—56, 104, 141, 195—205
  c-fos gene and, 199
  c-myc gene and, 199—200
  as competence factor, 200—201
  c-src gene and, 200
  in epidermal growth factor action, 197—198
  extranuclear effects of, 198—199
  functions of, 196
  insulin and, 93

mechanism of action of, 197—201
in neoplastic cells, 203—204
in nonproto-oncogene gene expression, 200
nuclear effects of, 199—201
sis gene and, 201—203
structure of, 195
in transformed cells, 203—204
transforming growth factor and, 150—151
Platelet-derived growth factor (PDGF) gene, 3, 11
Platelet-derived growth factor (PDGF) receptor, 49, 195—198, 201—202, 204
Platelet-derived permeability factor, 195
Platelet factor-4, 195
Platelets, 48, 50, 149, 151, 203, 205
PMA, see Phorbol 12-myristate 13-acetate
Polyacetate, 123
Polyamines, 5, 30, 52, 94, 229
Polykaryon, 6
Polyoma virus, 18, 33, 126
Polyoma virus middle-T antigen, 52
Polyphosphoinositide phosphodiesterase, 46
Polytene chromosome, 35
Postmitotic phase, 4
Post-transductional event, 27
Potassium, 30, 48
pp60$^{c\text{-}src}$, 52, 58—59, 200
pp60^{src}, 59, 62
pp60$^{v\text{-}src}$, 44, 52—53, 56—57, 61—62
  epidermal growth factor and, 120—121, 125
  hematopoietic growth factors and, 173, 175
  insulin and, 89, 92
  platelet-derived growth factor and, 200
  steroid hormones and, 228, 230
  transforming growth factor and, 149
Pre-AIDS syndrome, 168
Pre-DNA-synthetic phase, 4
Prenatal development, 148
Prepro-epidermal growth factor, 112
Preproinsulin, 86
Preproinsulin gene I, 86
Preproinsulin gene II, 86
Preprolactin, 5
Priess cells, 59
Progenitor cells, 158, 177
Progesterone, 94, 230, 233
Progesterone receptor, 121, 125, 228
Progestins, 122
Progression factor, 5, 171, 196, 200
Proinsulin, 86
Prolactin, 2, 141, 229
Prolactin gene, 31, 227
Proliferating cell nuclear antigen (PCNA), 6
Proliferin, 5
Proliferin-related protein (PRP), 5
Promoter, 30, 114, 172
Propylthiouracil, 221
Prostaglandins, 30, 33, 113, 147, 198, 200
Prostate, 228—229
Prostate cancer, 228—229
Prostate growth factor, 2
Protamines, 125

Protamine sulfate, 196
Protease, lysosomal, 28
Protease-activated kinase II (PAK-II), 61, 125
Protein kinase, 29, 32
  calcium/calmodulin-dependent, 30, 43, 45
  calcium-phospholipid dependent, see Protein kinase C
  cAMP-dependent, 30—36, 50, 56, 60—63, 143
    catalytic subunit of, 34
    epidermal growth factor and, 115, 118, 121, 125
    regulatory actions of, 34
    steroid hormones and, 228
  cGMP-dependent, 30, 35
  nontyrosine-specific, 60—62
  polyamine-dependent, 30
  tyrosine-specific, 49, 63, 147, 170
    epidermal growth factor and, 115, 119—120
    insulin and, 92
    in malignant cells, 58—60
    in normal cells, 57—60
    platelet-derived growth factor and, 196, 198
Protein kinase C, 16, 30, 43, 46—55, 60, 62, 142
  cell proliferation and, 47—48
  DNA synthesis and, 47—48
  epidermal growth factor and, 118, 124, 126, 128
  hematopoietic growth factors and, 160, 171, 179
  insulin and, 89, 91—92
  interleukin-2 and, 166—167
  phosphoinositides and, 46—55
  platelet-derived growth factor and, 198—200
  transferrin and, 213—214
Protein kinase modulator, 35
Protein phosphatase, phosphotyrosine-specific, 60
Proteins
  phosphorylation of, 30, 33, 55—63, 113
    cell differentiation and, 62—63
    cell proliferation and, 62—63
    epidermal growth factor and, 124—125
    insulin and, 92—93
    mitogenesis and, 63
    neoplastic transformation and, 63—64
    nerve growth factor and, 143
  synthesis of, 92—93, 161, 198
Prothoracicotropic hormone, 86
Proto-oncogenes, see also c-*onc* gene, 8—11
  cancer and, 9—10
  chromosomal assignment of, 9—11
  expression of
    calcitriol and, 234
    cAMP and, 35
    colony stimulating factor 1 and, 171
    colony stimulating factor 2 and, 172—173
    epidermal growth factor and, 127—128
    estrogens and, 230
    growth factors and, 14—18
    in hematic neoplastic cells, 179—181
    hormones and, 15—18
    interferon and, 175
    interleukin-2 and, 167—168
  protein products of, 9

Protoporphyrin IX, 213
PRP, see Proliferin-related protein
PTH, see Parathyroid hormone
PTR1 RNA, 18, 126
Pulmonary cancer, 121
Putrescine, 30, 229
PY815 mastocytoma cells, 35
Pyrogen, endogenous, 161

Q

5q-syndrome, 170, 172
Quiescent state, 3—4

R

Radioiodine, 221
raf gene, 8
Raji cells, 59
RAS-1 gene, 39
RAS-2 gene, 39
ras gene, 128
ras proteins, 37—43
Rat-1 cells, 150
Receptor, 1, 15, 27—29, 31, 51
  androgen, 228
  beta-adrenergic, 36
  bombesin, 13
  calcitriol, 233—234
  colony-stimulating factor 1, 159, 170—171
  epidermal growth factor, see Epidermal growth factor receptor
  estradiol, 228
  estrogen, see Estrogen receptor
  glucagon, 106
  glucocorticoid, see Glucocorticoid receptor
  G proteins and, 36
  granulocyte-colony stimulating factor, 173
  growth factor, 28—29
  hormone, 28—29
  IGF-I, 49
  insulin, see Insulin receptor
  insulin-like growth factor, see Insulin-like growth factor receptor
  interferon, see Interferon receptor
  interleukin-1, 161
  interleukin-2, see Interleukin-2 receptor
  internalization of, 28
  intracellular, 28
  low density lipoprotein, 112, 178
  in microorganisms, 28
  mobility of, 28
  nuclear, 29
  phorbol ester, 50—51, 124
  platelet-derived growth factor, see Platelet-derived growth factor receptor
  progesterone, 121, 125, 228
  regulation of expression of, 29
  spare, 88
  steroid hormone, 27, 227—228
  thrombin, 48
  thyroid hormone, see Thyroid hormone receptor
  thyroid stimulating hormone, 32
  transferrin, see Transferrin receptor
  transforming growth factor, see Transforming growth factor receptor
rel gene, 8
Renal carcinoma cells, 148
Renal failure, 177
Repressor protein, 163
Reticulocytes, 213
Reticuloendothelial system, 211
Retinoblastoma-derived growth factor, 60
Retinoic acid, 17—18, 51, 54, 95, 151, 179
Retrovirus, 147
  acute transforming, 8—9, 122—123
  chronic transforming, 8—9, 232
Reuber H35 hepatoma cells, 33
Rhodopsin, 36—38
Rhodospirillum rubrum, 58
Ribosomal protein S6, 48, 51, 61—62, 143
  epidermal growth factor and, 125
  insulin and, 92
  platelet-derived growth factor and, 198
  steroid hormones and, 230
Ribosomal protein S10, 62
R-*myc* gene, 171
RNA polymerase II, 232
RNA synthesis, 88, 92, 126, 143, 230
Rochester URII avian sarcoma virus (URII-ASV), 8, 52, 63
ros gene, 8, 52—53
Rous-associated virus type 1, 119
Rous sarcoma virus (RSV), 8, 12, 18, 33, 44, 53, 56, 59, 63
  epidermal growth factor and, 120
  hematopoietic growth factors and, 173, 175
  steroid hormones and, 233
RPMI-1788 lymphoblastoid cells, 176
RSV, see Rous sarcoma virus

S

S49 thymoma cells, 231
S6 kinase, 92, 125
Saccharomyces cerevisiae, 28, 34, 39, 58, 62, 227
Salivary gland adenocarcinoma, 120
Salivary gland epithelial cells, 151
Sarcoma, 121
  osteogenic, 204
Sarcoma cells, 202
Sarcoma growth factor (SGF), 148
Sea urchin eggs, 48
Second messenger, see Mediator
Serum, 6, 15, 51, 61, 63, 211
Sézary leukemic T-cells, 165
SF268 glioblastoma cells, 122
SGF, see Sarcoma growth factor
Simian sarcoma virus (SSV), 8, 201—202, 204

Simian virus 40 (SV40), 6, 12, 18, 33
  epidermal growth factor and, 122—123
  platelet-derived growth factor and, 199, 203—204
  thyroid hormones and, 219
Simian virus 40 (SV40) large-T antigen, 86
SIRS, see Soluble immune response suppressor
sis gene, 8, 201—203
SK-hep-1 hepatic adenocarcinoma cells, 161
ski gene, 8
Skin cancer, 94
Skin fibroblasts, 28, 93, 105, 128, 196
SK-MG-3 astrocytoma cells, 116, 122
SKV 770 AV, see SKV 770 avian virus
SKV 770 avian virus (SKV 770 AV), 8
Slime mold, see *Dictostelium discoideum*
Small cell lung cancer, 12—13
Smooth muscle cells, 104, 142, 174, 195—196, 203, 233
Snake venom, 142
Snyder-Theilen feline sarcoma virus (ST-FeSV), 8, 149
Sodium, 30, 48
Sodium/hydrogen antiport, 30, 48, 167, 197
Sodium/hydrogen exchange, 50, 143
Sodium/potassium exchange, 52
Sodium/potassium pump, 48
Soluble immune response suppressor (SIRS), 7
Somatomedin A, see Insulin-like growth factor II
Somatomedin C, see Insulin-like growth factor I
Somatotropin, see Growth hormone
Spare receptor, 88
Spectrin-actin complex, 34
Spermidine, 30, 229
Spermine, 30, 229
S phase, 1—6, 15, 18, 30, 35, 42—48
  hematopoietic growth factors and, 165, 168, 171
  insulin and, 94
  platelet-derived growth factor and, 199
  steroid hormones and, 231
  thyroid hormones and, 219, 222
Spiny dogfish, 227
Spleen, 59, 178
Splenocytes, 167
Squalus acanthias, 227
Squamous cell carcinoma, 94, 114
src gene, 8, 52—53, 56, 143
src gene family, 34, 61, 89, 115, 120—121, 128
SSV, see Simian sarcoma virus
Stem cells, 157—158, 160, 165
Steroid hormone(s), 3, 29, 227—234
Steroid hormone receptor, 27, 227—228
ST-FeSV, see Snyder-Theilen feline sarcoma virus
Stress fiber, 199
Stress protein family, 50
Submaxillary gland, 111—114, 128—129, 142
Suramin, 204
SV28 kidney cells, 141, 204
SV40, see Simian virus 40
Swiss 3T3 cells, 6, 16, 49, 55, 197
  TNR29 variant line of, 128
Swiss 3T3 fibroblasts, 63, 219

Swiss-Webster 3T3 cells, 115
Systemic lupus erythematosus, 166

T

T3, see Triiodothyronine
3T3 fibroblasts, 12, 15, 49, 126
3T3-L1 adipocytes, 92
T-3 monoclonal antibody, 162—163
3T3-TNR9 cells, 51, 63
T4, see Thyroxine
T47D breast cancer cells, 60, 106, 122, 203—204
Tac antigen, 164—167, 179
TAF, see Tumor angiogenesis factor
T-cell growth factor (TCGF), see Interleukin-2
T-cell leukemia/lymphoma, 163
T-cell malignancy, 166
T-cell receptor, antigen-specific, 162—163
T-cells, 54, 58—59, 157—168, 173, 178, 214
TCGF, see Interleukin-2
Teleocidin, 50
Testosterone, 51, 117, 177, 228, 233
12-*O*-Tetradecanoyl-phorbol 13-acetate (TPA), 49—53, 142
  epidermal growth factor and, 123—124, 128
  hematopoietic growth factors and, 167, 170—171, 179
Tetrahymena, 28, 86, 196
TGF, see Transforming growth factor
Thrombin, 48
Thrombin receptor, 48
beta-Thromboglobulin, 195
Thymocytes, 161, 173
Thymus, 165
Thyroglobulin, 41—42, 127, 219, 221
Thyroglobulin gene, 219
Thyroid, dysfunction of, 220
Thyroid cells, 15, 41
Thyroid hormone(s), 2, 29, 89, 122, 219—222, 230
Thyroid hormone receptor, 27, 219—220
Thyroid-stimulating hormone (TSH, thyrotropin), 2, 15, 32, 122, 141
Thyroid-stimulating hormone (TSH, thyrotropin) receptor, 32
Thyroid tumor, 32
Thyrotropin, see Thyroid-stimulating hormone
Thyroxine (T4), 219—222
TIF, see Tumor-inhibitory factor
Tissue repair, 47—48, 161, 174, 196
TLCK, see alpha-*p*-Tosyl-L-lysine chloromethyl ketone
T-lymphocytes, 15, 58, 105, 158—161, 166, 199
T-lymphoma cells, 14, 59
tms locus, 14
TNF-α, see Tumor necrosis factor α
TNF-β, see Tumor necrosis factor β
alpha-*p*-Tosyl-L-lysine chloromethyl ketone (TLCK), 59
TPA, see 12-*O*-Tetradecanoyl-phorbol 13-acetate
Transcription, 30—31, 35, 51, 54, 62

platelet-derived growth factor and, 199—201
  steroid hormones and, 231—232
  thyroid hormones and, 220
Transducin, 36—37
Transducing signal, intracellular, 27
Transduction mechanisms, 27
  postreceptor, 29—30
Transferrin, 3, 28, 165, 211—215
Transferrin gene, 3, 211—212
Transferrin-like proteins, 211
Transferrin receptor, 49, 124, 165, 178—179, 211
  expression on cell surface, 213—214
  in neoplastic cells, 214
  in normal cells, 214
  structure of, 213
Transferrin receptor gene, 3, 212
Transformation
  cAMP and, 32—33
  radiation-induced, 128
  sis-induced, 203
  transforming growth factor beta in, 150—151
Transformed cells, 1, 4, 12, 94—95
  insulin-like growth factor receptor in, 105—107
  insulin receptor in, 105—107
  platelet-derived growth factor in, 203—204
Transforming growth factor (TGF), 12—13, 56, 147—152
Transforming growth factor α (TGF-α), 2, 7, 111, 148—149
Transforming growth factor α (TGF-α) gene, 3, 11, 148
Transforming growth factor α (TGF-α) receptor, 149—150
Transforming growth factor α-related polypeptides, 148
Transforming growth factor (β (TGF-β), 2, 7, 18, 122—123, 205
  in cell proliferation, 150—151
  mechanism of action of, 149—150
  structure of, 149
  in transformation, 150—151
Transforming growth factor β (TGF-β) gene, 3
Transforming growth factor β (TGF-β) receptor, 149—150
Transmembrane signaling, 46
Transposable element, 126
Trifluoperazine, 126
Triiodothyronine (T3), 219—222
Triose phosphate isomerase, 126
Trophoblast cells, 214
Troponin-C, 44
Trypanosoma, 18
TSH, see Thyroid-stimulating hormone
Tubulin, 93, 125
Tumor
  animal, 12—14
  plant, 14—15
  regression of, 33, 166, 176
Tumor angiogenesis factor (TAF), 2, 141
Tumorigenicity, 12, 42
Tumorigenic processes, 56

Tumor-inhibitory factor (TIF), 7
Tumor-inhibitory factor 1 (TIF-1), 7
Tumor-inhibitory factor 2 (TIF-2), 7
Tumor necrosis factor α (TNF-α), 3, 176
Tumor necrosis factor α (TNF-α) gene, 3, 11, 176
Tumor necrosis factor α (TNF-α) receptor, 176
Tumor necrosis factor β (TNF-β, lymphotoxin), 3
Tumor necrosis factor β (TNF-β, lymphotoxin) gene, 3, 11
Tumor progression, 63, 148
Tumor promoter, 42, 49—52, 63, 166
  c-*fms*/CSF-1 receptor expression and, 170—171
  epidermal growth factor receptor and, 123—124
Tyrosine aminotransferase, 91, 93, 231
Tyrosine hydroxylase, 143

U

U-937 monocytic cells, 17, 54, 90, 95, 106—107, 175
URII-ASV, see Rochester URII avian sarcoma virus
Urine, 151, 169
URO, see Epidermal growth factor
Urogastrone, see Epidermal growth factor
Urogenital system, 122
Uterine cancer, 94
Uterus, 229

V

v-*abl* gene, 9, 13, 55—56, 61
  epidermal growth factor and, 115
  hematopoietic growth factors and, 160, 180—181
Vaccinia virus, 151
Vaccinia virus growth factor, 111
Vaginal cancer, 94
Vanadate, 60
Vascular function, 113
Vasopressin, 49
v-*erb* gene, 119, 227
v-*erb*-A gene, 119
v-*erb*-B gene, 56, 89, 115, 119—120
Vesicular stomatitis virus, 58
v-*ets* gene, 13
v-*fes* gene, 34, 56, 115
v-*fgr* gene, 56
v-*fms* gene, 17, 56, 61, 115
v-*fps* gene, 34, 56, 115, 123, 125
v-H-*ras* gene, 39, 42, 222
  hematopoietic growth factors and, 175, 178
  steroid hormones and, 230, 233
  transforming growth factor and, 160
Vinculin, 48, 199
Viral oncogens, see v-*onc* genes
Visual system, 36—38
Vitamin D, 233
v-K-*ras* gene, 12—13, 33, 39—42, 178
v-*mht* gene, see v-*raf* gene
v-*mil* gene, 19, 53, see v-*raf* gene

v-*mos* gene, 9, 33—34, 56, 61
  epidermal growth factor and, 113
  hematopoietic growth factors and, 175
  steroid hormones and, 233
v-*myb* gene, 9, 13, 173
v-*myc* gene, 13, 54, 160, 173, 179
v-*onc* protein products, 13—14
v-*onc*-transformed cells, growth factors in, 18
v-*raf* gene, 53, 56, 61
v-*ras* gene, 9, 13, 39—42, 93
  epidermal growth factor and, 125—126, 128
v-*ros* gene, 52—53, 56, 89
v-*sis* gene, 201—203
v-*src* gene, 12—13, 19, 34, 52—56, 59
  epidermal growth factor and, 115, 120—121, 123—126
  steroid hormones and, 230
v-*src* gene family, 173
v-*srs* gene, 53
Vulval cancer, 94
v-*yes* gene, 34, 56, 61, 115

W

W7, 126
W138 fibroblasts, 123
Walker carcinoma 256, 221
WEHI-3B myelomonocytic leukemia cells, 159—160, 172—173
Wheat germ agglutinin, 28
WI-38 fibroblasts, 142, 198
Wilms' tumor, 85, 103—104
Wound healing, 47—48, 113, 147, 196

Y

Y-79 retinoblastoma cells, 60
YAC-1 cells, 58
Yamaguchi avian sarcoma virus (Y-ASV), 8
Y-ASV, see Yamaguchi avian sarcoma virus
Yeast, see *Saccharomyces cerevisiae*
yes gene, 8, 120